Search Methods in Artificial Intelligence

Search Methods in Artificial Intelligence is designed to provide in-depth knowledge on how search plays a fundamental role in problem solving. The book is meant for undergraduate and graduate students pursuing courses in computer science and artificial intelligence. Starting with basic search, it covers a variety of complex algorithms designed for different kinds of problems. It demonstrates that search is all-pervasive in artificial intelligence and equips the reader with relevant skills.

The text begins with an introduction to search spaces that confront intelligent agents. It illustrates how basic algorithms like depth first search and breadth first search run into exponentially growing spaces. Discussions on heuristic search follow along with stochastic local search, algorithm A*, and problem decomposition. The role of search in playing board games, deduction in logic, automated planning, and machine learning is described next. The book concludes with a coverage of constraint satisfaction.

Deepak Khemani has been actively working in the field of artificial intelligence (AI) for over four decades – first as a student at Indian Institute of Technology (IIT) Bombay and then as a Professor in the Department of Computer Science and Engineering at IIT Madras. Currently he is Professor at Plaksha University, Mohali. He has three well-received courses on AI on SWAYAM, a MOOC platform launched by the Government of India. He is also the author of *A First Course in Artificial Intelligence* (2013).

Search Methods in Artificial Intelligence

Deepak Khemani

Shaftesbury Road, Cambridge CB2 8EA, United Kingdom

One Liberty Plaza, 20th Floor, New York, NY 10006, USA

477 Williamstown Road, Port Melbourne, VIC 3207, Australia

314–321, 3rd Floor, Plot 3, Splendor Forum, Jasola District Centre, New Delhi – 110025, India

103 Penang Road, #05–06/07, Visioncrest Commercial, Singapore 238467

Cambridge University Press is part of Cambridge University Press & Assessment, a department of the University of Cambridge.

We share the University's mission to contribute to society through the pursuit of education, learning and research at the highest international levels of excellence.

www.cambridge.org
Information on this title: www.cambridge.org/9781009284325

© Deepak Khemani 2024

This publication is in copyright. Subject to statutory exception and to the provisions of relevant collective licensing agreements, no reproduction of any part may take place without the written permission of Cambridge University Press & Assessment.

First published 2024

Printed in India by Nutech Print Services, New Delhi 110020

A catalogue record for this publication is available from the British Library

ISBN 978-1-009-28432-5 Hardback
ISBN 978-1-009-28433-2 Paperback

Cambridge University Press & Assessment has no responsibility for the persistence or accuracy of URLs for external or third-party internet websites referred to in this publication and does not guarantee that any content on such websites is, or will remain, accurate or appropriate.

Contents

Preface xi
Acknowledgements xiii

1 Introduction 1
  1.1 Can Machines Think? 1
  1.2 Problem Solving 4
  1.3 Neural Networks 6
    1.3.1 Deep neural networks 8
  1.4 Symbolic AI 10
    1.4.1 Symbols, language, and knowledge 10
    1.4.2 Symbol systems 12
    1.4.3 An architecture for cognition 13
  1.5 The Core of Intelligence 14
    1.5.1 Remember the past and learn from it 15
    1.5.2 Understand and represent the present 18
    1.5.3 Imagine the future and shape it 23
  A Note for the Reader 26
  Exercises 26

2 Search Spaces 29
  2.1 The State Space 30
    2.1.1 Generate and test 31
  2.2 Search Spaces 32
  2.3 Configuration Problems 33
    2.3.1 The map colouring problem 33
    2.3.2 The N-queens problem 34
    2.3.3 The SAT problem 35
  2.4 Planning Problems 35
    2.4.1 The 8-puzzle 36
    2.4.2 River crossing puzzles 37
    2.4.3 The water jug problem 38
    2.4.4 The travelling salesman problem 39
  2.5 The Solution Space 42
    2.5.1 Constructive methods 42
    2.5.2 Perturbative methods 43
  Summary 44
  Exercises 44

3 Blind Search — 47
- 3.1 Search Trees — 48
- 3.2 Depth First Search — 49
  - 3.2.1 How to stop going around in circles — 51
  - 3.2.2 Reconstructing the path — 53
  - 3.2.3 The complete algorithm — 55
  - 3.2.4 Backtracking in DFS — 57
- 3.3 Breadth First Search — 58
- 3.4 Comparing DFS and BFS — 60
  - 3.4.1 Space complexity — 61
  - 3.4.2 Time complexity — 62
  - 3.4.3 Quality of solution — 64
  - 3.4.4 Completeness — 64
- 3.5 Depth First Iterative Deepening — 64
  - 3.5.1 Depth bounded depth first search — 65
  - 3.5.2 Depth first iterative deepening — 66
  - 3.5.3 Space complexity — 67
  - 3.5.4 Time complexity — 67
  - 3.5.5 Quality of solution — 67
  - 3.5.6 Completeness — 69
- 3.6 Uninformed Search — 70
- *Summary* — 71
- *Exercises* — 72

4 Heuristic Search — 75
- 4.1 Heuristic Functions — 76
  - 4.1.1 Map colouring — 77
  - 4.1.2 SAT — 77
  - 4.1.3 The 8-puzzle — 77
  - 4.1.4 Route finding — 78
  - 4.1.5 Travelling salesperson problem — 80
- 4.2 Best First Search — 81
  - 4.2.1 Quality of solution — 84
  - 4.2.2 Completeness — 85
  - 4.2.3 Space complexity — 85
  - 4.2.4 Time complexity — 86
- 4.3 Local Search Methods — 86
  - 4.3.1 An optimization problem — 87
  - 4.3.2 Hill climbing — 88
  - 4.3.3 Completeness — 89
  - 4.3.4 Quality of solution — 89
  - 4.3.5 Time complexity — 89
  - 4.3.6 Space complexity — 90
- 4.4 Heuristic Search Terrains — 90
  - 4.4.1 Hill climbing in the blocks world domain — 90
  - 4.4.2 Heuristic functions — 91
  - 4.4.3 The SAT landscape — 95

	4.5	Neighbourhood Functions		96
		4.5.1	Neighbourhood functions for the TSP	96
		4.5.2	Neighbourhood functions for SAT	98
		4.5.3	Variable neighbourhood descent	99
		4.5.4	Beam search	101
	4.6	Escaping Local Optima		103
		4.6.1	Exploration versus exploitation	104
		4.6.2	Tabu search	104
		4.6.3	Iterated hill climbing	107
	Summary			108
	Exercises			108
5	**Stochastic Local Search**			**115**
	5.1	Climbing Mount Improbable		116
		5.1.1	Random walk: pure exploration	117
		5.1.2	Stochastic hill climbing: calibrated randomness	117
		5.1.3	Simulated annealing: controlled randomness	119
	5.2	Evolution: Survival of the Fittest		122
		5.2.1	Genetic algorithms: churning in a population of candidates	125
		5.2.2	TSP: Representations and crossover operators	130
			5.2.2.1 TSP: path representation	131
			5.2.2.2 TSP: adjacency representation	134
			5.2.2.3 TSP: ordinal representation	136
	5.3	Swarm Intelligence: The Power of Many		138
		5.3.1	Ant colony optimization	141
	Summary			144
	Exercises			144
6	**Algorithm A* and Variations**			**147**
	6.1	Branch & Bound		148
		6.1.1	Performance of B&B	152
		6.1.2	B&B on the TSP	153
		6.1.3	Higher estimates are better estimates	156
	6.2	Algorithm A*		156
		6.2.1	Dijkstra's algorithm	156
		6.2.2	A*	158
		6.2.3	A* is admissible	162
		6.2.4	Proof of admissibility	163
		6.2.5	Higher is better	164
		6.2.6	The monotone condition	166
		6.2.7	Performance of A*	167
		6.2.8	Weighted A*	167
	6.3	Space Saving Versions of A*		169
		6.3.1	Iterative deepening A*	169
		6.3.2	Recursive best first search	170
		6.3.3	Sequence alignment	171

		6.3.4	Pruning CLOSED	173
			6.3.4.1 Divide and conquer frontier search	173
			6.3.4.2 Smart memory graph search	175
		6.3.5	Beam stack search	176
		6.3.6	Pruning OPEN and CLOSED	178

Summary 179
Exercises 180

7 Problem Decomposition 185

7.1 Pattern Directed Inference Systems 185
7.2 Rule Based Production Systems 189
    7.2.1 The working memory in OPS5 190
    7.2.2 Patterns in rules 191
    7.2.3 Actions in rules 192
    7.2.4 The inference engine 194
    7.2.5 Conflict resolution strategies 195
    7.2.6 The Rete net 197
    7.2.7 The Rete algorithm 201
7.3 Problem Decomposition with And–Or Graphs 205
    7.3.1 DENDRAL 206
    7.3.2 Symbolic integration 209
    7.3.3 Algorithm AO* 211

Summary 216
Exercises 216

8 Chess and Other Games 221

8.1 Game Theory 223
    8.1.1 The prisoner's dilemma 224
    8.1.2 Types of games 225
8.2 Board Games 227
    8.2.1 Strategies 229
    8.2.2 Limited lookahead with the evaluation function 231
    8.2.3 Algorithm Minimax 235
    8.2.4 AlphaBeta pruning 236
    8.2.5 Algorithm SSS*: best first search 239
    8.2.6 Algorithm SSS*: an iterative version 243
    8.2.7 AlphaGo, AlphaGo Zero, and AlphaZero 246
8.3 Backgammon and Scrabble: Elements of Chance 247
8.4 Contract Bridge: A Challenge for AI 250

Summary 258
Exercises 258

9 Automated Planning 263

9.1 Representation 264
    9.1.1 Time and change 264
9.2 Simple Planning Domains 266
    9.2.1 STRIPS 266
    9.2.2 The blocks world domain 268

	9.3	State Space Planning	272
		9.3.1 Forward state space planning	273
		9.3.2 Domain independent heuristics	274
		9.3.3 Backward state space planning	277
	9.4	Goal Stack Planning	280
		9.4.1 Linear planning	281
		9.4.2 Sussman anomaly	284
	9.5	Partial Order Planning	286
		9.5.1 A two armed robot	293
	9.6	Algorithm Graphplan	295
		9.6.1 The planning graph	295
		9.6.2 Heuristics and solutions from the planning graph	300
	9.7	Planning as Satisfiability	302
		9.7.1 Direct encoding	303
		9.7.2 The planning graph as SAT	306
	9.8	Richer Planning Domains	307
		9.8.1 Durative actions	308
		9.8.2 Metric domains	310
		9.8.3 Conditional effects	310
		9.8.4 Contingent planning	311
		9.8.5 Trajectory constraints and preferences	312
		9.8.6 Coordination in multi-agent systems	312
		9.8.7 Epistemic planning	313
	Summary		315
	Exercises		315
10	**Deduction as Search**		**319**
	10.1	Logical Connectives	320
		10.1.1 Truth tables	321
	10.2	Entailment and Proof	323
		10.2.1 Soundness	324
		10.2.2 Completeness	326
	10.3	First Order Logic	328
		10.3.1 Terms and domains	329
		10.3.2 Atomic formulas	330
		10.3.3 Quantifiers, formulas, and sentences	331
	10.4	Deduction in FOL	333
		10.4.1 Implicit quantifier form	335
		10.4.2 Unification	337
		10.4.3 Forward chaining	339
		10.4.4 Backward chaining and deductive retrieval	341
		10.4.5 Prolog	343
		10.4.6 Incompleteness of forward and backward chaining	346
		10.4.7 The resolution refutation method	347
		10.4.8 FOL with equality	351

10.5	The Family of Logics		355
	10.5.1 Horn clause logic and Prolog		356
	10.5.2 Description logics		357
	10.5.3 Default reasoning		359
	10.5.4 Event calculus		360
	10.5.5 Epistemic logic		361
Summary			362
Exercises			362

11 Search in Machine Learning by Sutanu Chakraborti — 367

11.1	Decision Trees	371
11.2	k-Nearest Neighbour	376
11.3	Bayesian Classification	378
11.4	Artificial Neural Networks	382
11.5	K-MEANS Clustering	389
Summary		391
Exercises		391

12 Constraint Satisfaction — 393

12.1	Constraints: Clearing the Fog		394
	12.1.1 The map colouring problem		394
	12.1.2 The N-queens puzzle		395
12.2	Algorithm BACKTRACKING		398
	12.2.1 Static variable ordering		400
	12.2.2 Dynamic variable ordering		402
12.3	Constraint Propagation		403
	12.3.1 Arc consistency		404
	12.3.2 The Waltz algorithm		408
	12.3.3 Path consistency		411
	12.3.4 i-Consistency		414
	12.3.5 Directional consistency		415
12.4	Lookahead Search		417
	12.4.1 Algorithm forward checking		419
	12.4.2 Algorithm DAC-Lookahead		422
	12.4.3 Algorithm AC-Lookahead		425
12.5	Informed Backtracking		426
	12.5.1 Gaschnig's backjumping		427
	12.5.2 Graph based backjumping		429
	12.5.3 Conflict directed backjumping		432
Summary			434
Exercises			435

***Appendix: Algorithm and Pseudocode Conventions** by S. Baskaran* — **441**

References — **449**

Index — **461**

Preface

This book is meant for the serious practitioner-to-be of constructing intelligent machines. Machines that are aware of the world around them, that have goals to achieve, and the ability to imagine the future and make appropriate choices to achieve those goals. It is an introduction to a fundamental building block of artificial intelligence (AI). As the book shows, search is central to intelligence.

Clearly AI is not one monolithic algorithm but a collection of processes working in tandem, an idea espoused by Marvin Minsky in his book *The Society of Mind* (1986). Human problem solving has three critical components. The ability to make use of experiences stored in memory; the ability to reason and make inferences from what one knows; and the ability to search through the space of possibilities. This book focuses on the last of these. In the real world we sense the world using vision, sound, touch, and smell. An autonomous agent will need to be able to do so as well. Language, and the written word, is perhaps a distinguishing feature of the human species. It is the key to communication which means that human knowledge becomes pervasive and is shared with future generations. The development of mathematical sciences has sharpened our understanding of the world and allows us to compute probabilities over choices to take calculated risks. All these abilities and more are needed by an autonomous agent.

Can one massive neural network be the embodiment of AI? Certainly, the human brain as a seat of intelligence suggests that. Everything we humans do has its origin in activity in our brains, which we call the mind. Perched on the banks of a stream in the mountains we perceive the world around us and derive a sense of joy and well-being. In a fit of contented creativity, we may pen an essay or a poem using our faculty of language. We may call a friend on the phone and describe the scene around us, allowing the friend to visualize the serene surroundings. She may reflect upon her own experiences and recall a holiday she had on the beach. You might start humming your favourite song and then be suddenly jolted out of your reverie remembering that friends are coming over for dinner. You get up and head towards your home with cooking plans brewing in your head.

So, in principle at least one can imagine a massive neural network that could do all the above. But how would it be implemented? What kind of a training process would instil all such knowledge and memories in the neural brain? Human beings go through a lifetime of learning. A human baby, unlike a fawn, is an utterly helpless creature and needs to be nurtured for years. Taught in schools, influenced by peer groups, moulded through culture and religion, coached in sports. Every human is said to be unique, even identical twins. We celebrate this diversity, even when it is sometimes a source of crime and conflict. Are we ready for idiosyncratic machines? Or do we aim for identical assembly line robots? But what or who would they be like? And what about issues of fairness? And generation of harmful or misleading content?

The twenty-first century has seen an explosion in machine learning as exemplified by deep neural networks which outperform humans on many classification tasks, and large language models that can generate an essay, a college application, or a poem in a jiffy. Massive computing power and humungous amounts of data have made this possible. It has been very impressive, but has it peaked? Do we need to move on and seek another path to the Holy Grail, machines which autonomously solve problems for us? Instead of blanket ingestion of all data on the internet, perhaps we need to build machines which learn from human expertise to become experts in specific domains. And do useful things for us.

This book is a step in that direction. It is designed to be a complete guide to one specific aspect of problem solving – the use of search methods. It is intended to be one in-depth module for the task of building AI, and its contents can be covered in a one semester course. We begin by learning to walk with small problems, and gradually build a repertoire of search algorithms that would allow us to navigate the high seas and vast deserts. The algorithms are general purpose, but our representations are tailormade for the individual domains. We urge the interested reader to implement the algorithms described here and develop a suite of search algorithms that can be used to solve specific problems.

One common feature in all these algorithms is that they operate on symbolic data, where symbols stand for things meaningful to us, and algorithms operate upon them. This approach is, as *hypothesized* by Herbert Simon and Alan Newell in 1976, both necessary and sufficient to create AI.

Maybe one day these many algorithms and the different problems they solve will come together in an integrated entity as a step towards artificial general intelligence. But that will need advances in knowledge representation where different domains and problems can be uniformly expressed in a common language. There is work still ahead for us.

Acknowledgements

Several people have contributed in myriad ways to this book, some directly and some indirectly. Many students in my class, both online and offline, have triggered a thought process by asking incisive questions and making insightful observations. I gratefully acknowledge all of them collectively. They have made the job of teaching and learning rewarding.

Baskaran Sankaranarayanan has been involved with my courses over the last few years, answering student queries, helping with question papers and figures, and most importantly by standardizing the way in algorithms are written in pseudo code. He has written the appendix in this book on algorithm style, and the algorithms in the book conform to that style.

Sutanu Chakraborti has been a long-time collaborator working in AI. He wrote a chapter on natural language processing in my previous book, *A First Course in Artificial Intelligence*. In this book he has written one chapter on machine learning, despite a pressing schedule.

I am indebted to the following who have read and commented upon parts of the book – Nitin Dhiman, Aditi Khemani, Kamal Lodaya, Adwait Pramod Parsodkar, Devika Sethi, and Shashank Shekhar.

I am grateful to the team at CUP for their constant support from the very beginning. Vaishali Thapliyal took up my book proposal with gusto and got reviews from external reviewers expeditiously, yielding some very valuable suggestions and feedback. When the manuscript was ready Vikash Tiwari and Ankush Kumar initiated the production process immediately. Aniruddha De and Karan Gupta have done an excellent job with proofreading and copy editing, ironing out many discrepancies and bringing uniformity to the writing style.

Finally, I would like to thank the friends and family who have supported the book writing in many ways.

CHAPTER 1

Introduction

We will adopt the overall goal of artificial intelligence (AI) to be 'to build machines with minds, in the full and literal sense' as prescribed by the Canadian philosopher John Haugeland (1985).

Not to create machines with a clever imitation of human-like intelligence. Or machines that exhibit behaviours that would be considered intelligent if done by humans – but to build machines that reason.

This book focuses on search methods for problem solving. We expect the user to define the goals to be achieved and the domain description, including the moves available with the machine. The machine then finds a solution employing first principles methods based on search. A process of trial and error. The ability to explore different options is fundamental to thinking.

As we describe subsequently, such methods are just amongst the many in the armoury of an intelligent agent. Understanding and representing the world, learning from past experiences, and communicating with natural language are other equally important abilities, but beyond the scope of *this* book. We also do not assume that the agent has meta-level abilities of being self-aware and having goals of its own. While these have a philosophical value, our goal is to make machines do something useful, with as general a problem solving approach as possible.

This and other definitions of what AI is do not prescribe *how to test* if a machine is intelligent. In fact, there is no clear-cut universally accepted definition of intelligence. To put an end to the endless debates on machine intelligence that ensued, the brilliant scientist Alan Turing proposed a behavioural test.

1.1 Can Machines Think?

Ever since the possibility of building intelligent machines arose, there have been raging debates on whether machine intelligence is possible or not. All kinds of arguments have been put forth both for and against the possibility. It was perhaps to put an end to these arguments that Alan Turing (1950) proposed his famous *imitation game*, which we now call the *Turing Test*. The test is simply this: if a machine interacts with a human using text messages and can fool human judges a sufficiently large fraction of times that they are chatting with another human, then we can say that the machine has passed the test and is intelligent.

Since then, many programs have produced text based interactions that are convincingly human-like, for example, *ChatGPT*[1] being one of the latest. Advances in machine learning algorithms for building language models from large amounts of training data have enabled machines to churn out remarkably well structured impressive text. Humans are quite willing to believe that if it talks like a human, then it must think like a human. Even when the very first chat program *Eliza* threw back user sentences with an interrogative twist, its creator Edward Weizenbaum was shocked to discover that his secretary was confiding her personal woes to the program (Weizenbaum, 1966). Pamela McCorduck (2004) has observed in *Machines Who Think* that in medieval Europe people were willing to ascribe intelligence to mechanical toy statues that could nod or shake their head in response to a question.

Clearly relying on human impressions based on interaction in natural language is not the best way of determining whether a machine is intelligent or not. With more and more machines becoming good at generating text rivalling that produced by humans, a need is being felt for something that delves deeper and tests whether the machine is actually reasoning when answering questions.

Hector Levesque and colleagues have proposed a new test of intelligence which they call the *Winograd Schema Challenge*, after Terry Winograd who first suggested it (Levesque et al., 2012; Levesque, 2017). The idea is that the test cannot be answered by having a large language model or access to the internet but would need common sense knowledge about the world. The test subject is given a sentence that refers to two entities of the same kind and a pronoun that could refer to either one of them. The question is which one, and the task is called anaphora resolution. The ambiguity can easily be resolved by humans using common sense knowledge. The strategy is to have two variations of the sentence, each having a different word or a phrase that leads to different anaphora resolution. One of the versions is presented to the subject with a question about what the pronoun refers to. Guesswork on a series of such questions is only expected to produce about half the correct answers, whereas a knowledgeable (read intelligent) agent would do much better. The following is the example attributed to Winograd (1972).

- The town councillors refused to give the angry demonstrators a permit because they feared violence. Who feared violence?
  - **(a)** The town councillors
  - **(b)** The angry demonstrators
- The town councillors refused to give the angry demonstrators a permit because they advocated violence. Who advocated violence?
  - **(a)** The town councillors
  - **(b)** The angry demonstrators

In both cases, two options are given to the subject who has to choose one of the two. Here are two more examples of the Winograd Schema Challenge, with two sets of sentences, each one of which is presented and followed by a question.

- The trophy doesn't fit in the brown suitcase because it's too big. What is too big?
  - **(a)** the trophy
  - **(b)** the suitcase

---

[1] ChatGPT: Optimizing Language Models for Dialogue. https://openai.com/blog/chatgpt/, accessed December 2022.

- The trophy doesn't fit in the brown suitcase because it's too small. What is too small?
  (a) the trophy
  (b) the suitcase

The following sentence is from the First Winograd Challenge at the International Joint Conference on AI in 2016 (Davis et al., 2017).

- John took the water bottle out of the backpack so that it would be lighter.
- John took the water bottle out of the backpack so that it would be handy.

What does 'it' refer to? Again, two options are given to the subject who is asked to choose one.

The authors report that the Winograd Schema Test was preceded by a pronoun disambiguation test in a single sentence, with examples chosen from naturally occurring text. Only those programs that did well in the first test were allowed to advance to the Winograd Schema Test. Here is an example from their paper which has been taken from the story 'Sylvester and the Magic Pebble'.

- The donkey wished a wart on its hind leg would disappear, and it did.

What vanished? The important thing is that such problems can be solved only if the subject is well versed with sufficient common sense knowledge about the world and also the structure of language.

A question one might ask is why should a test of intelligence be language based? After all, intelligence manifests itself in other ways as well. Could one of these also be an indicator of intelligence?

One area that has been proposed is in the arts, where creativity is the driving force. Computer generated art has time and again come to the limelight. Many artworks by *AARON*, the drawing artist created by Harold Cohen (1928–2016), have been demonstrated at AI conferences over the years (Cohen, 2016). A slew of text-to-image AI systems including DALL-E, Midjourney, and Stable Diffusion have all been released for public use recently.

Erik Belgum and colleagues have proposed a Turing Test for musical intelligence (Belgum et al., 1989). In the fall of 1997, Douglas Hofstadter organized a series of five public symposia centred on the burning question 'Are Computers Approaching Human-Level Creativity?' at Indiana University. This fourth symposium was about a particular computer program, David Cope's *EMI* (Experiments in Musical Intelligence) as a composer of music in the style of various classical composers (Cope, 2004). A two-hour concert took place in which compositions written by EMI and compositions written by famous human composers were performed without identification, and the audience was asked to vote for which pieces they thought were human-composed and which were computer-composed. Subsequently, David Coco-Pope published an article written by a computer program *EWI* (Experiments in Written Intelligence) in the style of Hofstadter, grudgingly conceded by Hofstadter himself at the end of the article (Hofstadter, 2009).

After the 2011 spectacular win by IBM's program *Watson* in the game of *Jeopardy* over two players who were widely considered to be the best that the game had seen, the company unveiled a program *Chef Watson* with the following claim – 'In our application, a computationally creative computer can automatically design and discover culinary recipes that are flavorful, healthy, and novel!'[2] The market is now abuzz with robots that can cook for you, for example, as reported in Cain (2022).

Recently, when DeepMind's *AlphaGo* program beat the reigning world champion Lee Sedol in the oriental game of *go*, the entire world sat up and took notice (Silver et al., 2016). This followed an equally impressive win almost twenty years earlier in 1997 when IBM's *Deep Blue* program beat the then world champion Garry Kasparov in the game of chess (Campbell et al., 2002). Both the games are two person board games in which programs can search game trees as described in Chapter 8. The challenge in these games is to search the huge trees that present themselves. While chess is played on an 8×8 board, *go* is played on a 19×19 board, which generates a much larger game tree. And yet a combination of machine learning and selective search proved invincible. Both these games are conceptually simple even though the search trees are large. In the author's opinion, only when a computer program can play the game of contract bridge at the level described in Ottlik and Kelsey (1983) can we legitimately stake a claim to have created an AI.

Meanwhile, one should perhaps take a cue from Alan Turing himself, move away from the bickering, and get on with the design and implementation of autonomous machines who[3] do useful things for us. In the summer of 1956, John McCarthy and Marvin Minsky had organized the Dartmouth Conference with the following stated goal – 'The study is to proceed on the basis of the conjecture that every aspect of learning or any other feature of intelligence can in principle be so precisely described that a machine can be made to simulate it' (McCorduck, 2004). *That* is the spirit of our quest for AI.

1.2 Problem Solving

Our quest is for a machine that is autonomous and whose behaviour is goal directed. Whatever it does, it should do autonomously. We imagine a scenario in which the machine is an agent to serve the goals given it to it by a *user*. Current applications take a short horizon view of achieving specific goals, though we can imagine a persistent agent engaging with the human over long periods, perhaps even the user's lifetime. We ignore the doomsday scenarios in which machines overcome and subjugate humans, though this idea has been fashionable amongst certain sections of science writers. That so-called *singularity* is not even on the horizon (Larson, 2021).

We want our machines to solve problems for us. Given a set of goals, the machine must engage with the world to achieve those goals. The goals may be short term or long term, and the world in which the problem solving agent operates may be changing, even in the simplest case when the agent is the only one acting and effecting the change. The agent must sense its environment, deliberate over its goals, and act in the domain. The agent must not just be reactive, operating in a hard-coded *stimulus–response* cycle, but should be able to act flexibly in a *sense–deliberate–act* cycle autonomously. A schematic of an autonomous agent is shown in Figure 1.1.

In all life forms, deliberation happens in the brain. Incoming sensory data is processed in the context of what the creature already knows. Our understanding of the animal brain is that it is a collection of a large number of very simple processing components called *neurons*. Each

---

[2] https://researcher.watson.ibm.com/researcher/view_group.php?id = 5077, accessed December 2022.
[3] In the style of *Machines Who Think* by Pamela McCorduck.

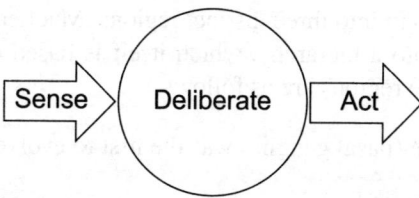

Figure 1.1 An autonomous agent operates in a three stage cycle. It receives input from its sensory mechanism, it deliberates over the inputs and its goals, and acts in the world.

neuron is connected to many other neurons and each connection has a weight that evolves with experience. This changing of weights is associated with the process of learning.

The neurons at the sensing end of the brain accept information coming in from various senses like sight, sound, smell, taste, and touch. The general model of processing is that once a neuron is activated, it sends a signal down its principal nerve called the axon, which distributes the signal to other connected neurons. The weights of the connections determine which connected neurons receive how much of the signal. Eventually the signals reach the neurons at the output end, sending signals down the motor neurons that activate muscles that produce sounds from the mouth and movement of the limbs.

Some simple creatures may be just reactive, recognizing food or prey and triggering appropriate actions, but as we move up the hierarchy, there may be more complex processing happening in the brain, involving memory (in Greek mythology, the dog Argos recognizes Odysseus at once when humans could not), planning (monkeys in Japan have been known to season their food with salt water), and reasoning (remember all those experiments with mice in mazes and Pavlov's dog). Whatever the cognitive capability of the creature, our view of their brains can be captured as shown in Figure 1.2.

Different life forms have differently sized brains relative to the sizes of their bodies. Earlier life forms had simple brains often referred to as the reptilian brain. In the 1960s, the American neuroscientist Paul MacLean (1990) formulated the *Triune Brain* model, which is based on

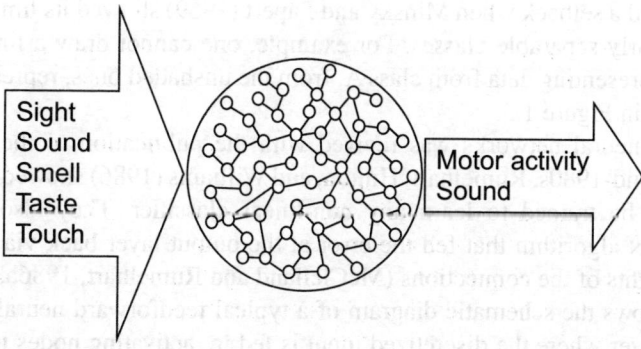

Figure 1.2 The neural animal brain. All life forms represent knowledge in the form of weights of connections between neurons in their brain and body. The numbers do not mean anything to us, and we say that the representation is sub-symbolic.

the division of the human brain into three distinct regions. MacLean's model suggests that the human brain is organized into a hierarchy, which itself is based on an evolutionary view of brain development. The three regions are as follows:

1. Reptilian or primal brain (basal ganglia) was the first to evolve and is the one in charge of our primal instincts.
2. Paleomammalian or emotional brain (limbic system) was the next to evolve and handles our emotions.
3. Neomammalian or rational brain (neocortex) which is responsible for what we call as thinking.

According to MacLean, the hierarchical organization of the human brain represents the gradual acquisition of the brain structures through evolution. The human brain, considered by many to be the most complex piece of matter in the universe, is made up of a cerebrum, the brain stem, and the cerebellum. The cerebrum is considered to be the seat of thought and, in humans, comprises two halves, each having an inner white core and an outer cerebral cortex made up of grey matter.

It is generally believed that the larger the brain, the greater the cognitive abilities of the owner.

1.3 Neural Networks

A neuron is a simple device that computes a simple function of the inputs it receives. Collections of interconnected neurons can do complex computations. Insights into animal brains have prompted many researchers to pursue the path of creating *artificial neural networks* (ANNs).

An ANN is a computational model that can be trained to perform certain tasks by repeatedly showing a stimulus and the expected response. It is best suited for the classification task. The earliest neural network was the *perceptron* (McCulloch and Pitts, 1943; Rosenblatt, 1958) which had one layer of neurons and could serve as a binary linear classifier. That is, whenever two classes in some space could be separated by a line or a plane in appropriate dimensions, the perceptron could be trained to learn the position and the orientation of the separator. Research in this area suffered a setback when Minsky and Papert (1969) showed its limitations – it could only classify linearly separable classes. For example, one cannot draw a line to separate the shaded circles, representing data from class A, from the unshaded ones, representing data from class B, as shown in Figure 1.3.

The work in neural networks was revived with the publication of the BACKPROPAGATION algorithm. In the mid-1980s, Rumelhart, Hinton, and Williams (1986) showed that a *multi-layer perceptron* could be trained to learn any non-linear classifier. They also popularized the BACKPROPAGATION algorithm that fed the error at the output layer back via the hidden layer, adjusting the weights of the connections (McClelland and Rumelhart, 1986a, 1986b).

Figure 1.4 shows the schematic diagram of a typical feedforward neural network. On the left is the input layer where the discretized input is fed in, activating nodes in the layer. Then, activation spreads from the left to the output layer on the right via the nodes in the hidden layer. In Figure 1.4, there are five output nodes, which could stand for five class labels. In the simplest case, when the network has learned to classify the input, which could be an image, one output

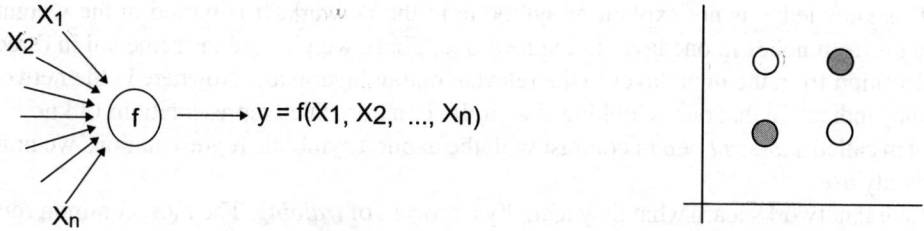

Figure 1.3 A neuron is a simple processing device that receives signals and generates an impulse as shown on the left. On the right is an example of a classification problem in which no line can be drawn to separate the shaded circle from the unshaded ones.

node is activated, indicating the class label. What the neural network has learnt is the *association* between the pattern of activation in the input layer and the class label at the output layer.

The fact that the input may be an image of a scene is only in the mind of the user, as is the name given to the class label. In the figure, the names are five animals, but the neural network has no idea that one is talking of animals, or a particular animal like a horse or a bear. It just *knows* which label to activate with a given image.

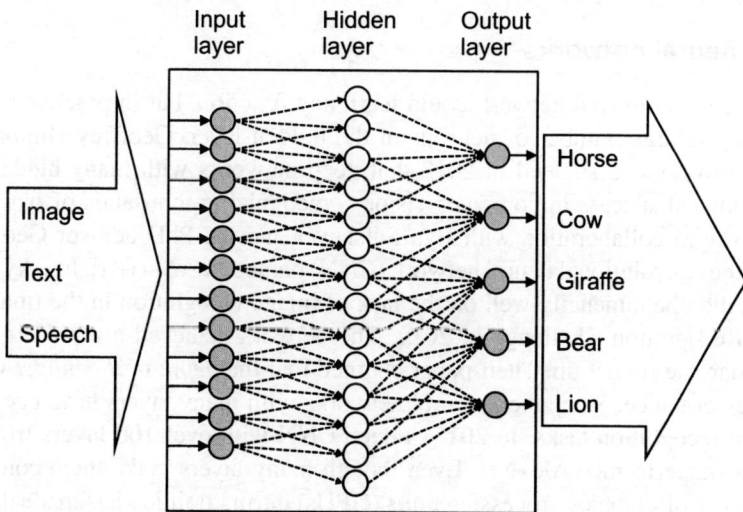

Figure 1.4 A feedforward artificial neural network learns a function from the input space to the output classes. Learning happens via the adjustment of edge weights. The labels of the output classes are meaningful only to the user.

This knowledge is not explicit or symbolic in the network. It is buried in the weights of the edges from nodes in one layer to the next one. These weights are instrumental in directing the activation from the input layer to the relevant output layer node. Nowhere in the network is there any indication that one is looking at a giraffe or a lion. Such representations of knowledge are often called *sub-symbolic* in contrast with the explicit symbolic representations we humans commonly use.

Neural networks learn what they learn by a process of *training*. The most common form of training is called *supervised learning*, in which a user presents input patterns to the network, and for each input shows what the output label should be. Every time an input pattern is presented, the network makes its own decision of what the activation value of the class label is. For example, if a bear is shown to the network, it might compute the output values as [0.2, 0.1, 0.0, 0.4, 0.3] when the expected output is [0, 0, 0, 1, 0], indicating that it is the fourth node (the bear). The error in the actual output defines a loss function that BACKPROP (as it is also known) aims to minimize. Most variations of the algorithm compute the gradient of the loss function with respect to the weights and do a small change in the edge weights in each cycle to reduce the loss. This can be viewed as *gradient descent*, an algorithm we look at later in the book.

The other forms of learning that are popular are *unsupervised* learning in which algorithms can learn to identify clusters in data, and *reinforcement* learning in which feedback from the world is used to adjust relevant weights. Reinforcement learning has achieved great success in game playing programs that learn how to play by playing hundreds of thousands of games against themselves, learning how to evaluate board positions from the outcomes of the games.

1.3.1 Deep neural networks

In principle the three layered network could learn any function, but in practice it was hard to do so, requiring a large number of neurons in the hidden layer. Geoffrey Hinton persevered with neural networks and showed in 2012 that deep networks with many hidden layers can achieve phenomenal success in computer vision – recognizing thousands of *types* of objects. Alex Krizhevsky in collaboration with Ilya Sutskever and his PhD advisor Geoffrey Hinton implemented the convolutional neural network (CNN) named *AlexNet* (Krizhevsky et al., 2017). This program did phenomenally well on the task of image recognition in the ImageNet Large Scale Visual Recognition Challenge in 2012. The network achieved a top-5 error of 15.3%, much better than the runner up. Their paper claimed that the *depth of the model* was essential for its high performance. Since then, neural networks with many layers have been doing very well in pattern recognition tasks. In 2015, a deep CNN with over 100 layers from Microsoft Research Asia outperformed AlexNet. Even though many layers make them computationally expensive, the use of graphics processing units (GPUs) during training has made them feasible. Figure 1.5 shows a schematic of a deep neural network.

The development of newer architectures and newer algorithms was instrumental in the spurt of interest in deep neural networks. Equally responsible perhaps was the explosion in the amount of data available on the internet, for example, the millions of images with captions uploaded by users, along with rapid advances in the computing power available. In 2018, three scientists, Geoffrey Hinton, Yann LeCun, and Yoshua Bengio, were jointly awarded the Turing Award for their work in this area. Deep networks got further impetus with the availability

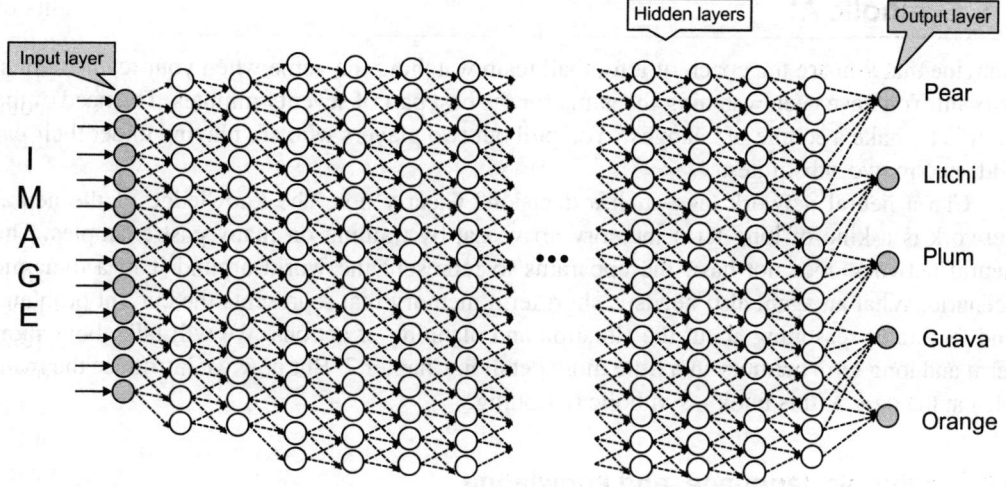

Figure 1.5 The schematic diagram of a deep neural network.

of open source software like *Tensorflow*[4] from Google that makes the task of implementing machine learning models easier for researchers.

More recently, *generative neural networks* have been successfully deployed for language generation and even creating paintings, for example, from OpenAI.[5] Generative models embody a form of unsupervised learning from large amounts of data, and are then trained to *generate data* like the one the algorithms were trained on. After having been fed with millions of images and text and their associated captions, they have now learnt to generate similar pictures or stories from similar text commands. Programs like *ChatGPT, Imagen,* and *DALL-E* have created quite a flurry amongst many users on the internet.

Deep neural networks are very good at the task of pattern recognition. Qualitatively, they are no different from the earlier networks, but in terms of performance they are far superior. The main task they are very good at is classification, a task that some researchers have commented is accomplished 'in the blink of an eye' by all life forms (Darwiche, 2018). The question one might ask is what after that?

Both in the case of generative models and deep neural network based classification, one must remember that the programs are throwing back at us whatever data has been fed to them. They do not *understand* what they are writing or drawing even though there is some correlation between the input query or command and the output generated.

For understanding and acting upon such perceived data, one needs to create models of the world to reason with. This is best done by explicit symbolic representations, which have the added benefit that they can contribute to explanations.

---

[4] https://developers.google.com/machine-learning/crash-course/first-steps-with-tensorflow/toolkit, accessed December 2022.

[5] https://openai.com/blog/generative-models/, accessed December 2022.

1.4 Symbolic AI

Imagine that you are the coach of a football team watching the game when your team is down two–nil. You have been watching the game for the best part of seventy minutes. The need of the hour is to make a couple of changes.[6] You pull out two players who are playing under their par and send in two substitutes.

Can a neural network take such a decision? Clearly not. The knowledge of the neural network is a kind of long term memory arrived at by training on many past examples. The neural network does not have the apparatus to represent the world around it in a dynamic scenario. What an agent also needs is short term memory that represents the current problem, and facilitates reasoning about the situation and planning of actions. We will talk about short term and long term memory in a little more detail in Chapter 7. But now we introduce the main idea at the core of this book – symbolic reasoning.

1.4.1 Symbols, language, and knowledge

Arguably, humankind broke away from the rest of the animal world with the development of language. The ability to give *names* to concepts combined with a shared understanding of what words mean not only has been a boon for communication (remember the boy who shouted wolf?) but has also provided a basis for representing complex concepts.

The core of *language*, whether spoken or written, is the *symbol*. A symbol is a perceptible something that stands for something else. The study of how signs and symbols are created and how they are interpreted is called *semiotics*. We are all familiar with road signs indicating the presence of schools, crossings, restaurants, U-turns, and so on. Most commercial activities are promoted using logos of companies which too stand for the company. Biosemiotics is the study of how complex behaviour emerges when simple systems interact with each other through signs. The pheromone trails left by ants for other ants to follow and the waggle dance of the honey bees to convey the location of food source to fellow bees are examples. The key feature is the use of words, behaviours, and shapes, collectively known as *semiosis*, as a means of transmitting meaningful information encoded in symbols and decoded by the receiver.

Human languages have evolved to describe what we see and perceive. The simplest kinds of names were probably just atomic but gradually we learnt to combine words to devise compound names, for example, *der liegestuhl* (the lounge chair) in German and *Himalaya* (the abode of snow) in Sanskrit. But at the simplest atomic level, a word, whether a noun, adjective, adverb, or verb, simply stands for something. Once many years ago a curious four-year-old had asked me: 'Why is a (ceiling) fan called a fan, and not something else?' The answer perhaps is that words acquire meaning via wide social agreement and also derive from related words. Words vary over regions, sometimes gradually and sometimes abruptly. The English word 'potato' corresponds to 'patata' in Sindhi, and 'batata' in Marathi. But some languages have a radically different name, 'alu' or 'aloo', for it. A look at the names of numbers across different languages also reveals a remarkable similarity that is unlikely to be sheer coincidence. Most languages have names starting with 's' for the equivalent of the number six, which is *sechs*

---

[6] Even as I write this, France has scored two goals in two minutes to draw level with Argentina in the FIFA World Cup final of 2022.

in German, *seks* in Norwegian, *seis* in Spanish and Portuguese, *shash* in Sanskrit, and *sitta* in Arabic. Likewise, the number seven is *saat* in Hindi, *saith* in Welsh, *sieben* in German, *sedam* in Serbian, *septem* in Latin, and *sapta* in Sanskrit. At the same time, the same word may mean different things in different languages, much to the consternation of uninformed travellers, who may not realize that *gift* in German means poison or a toxin, and *helmet* in Finnish means pearls. In addition, the diversity in the world across regions results in communities having fine-grained words indicating small differences in what *they* encounter in their lives. Nordic countries have a multitude of words for different kinds of snow. A Swede may use *Kramsnö* for squeezy snow, perfect for making snowballs, and *Lappvante* for thick, falling snow amongst the many words that residents of Kerala may club into one word, *snow*. At the same time, the people in Kerala have different names for a variety of what the Scandinavian countries might just refer to as a *banana*. *Ethapazham*, for example, is the name of the longest banana available, *chenkadali* is the red banana, and *poovan* a small banana.

Observe that when we talk of trees and fruits and animals, we talk of them *as we perceive* them. There is a process of *reification* or *abstraction* that happens here. A human body is made up of about 10^{27} atoms, but we do not think of it at that level of detail. We cannot. We think of a person as an *individual* and think about the body parts as individual entities. The atoms that are part of our body are anyway transient, but our notion of the self is persistent. In his book titled *Creation*, author Steve Grand (2001) highlights the fact that when we *perceive a stationary cloud* atop a mountain pass, it is really moist wind blowing over it with water molecules condensing as they reach the top and becoming visible even as they flow on. The cloud, like our own body, is in our mind. The concepts that we form in our heads are often at a convenient level of aggregation. In any case, humans started off by giving names to what we see and what we do. Our visual perception system has a finite field of vision. A fascinating chronicle on the sizes of objects in the universe, *The Powers of Ten*, lists different physical entities that exist at different scales (Morrison et al., 1986). In the book, and a short movie of the same name, the authors zoom out from human-size objects to the very ends of our universe, and them zoom back in and onwards onto the subatomic level. Our human perception is limited from about 10^{-4} m where we can see a pollen grain shining in a ray of sunlight, to larger objects – a mustard seed (10^{-3} m), a fingernail (10^{-2} m), a sunbird (10^{-1} m), a child (10^0 m), a small tree (10^1 m), a pond (10^2 m), the Golden Gate bridge (10^3 m), and a small town seen from a hill (10^4 m). We find it easy to give names for objects at these scales. For larger or smaller scales, we have to rely on science to inform us. We know that the diameter of solar system is 11,826,600,000 km, and the diameter of the Milky Way is about 100,000 light years across. We know that a virus is about 10^{-7} m, and the size of the carbon atom nucleus is about 10^{-14} m. All this is secondary knowledge derived from our scientific endeavour, even though we often cannot visualize very large or very small distances at the extreme scales. Quantum mechanics has further obfuscated our understanding of the world. Marcelo Gleiser (2022) writes that quantum physics has redefined our understanding of matter. In the 1920s, the wave–particle duality of light was extended to include all material objects, from electrons to you. Cutting-edge experiments now explore how biological macromolecules can behave as both particle and wave.

When we talk of the spoken word, we think of it as a symbol that stands for something. Symbols take on a life of their own when they are represented by tangible marks, which are not transitory like sounds, but have a degree of permanency associated with them.

The earliest humans known to engrave symbols on clay were the Sumerians in ancient Mesopotamia, which is often known as the cradle of civilization. The first engravings were pictographs, but soon evolved into more abstract entities like symbols from an alphabet. The earliest form of writing was cuneiform writing.

> First developed around 3200 B.C. by Sumerian scribes in the ancient city-state of Uruk, in present-day Iraq, as a means of recording transactions, cuneiform writing was created by using a reed stylus to make wedge-shaped indentations in clay tablets. Cuneiform as a robust writing tradition endured 3,000 years. The script – not itself a language – was used by scribes of multiple cultures over that time to write a number of languages other than Sumerian, most notably Akkadian, a Semitic language that was the lingua franca of the Assyrian and Babylonian Empires.[7]

It was replaced by alphabetic writing sometime after the first century AD. The breakthrough came when symbols were not only employed as images representing objects and events like a hunt, but abstract entities like sounds. A set of symbols forms an alphabet. Alphabetic symbols could now come together to form words, and words could form sentences. The spoken word became the written word. Different natural languages evolved in many regions of the world. The common theme was writing.

The faculty of language in turn created a mechanism of knowledge dissemination. Starting with stories in the oral tradition, the invention of writing made it possible for us to leave a permanent imprint for anyone to read at any time in any place. The invention of the internet made all this information available for everyone instantaneously.

The basis of the written word was the idea of symbols.

1.4.2 Symbol systems

While it is true that ANNs have knowledge about the world encoded in the weights, it is not knowledge accessible to the user. Humans beings too have knowledge encoded into our neural brains, but we have somehow evolved the ability of representing and reasoning with symbols. We have the ability to model the world around us in our heads and describe what we know in natural language.

Herbert Simon and Alan Newell (1976) proposed that the ability to represent symbolic knowledge and reason with it is sufficient and necessary for intelligence – 'A physical symbol system has the necessary and sufficient means for general intelligent action'. This is known as the *physical symbol system hypothesis*.

- Symbol: A perceptible something that stands for something else. For example, alphabet symbols, numerals, road signs, musical notation.
- Symbol System: A collection of symbols – a pattern. For example, words, arrays, lists, even a tune.

---

[7] https://www.archaeology.org/issues/213-1605/features/4326-cuneiform-the-world-s-oldest-writing, accessed September 2022

- Physical Symbol System: That obeys laws of some kind, a formal system. For example, long division, computing with an abacus, an algorithm that operates on a data structure (which is a symbol system).

The idea of symbolic reasoning goes back to olden times. John Haugeland (1985) traces the evolution of the *idea* of thinking being symbolic to medieval Europe, reproduced here: Galileo Galilei (1564–1642) in *The Assayer* (published 1623) says that 'tastes, odors, colors, and so on are no more than mere names so far as the object in which we locate them are concerned, and that they reside in consciousness'. Further, that 'philosophy is written in this grand book, the universe ... It is written in the language of mathematics, and its characters are triangles, circles, and other geometric figures'. Galileo Galilei gave us what we call the laws of motion, and his explanations were expressed in geometry. The English philosopher Thomas Hobbes (1588–1679) first put forward the view that thinking itself is the manipulation of symbols. Galileo had said that all reality is mathematical in the sense that everything is made up of particles, and our sensing of smell or taste was how we reacted to those particles. Hobbes extended this notion to say that thought too was made up of (expressed in) particles which the thinker manipulated. However, he had no answer to the question of how a symbol can *mean* anything. In *De Corpore*, Hobbes first describes the view that reasoning is computation early in Chapter 1. 'By reasoning', he says, 'I understand computation.' Hobbes was influenced by Galileo. Just as geometry could represent motion, thinking could be done by manipulation of mental symbols. René Descartes (1596–1650) further extended the idea by saying that 'thoughts themselves are symbolic representations'. Descartes was the first to clarify that a symbol and what it symbolizes are two different things, but then he ran into the *mind–body dualism*. If reasoning is the manipulation of meaningful symbols according to rational rules, then who is manipulating the symbols? It can be either mechanical or meaningful, but how can it be both? How can a mechanical manipulator pay attention to meaning? These are some of the questions we are still to find answers for.

1.4.3 An architecture for cognition

It has become clear that intelligence cannot be manifested by a single algorithm or a single representation. In his influential book *The Society of Mind*, Marvin Minsky (1986) emphatically argues that a mind is a society of many processes working together and in tandem. We have already hinted that ANNs are good at certain tasks like pattern recognition, but are not easy to adapt to acting autonomously in the world managing dynamic information. We highlight the different aspects of intelligence in the next section based on the proposed architecture shown in Figure 1.6.

As one can see, this is a refinement of Figure 1.1 with *a society of algorithms* coming together to constitute an autonomous intelligent agent. AI will not be a monolithic hammer. It will be a delicate orchestra of many different processes working together, each playing its own part.

The inputs that the agent works with are the same with the different senses feeding information. Much of this information is processed by neural networks that serve as signal to symbol transducers. In practice ANNs work with discretized signals, which technically are symbols themselves. However, we prefer to think of the input as being at the signal level, because it is the raw data that the machine has to work with, for example, an image represented as an array of pixels each having a discrete value stored in it. Only when a program processes

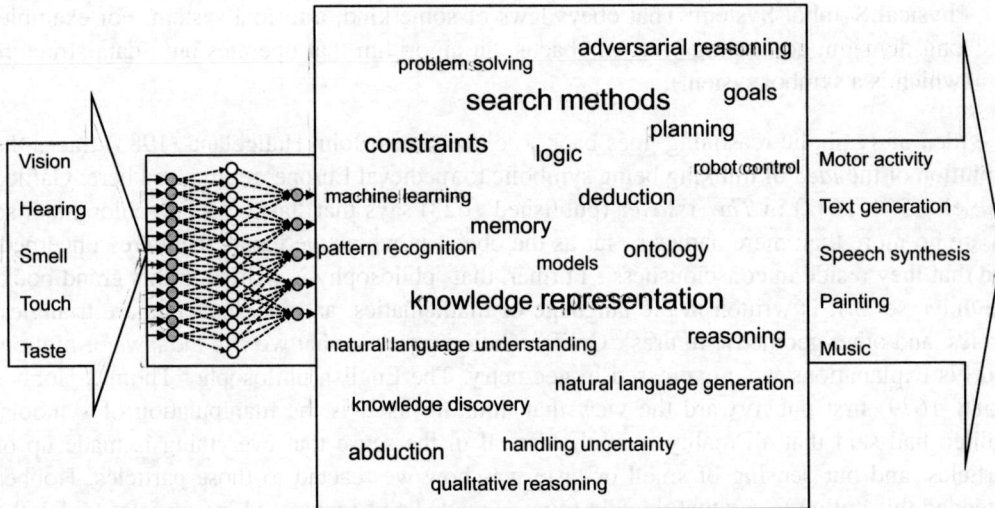

Figure 1.6 An architecture for cognition. In classical AI, an intelligent agent senses the world around it and maps it to a symbolic representation making inferences, and planning for its goals.

these pixels and extracts information about the objects in the image do we say that we have a symbolic representation. Early work on pattern recognition was syntactic in nature, for example, as described in Gonzalez and Thomason (1978). One would extract edges in the image and apply grammar rules to combine edges to (say) recognize handwritten characters. Processing complex images was not feasible at all. However, neural networks have proven to be excellent at processing images and recognizing patterns and individuals.

In our proposed architecture, the task of deliberation is done using symbolic reasoning. As shown in Figure 1.6, the deliberation phase may invoke many different algorithms. We outline the different process in the next section.

1.5 The Core of Intelligence

Given that it is hard to define what intelligence is, it is easier to adopt a behavioural characterization of intelligence – if it behaves intelligently, then it must be intelligent. Alan Turing's *imitation game* adopted the approach of conversational interaction with the idea that if it can talk intelligently, it must be intelligent. We judge a machine to be intelligent if *we* are convinced that it is behaving intelligently.

A question that has often been asked about machine intelligence is that of meaning. How could a symbol *mean* anything to a machine? We have observed that Hobbes was faced with this when he said that thinking was the manipulation of symbols. The Scottish philosopher David Hume (1711–1776) did away with the notion of meaning altogether. An admirer of Isaac Newton, Hume proposed that just like the heavenly bodies moved based on the laws of physics, impressions, and ideas were (like) the basic particles to which all mental forces and operations applied. He did not question why they did so but was satisfied by the empirical observation that they did do so.

The question of meaning was also raised by the American philosopher John Searle (1932–) with his Chinese room thought experiment in which a native English speaker locked inside a room with boxes of symbols and a set of instructions on how to manipulate them could answer questions in the Chinese language slipped below on pieces of paper without understanding a word of the Chinese language.

The digital computer manipulates symbols based on a set of instructions given to it. Does it understand the meaning of the symbols that it is manipulating? If it is adding two numbers, does it know that it is adding numbers? Do all of us understand what a number is (McCulloch, 1961)? Or if it is beating a world champion in the game of chess, does it even know that it is playing chess, or what winning is?

We will sidestep these questions on meaning and focus instead on utility and meaningful action. Build machines that operate in a purposeful goal directed manner.

In this book we assume that our goal is to build machines that autonomously solve problems for us, and that the goals of the machines are the goals we have given them to solve. Given a problem to solve, and given a set of operators in its repertoire, the task of the problem solver is to choose actions that will achieve the goal.

At the core is the ability to create a model of the world in which the agent is operating, and reason about its goals, plans, and actions with the representation. The model of the world is the base for all cognitive activity. This model contains the memories of the agent, lessons learnt, and the representation of the world in which it operates. The agent needs the ability to imagine worlds that are not immediately perceptible, or which the agent may desire to create.

Broadly speaking, there are three kinds of processes that come together to solve a problem, and which form the core of intelligent behaviour.

1.5.1 Remember the past and learn from it

Problem solving refers to the activity of making, and acting upon, decisions to transform a given situation to a desired one. The ability to solve problems is perhaps the prime hallmark of intelligent behaviour. Very often we look at a problem and a solution comes to our mind. This happens because we humans have memory, and we draw upon our past experiences, and that of others via language. We ascribe lack of intelligence to people who make the same mistakes repeatedly. And we believe that experts are those who learn even from mistakes of others. Figure 1.7 shows the learning cycle an agent may go through during problem solving. Different mechanisms for exploiting memory may have different forms of learning.

One approach that has focussed on storing past experiences is *case based reasoning* (CBR). CBR has been surprisingly effective in many industrial applications (Watson, 1997, 2002). The strategy behind CBR is simple.

- Similar problems have similar solutions, is the adage behind memory based reasoning.
- And, importantly, problems are indeed often similar.

In the simplest form, CBR maintains a *case base* of *problem solution* pairs <p, s>. The problem part of a case is a description of the problem that the case solves. The description may be attribute value pairs or it could be in natural language text. The following is the 4R cycle that CBR follows.

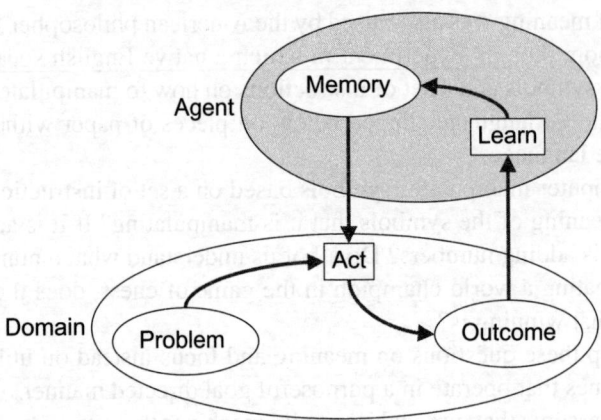

Figure 1.7 A memory based agent employs the knowledge stored in its memory to solve a problem in the domain. Based on the outcome of each instance of solving a problem, the agent refines its knowledge and improves over time.

- Retrieve: When the agent encounters a new problem, it searches the case base for the most similar problem. Often more than one case is retrieved in the style of *k nearest neighbours* retrieval.
- Reuse: The solution that is retrieved along with the case is adapted to the current problem. This could involve adjusting some parameters that are different in the current description and the retrieved one and adjusting the solution part.
- Revise: If the solution does not work, tweak it. This could involve human intervention.
- Retain: If the revised solution is significantly different, add it to the case base. The next time a similar problem shows up, this could be useful.

CBR has been particularly useful in domains that are not well modelled and where the problem solving knowledge is more *experiential* than *analytical*. One of the earliest successes was the *Clavier* system developed to cure aircraft parts at Lockheed Missiles and Space Company in Sunnyvale, California (Watson, 1997). The task involved placing parts made of composite materials that kept changing in an autoclave, which is an expensive resource. The quality of the product depended on where it was placed on the tray, and operators were essentially following similar layouts from the past. Curing is an unrewarding art, rather than a science, but Clavier reduced the discrepancy reports considerably as its case base grew from an initial twenty to several thousands. Figure 1.8 shows a schematic of a CBR system employed for knowledge management in a manufacturing setting (Khemani et al., 2002).

CBR is a form of instance based learning (see also Chapter 11) in which the system memorizes past experiences and remembers them. Another approach is to assimilate the knowledge accrued from experience into compact structures that can be used. *Neural*

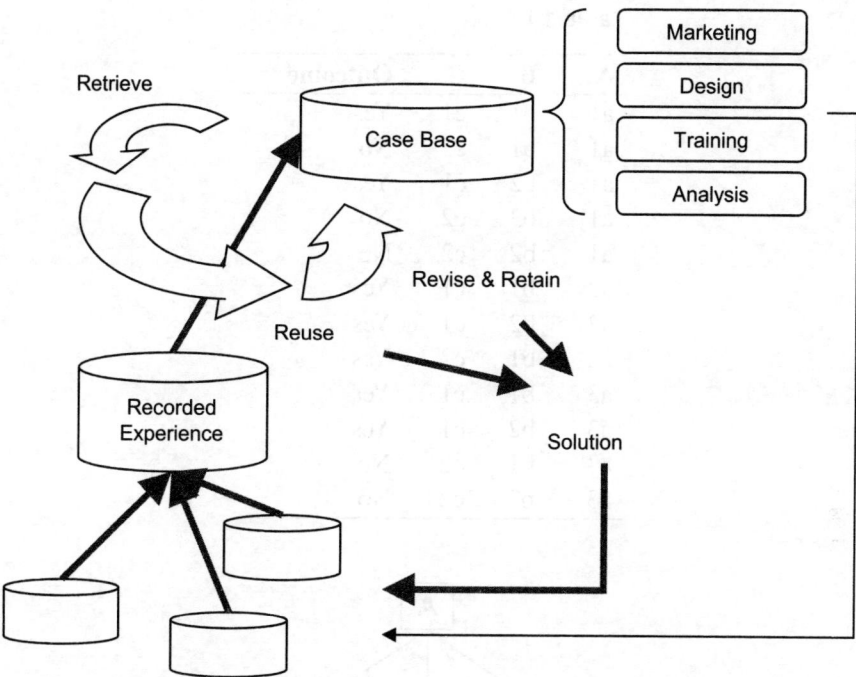

Figure 1.8 A CBR system in the manufacturing industry. The data recorded from different locations in the shop floor is assimilated into a case base. The resulting system can have multiple applications. The CBR is essentially a tool for knowledge management.

networks are examples of such learning, but there have also been more explainable structures like *decision trees*, where attributes and their values used for classification can be read from the path in the tree. Consider a small data set shown in Table 1.1 for the sake of illustration. There are three attributes A, B, and C in this data set with values {a1, a2, a3}, {b1, b2, b3}, and {c1, c2} respectively. There are two class labels 'Yes' and 'No' in each row in the table.

One could of course use CBR for prediction given a new problem in which the values of the three attributes are given. But for such well defined domains, it is convenient to build a *decision tree*. A decision tree is a discrimination tree that tests for the value of one variable at the root node, and then traverses an appropriate branch to test the value of another variable. The algorithm for constructing a decision tree with nominal attribute values is the well known *ID3* algorithm (Mitchell, 1997). The basic idea behind the algorithm is to choose that attribute that separates the two classes as best as possible. A decision tree for the data in Table 1.1 is shown in Figure 1.9.

When a new record of the values for A, B, and C comes in, it is dropped down the tree. At each node, the value of some attribute is tested and the record follows an appropriate branch. Leaves in the tree are labelled with class information. Observe that other trees may be possible, testing a different attribute at each stage. The ID3 algorithm is designed to build short trees.

Table 1.1

A	B	C	Outcome
a1	b1	c1	Yes
a1	b1	c2	No
a1	b2	c1	Yes
a1	b3	c2	No
a1	b2	c2	No
a2	b1	c1	Yes
a2	b2	c1	Yes
a2	b1	c2	Yes
a3	b1	c1	Yes
a3	b2	c1	Yes
a3	b1	c2	No
a3	b2	c2	No

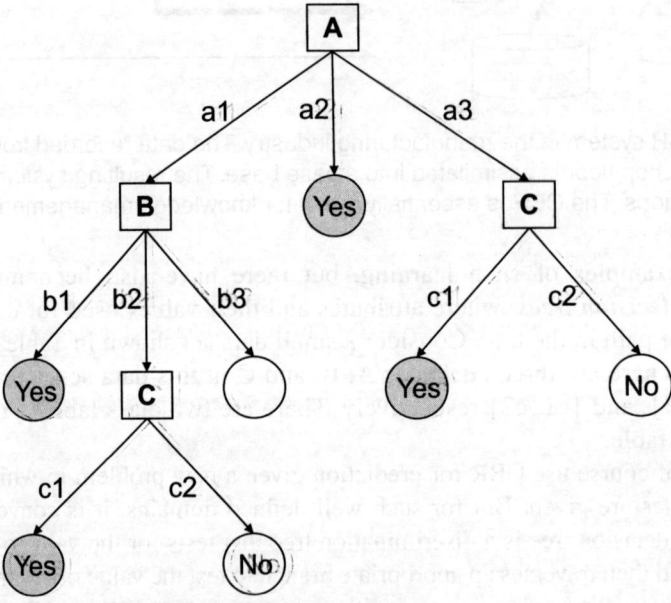

Figure 1.9 A decision tree based on the data from Table 1.1. The root nodes tests for the attribute A and traverses the appropriate branch. The leaf nodes have the class labels.

1.5.2 Understand and represent the present

The world around us and including us operates according to and can, in principle, be explained by the fundamental laws of physics. Nothing else is needed. But we do not think of the world around us as swarms of particles. This is simply because there are far too many of them, even to describe a grain of rice.

We the thinking creatures create our own worlds in our minds. And it is only our own creation that is meaningful to us. We create *categories* in our heads, for example, a human, a fox, a river, and a tree. We define an *ontology* in terms of which we represent the world. Every domain of study from physics to chemistry, to biology, to economics defines its own ontology. Or its own *terminology*. While the notion of an ontology has roots in philosophy, it has found a formal definition in computer science, as an *explicit specification of a conceptualization* (Gruber, 1993).

Behind every word of a language there sits a concept, and a knowledgeable agent relates that concept to others, within an ontology. For example, we associate the word 'banana' with a particular kind of fruit, growing on a particular kind of a single stem tree, which has a skin that can be peeled off before eating, and which has leaves that can be used to serve a meal on. Likewise, with verbs we can conjure up actions mentally, for example, jogging. In any case, whenever we use words in a language, they just stand for some concepts that an agent may have in its knowledge representation scheme. Roger Schank and his group at Yale university showed that the moment one talks of a person going into a restaurant one needs to retrieve all that one knows about what typically happens in a restaurant to make sense of the conversation (Schank and Abelson, 1977; Schank and Riesbeck, 1981). At around the same time, Marvin Minsky (1975) published his idea of *frames* for knowledge representation, which eventually led to the ideas of object oriented programming.

With the advent of the internet, when programs could talk to each other, defining ontologies gained prominence. Computational ontologies are a means to formally model the structure of a system, that is the relevant *entities* and *relations* that emerge from its observation, and which are useful to our purposes. The backbone of an ontology consists of a generalization/specialization hierarchy of concepts, that is a *taxonomy* (Guarino et al., 2009).

Figure 1.10 shows a snippet of a sample ontology represented as a frame system. The shaded squares are abstract frames, corresponding to concepts in an ontology. The *IS-A* slot in a concept defines an abstraction hierarchy, for example, 'a dog is a mammal'. The unshaded nodes represent individuals or instances of concepts. An ontology may have other kinds of links, for example, the fact that 'Ted is a dog owned by Socrates who is a human'.

The idea of *semantic networks* was already well developed. A semantic network is a graphical model in which nodes representing concepts are connected with labelled edges representing relations. Early work on semantic nets was motivated by natural language processing. Ross Quinlan is often credited with crystalizing the idea (Quillian, 1967, 1968). Subsequently, the idea of semantic nets evolved into the idea of *knowledge graphs*, which were semantic networks spread over the internet. The idea of the *Semantic Web* evolved in the twenty-first century. In 2012, Google adopted the term 'knowledge graph' (Singhal, 2012).

A knowledge graph is a collection of nodes and named edges. We can create an *abstract type* called *event* and describe the rest of the relations for an *instance* of that event. It has become common to express these as *triples* <subject, predicate, object> or <subject, property, value>. For example, here is an incident of kids fighting: 'Divya hit Atul with a stick yesterday afternoon in a park.'

    (Hitting_event, Instance, $e45$)   : $e45$ is an instance of Hitting_event
    ($e45$, Actor, divya54)   : The actor of $e45$ is Divya
    ($e45$, Object, atul81)   : The object of $e45$ is Atul

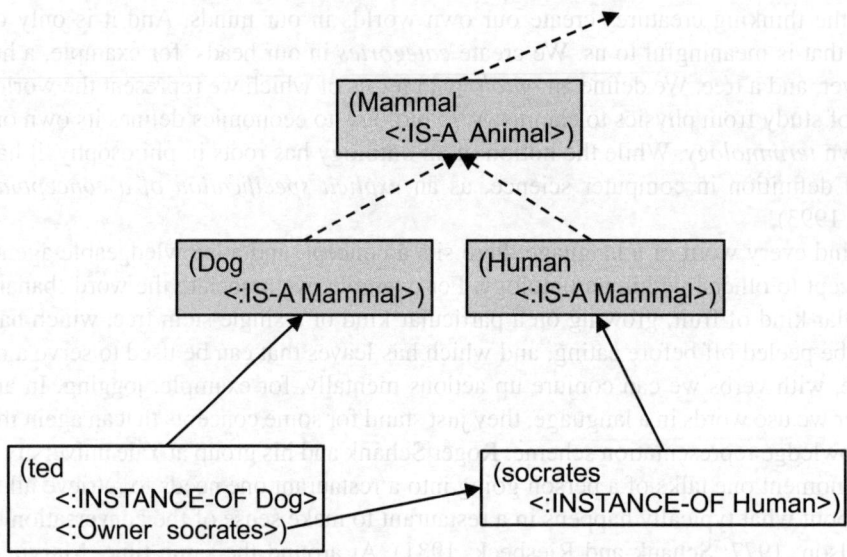

Figure 1.10 A snapshot of an ontology captured in a frame system. The shaded nodes are abstract concepts, and the unshaded ones are instances of concepts. An ontology may define an abstraction hierarchy as well as other kinds of relations.

(*e45*, Date, March_21) : The date of *e45* is March 21
(*e45*, Loc, park28) : The location of *e45* is the park
(*e45*, Instrument, stick14) : The Instrument of *e45* is the stick

Figure 1.11 shows how information about musical performances and poetry could be represented as a knowledge graph.

Representation is only one side of the coin. *Reasoning* is the other.

We are never privy to everything there is to know. We have partial knowledge of the world and can try and fill in by making *inferences*. The process of making inferences is called *reasoning*. There are three kinds of inferences. All are in some way connected with a logical relation often captured as 'IF antecedent THEN consequent', expressed as a sentence in some logic. When the sentence is true, then it often expresses a causal connection from the antecedent to the consequent. But as Judea Pearl (2019) has shown to us, there can be confusion between causality and correlation, which has been exploited by the tobacco industry to contest the connection between smoking and cancer.

- Deduction: From a given set of facts, infer another fact that is necessarily true. Deduction is the bread and butter of logic. We study deduction in Chapter 10. Deduction is sound because it goes from the antecedent to the consequent. For example, the statement 'If X is a trapezium then X is a quadrilateral' is true by definition. So if anyone has drawn a trapezium, then she has drawn a quadrilateral. Deduction only makes explicit what is implicit in the knowledge base.

Introduction | 21

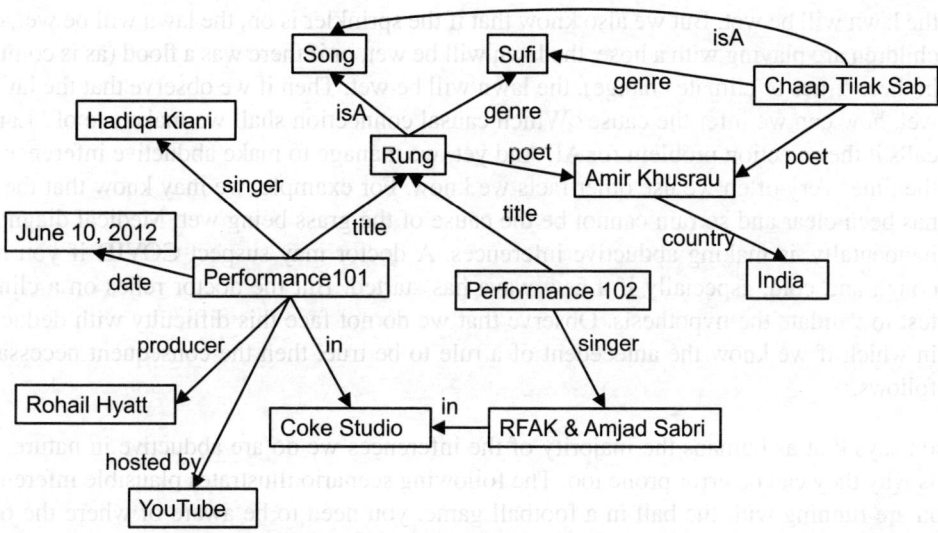

Figure 1.11 A triple subject, predicate, object store can store heterogeneous information in a knowledge graph, with nodes connected by directed edges. Each labelled edge goes from subject to object and is labelled by the property or the predicate. The figure shows a snippet of a knowledge graph relating to music and poetry.

- Induction: From a given sets of facts, infer a new fact. Also known as generalization. Induction can create new knowledge. Recognizing that a number of entities in the domain share some common property and asserting that as a general statement. For example, from the observations,

    The peepul leaf is green.
    The tamarind leaf is green.
    The neem leaf is green.
    The mango leaf is green.

  one can conclude,

    All leaves are green.

  Induction or generalization is the basis of machine learning. Such conclusions are, however, not sound, which means they are not necessarily true. One may know of a plant that does not have green leaves, for example, the Japanese maple. Nevertheless, the ability to generalize has tremendous use in practice.

- Abduction: From a given set of facts, infer another fact that is possibly true. For example, if one has cough and fever, one might hypothesize that one has COVID. But one might have a cough and fever from other causes as well. One cannot say with certainty what the cause is. Eric Larson (2021) explains this with a well known example. We know that if it rains, the lawn will be wet. But we also know that if the sprinkler is on, the lawn will be wet, or if children are playing with a hose, the lawn will be wet, or if there was a flood (as is common in these times of climate change), the lawn will be wet. Then if we observe that the lawn is wet, how can we infer the cause? Which causal connection shall we make use of? Larson calls it the selection problem for AI. And yet, we manage to make abductive inferences all the time. Very often we use other facts we know. For example, we may know that the sky has been clear and so rain cannot be the cause of the grass being wet. Medical diagnosis, incidentally, is making abductive inferences. A doctor may suspect COVID if you have cough and cold, especially if a new wave has started. But the doctor relies on a clinical test to validate the hypothesis. Observe that we do not face this difficulty with deduction in which if we know the antecedent of a rule to be true, then the consequent necessarily follows.

Larson says that as humans the majority of the inferences we do are abductive in nature, and that is why they can be error prone too. The following scenario illustrates plausible inferences. If you are running with the ball in a football game, you need to be aware of where the other players are and what they intend to do. This inference of intention comes from background knowledge about the strategies and tactics used by the team. You should be able to imagine that if you kick the ball to where your teammate should be running to, then he would have a better shot at the goal. The opponents no doubt are thinking about it too. Why is the opposing team player running towards that spot? Making inferences is the basic cognitive act for intelligent minds and we are constantly making inferences.

Another example is the work done by Roger Schank and his group with stereotypical situations knowledge which is instrumental in generating expectations about what must have happened and what to expect. If we hear that 'John went to a restaurant. He ordered a masala dosa. He left satisfied,' we can imagine what must have happened because we have knowledge about how restaurants function, even though the story is cryptic. We know that he must eaten the dosa, and must have paid for it, because that is the normal behaviour in a restaurant.

In summary, the agent must be able to *reason* with what it knows to infer what is implicit (deduction) or even what is unknown (induction) to create new knowledge. It must be able to hypothesize connections between facts and events (abduction) to anticipate what is happening in the world around it, and what other agents are up to. It must be able to recognize intentions and plans of collaborators and adversaries, make its own plans, evaluate and choose the best ones, execute them, monitor them as they are executed, replan if necessary or take advantage of an unexpected opportunity. It must also be able to use the science of probability to judge which of its possible decisions is most likely to succeed.

The previous two subsections have described in a nutshell those aspects of AI that would each need a complete book for any justice to be done. We have briefly dwelt upon these to highlight the fact that they are necessary for building intelligent systems, along with other processes shown in Figure 1.6.

While all these are necessary, we now come to the subject matter of this book – *solving problems by first principles* by projecting decisions into the future to tease out those that solve the problem. The search methods that we study in this book arrive at solutions by trying out different options available to the agent. In the following section we outline the contents of this book.

1.5.3 Imagine the future and shape it

We have seen that having a memory enables an agent to capture and store experiences that can facilitate solving problems. Memory based reasoning allows us to reuse solutions from the past. Why would one want to reinvent the wheel every time? In addition, the faculty of language allows agents to share knowledge and experience allowing new researchers to 'stand on the shoulders of giants' (Hawking, 2003).

But the world we live in is not always so simple. For every nugget of wisdom, we can find one that says the exact opposite.[8] The Greek philosopher Heraclitus of Ephesus (535–475 BC) believed that change is a constant and famously said – *No man can step into the same river twice*. Problems are often similar, but not always. When faced with a new problem, the intelligent agent needs to solve it by first principles, *ab initio*. This book is devoted to such an approach to solving problems, in which the agent searches for solutions by projecting proposed actions into the future.

The ability to model the world is central to intelligent behaviour. This model must include not only the description of the world but also the moves available to the agent to achieve the goals it has. The algorithms that we study allow the agent to *imagine* the consequences of the actions and decisions and try out various combinations in a *mental simulation* to select the sequences that would work. The ability to imagine is perhaps the single most important characteristic of intelligence.

We now describe the chapterwise contents of this book.

Chapter 2 introduces the basic machinery of search. We begin by defining the state space, which is the space of all possible states. We define a neighbourhood function, MoveGen(*state*), that specifies the moves that an agent can make in any given state. The task of the search algorithm is to find a sequence of such moves from a given *start* state to a desired *goal* state. A function called GoalTest(*state*) inspects a state and identifies whether it is a goal state or not. The search algorithm begins at the *start* state, uses to MoveGen function to generate and navigate the state space. Then it either returns the solution or reports failure if it has exhausted all possibilities. In the process, it has generated and explored a part of the state space in the form of a *search tree*.

We also introduce the notion of a *solution space*, which is a space of *candidate solutions*. Instead of moves in the state space from one state to another, the neighbourhood function in the solution space is *perturbation* of candidates in search of the solution.

Chapter 3 introduces the simplest search strategies, *depth first search* and *breadth first search*, familiar to the students of graph algorithms. The difference in AI search is that the graphs are generated on the fly. Apart from time and space complexity, we also look at completeness and the quality of solution found. We compare the two algorithms and devise a new algorithm, *depth first iterative deepening*, that combines the best features of both. In all three cases, we recognize that the search spaces grow exponentially due to *combinatorial explosion (CombEx)*. This leads us to explore various approaches to mitigate the effects of CombEx in the chapters that follow.

---

[8] Consider the two sayings – 'Out of sight, out of mind' versus 'Absence makes the heart grow fonder'.

Chapter 4 introduces *informed search* aimed to guide the search towards the goal instead of the *blind* or *uninformed* search methods of Chapter 3. We introduce the idea of a *heuristic* function $h(n)$ that looks at a state n and computes an *estimate* of its *distance* to the goal state. Search that employs such a heuristic function is called *heuristic search*. From the set of all available candidates, the algorithm *best first search* picks that node that appears to be closest to the goal. We show that if a solution exists, the algorithm will find it. That is, it is complete for finite state spaces. The performance of the algorithm depends upon the quality of heuristic estimate, and it has been empirically found that most implementations still need exponential time and space.

In an effort to save on space, we resort to *local* search. The algorithm *hill climbing* burns its bridges and considers only the neighbours of the current state (or candidate solution). Instead of using the GOALTEST function, it looks for an optimum value of the heuristic function. While it does result in reduced complexity, it loses out on completeness. It does not guarantee finding a solution and can get stuck in a *local optimum*. Then begins the quest for variations in local search more likely to succeed in finding a solution. We look at the algorithm *beam search* that explores more than one path, and at *tabu search* that allows a search algorithm to get off a local optimum and continue exploration. We also introduce an aspect of stochastic search with *iterated hill climbing*.

Chapter 5 is devoted to stochastic local search methods. All the algorithms in the chapter draw inspiration from processes in nature. *Simulated annealing* begins with randomized moves and gradually makes them deterministic, reminiscent of the annealing process used to form materials in the lowest energy states. *Genetic algorithms* mimic the process of survival of the fittest in natural selection and mix and churn the components available in a population of candidates. We look at how genetic algorithms solve the travelling salesperson problem. Finally, we introduce the *ant colony optimization* algorithm that draws inspiration from how ants communicate with each other via pheromone trails and collaborate to find shortest paths. All the three algorithms studied are popular in the optimization community.

The algorithms studied so far do not guarantee an optimal solution (except breadth first search). Chapter 6 introduces the well known algorithm $A*$ that employs heuristic search and also guarantees an optimal solution even for infinite state spaces. We say that the algorithm is *admissible* and we present a proof of its admissibility. We introduce the problem of sequence alignment from bioinformatics where A* is applicable. We then look at space saving variations of A* that can solve much bigger problems than A* can.

Chapter 7 looks at problem decomposition and how problems can be solved in parts. We begin by looking at *pattern directed inference systems* in which an algorithm looks for patterns in the given state and triggers appropriate actions. The production system architecture lays the foundations of building *rule based expert systems*. We present the RETE algorithm which is an efficient implementation that is used in many *business rule management systems*. Such systems also serve as a vehicle of declarative programming exemplified by the language *OPS5* in which the user just writes the rules and an *inference engine* decides which rules to apply to what data.

We then look at a goal directed approach to breaking down problems into subproblems with *And–Or* graphs. The idea is to decompose the problem into simpler subproblems and continue the process till the reduced problems are primitive problems with trivial solutions. We present the algorithm AO* that can be used to find an optimal least cost decomposition strategy.

In Chapter 8, we turn to games as an example of adversarial reasoning. We introduce games that are popular amongst humans and computer programmers, and focus on two person board games like chess. We begin with the idea of the minimax value that is the Nash equilibrium for a game. Games like chess have game trees that are too large to be analysed completely though, and one needs to introduce the notion of an *evaluation function* that allows the *minimax algorithm* to find a move by limited lookahead. Then we look at efficient versions, *alpha beta pruning* and *SSS**, that do progressively more pruning of the game tree. We conclude with some commentary on the program AlphaGo that created a sensation by beating the *go* champion in 2016, and on the games backgammon and scrabble. Finally, we introduce contract bridge as an open challenge.

Chapter 9 is devoted to automated planning that has emerged as an independent area of research, with its own representation and move generation schemes. The community has devised an array of domain description languages allowing for richer descriptions and actions. A distinctive feature of planning domains is that the planning operators have an inbuilt arrow of time. When an action that is applicable is applied to a given state in a forward direction, it always results in another state. Searching backwards from goal descriptions is desirable because of low branching factor but runs into inconsistencies that require double checking the plan found. We look at a variation called *goal stack planning* that gets around this problem. We also look at *plan space planning* that searches in the solution space, allowing in principle separation of action selection from action scheduling. Next, we look at two stage approaches that first encode the planning problem into an intermediate representation and then solve that. Our focus in this chapter is on the algorithms that work in the simplest planning domains called *STRIPS* domains, but we do describe the more expressive planning domains that the community has engaged with.

Chapter 10 takes a quick look into the rich world of representation and reasoning. We look at the notions of *entailment* and *proof*, and at how deduction involves search. Our focus is on formal reasoning as has evolved in the logic community with emphasis on sound and complete theorem proving. We confine ourselves to *first order logic* that occupies a major part of the representation landscape in computer science. Our goal is to show how search is instrumental in reasoning. Like in planning, we explore both *forward reasoning* from facts to conclusions and *backward goal directed reasoning* for theorems to be proved. Both turn out to be incomplete, in the sense that given a knowledge base, there can be statements that are entailed, but the two algorithms are unable to find a proof for them. We then introduce the well known *resolution refutation method* that is both sound and complete. We also introduce the idea of logic as a programming language and also look at some more expressive languages briefly.

Chapter 11, written by Sutanu Chakraborti, takes a brief look at *machine learning*, highlighting the fact that the process of learning or training involves search in the space of possible models. A model encapsulates a function from the input to the output and usually computes the output quickly. This is because the relation between the input and output is acquired and made explicit in the learning phase. Models may go through extensive training phase, but the process of using them is quick. We look at some examples of training, including Bayesian classification, *k* nearest neighbours, decision trees, and neural networks. We also look at the K-means algorithm as an example of unsupervised learning.

Finally, Chapter 12 looks at *constraint satisfaction* which offers the tantalizing possibility of integrating the different kinds of processes needed for intelligence into one. We study finite domain *constraint satisfaction problems*, and show how search and reasoning can be combined, and look at some algorithms where reasoning is effective in reducing the search effort. The most attractive feature of constraints is that they offer a unifying formalism for representation, when solutions can be found by general purpose methods. Eugene Freuder (1997), one of the founding figures in constraint programming, has said – 'Constraint Programming represents one of the closest approaches computer science has yet made to the Holy Grail of programming: the user states the problem, the computer solves it'.

A Note for the Reader

The proof of the pudding is in the eating. It is all very well to speculate upon the nature of intelligence and whether machines will ever be able to think or not. There is no compelling argument that they will not be able to do so. At the same time, there is no convincing demonstration of artificial general intelligence so far. Yet machines continue to excel in individual tasks doing many useful things, from diagnosing diseases from images and radiographs, finding routes for us in a new city, controlling an autonomous robot on Mars, and beating us humans on all board games. All this is done with algorithms to tackle specific problems. Will they all come together as one entity that will solve all problems that have been solved independently? A key stumbling block is going to be knowledge representation. Can we create a general representation scheme in which all problems can be posed and solved? Then the machine will be capable of analogical thinking across diverse domains, and perhaps spot connections hitherto unknown. If we can do that, then we would have surpassed humans on this aspect, because no individual human is a master of all. All of us professionals in the modern world work with our narrow domain ontologies and are specialists in our areas of expertise. The frontiers of knowledge have expanded so much that the days of the Renaissance men like Leonardo da Vinci are over.

Meanwhile, the student of AI must acquire the tools of the trade. The building blocks. Understand the algorithms that are beneath the hood of a problem solver and learn to implement them. The crux of this book is a collection of 50 odd algorithms addressing different aspects of problem solving. These algorithms are described in pseudo code that may need getting used to. S Baskaran, an expert programmer, has put together *an appendix* that is at the end of the book. We urge the reader to study that before embarking upon the book. It should go a long way in your quest of a suite of programs for sophisticated, interesting, and useful applications.

Exercises

1. Alan Turing prescribed the Imitation Game as a test of whether machines can think or not. We call the test the Turing Test. Discuss the merits and demerits of the Turing Test. Is the ability to chat intelligently a sufficient indicator of intelligence? What is the role of world knowledge in a meaningful conversation? How should a machine react to a topic it does not know about and still convince the judge that he is chatting with a human? Should a machine introduce errors or delays intentionally, for example, when given a massive arithmetic problem?

2. Devise three sets of questions for the Winograd Schema Challenge that would require world knowledge to answer correctly.
3. Natural language is notoriously ambiguous, a fact that has been widely exploited to create punch lines that surprise the listener. For example, 'Time flies like an arrow, fruit flies like a banana' sometimes attributed to Groucho Marx. When humans parse language, they start building a semantic picture of what they are listening to. Garden path sentences force the listener to abandon an initial likely interpretation after hearing the complete sentence. For example, 'The old man's glasses were filled with sherry'. Given the sentence 'She shot the girl with the rifle', how would a computer chat program answer the following question 'Who had the rifle?'
4. In Chapter 8, we discuss games as models of rational behaviours aimed at maximizing the agent's own payoff. While this is an economic model that explains why individuals, corporates, and nations behave the way they do, what does it say about the collective intelligence of humankind whose focus is on arms manufacture, sale, and use, even while climate change looms upon us?

CHAPTER 2
Search Spaces

In this chapter we lay the foundations of problem solving using first principles. The first principles approach requires that the agent represent the domain in some way and investigate the consequences of its actions by simulating the actions on these representations. The representations are often referred to as *models* of the domain and the simulations as *search*. This approach is also known as *model based reasoning*, as opposed to problem solving using *memory* or *knowledge*, which, incidentally, has its own requirements of searching over representations, but at a sub-problem solving retrieval level.

We begin with the notion of a *state space* and then look at the notion of *search spaces* from the perspective of search algorithms. We characterize problems as *planning problems* and *configuration problems*, and the corresponding search spaces that are natural to them. We also present two iconic problems, the Boolean satisfiability problem (SAT) and the travelling salesman problem (TSP), among others.

In this chapter we lay the foundations of the search spaces that an agent would explore.

First, we imagine the space of possibilities. Next, we look at a mechanism to navigate this space. And then in the chapters that follow we figure out what search strategy an algorithm can use to do so efficiently.

Our focus is on creating domain independent solvers, or agents, which can be used to solve a variety of problems. We expect that the users of our solvers will implement some domain specific functions[1] in a specified form that will create the domain specific search space for our domain independent algorithms to search in. In effect, these domain specific functions create the space, which our algorithm will view as a graph over which to search. But the graph is not supplied to the search algorithm upfront. Rather, it is constructed on the fly during search. This is done by the user supplied *neighbourhood function* that links a node in this graph to its neighbours, generating them when invoked. The neighbourhood function takes a node as an input and computes, or returns, the set of neighbours in the abstract graph for the search algorithm to search in.

---

[1] Throughout this book, we will use the word 'function' as a synonym for a program. The function accepts the input as an argument and *returns* the output it computes. This is in accordance with the style of functional programming.

2.1 The State Space

The *state space* is the space of all possible states. A *state* is a description of the world in some language and corresponds to a node in the *state space*. The problem solver begins in a given or *start* state and has some desired state as a *goal* state. The desired state can be a single state that is completely specified, or it may be a set of states described by a common property.

The agent has access to a neighbourhood function, which defines all the *moves* that can be made to transform any state in some way. We will use the name MOVEGEN function for the neighbourhood function, to signify that it *generates the moves* in a given state. It takes a state as input and returns the set of its neighbours. The MOVEGEN function will have to be supplied by the user for our solvers. We will also assume that the user has defined a GOALTEST function that takes a state as input and determines whether it is a goal state or not. This will provide a termination criterion for the search algorithm.

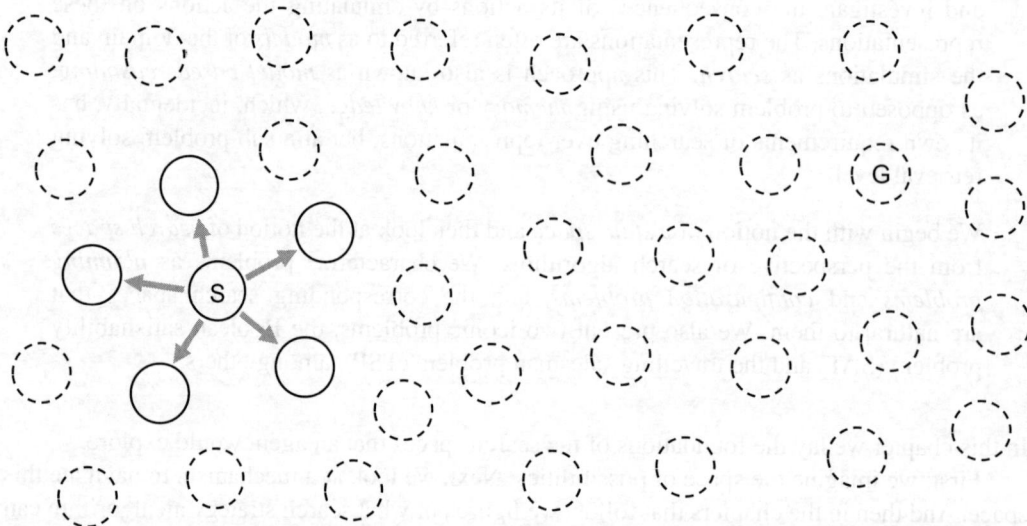

Figure 2.1 A state space is the set of possible states. Each state can be seen as a node in an implicit graph. A neighbourhood function we call MOVEGEN(N) takes a node as input and returns the neighbours of N that can be reached in one move. We call the given state as the start state S and a desired state as a goal state G. There may be more than one goal states. The figure shows the neighbours of the state S generated by MOVEGEN. The graph is generated on the fly by repeated calls to MOVEGEN.

The state space can be seen as a graph. However, it is a graph that is implicit and defined by the MOVEGEN function, as depicted in Figure 2.1. Observe that we have shown the edges in this figure as directed edges. This is only to highlight the exploration away from the start state S. In practice, the edges in a state space may be directed when the moves are not reversible, or undirected when they are. For example, in the domain of cooking, chopping vegetables cannot be undone, but in the river crossing problem described later, one could simply row the boat back. In some problems some moves may be reversible while others are not, for example, the water jug problem described later. The MoveGen function captures all these variations. For any node as input, it returns the set of neighbour nodes reachable in one move.

MoveGen(N)
  Input: a node N
  Output: the neighbours of node N

For the moment, we will not *name* the moves, and simply rely on the MoveGen function to provide the resulting states. Later, in Chapter 9, we will look at how the automated planning community explicitly represents actions and reasons with them.

2.1.1 Generate and test

A search algorithm navigates the state space generated by the MoveGen function. It will terminate when it has reached the goal state. We require the user to supply another function, GoalTest, that will identify a goal state.

GoalTest(N)
  Input: a node N
  Output: *true* if N is a goal state, and *false* otherwise

The GoalTest function is a predicate that returns *true* if the input is a goal state. The high level search algorithm maintains a set of candidate states, which is traditionally called *OPEN*, and repeatedly picks some node N from OPEN, till it picks a goal node. OPEN is initialized to the start state S. It returns the goal state if it finds one, else it returns *fail*. That happens if there is no way of reaching the goal state. In graph theory terms, there is no path to the goal state.

Algorithm 2.1 Algorithm *SimpleSearch* picks some node N from OPEN. If N is the goal, it terminates with N, else it calls MoveGen and adds the neighbours of N to OPEN.

SimpleSearch()
  1    $OPEN \leftarrow \{S\}$
  2   **while** *OPEN* is not empty
  3   **do** pick *some* node N from OPEN
  4         $OPEN \leftarrow OPEN - \{N\}$
  5   **if** $GoalTest(N) = TRUE$
  6   **then** return N
  7   **else** $OPEN \leftarrow OPEN \cup MoveGen(N)$
  8   **return** FAILURE

This problem solving strategy is also known as *generate and test*. It embodies the strategy of navigating the space, in some order, generating candidates one by one, and testing each for being a goal state. As one can imagine, the choice of *which* node to pick from OPEN will determine how quickly the algorithm terminates. This choice will be addressed in the next few chapters.

2.2 Search Spaces

The state space is the world of possible states. A search space depicts how a search algorithm explores these possibilities. The high level algorithm described earlier does not specify *which* node to pick up for inspection. The simple search algorithm begins at the start node and repeatedly picks *some* node from OPEN, and adds more nodes to OPEN at each stage. The various choices it can make define its search space. Figure 2.2 shows the search space corresponding to a tiny state space.

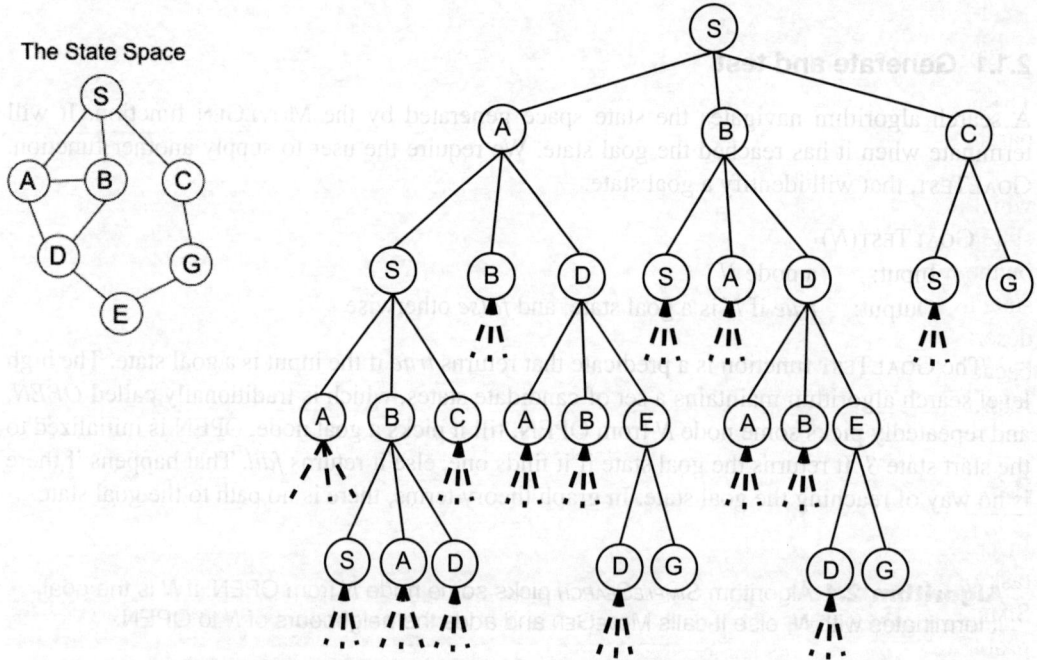

Figure 2.2 Given a tiny state space with a start state *S* and a goal state *G*, the simple search algorithm defines a search space which is a tree. Starting with *S*, it can go down any branch in search of a goal node. Observe that the tree has some infinite branches.

As the simple search algorithm does not specify which candidate to pick for inspection next, it could go down any path in the search tree. Observe that some of these paths are of infinite length, wherein the algorithm traverses cycles in the state space. In the following chapters we will look at different strategies for exploring this search space and study their properties.

We will evaluate the performance of our search algorithms on the following four criteria.

- Completeness. An algorithm is said to be complete if it is guaranteed to find a goal state if one is reachable. We also call such algorithms as being *systematic*. By this we mean that the algorithm searches the entire space and returns *fail* only if a goal state is not reachable.

- Quality of solution. We may optionally specify a quality measure for the algorithm. We begin with the length of the path found as a measure of quality. Later, we will associate edge costs with each move, and then the total cost of the path will be the measure. This will typically be in the case for *planning* problems (described later) where we can associate a cost with the solution found.
- Space complexity. This looks at the amount of space the algorithm requires to execute. We will see that this will be a critical measure, as the number of candidates in the search space often grows exponentially.
- Time complexity. This describes the amount of time needed by the algorithm, measured by the number of candidates inspected. The most desirable complexity will be linear in path length, but one will have to often contend with exponential complexity.

The next few chapters will occupy us with finding increasingly better algorithms on the above parameters, often making a tradeoff on one to improve upon another.

We will now identify two distinct kinds of problems. Planning problems are problems in which the sequence of moves to the goal state is of interest. Often in such problems the goal state is clearly specified, and one needs to find a path to a goal state. Configuration problems, on the other hand, have only an abstract description of the desired state, and the task is to find a state that conforms to the description. The path to the goal state is generally of no interest. We describe both kinds of problems with some examples. We begin with configuration problems, because the high level algorithm described earlier is more suited to such problems, since it only returns the goal state.

2.3 Configuration Problems

In configuration problems we describe the desired or goal state in some manner, and the task of the problem solver is to find a state that fits the description. Given that the algorithm SIMPLESEARCH returns the goal state when it finds one, it can be employed to solve configuration problems directly. Here are some examples of configuration problems.

2.3.1 The map colouring problem

The map colouring problem has been of interest to mathematicians for quite a while. The task is to colour the regions of a map such that adjacent regions have different colours. One could even specify which colours are allowed for each region.

A map can be represented as a graph where each node is a region, and an edge connects two adjacent regions. A map that can be drawn on 2-dimensional surfaces is represented as a planar graph, in which edges do not cross each other. It was conjectured in the nineteenth century that four colours suffice to colour any planar map. This is known as the four colour theorem. The proof was found in 1976 with the help of an automatic theorem prover.

Figure 2.3 shows an example of a map colouring problem with six regions {A, B, C, D, E, F} along with the allowed colours on the left. A search algorithm would try out the different colours for each node and terminate when all regions have been assigned a colour satisfying the conditions. The map on the right shows a solution.

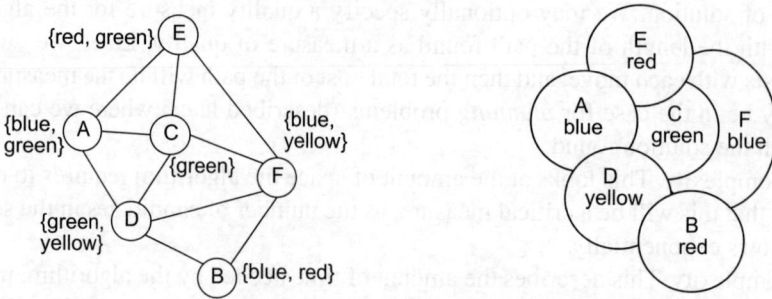

Figure 2.3 A map colouring problem (*left*) represented as a planar graph, and a solution (*right*). Each node in the graph is a region and the colours next to it are the set of allowed colours for that region. An edge between two nodes exists if the two regions have a common boundary.

2.3.2 The N-queens problem

Given an $N \times N$ chessboard, the task is to place N queens on the chessboard so that no queen attacks another queen. A queen, as per the rules of chess, attacks another piece if the two are in the same row, column, or diagonal.

The 8-queens problem refers to the specific case of a standard 8×8 chessboard, and the reader is encouraged to try solving the problem on a physical board. There are 92 solutions, so finding one should not be too hard, even though there are $^{64}C_8 = 4{,}426{,}165{,}368$ arrangements of the eight queens possible. Three of the solutions are shown in Figure 2.4.

In Chapter 12, we shall study how search can be combined with reasoning in a constraint based formulation, and we will revisit the *N*-queens and map colouring problems.

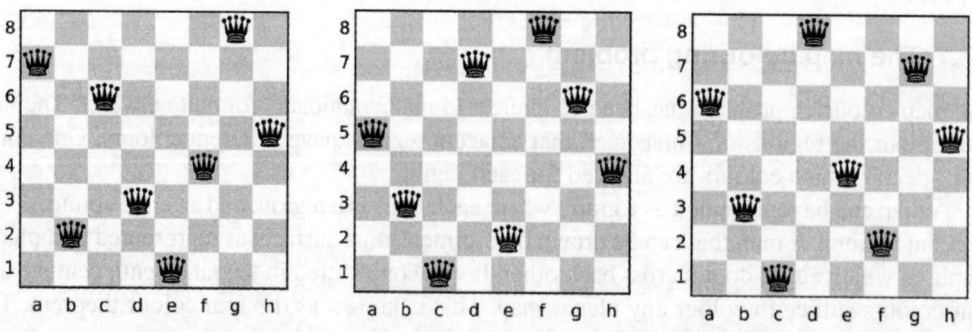

Figure 2.4 Three solutions for an 8-queens problem. A board with eight queens may be represented by a list signifying the column for each queen. This is possible because each column can have only one queen. The solution on the left is represented by (*e, b, d, f, h, c, a, g*) where *e* is the column for the queen in row 1, *b* is the column for the queen in row 2, and so on.

2.3.3 The SAT problem

Perhaps the most famous configuration problem is the SAT, or *Boolean satisfiability*, problem. The SAT problem is as follows. Given a set of propositional (or Boolean) variables and a propositional formula (or a Boolean expression) on those variables, one has to assign a value *true* or *false* to each propositional variable that makes the formula true. Equivalently, one has to assign 1 or 0 to each Boolean variable so that the expression evaluates to 1. Given N variables, a candidate can be represented as an N-bit binary string. For example, given the variables $\{a, b, c, d, e\}$, the string 10011 represents the assignment $a = 1, b = 0, c = 0, d = 1$ and $e = 1$.

We often study SAT by looking at expressions in conjunctive normal form (CNF), since any formula can be converted to CNF. We will keep coming back to SAT problems as we investigate different approaches to solve them. The following is an example SAT problem on five variables $\{a, b, c, d, e\}$ in a CNF form having six clauses. We presume that the reader is familiar with the semantics of the three logical operators used here. The reader is encouraged to find a solution.

$$(b \vee \neg c) \wedge (c \vee \neg d) \wedge (\neg b) \wedge (\neg a \vee \neg e) \wedge (e \vee \neg c) \wedge (\neg c \vee \neg d)$$

The SAT problem has been well studied and finds many applications. Many problems can be formulated as SAT and then solved by one of the general-purpose SAT solvers of the kind we will study. This exemplifies the spirit of general or weak methods, in which we seek to explore general-purpose problem solving methods, which can then be applied to different individual problems.

The SAT problem is also an epitome of exponential complexity. SAT was one of the earliest problems to be proven NP-complete, which is a complexity class of problems that can be solved in non-deterministic polynomial time (Cook, 1971). Solving the SAT problem by brute force can be unviable when the number of variables is large. A formula with 100 variables will have 2^{100} or about 10^{30} candidates. Even if we could inspect a million candidates per second, we would need 3×10^{14} centuries or so. Clearly that is in the realm of the impossible as far as humankind is concerned. Further, it is believed that NP-complete problems do not have algorithms whose worst case running time is better than exponential in the input size.

One often looks at specialized classes of SAT formulas labelled as k-SAT, in which each clause has at most k literals. It has been shown that 3-SAT is NP-complete. On the other hand, 2-SAT is solvable in polynomial time. For k-SAT, complexity is measured in terms of the size of the formula, which in turn is at most polynomial in the number of variables.

2.4 Planning Problems

Planning problems are those in which the goal state, or a partial description of it, is given to us, and the task is to find a sequence of moves, or a *plan*, to transform the given state into the desired state. Since our representation does not have named moves, we can represent the plan with a sequence of states that are on the path to the goal state. Thus, a plan with N moves may be represented as $<S, S_1, S_2, ..., S_{N-1}, G>$, where S is the start state, G is the goal state, and each S_i is a state on the path. Now, our algorithm SIMPLESEARCH will not do the job, because it will only return the goal state G. We will address this shortcoming in Chapter 3 when we start looking at search algorithms in greater detail. We will also look at representations used by the planning community, in which named actions are represented explicitly, in Chapter 9. Meanwhile, we look at some examples of planning problems.

2.4.1 The 8-puzzle

The 8-puzzle is a one player game popular with children. It was created by Sam Loyd (1841–1911) in the 1870s and called a sliding tile puzzle. In the 8-puzzle, a set of eight tiles are placed in a 3 × 3 grid, with one location being blank, which we can refer to as the blank tile. Any tile adjacent to the blank tile can slide into its place, thus making a move in the state space. Given a jumbled up puzzle, the task is to find a sequence of moves that transforms the puzzle into a desired state. Figure 2.5 shows an instance of the 8-puzzle.

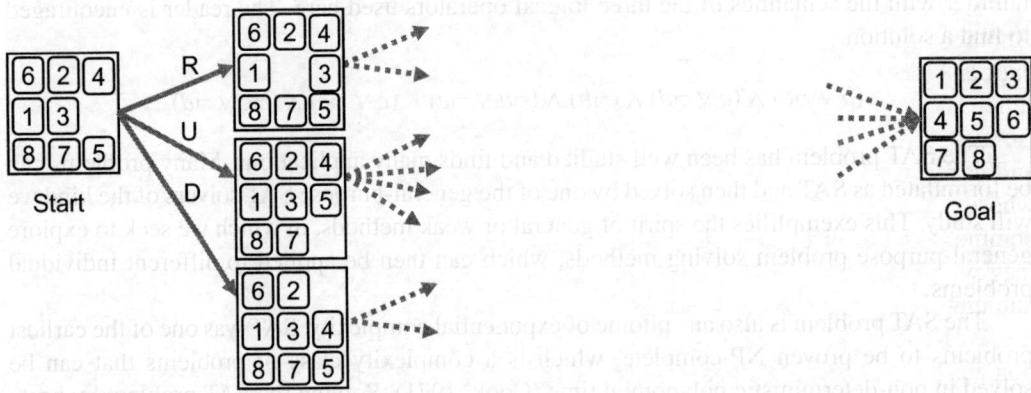

Figure 2.5 The 8-puzzle consists of eight tiles on a 3 × 3 grid. A tile can slide into an adjacent location if it is empty. A move is labelled R if a tile moves right, and likewise for up (U), down (D), and left (L). *Planning* algorithms use *named* moves.

The directed arrows signify the direction of search. In the actual puzzle, the moves are reversible, so the arrows should be bidirectional, but are not depicted in the figure for simplicity. Interestingly, not every starting position leads to the given goal position. This is because the set of possible states is disjoint, having two partitions. Two states that differ only in the position of two adjacent tiles belong to different partitions, and there is no path (sequence of moves) from one to the other. Exploiting this property for popularizing his puzzle, Sam Lloyd offered a prize of USD 1,000, a lot of money in those days, for solving such an instance of the problem. No one could claim the prize though, because there was no solution for the published instance!

One of the earliest textbooks on artificial intelligence (AI) (Nilsson, 1971) used this puzzle extensively, comparing different heuristics for performance. We shall look at heuristic search in Chapter 4. This puzzle, like its 3-dimensional cousin, the Rubik's cube, exemplifies problems with *non-serializable* subgoals. That is, if one chooses an order of bringing tiles into their destined place, for example, the top row first, then while solving subsequent subgoals one is forced to disrupt the earlier partial solution. Solvers of the Rubik's cube must be familiar with this phenomenon. The trick we employ as humans is to learn whole sequences of moves, called macros, or macro operators, for solving each subgoal, while turning a blind eye to the disruptions in the intermediate states. The researcher Richard Korf explored ways of learning such operators for his doctoral thesis (Korf, 1985).

A larger version of the 8-puzzle is the 15-puzzle, on a 4×4 grid. While the 8-puzzle has $9! = 362,880$ states (in two partitions), the 15-puzzle has about 10^{13} or 10 trillion states. Searching through these is indeed a formidable task, but people do find solutions. Searching for the shortest path solution, however, is harder, because the average length of the shortest path is 53, the longest being 80. It was only in 2005 that Korf and Schultze (2005) reported a large-scale parallel breadth first search running for over 52 hours to find the shortest path. We will look at breadth first search in Chapter 3. We will also look at the puzzle again in Chapter 4 on heuristic search.

2.4.2 River crossing puzzles

Another popular class of puzzles are river crossing puzzles. These puzzles involve multiple entities needing to cross a river, using a small boat, typically allowing two passengers at a time. The challenge comes because of conditions on combinations of entities allowed to be isolated on a bank, for the safety of some of them. An example is the well known missionaries and cannibals puzzle, in which there are three of each wanting to cross the river. Based on whose perspective you see the problem from, the conditions can be one of the following. The commonly known version says that if the cannibals outnumber the missionaries on either bank, they will make a meal of them. In a variation, known only to cannibals, if the missionaries outnumber the cannibals they would convert them.

We look at another variation, in which a man needs to transport a goat, a cabbage, and a lion across the river without ever leaving the cabbage alone with the goat, or the goat in turn with the lion, for fear of them being eaten. We explore the state representation and the move generation for this problem.

Let us represent the prospective passengers with the symbols G (goat), L (lion), C (cabbage), and B (boat). The man is implicitly where the boat is. In addition, we need to represent on which bank the elements are. Figure 2.6 shows three possible list representations for the problem. Our lists are represented in square brackets.

In the first, we represent two lists: one for the left bank and the other for the right bank. The start state is [[G L C B] []] indicating that all four elements are on the left bank. The list [[G L] [C B]] says that the goat and the lion are on the left bank and the boat and the cabbage are on the right bank. Note that this is a disallowed state as the lion would eat the goat. Should this be pruned by the MOVEGEN function or left to the search algorithm to avoid? The goal state is represented by [[] [G L C B]].

The second representation simply lists the objects on the left bank, since the others must be on the right. The start state is [G L C B]. The list [G L] says that the goat and the lion are on the left bank and the boat and the cabbage are on the right bank. The goal state is represented by [].

The third representation has one list of all elements on the bank where the boat is. Here Left or Right is mentioned instead of the boat. The start state is [G L C Left]. The list [C B Right] says that the goat and the lion are on the left bank and the boat and the cabbage are on the right bank. The goal state is represented by [G C L Right].

38 | Search Methods in Artificial Intelligence

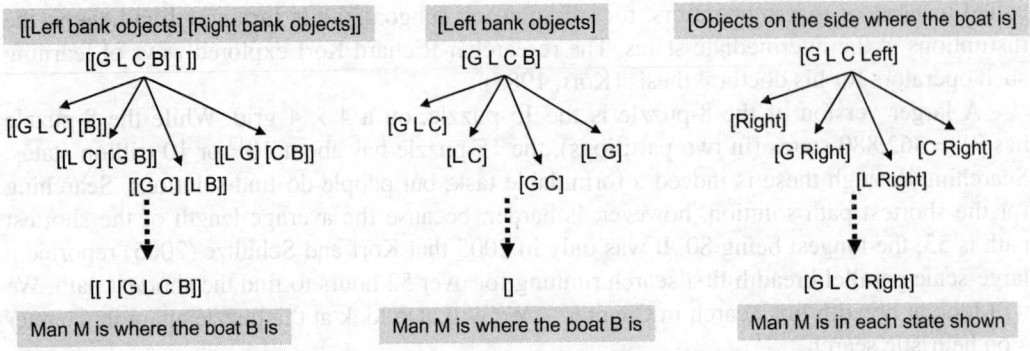

Figure 2.6 Three representations for the man, goat, lion, and cabbage problem. The one on the left has two lists, one for each bank. The representation in the middle has only elements on the left bank. The representation on the right has a list of elements on the bank where the boat is, along with the label specifying which bank it is. The reader is encouraged to write the MOVEGEN functions for these representations.

The reader is encouraged to write the MOVEGEN and GOALTEST functions for the above representations.

A note about representation would not be out of place here. Very often we choose a representation designed for a specific implementation, with the semantics being only in the mind of the programmer. This can have two drawbacks. One, the representation may not be suitable for another problem or implementation. Two, the representation may not make sense to someone else who looks at the program. On both counts, it should be good practice to choose clear and meaningful representations that are interpretable by other humans or programs.

2.4.3 The water jug problem

Assume that you have three jugs, with capacities 3, 5, and 8 litres. The 8-litre jug is filled with water, and the other two jugs are empty. You are required to measure out 4 litres of water, while conserving the total amount of water. The only moves one can make are adding water to a jug till it is filled to the brim or emptying one jug into another if it has the capacity to accept the volume of water. What is your plan?

Let the representation be a triple denoting the quantities in the 8-, 5-, and 3-litre jugs, respectively. Figure 2.7 depicts the entire state space and the possible moves. The start state is (8, 0, 0). The goal state can be any one of (1, 4, 3), (4, 4, 0), and (4, 1, 3). Observe that some moves are reversible, while others are not. For example, one can undo the move from (3, 2, 3) to (3, 5, 0) but not from (3, 2, 3) to (5, 0, 3).

Observe that only states with measurable amounts of water are part of the state space. For example, while the state (6, 1, 1) is physically possible, there is no way of measuring these quantities with the set of moves available. Likewise, the state (4, 3, 1) cannot be reached given the moves.

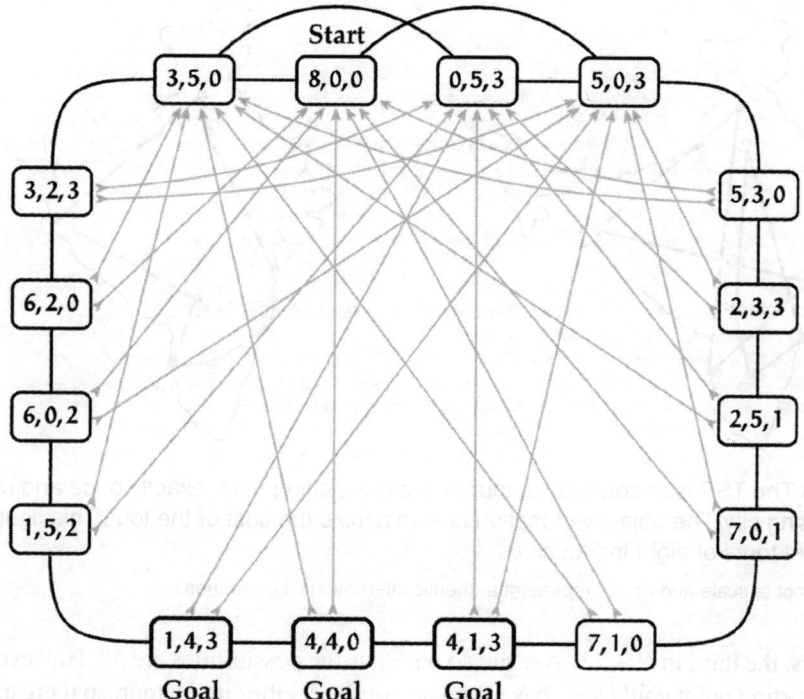

Figure 2.7 The states reachable from the start (8, 0, 0) with three jugs of capacity 8, 5, and 3 litres. Undirected edges on the periphery represent reversible moves. The figure also identifies three goal states in which 4 litres of water has been measured.

2.4.4 The travelling salesman problem

Many consider the travelling salesman[2] problem (TSP) to be the holy grail of computer science. This is because it is one of the hardest to solve in the general case. The problem is as follows.

Every week, or every month, a travelling salesperson embarks upon a tour of a region, starting off from home, visiting every city on her itinerary exactly once, and returning home eventually. She needs to find the optimal order of visiting cities such that the total cost of the tour is minimized. The cost could be in terms of distances travelled, the price of fuel, the time taken from one city to another, or the price of train or flight tickets. Mathematically, the tour of the salesperson is called a Hamiltonian cycle, after the Irish mathematician W.R. Hamilton who first described the problem in the early nineteenth century.

We state the problem by specifying the cost between every pair of cities as a number. We assume that a completely connected graph is given to us as input. If any pair of cities is not directly connected on the map, we add an edge with a prohibitively high cost. Figure 2.8 illustrates two possible tours for some cities in India.

The number of possible TSP tours grows factorially with the number of cities. Given N cities, a brute force tour construction method could pick the first city in N ways, the second in

---

[2] In the days when this problem was posed perhaps only men did this work.

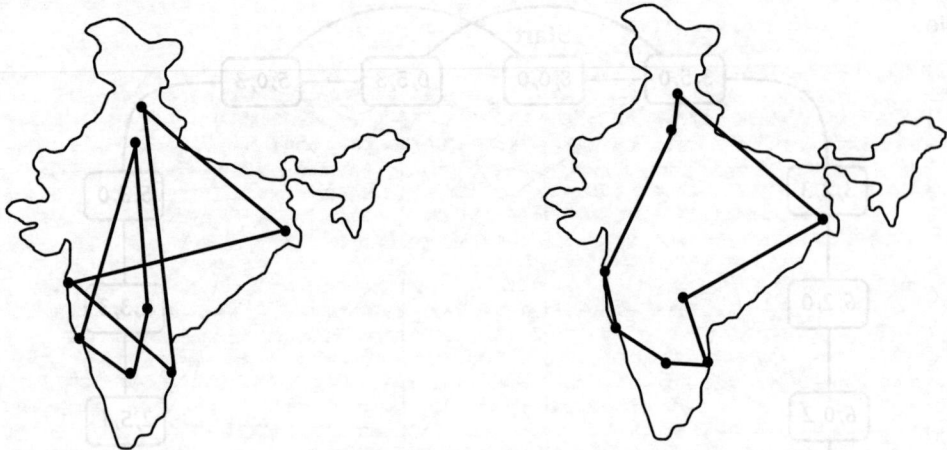

Figure 2.8 The TSP is to construct a tour of *N* cities visiting each exactly once and returning to the starting city. The objective function is to minimize the cost of the tour. This figure shows two different tours of eight Indian cities.

Note: Maps not to scale and do not represent authentic international boundaries.

$N - 1$ ways, the third in $N - 2$ ways, and so on. Thus the possibilities are $N!$. However, for each tour, the starting point could well have been any of the N cities in the tour, so there are in effect $(N - 1)!$ distinct tours. Also, for every tour, travelling in the opposite direction would result in the same cost, assuming the costs are symmetric. In that case, one can say that there are $(N - 1)!/2$ distinct tours. But since most algorithms cannot always identify such duplications, we often say that the problem is of complexity $N!$.

Now the factorial function grows much faster than the exponential function. Let us compare an N variable SAT problem with an N-city TSP. Table 2.1 compares the number of candidate solutions for the two problems as N increases. Observe the ratio between the two. The TSP problem does indeed grow much faster!

Let us look at the value 2^{100} which is the number of candidates for a 100-variable SAT problem. This number is 267,650,600,228,229,401,496,703,205,376. In the US number naming system, this is 1 nonillion, 267 octillion, 650 septillion, 600 sextillion, 228 quintillion, 229 quadrillion, 401 trillion, 496 billion, 703 million, 205 thousand, 376. This is about 10^{30}.

The number in every row for SAT is double the number in the previous row. If one were to take a sheet of paper that is 0.1 millimetre thick and double the thickness (by folding it) one hundred times, the resulting stack would be 13.4 billion light years tall. It would reach from Earth to beyond the most distant galaxy we can see with the most powerful telescopes – almost to the edge of the observable universe. A 100-variable SAT is hard enough. But 100! is a much bigger number. The following output from a simple Lisp[3] program shows the number.

---

[3] Lisp has a built-in feature to handle large numbers. It has long been a favourite language of AI researchers, primarily because it allows dynamic structures to be built naturally, and its functional style allows the creation of domain specific operators easily. It has been on a little bit of decline with the advent of newer languages, and a diminishing community makes it daunting for new entrants to try their hand at it now.

Table 2.1. Comparing SAT and TSP state spaces

N	Candidates in SAT: 2^N	Candidates in TSP: $N!$	Ratio
1	2	1	0.5
2	4	2	0.5
3	8	6	0.75
4	16	24	1.5
5	32	120	3.75
6	64	720	11.25
7	128	5,040	39.38
8	256	40,320	157.5
50	1,125,899,906,842,624	$3.041409320 \times 10^{64}$	3×10^{49}
100	$1.267650600 \times 10^{30}$	$9.332621544 \times 10^{157}$	10^{127}

```
(factorial 100)
  →
9332621544394415268169923885626670049071596826438162146856
92963895217599993229915608941463976156518286253697920827223
75825118521091686400000000000000000000000000
```

This number is larger than 10^{157}, and clearly it is impossible to inspect all possible tours in a 100-city problem. Inspecting all candidates is not going to be an approach, and the TSP problem has been shown to be NP-hard (Gary and Johnson, 1979). Exact solutions are hard to find for a given large problem, and thus it makes it difficult to evaluate an algorithm. A library of TSP problems *TSPLIB*[4] (Reinelt, 2013) with exact solutions is available on the web. An exact solution for 15,112 German cities from TSPLIB was found in 2001 using the cutting-plane method proposed by George Dantzig, Ray Fulkerson, and Selmer Johnson in 1954, based on linear programming (Dantzig et al., 1954). The computations were performed on a network of 110 processors. In May 2004, the TSP of visiting all 24,978 cities in Sweden was solved: a tour approximately 72,500 kilometres long was found, and it was proven that no shorter tour exists. In March 2005, the TSP of visiting all 33,810 points in a circuit board was solved using Concorde (Applegate et al., 2007): a tour of length 66,048,945 units was found, and it was proven that no shorter tour exists.

One reason why TSP is much harder than SAT is that it is an *optimization* problem. We are looking for the lowest cost tour. In SAT, when we inspect a candidate, we know whether it is a solution or not. On the other hand, looking at a valid TSP tour, there is no way of knowing whether it is the optimal in the general case. Chapter 6 describes the BRANCH&BOUND algorithm that can find the optimal tour without having to search the entire space, and in Chapter 4 we will study heuristic search algorithms that can give very good solutions. Under certain conditions simpler algorithms work well. For example, when the cost function behaves like a metric distance function that satisfies the triangle inequality. The triangle inequality says that the sum of two sides of a triangle is less than the third side. Human beings are often satisfied

---

[4] http://comopt.ifi.uni-heidelberg.de/software/TSPLIB95/, accessed June 2022.

with good solutions in lieu of optimal ones, especially when the cost of finding them is much lower. It is said that we are *satisficers* and not *optimizers*.

Stochastic local search (SLS) methods (Hoos and Stutzle, 2005) can find very good solutions quite quickly. We will study such methods in Chapter 5. For example, for randomly generated problems of 25 million cities, a solution quality within 0.3% of the estimated optimal solution was found in 8 CPU days on an IBM RS6000 machine (Applegate et al., 2003). More results on performance can be obtained from the website for the DIMACS (the Center for Discrete Mathematics and Theoretical Computer Science, http://dimacs.rutgers.edu/) implementation challenge on TSP (Johnson et al., 2003; Applegate et al., 2007). Some more interesting TSP problems available (Applegate et al., 2007) are as follows: The *World TSP* – a 1,904,711-city TSP consisting of all locations in the world that are registered as populated cities or towns, as well as several research bases in Antarctica; *National TSP Collection* – a set of 27 problems, ranging in size from 28 cities in Western Sahara to 71,009 cities in China. Many of these instances remain unsolved, providing a challenge for new TSP codes; and *VLSI TSP Collection* – a set of 102 problems based on VLSI data sets from the University of Bonn. The problems range in size from 131 cities to 744,710 cities.

TSP problems arise in many applications (Johnson, 1990), for example, circuit drilling boards (Litke, 1984), where the drill has to travel over all the hole locations, X-ray crystallography (Bland and Shallcross, 1989), genome sequencing (Agarwala et al., 2000), and VLSI fabrications (Korte, 1990). These can give rise to problems with thousands of cities, with the last one reporting 1.2 million cities. Many of these problems are what are known as *Euclidean TSPs*, in which the distance between two nodes (cities) is the Euclidean distance. One can devise approximation algorithms that work in polynomial time. Arora (1998) reports that in general, for any $c > 0$, there is a polynomial-time algorithm that finds a tour of length at most $(1 + 1/c)$ times the optimal for geometric instances of TSP, which is a more general case of a Euclidean TSP. Special cases of TSPs can be solved easily. For example, if all the cities are known to lie on the perimeter of a convex polygon, a simple greedy algorithm *TSP-NearestNeighbour* works well.

Interestingly, TSP can be seen both as a planning problem and a configuration problem, hinting that the two are just different ways of looking at a problem. As a planning problem, we can think of solving it by constructing a tour by moving from one city to another. As a configuration problem, we can think of the solution as a particular ordering of the cities. Algorithms that take the second view are said to operate in the solution space.

2.5 The Solution Space

When we look at problems like TSP, SAT, and *N*-queens, we can think of two approaches to solving them.

2.5.1 Constructive methods

The first approach, which is called a *constructive* approach, synthesizes the solution piece by piece. In a way, this is consistent with the state space representation, where a move takes you one step closer to a solution. The solution could be a plan for a planning problem, where constructive approaches are natural and intuitive, and we construct the plan move by move.

But constructive methods can also be used for configuration problems, where again the solution is synthesized piece by piece. In the TSP, we can synthesize the tour one edge at a time. In the map colouring problem, we assign a colour region by region. In the N-queens problem, we can imagine, or represent, an empty board and place the queens one by one. In the SAT problem, we can assign values to variables one at a time.

2.5.2 Perturbative methods

We can also think of a search space which contains candidate solutions. We call this space as the solution space or plan space as opposed to the state space. It is characterized by the fact that when the search algorithm finds the goal node, it already has the solution. Every entity in this space is a candidate solution. Moves in this space are perturbations to candidates, resulting in other candidates.

In the SAT problems on N variables, the candidates can be N-bit strings. For example, on five variables $\{a, b, c, d, e\}$ mentioned earlier in this chapter, a candidate can be 11111, signifying $a = 1, b = 1, ..., e = 1$. One MoveGen function, or neighbourhood function, could be to flip one bit. This is depicted in Figure 2.9.

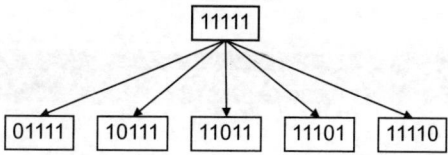

Figure 2.9 A flip-one-bit MoveGen function for SAT flips one bit of a candidate. For a 5-variable SAT problem, it produces a set of five neighbours.

However, there can be other neighbourhood functions as well. For example, one could flip any two bits, or flip one or two bits, and so on. We shall explore the implication of choosing different neighbourhood functions in Chapter 4. Meanwhile, here is a point to ponder. Why not allow the neighbourhood function to change any number of bits? Then the solution would be just one move away!

In the TSP problem, one representation of the candidates, known as the path representation, is to list the cities in the order they are visited. For example, for the five cities Chennai, Bangalore, Goa, Delhi, and Mandi depicted by the initial letters, a candidate tour could be [B, M, G, C, D]. One perturbation function could be a 2-city exchange in which a new permutation is created by swapping some two cities. We shall also explore different neighbourhood functions in Chapter 4.

In Chapter 9, we will look at plan space methods for planning problems. In the plan space, each node is a candidate plan, and the search is for those candidates that are solution plans. The neighbourhood functions there will be a little different from the simple perturbation that is possible for configuration problems like SAT and TSP.

Summary

In this chapter we have laid the foundations of problem solving by search. Search is a first principles approach in which the problem solver simulates the effects of its intended actions, and by a process of searching through various possibilities, arrives at a solution.

First principles methods are needed to solve problems in the first place. When faced with a new problem, this is the only approach to finding a solution. But we do not employ them all the time. For problems that are similar to the ones encountered earlier, making use of memory and learning is much more effective. Humankind has even learned to employ social memory through language, telling stories and writing books for others to read and benefit.

A search method has to wade through a sea of possibilities, as the number of choices grows exponentially. We have named this adversary *CombEx* (Khemani, 2013) for combinatorial explosion and likened it to the mythical Hydra that Hercules had to battle. Hydra would sprout new heads every time Hercules cut off one. Our search spaces also multiply as we inspect a candidate and delve deeper. And yet we are encouraged by the story that Hercules was able to triumph over Hydra.

In the rest of this book we shall study different ways of exploring search spaces and different kinds of problems that can be formulated as search problems. In the next chapter we begin by setting up the basic mechanisms for search, and in the following chapters we will investigate how to battle CombEx.

Exercises

1. Modify the SIMPLESEARCH algorithm to store the path information in the nodes of the search tree. Hint: it is natural to store the reverse of the path as a list.
2. [Baskaran] Consider a 5–3–2 water jug puzzle with three jugs *a*, *b*, and *c* of capacity 5, 3, 2 litres, respectively. A state is represented as a 3-digit string *ABC*, where *A*, *B*, and *C* denote the volume of water present in the jugs *a*, *b*, and *c*, respectively. For example, 302 denotes a state where jug *a* has 3 litres of water, jug *b* is empty, and jug *c* has 2 litres of water. Let 500 be the start state and 131 be the goal state; find the shortest path from start to goal.

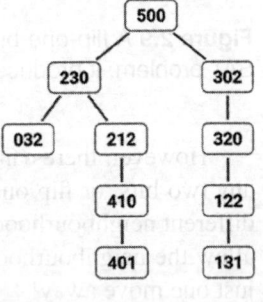

3. Can you think of a map that can be drawn on a plane such that two colours are sufficient to colour the map?
4. Can you think of a map in which four colours are the minimum needed to colour the regions, and where three would not be enough?
5. The chapter has some solutions for the 8-queens problem. Try solving the *N*-queens problem with the following values of *N*: 2, 3, 4, 5, 6, and 7. Do all of them have solutions?
6. Consider the knight's tour problem on a standard 8 × 8 chessboard. The knight can begin on any square and must visit every other square on the board exactly once. Draw a chessboard and fill in the numbers 1–64 which stand for the order in which the knight visits the squares.
7. Given a set of three variables {*a*, *b*, *c*}, construct a SAT problem which has no solution.

8. Given a set of three variables $\{a, b, c\}$, construct a SAT problem such that every assignment is a solution.
9. Of the eight locations in the 8-puzzle, a blank in the centre allows four moves, a blank on the side allows three moves, and a blank in a corner allows two moves. Design a representation and a MoveGen function that works with the representation.
10. Given the representation of the water jug problem as a triple, devise the MoveGen function.
11. Given the following representation for the man, goat, lion, and cabbage river crossing problem, write the MoveGen and GoalTest functions. The representation is a list of two lists: one with the elements on the left bank and the other with elements on the right bank. The start state is [[G L C B] []] indicating that all four elements are on the left bank. Observe that B stands for the boat, and the assumption is that the man is where the boat is, since only he can row the boat.
12. Write the MoveGen and GoalTest functions when the representation is only one list, with the elements on the left bank. The corresponding start state is [G L C B].
13. Write the MoveGen and GoalTest functions with the following representation: there is only one list, with the elements on the side where the boat is, along with the side. The corresponding start state is [G L C Left].
14. In the perturbative move described in the chapter for SAT, one is allowed to flip one bit in the candidate. If there are N variables, then this results in N neighbours being generated. What would the neighbourhood size be if the move was to flip two bits? And if one could flip one or two bits?
15. In the SAT problem with N variables, if we flip one bit, the path to the solution in the worst case would be of length N, when each bit has to be flipped. If the neighbourhood function allowed one to change *any* number of bits, then the solution would be only one move away. Comment on this neighbourhood function.

CHAPTER 3

Blind Search

In this chapter we introduce the basic machinery needed for search. We devise algorithms for navigating the implicit search space and look at their properties. One distinctive feature of the algorithms in this chapter is that they are all *blind* or *uninformed*. This means that the way the algorithms search the space is always the same irrespective of the problem instance being solved.

We look at a few variations and analyse them on the four parameters we defined in the last chapter: completeness, quality of solution, time complexity, and space complexity. We observe that complexity becomes a stumbling block, as our principal foe *CombEx* inevitably rears its head. We end by making a case for different approaches to fight CombEx in the chapters that follow.

In the last chapter we looked at the notion of search spaces. Search spaces, as shown in Figure 2.2, are trees corresponding to the different traversals possible in the state space or the solution space. In this chapter we begin by constructing the machinery, viz. algorithms, for navigating this space. We begin our study with the corresponding tiny state space shown in Figure 3.1.

The tiny search problem has seven nodes, including the start node S, the goal node G, and five other nodes named A, B, C, D, and E. Without any loss of generality, let us assume that the nodes are states in a state space. The algorithms apply to the solution space as well. The left side of the figure describes the *MoveGen* function with the notation Node → (list of neighbours). On the right side is the corresponding graph which, remember, is implicit and not given upfront. The algorithm itself works with the MoveGen function and also the *GoalTest* function. The latter, for this example, simply knows that state G is the goal node. For configuration problems like the N-queens, it will need to inspect the node given as the argument.

The search space that an algorithm explores is implicit. It is generated on the fly by the MoveGen function, as described in Algorithm 2.1. The candidates generated are added to what is traditionally called *OPEN*, from where they are picked one by one for inspection. In this chapter we represent OPEN as a *list* data structure. We also replace the nondeterministic move of picking *some* candidate from OPEN with the operation of picking the node at the head of the list. By doing so, we shift the onus of devising a strategy for picking the next move to the way we *maintain* OPEN. The modified version that stores OPEN as a list is described in Algorithm 3.1.

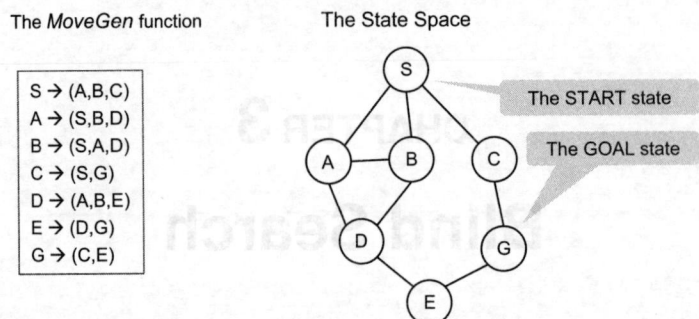

Figure 3.1 A tiny search problem. The figure on the left shows the *MoveGen* function in the format Node → (List of neighbours). The figure on the right shows the corresponding state space. Observe that all moves are bidirectional.

Algorithm 3.1 Algorithm SIMPLESEARCH picks node *N* from the head of OPEN. If *N* is the goal, it terminates with *N*, else it calls MOVEGEN and adds the neighbours of *N* to OPEN in some manner. The way this is done determines the behaviour of the search algorithm.

SIMPLESEARCH(S)
1   OPEN ← [S]
2   **while** *OPEN* **is not empty**
3   *N* ← **head** OPEN
4   OPEN ← **tail** OPEN
5   **if** GOALTEST(*N*) = TRUE
6   **then**   **return** *N*
7   **else**   Combine *MOVEGEN(N)* with OPEN
8   **return** *FAILURE*

3.1 Search Trees

The candidate that the search algorithm picks from OPEN determines its behaviour. Algorithm SIMPLESEARCH picks the node at the head of OPEN. The question now is where the new nodes generated by MOVEGEN are added. The search space the algorithm will explore in some order is given in Figure 3.2. Note that this is a tree, which we will call the *search tree*. The order in which children are generated is the same as in the MOVEGEN function. Given that the algorithm is searching for the goal *G*, any branch in which *G* occurs ends there, as shown in the figure with shaded nodes.

A search algorithm wanders through the state space in the quest of a goal node. In general, the algorithm could traverse the search tree in any order and may even dive down some branch

Blind Search | 49

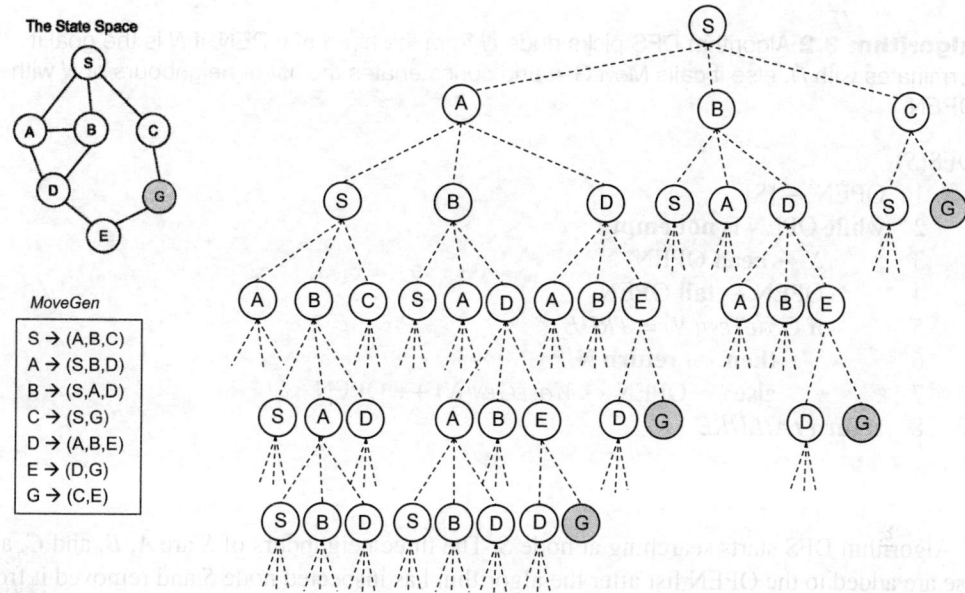

Figure 3.2 A part of the *search tree* for the tiny search problem of Figure 3.1. For each node, the children are ordered as in the list given by MoveGen. Every path in the search tree continues till it encounters the goal node G. Observe that some paths are never-ending.

in the above tree. The actual behaviour would depend on how OPEN is organized in the implementation. Different algorithms do their own thing and explore the space in different ways. The way they do so is captured in the corresponding search tree, which may be a subtree of the above tree.

The search tree depicts the space as viewed by a given search algorithm. Each variation of the algorithm generates its own search tree. Given that we have decided to extract the next candidate from the head of the OPEN list, we need to decide how to add the new nodes to OPEN. We begin by treating OPEN as a stack data structure. The resulting search is known as *depth first search*.

3.2 Depth First Search

When new nodes are added to the head of the stack, the newest nodes are the first ones to be inspected. The stack data structure is characterized by the last-in-first-out (LIFO) behaviour. Later in this chapter we shall look at the case where OPEN is treated as a queue[1] with the first-in-first-out (FIFO) behaviour. Algorithm 3.2 incorporates this change with Line 7 in Algorithm 3.1 so that OPEN behaves like a stack. We call the revised algorithm *DFS*.

---

[1] Thanks to the Coronavirus all of us have learnt to form queues at grocery stores.

50 | Search Methods in Artificial Intelligence

> **Algorithm 3.2** Algorithm DFS picks node *N* from the head of OPEN. If *N* is the goal it terminates with *N*, else it calls MoveGen and concatenates the list of neighbours of *N* with OPEN.
>
> DFS(*S*)
> 1   OPEN ← [S]
> 2   **while** OPEN is **not empty**
> 3   *N* ← **head** OPEN
> 4   OPEN ← **tail** OPEN
> 5   **if** GoalTest(*N*) = TRUE
> 6   **then**   **return** *N*
> 7   **else**   OPEN ← MoveGen(*N*) ++ OPEN
> 8   **return** FAILURE

Algorithm DFS starts searching at node *S*. The three neighbours of *S* are *A*, *B*, and *C*, and these are added to the OPEN list after the algorithm has inspected node *S* and removed it from OPEN. The new nodes are added in the order generated by the MoveGen function, with node *A* ending up at the head of OPEN, or at top of the stack. DFS picks *A* next. The search tree explored by the algorithm is the entire tree shown earlier and repeated in Figure 3.3. The figure also shows on the right how OPEN evolves after each step. Remember that at each stage the next node is picked from the *head* of OPEN.

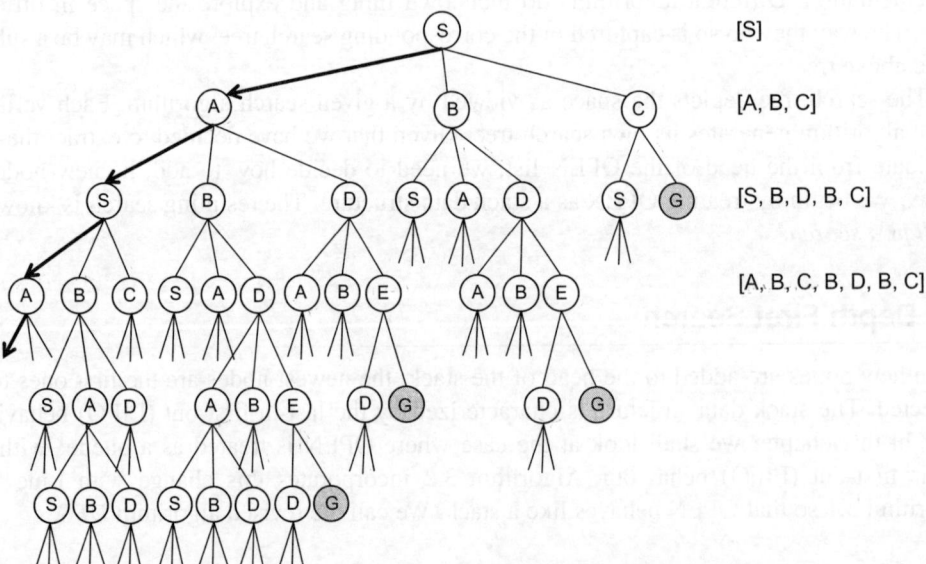

Figure 3.3 The entire search tree is available for algorithm DFS, which is the simplest instance of depth first search. It dives down the never-ending leftmost path and goes into an infinite loop. The OPEN list is shown on the right at each level, with the algorithm always picking the node at the head at each stage.

Algorithm DFS dives headlong into the first branch it sees. The fact that it does so on the leftmost branch is only because the first node in the list returned by MoveGen always ends up in at the head of OPEN. OPEN behaves like a stack with the LIFO property. The behaviour of depth first search can thus be seen to embody the *newest* nodes first strategy. In the search tree that translates to *deepest* nodes first, giving the algorithm its name.

Even though the state space is finite, the search tree is infinite. This is another way of saying that a search algorithm could well go in cycles, moving around in the state space without moving towards the goal. As can be seen, the search simply oscillates between nodes S and A, and will keep doing so till the machine crashes. In the river crossing problem this would simply mean repeatedly going over to the other bank (with the goat) and coming back. For the 8-puzzle this would mean moving the same tile back and forth.

Observe that in the search tree in Figure 3.3, there are several occurrences of the goal node G, each with a different path leading to it. Will some search algorithm find one?

There are two issues with algorithm DFS. First, the problem of looping indefinitely as seen above. And second, even if the algorithm were to find the goal, it only returns the goal node (Line 6) and *not* the path that we seek. We resolve the two lacunae one by one.

3.2.1 How to stop going around in circles

How does one stop the algorithm from going round and round in cycles?

Let us look at some Greek mythology, in which the Labyrinth was an elaborate, confusing structure built by the legendary Daedalus for King Minos of Crete at Knossos (Stieger, 2014). Minos used it to hold the Minotaur, a monster eventually killed by the hero Theseus. The Labyrinth, or the maze, was designed so that one found it very difficult to find the way out once inside, because one kept going round in circles. In the Indian city of Lucknow, one can find a maze too, in the Bara Imambara, built by the Nawab of Awadh in 1784. It is a 3-dimensional labyrinth with 489 identical doorways and is referred to as the Bhul Bhulaiya, which in Hindi roughly means a place where you forget where you came from.

The idea of a maze exemplifies the problem faced by a search algorithm, which has access only to the immediate neighbours, and needs to somehow find a path to the goal. There is lack of a global perspective of the kind humans have when, for example, poring over a city map. However, when on the ground at a crossroads, one often does not have a sense of direction looking only at the immediate neighbourhood. Similar experiences have been narrated by trekkers in the mountains, who can see the destination, and the path to it, from the peak, only to get confused by diverging trails lower down in the forest.

Let us return to the Greek story. Theseus was a son of King Aegeus of Athens. He travelled to Crete, where Ariadne, King Minos' daughter, fell in love with him. She offered to help him conquer the Labyrinth and kill the Minotaur if he would marry her and take her away from Crete. He agreed and Ariadne gave him a ball of red thread, which Theseus unrolled as he explored the labyrinth. The thread allowed him to find his way out of the labyrinth.

This approach of exploring a problem *systematically* is known as *Ariadne's thread* in the logic community: 'The key idea is to create and maintain a record of the available and exhausted options. This record is referred to as the 'thread'. Maintaining such a record allows us to explore a search space systematically.

We implement this idea by maintaining another list apart from OPEN, that stores the nodes already inspected. This list is traditionally called *CLOSED*. Every time we remove a

52 | Search Methods in Artificial Intelligence

node *N* from OPEN for inspection, we add it to CLOSED. And before adding the neighbours of *N* generated by the MOVEGEN function, we remove those nodes that are already present in CLOSED. The corresponding search tree generated by the algorithm is shown in Figure 3.4 with solid edges. On the right, we show both OPEN and CLOSED as search progresses. The search is still depth first. Only that nodes already present in CLOSED are not generated again.

With this modification, our search algorithm does terminate by finding a path to the goal node. Moreover, the infinite search tree has transmogrified into a finite one. The perceptive reader would have noticed that the path found is not the shortest path. We will see below that the other option of treating OPEN as a queue does find the shortest path.

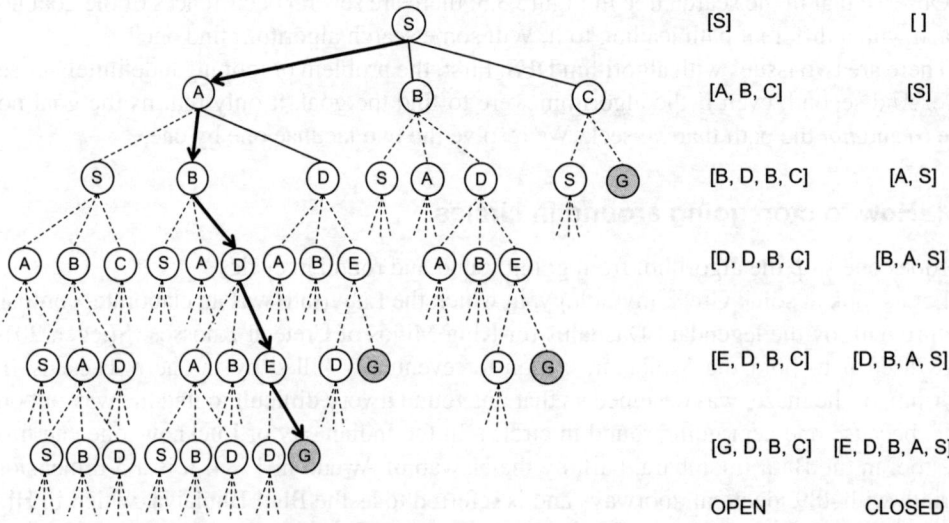

Figure 3.4 When nodes already on CLOSED are not added to OPEN, the search space accessible to DFS shrinks dramatically to the subtree with solid edges. It finds the path to the goal as shown. Note that the search tree depends on the left to right order of depth first search, which is why it does not go beyond the nodes *D*, *B*, and *C* on the right. All their neighbours would already be on CLOSED.

Looking at Figure 3.4, one can see that while nodes on CLOSED are not added again, nodes already on OPEN are again added. These are *B* and *C* as children of *S*, and *D* as a child of *A*. In fact, the path found to *G* is through the copy of *D* added later. If we filter out neighbours of a node that are already on OPEN, in addition to CLOSED, then we get an even smaller search tree as shown in Figure 3.5.

The resulting search tree includes each node from the state space exactly once. In this example it does include all nodes of the state space, but for a larger graph it may not have done so. Also, the path found by this variation is different from the one in Figure 3.4. The numbers in the search tree represent the order in which depth first search inspects the node till termination. We next turn our attention to modifying our algorithm to reconstruct and return the path found.

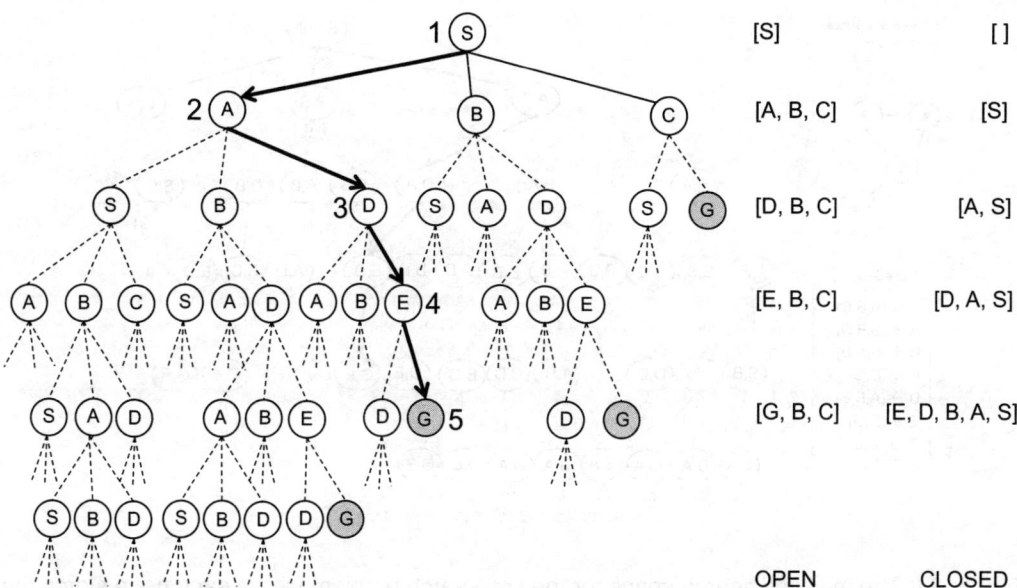

Figure 3.5 When only new nodes, which are neither on OPEN nor on CLOSED, are added to OPEN, the search tree shrinks even more. Now only one copy of every node exists in the search tree. Note that the path found is different too. The numbers next to nodes show the order in which the nodes are inspected.

3.2.2 Reconstructing the path

In configuration problems it suffices to find a goal node that matches the goal description. But in planning problems we require our search algorithm to return the path to the goal node. Again, we take recourse to Ariadne's thread which lays a trail of the path to the goal. We do this by modifying the node representation in the search tree, while the problem space remains as given to us.

One way of modifying the search tree would be to modify each node to represent the path from the start node. For example, node D in the search tree could be represented by [D, A, S], the reverse of the path from S to D. The GOALTEST function would then apply to the head of the search tree node, as would the MOVEGEN function. If GOALTEST succeeds, all one would need is to reverse the path already stored.

We adopt a different approach where each node in the search tree keeps track of the parent it was generated from. The node in the search tree, which we call a *nodePair* is a pair of two nodes, the node itself and its parent node. That is, *nodePair* = (Node, Parent). The corresponding search tree for our tiny example is shown in Figure 3.6.

Both the OPEN and the CLOSED lists now store *nodePairs* which are processed as described later in the revised depth first search algorithm. It will need to extract the first node in the *nodePair* to which the GOALTEST function is applied. When that succeeds, the revised algorithm will invoke a module we call RECONSTRUCTPATH. There are two arguments to RECONSTRUCTPATH. One is the *nodePair* in which the goal G is the first element. The other is CLOSED which has the memory thread. We illustrate the path reconstruction process

54 | Search Methods in Artificial Intelligence

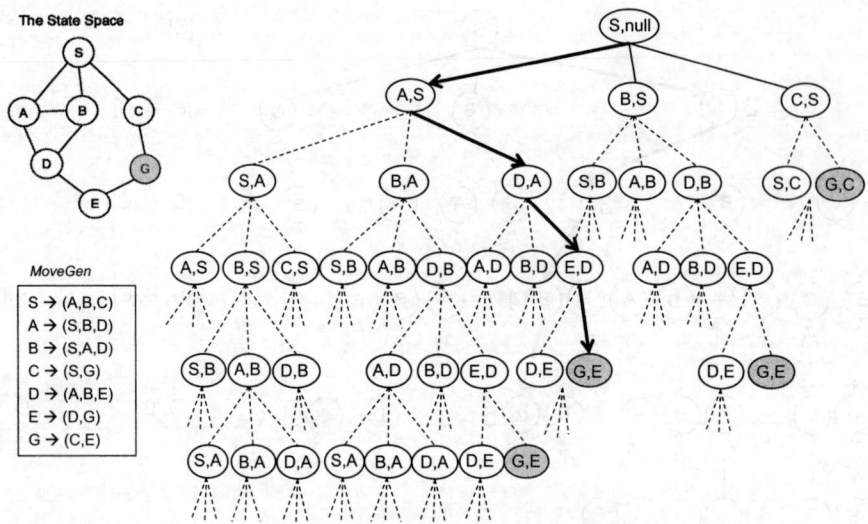

Figure 3.6 The modified search space for the tiny search problem where each node stores the parent too in a *nodePair*. The state space and the MoveGen function are repeated on the left. The figure also shows the path found by the depth first algorithm with only new nodes being added to OPEN.

with the tiny search problem. As shown in Figure 3.6, when GoalTest succeeds with the *nodePair* (*G,E*), CLOSED contains the following *nodePairs*: (*E,D*), (*D,A*), (*A,S*), and (*S*,null). The algorithm begins by initializing the path *P* to [*G*]. As long as the parent of the last node added to the path *P* is not *null*, it concatenates the parent to the path and looks for the parent of the parent in CLOSED. Starting with *E*, the following nodes are successively concatenated to the path: *D*, *A*, and *S*, at which the parent of *S* is *null* and the algorithm terminates with [*S, A, D, E, G*].

The algorithm is shown as Algorithm 3.3. It accepts a node pair of the form (goal, parent) and constructs the path by tracing back successive parents till it reaches the start node whose parent is *null*. Lines 1–4 show an ancillary function FindLink that retrieves the *nodePair* which has the parent of the node given as the first argument.

Algorithm 3.3. Algorithm ReconstructPath accepts the *nodePair* containing the goal node and constructs the path by tracing the parents via the *nodePairs* stored in CLOSED. It uses an ancillary function FindLink(node, CLOSED) which fetches the *nodePair* in which node is the first element.

ReconstructPath (nodePair, CLOSED)
1.   (node, parent) ← nodePair
2.   path ← node : [ ]
3.   **while** parent **is not null**
4.   path ← parent : path

5.   (_, parent) ← FINDLINK(parent, CLOSED)
6. **return** path

FINDLINK(node, CLOSED)
1.   **if** node = **first head** CLOSED
2.   **return head** CLOSED
3.   **else return** FINDLINK(node, **tail** CLOSED)

3.2.3 The complete algorithm

The revised depth first search algorithm *DFS* is shown as Algorithm 3.4 and the other ancillary functions it employs follow.

Algorithm DFS begins by initializing OPEN with the *nodePair* containing the start node S with the parent *null*. It also initializes CLOSED as the empty list. In practice for large problems we would prefer to represent CLOSED as a hash table, since the main task is to look for a particular *nodePair* in it. Then in Lines 3–12 it traverses the space in search of a *nodePair* whose first element N passes the GOALTEST function. If N turns out to be the goal node, it invokes the RECONSTRUCTPATH function (Lines 6, 7). For a configuration problem, one would return the goal node itself. Else it adds the *nodePair* to CLOSED and calls MOVEGEN with N (Lines 8, 9). In Line 10 it calls the *REMOVESEEN* function described later to prune nodes that are either in OPEN or CLOSED. For the ones that remain, it constructs *nodePairs* for each with N as the parent and appends them at the head of the tail of OPEN (Lines 11, 12), as a result of which OPEN behaves like a stack.

Algorithm 3.4. Algorithm DFS works with node pairs to keep track of the parent of each node. It treats OPEN like a stack. It removes any nodes generated earlier from the set of neighbours returned by MOVEGEN and for each adds the parent to make pairs before appending them to the tail of OPEN. For a planning problem, we call the module to reconstruct the path. For a configuration problem, we would only return the goal node when it is found.

DFS(S)
1 OPEN ← (S. **null**) : [ ]
2 CLOSED ← **empty list**
3 **while** OPEN **is not empty**
4   nodePair ← **head** OPEN
5   (N, _) ← nodePair
6   **if** GOALTEST(N) = TRUE
7   **return** RECOSTRUCTPATH(nodePair, CLOSED)
8   **else** CLOSED ← nodePair : CLOSED
9   children ← MOVEGEN(N)
10   newNodes ← REMOVESEEN(children, OPEN, CLOSED)
11   newPairs ← MAKEPAIRS(newNodes, N)
12   OPEN ← newPairs ++ (**tail** OPEN)
13 **return empty list**

There are two exit points for algorithm DFS. The first is in Line 7 when it finds the goal node. The second is in Line 13 when the while loop has ended, and it has no more nodes left in OPEN to inspect. It then returns the empty path. The latter exit happens if the state space is finite and there is no path to any goal node. If the state space itself were to be infinite then DFS could go down an infinite path, even if there was a finite path to a goal node.

The other ancillary functions are described below. When node *N* is not the goal node, its neighbours are generated by calling MOVEGEN (Line 9). Those neighbours that have already been inspected, in CLOSED, and those waiting to be inspected, in OPEN, are filtered out (Line 10). For each node in *nodeList*, it makes two calls to algorithm OCCURSIN, once to check if the node occurs in OPEN, and once to check for its presence in CLOSED. The algorithm REMOVESEEN is described below.

Algorithm 3.5. Algorithm REMOVESEEN accepts a list of nodes, and two lists of nodePairs, OPEN and CLOSED, and filters out those nodes in the nodeList that are present in either CLOSED or OPEN.

REMOVESEEN(nodeList, OPEN, CLOSED)
1 **if** nodeList is **empty**
2   **return empty list**
3 **else** node ← **head** nodeList
4   **if** OCCURSIN(node, OPEN) **or** OCCURSIN(node, CLOSED)
5   **return** REMOVESEEN(**tail** nodeList, OPEN, CLOSED)
6   **else return** node : REMOVESEEN(**tail** nodelist, OPEN, CLOSED)

The above program is a recursive program that steps through *nodeList*, removing a node if it is present in either CLOSED or OPEN (Lines 4, 5) or keeping it and recursively processing the tail of *nodeList* (Line 6). Algorithm REMOVESEEN in turn calls algorithm OCCURSIN to check if a given *node* occurs somewhere in the list *nodePairs* as a first element of a *pair*. Algorithm OCCURSIN is described below.

Algorithm 3.6. The procedure OCCURSIN checks for the presence of a node, as a first element of some nodePair in the list nodePairs

OCCURSIN(node, nodePairs)
1 **if** nodePairs **is empty**
2   **return** FALSE
3 **elseif** node = **first head** nodePairs
4   **return** TRUE
5 **else return** OCCURSIN(node, **tail** nodePairs)

The final piece of the jigsaw is the algorithm *MAKEPAIRS* that is called in DFS (Line 11) with a list of (new) nodes to be added to OPEN. In DFS the input to this is the nodes from which nodes generated earlier have been filtered by REMOVESEEN. MAKEPAIRS accepts this list and the parent they were generated from and forms pairs to be added to OPEN.

> **Algorithm 3.7.** MAKEPAIRS recurses down a list of nodes in nodeList, for each making a pair with the parent node. For example, MAKEPAIRS([A, B, C, D], S) = [(A,S), (B,S), (C,S), (D,S)]
>
> MAKEPAIRS(node List, parent)
> 1 **if** nodeList **is empty**
> 2   **return empty list**
> 3 **else return** (**head** nodeList, parent) : MAKEPAIRS(**tail** nodelist, parent)

3.2.4 Backtracking in DFS

When we talk about DFS, we also talk about backtracking. This did not manifest itself in the tiny search problem of Figure 3.1 where all branches in the search tree led to the goal. This is not the case in the following example in Figure 3.7 in which we have deleted node E from the state space of the tiny search problem. The figure also shows the corresponding search space, and with solid edges, the tree explored by DFS when only new nodes are added to OPEN.

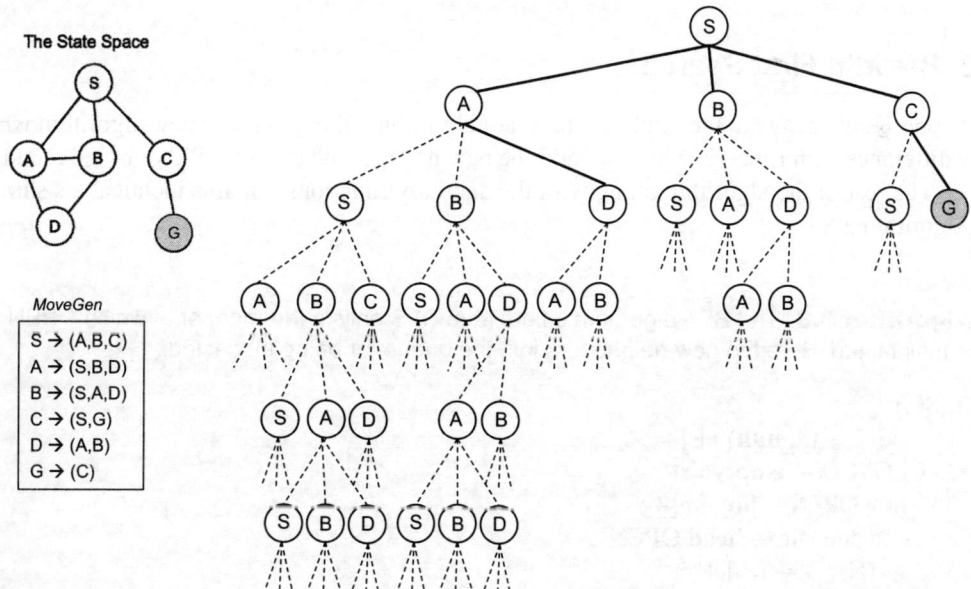

Figure 3.7 If we remove node E from the tiny search problem, then DFS is compelled to backtrack from the paths via A and B on the left in the search tree and find the path via node C on the right. When only new nodes are added, both DFS and BFS (described later) generate the same subtree shown in solid lines, but they explore it in a different order.

Backtracking happens when the algorithm reaches a dead end or has run out of options, and it backtracks to the parent to try the next option. In our implementation of DFS using a stack, this happens naturally. The reader is encouraged to simulate the progress of DFS on the above problem. DFS first goes down to D via node A. Node D is a dead end as it has no new neighbours, so after it is removed from OPEN it does not add any nodes to it. Node B naturally comes to the head of the list and likewise does not add anything to OPEN. Node C then becomes the next node and leads to the only path to the goal G. In general, when all options at some level are tried in a particular branch without adding new ones, the alternatives come to the fore automatically at the head of OPEN.

Depth first search admits the possibility of keeping only one copy of the state. Then instead of adding a neighbour state to OPEN, one can add two moves f_i and b_i, where f_i is a forward move and b_i is the corresponding backward move which undoes the forward move. When f_i is picked, then the state is modified to reflect the change. When MOVEGEN is applied to the resulting state it returns a list of forward and backward moves which are added to the head of OPEN. If there are no new moves to be made, then, as before, nothing is added and then b_i is the next move to be picked. That changes the state back to what it was before f_i was applied. Backtracking is thus explicit. The reader is encouraged to modify the program to work in this manner with only one state.

For the problem in Figure 3.7, DFS ends up inspecting the entire state space. The reader should verify that the order of inspecting them is S, A, D, B, C, G. The algorithm breadth first search (BFS) also inspects the entire space but in a different order. The reader should work that out after studying the algorithm BFS described below.

3.3 Breadth First Search

Only one small change to the implementation of algorithm DFS gives us a new algorithm BFS. The difference is in Line 12, where we add the new nodes at the rear of OPEN instead of at the front. The rest of the algorithm along with the ancillary functions remains identical, as shown in Algorithm 3.8.

Algorithm 3.8. The BFS algorithm differs from DFS only in the manner in which OPEN is maintained. Here the new neighbours join the queue for being inspected.

BFS(S)
1 OPEN ← (S, **null**) : [ ]
2 CLOSED ← **empty list**
3 **while** OPEN **is not empty**
4   nodePair ← **head** OPEN
5   (N, _) ← nodePair
6   **if** GOALTEST(N) = TRUE
7   **return** RECONSTRUCTPATH(nodePair, CLOSED)
8   **else** CLOSED ← nodePair : CLOSED
9   children ← MOVEGEN(N)
10   newNodes ← REMOVESEEN(children, OPEN, CLOSED)

```
11          newPairs ← MAKEPAIRS(newNodes, N)
12          OPEN ← (tail OPEN) ++ newPairs
13   return empty list
```

The new order of maintaining OPEN as a queue completely changes the behaviour of the search. The newest nodes first strategy led DFS to dive headlong into the search tree, and the state space. The FIFO, or newest nodes last, strategy of BFS forces new nodes to queue up to be inspected when their turn comes. Figure 3.8 shows the behaviour of BFS on the tiny problem of Figure 3.1. Nodes *A*, *B*, and *C*, the neighbours of *S*, are first in line. After *A* is inspected, its only new neighbour *D* is added, but to the rear of the queue. It is node *B* that is inspected after *A*. It has no new neighbours to add. Then search visits *C* and generates the goal node *G* as a child. Now, having finished with the first level in the search tree, the algorithm inspects *D* and then node *G*. When the path is reconstructed, it is the shortest path *S–C–G*. The order of visiting nodes, level by level, is also marked in Figure 3.8.

The conservative nature of BFS, to stick as close to the start node as possible, is an insurance against going into infinite loops. This is true even when we explore the entire search tree without filtering out any neighbours, as shown in the figure with dashed edges. This is a direct consequence of the level by level push into the search tree. There may be multiple paths to the goal node, as depicted by multiple occurrences of *G* in the complete search tree. Since BFS ventures incrementally away from the start node, it will find the shortest path when it first appears in some level. This is true even if the state space itself is infinite.

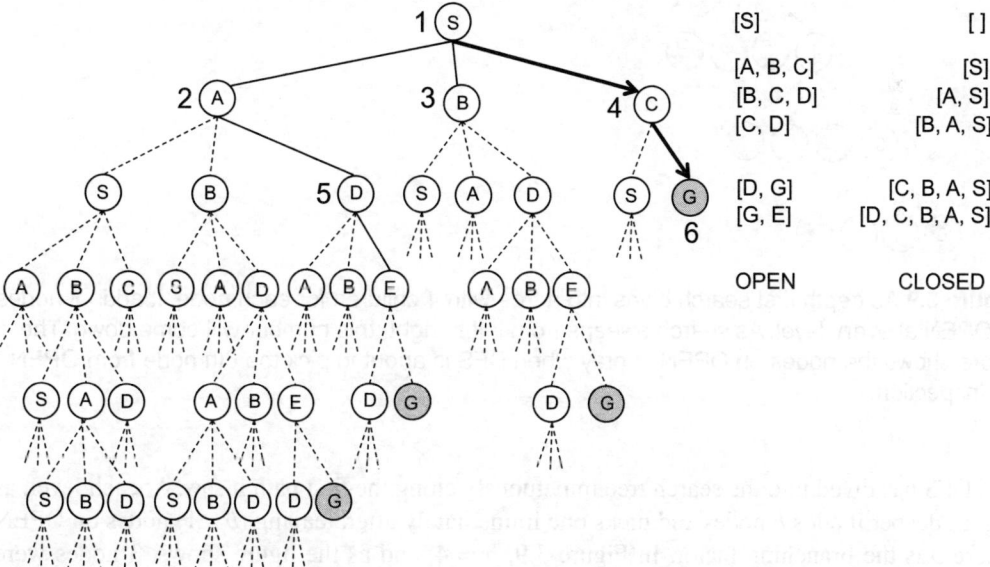

Figure 3.8 When OPEN is a queue, the behaviour of search is starkly different from that in DFS. Algorithm BFS goes down the search tree level by level as indicated by the numbers showing the order in which nodes are visited. As a consequence, it terminates with the shortest path to the goal *G*.

The reader is encouraged to explore the problem described in Figure 3.7 with the two algorithms DFS and BFS, in all their three avatars with new nodes in CLOSED being filtered out, new nodes both in CLOSED and OPEN being filtered, and when all nodes returned by MoveGen are added to OPEN.

3.4 Comparing DFS and BFS

There are four criteria we have stated for analysing a search algorithm – completeness, quality of solution, space complexity, and time complexity. We analyse and compare the two search algorithms based on these. The only difference between DFS and BFS is the way the OPEN list is maintained. Consider a search tree with a constant branching 4 for the sake of analysis. Let the root be at level 0. Then the first level has 4 nodes, the second level 16, the third 64, and so on. The total number of nodes grows exponentially with depth. Figure 3.9 shows the progress of DFS at the point when it is about to pick its 6th node.

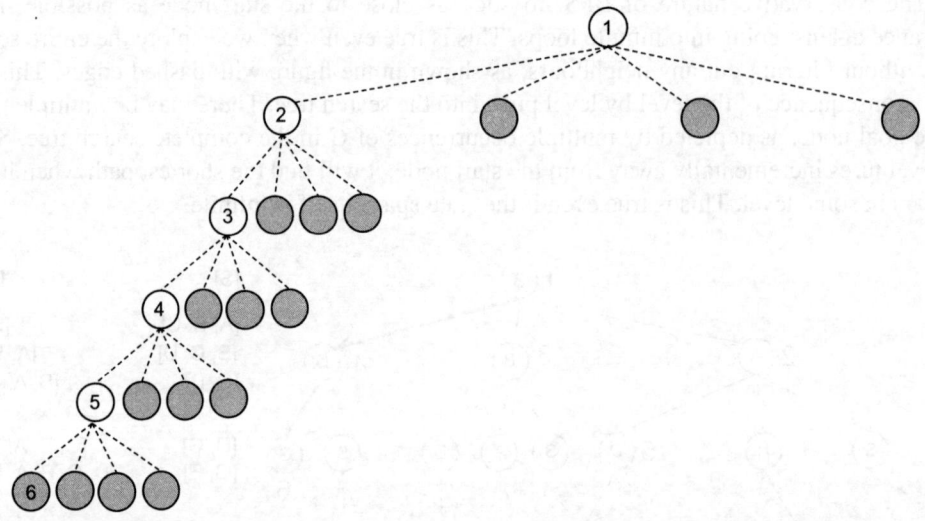

Figure 3.9 As depth first search dives into a tree with 4 children for each node, it adds 3 nodes to OPEN at every level. As search sweeps from left to right, this number will come down. The figure shows the nodes on OPEN in grey when DFS is about to pick the 6th node from OPEN for inspection.

DFS has dived into the search tree impetuously along the first path it saw. At each level, as it goes deeper it adds b nodes and picks one immediately after, leaving $(b-1)$ nodes on OPEN where b is the branching factor. In Figure 3.9, $b = 4$, and as the figure shows, 3 nodes were added to OPEN at each level, shown in grey.

BFS on the other hand has just reached level 2 when it picks the 6th node in the same search tree, as shown in Figure 3.10.

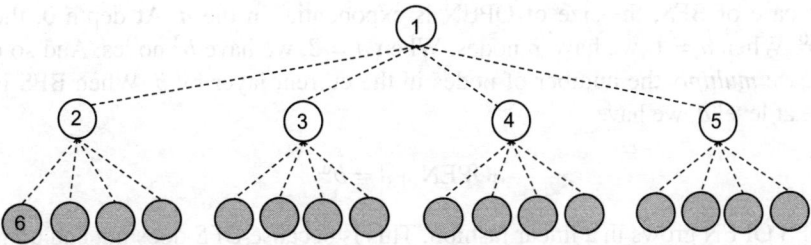

Figure 3.10 As breadth first search pushes into a tree with 4 children for each node, it multiplies the number of nodes in OPEN by 4 at every level. Thus OPEN grows exponentially with depth. The figure shows the nodes on OPEN in grey when BFS is just about to pick the 6th node from OPEN for inspection.

When it is about to pick the 6th node, BFS has finished inspecting all the nodes in level 1. In the process, it has added 4 nodes for each node it inspected at level 1. When it starts on level 2, it has 16 nodes in OPEN, which is 4^2 nodes. When it is about to pick the 6th node, there are 16 nodes on OPEN shown in grey in Figure 3.10. In general, as BFS enters level d, it is confronted with b^d nodes at level d where b is the branching factor.

3.4.1 Space complexity

Traditionally space complexity is estimated by the size of OPEN, which stores the candidate nodes yet to be inspected and defines the search frontier. From the analysis above DFS is a clear winner here. As search goes deeper into the tree, OPEN grows linearly with a constant number of nodes being added at every level. CLOSED on the other hand grows exponentially with the number of nodes being multiplied by a constant in each successive layer. Figure 3.11 depicts the two search frontiers in a schematic search tree.

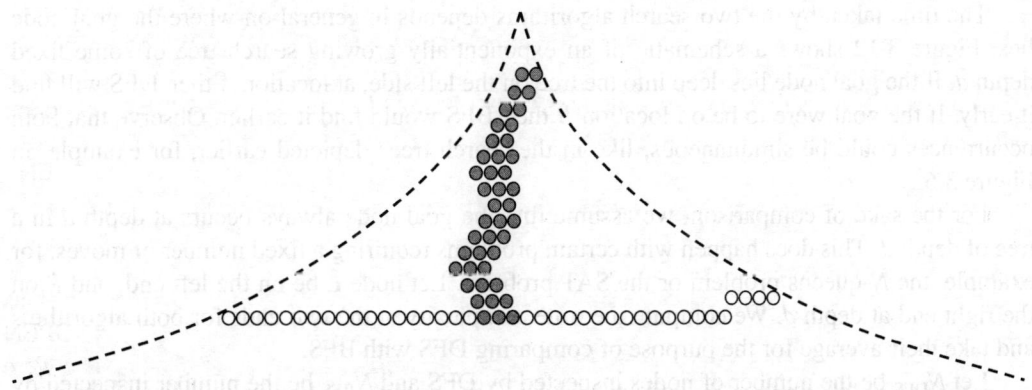

Figure 3.11 The search frontiers for DFS in grey and BFS in unshaded nodes. The outer envelope is meant to give a feel of exponential growth.

In the case of BFS, the size of OPEN is exponential in the d. At depth 0, there is one node, or b^0. When $d = 1$, we have b nodes. When $d = 2$, we have b^2 nodes. And so on. As we go deeper, we *multiply* the number of nodes in the current layer by b. When BFS is about to commence at level d, we have

$$|\text{OPEN}_{\text{BFS}}| = b^d.$$

For DFS OPEN grows in a linear fashion. This is because DFS does not end up generating b children for every node in the *current* layer, but only for the one node that it picks. We only add b nodes to OPEN as we go to the next layer. At depth 0, there is 1 node. At depth 1 we add b nodes. Of these, we remove the first for inspection, so in effect we have $(b - 1)$ nodes left. At every layer, we add b nodes and pick one for inspection. Counting the node that the algorithm is about the pick at depth d, we have

$$|\text{OPEN}_{\text{DFS}}| = d(b - 1) + 1$$

DFS thus scores over BFS on space complexity. The OPEN list grows only in a linear fashion as compared to the exponential growth of BFS.

Figure 3.11 also suggests a way of visualizing the progress of search. Assuming that the MoveGen function generates children from left to right, DFS dives down into the leftmost branch. Assuming that the tree is of finite depth, which would be the case for a finite state space, it may backtrack at some point, and explore the search tree sweeping it from left to right. BFS on the other hand pushes into the search tree layer by layer, with OPEN growing exponentially. In either case, the algorithm is oblivious of the goal, till the point it hits upon it.

3.4.2 Time complexity

Time complexity can be measured in terms of the number of times the GoalTest function is applied, or the number of nodes inspected by the algorithm. This is the size of CLOSED.

The time taken by the two search algorithms depends in general on where the goal node lies. Figure 3.12 shows a schematic of an exponentially growing search tree of some fixed depth d. If the goal node lies deep into the tree on the left side, at location A then DFS will find it early. If the goal were to be on location B then BFS would find it earlier. Observe that both occurrences could be simultaneous, like in the search trees depicted earlier, for example, in Figure 3.6.

For the sake of comparison, we assume that the goal node always occurs at depth d in a tree of depth d. This does happen with certain problems requiring a fixed number of moves, for example, the N-queens problem or the SAT problem. Let node L be on the left end, and R on the right end at depth d. We compute the time complexity at the two ends for both algorithms and take their average for the purpose of comparing DFS with BFS.

Let N_{DFS} be the number of nodes inspected by DFS and N_{BFS} be the number inspected by BFS.

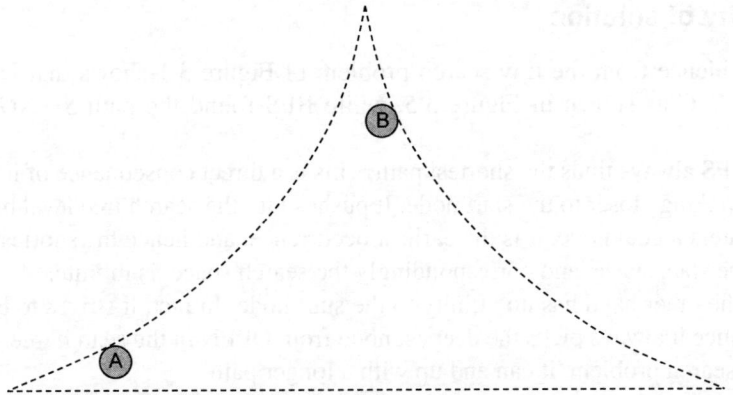

Figure 3.12 The time taken by search to find the goal node depends upon where the goal node lies in the search tree. If it is at location A deep into the tree on the left, then DFS will find it quickly while BFS will wade through all the levels before. If it is at location B on the right closer to the root, then BFS will find it early while DFS will sweep through the entire tree from left to right.

When the goal node is at L,

$$N_{DFS} = (d + 1)$$

the number of nodes down the leftmost branch.

$$N_{BFS} = (b^d - 1)/(b - 1) + 1$$

the number of nodes up to depth $(d - 1)$ plus 1 for the node at L.

When the goal node is at R,

$$N_{DFS} = (b^{d+1} - 1)/(b - 1)$$
$$N_{BFS} = (b^{d+1} - 1)/(b - 1)$$

the number of nodes in the entire search tree for both.

On average,

$$N_{DFS} = ((d + 1) + (b^{d+1} - 1)/(b - 1))/2 \cong b^d/2 \text{ for large } b$$
$$N_{BFS} = ((b^d - 1)/(b - 1) + 1 + (b^{d+1} - 1)/(b - 1))/2 \cong b^d(b + 1)/2(b - 1) \text{ for large } b$$

Thus the ratio of the time taken by BFS and DFS is

$$N_{BFS}/N_{DFS} = (b + 1)/(b - 1) \text{ for large } b$$

We observe that for large b, BFS has a slightly higher time complexity than DFS. But the important observation is that *both are exponential*.

3.4.3 Quality of solution

Anecdotal evidence from the tiny search problem of Figure 3.1 shows that DFS found the path S–A–D–E–G as shown in Figure 3.5, while BFS found the path S–C–G as shown in Figure 3.8.

In fact, BFS always finds the shortest path. This is a direct consequence of its conservative tendency of sticking closer to the start node. It pushes into the search tree level by level. When it first encounters a goal node, it is the earliest occurrence, and hence the shortest path. This is true even if the state space, and correspondingly the search space, is infinite.

DFS on the other hand has no affinity to the start node. In fact, it strives to be as far away as possible, since it always picks the deepest node from OPEN in the search tree. As illustrated with the tiny search problem, it can end up with a longer path.

3.4.4 Completeness

A corollary of the above property is that BFS is complete. If a path exists from the start state to the goal state, then BFS will find a path. This will happen even if the search tree is infinite, for example, as in Figure 3.3. Not only that, it will find the shortest path to the goal node. The moment a goal node appears in a new layer, it will be found. DFS on the other hand can go down an infinite path, again as in Figure 3.3, even though there is a path to the goal.

For finite state spaces with only previously unseen nodes being added, as in Algorithms 3.4 and 3.8, both algorithms are complete. If there is a path to the goal node, both will find one and will also terminate with failure if there is none. This is because, in every cycle, one node is moved from OPEN to CLOSED and eventually OPEN becomes empty. We also say that the search algorithms are *systematic* when they explore the entire finite state space, whatever the order they choose to do so in.

Most real world problems have finite state spaces, so we will consider both algorithms to be equal on this count. Infinite state spaces can come from the domain of mathematics, and we will perhaps need to be careful here. For example, in number theory, if we were to search for a counterexample to Fermat's Last Theorem (look for x, y and z such that $x^3 + y^3 = z^3$), the search will be futile, and no algorithm will terminate.

On the four criteria for analysing search algorithms, BFS did better on the quality measure, guaranteeing the shortest path, and DFS did better on space complexity, requiring only linear space to store OPEN. We next look at an algorithm that gives us the best of these two worlds, with surprisingly low extra cost.

3.5 Depth First Iterative Deepening

DFS is one of the most widely used algorithms for many problems. One place it has been used quite a bit is in Chess playing programs competing in tournaments. Chess tournaments allocate a certain amount of time, with the most common format allowing 90 minutes for the first 40 moves to each player. Each player has her own clock, with only one of the two ticking at any time. Each player is free to distribute their allocated time over the different moves, and as one can imagine, if you spend too much time over the initial moves, you have less for the remaining. Now, as we will study in Chapter 8, the most used algorithm is based on DFS. The programs look ahead a certain number of moves, called plies, before evaluating the board position. The farther you look ahead, the better the analysis. Chess programmers came up with the idea of doing a flexible amount of lookahead, based on the available time, and the idea of iterative deepening was born.

3.5.1 Depth bounded depth first search

Before we look at our target algorithm, let us look at a variation of DFS which operates with a depth bound. This algorithm, called *depth bounded depth first search* (DB-DFS), is essentially DFS but with a bound on the depth it cannot go beyond – a kind of *Lakshman rekha*, for those familiar with the epic of Ramayana. The situation is depicted in Figure 3.13.

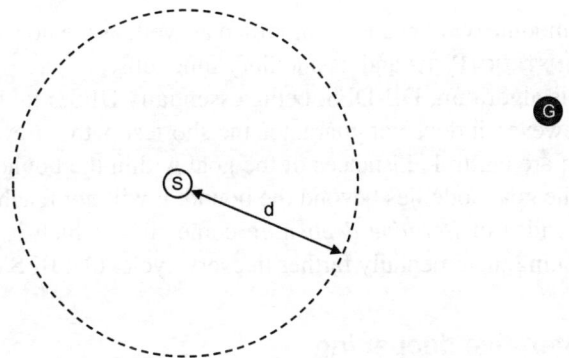

Figure 3.13 DB-DFS searches the state space in a depth first manner but does not venture beyond a boundary at depth *d*.

Algorithm 3.9 describes DB-DFS, which takes the depth bound as an argument, *depthBound*, and executes DFS only within the bound. It stores the depth information in the *nodePair*, which now in fact becomes a triple. But we retain its name. The root is at depth 0. DB-DFS invokes MOVEGEN only when the current node is within the depth bound.

> **Algorithm 3.9.** Algorithm DB-DFS is a modification to restrict DFS to a bound received as an argument. We extend the *nodePair* to store the depth value as a third parameter, even though we still call it a pair.
>
> DB-DFS(*S*, depthBound)
> 1   OPEN ← (S, **null**, 0) : [ ]
> 2   CLOSED ← **empty list**
> 3   **while** OPEN **is not empty**
> 4   nodePair ← **head** OPEN

```
5      (N, _, depth) ← nodePair
6      if GoalTest(N) = TRUE
7          return ReconstructPath(nodePair, CLOSED)
8      else CLOSED ← nodePair: CLOSED
9          if depth < depthBound
10             children ← MoveGen(N)
11             newNodes ← RemoveSeen(children, OPEN, CLOSED)
12             newPairs ← MakePairs(newNodes, N, depth + 1)
13             OPEN ← newPairs ++ tail OPEN
14         else OPEN ← tail OPEN
15 return empty list
```

The MakePairs module will need to be modified as well, as we now deal with triples. The same applies to ReconstructPath and its ancillary functions.

Let us analyse this algorithm. DB-DFS, being essentially DFS, requires only linear space for the OPEN list. However, it does not guarantee the shortest path, or even a path to the goal. What is more, if there are multiple instances of the goal within the bound, it may find the one farther away. And if the goal node lies beyond the bound, it will not reach it at all.

This is where the idea of *iterative deepening* comes in – which is to do a sequence of DB-DFS searches looking incrementally farther in every cycle, like BFS does.

3.5.2 Depth first iterative deepening

Depth first iterative deepening (DFID) starts with a depth bound 0 and starts doing DB-DFS. If it does not find the goal node, then it increments the bound and tries again. This process is repeated till a goal node is found in some cycle. The *high level* algorithm is given below.

Algorithm 3.10. DFID does a series of DB-DFSs with increasing depth bounds

```
DFID(start)
    1   depthBound ← 0
    2   while TRUE
    3       do    DB-DFS(start, depthBound)
    4             depthBound ← depthBound + 1
```

Can one get the best of both worlds? That is, the shortest path using linear space. The answer is yes, almost. A small incremental cost must be paid. We analyse the high level algorithm and highlight its positive features with a couple of problems. We also give some suggestions on how some lacunae can be addressed. We begin with the simpler arguments.

3.5.3 Space complexity

Beneath the hood of DFID is DFS which, as we have seen earlier, has linear space requirement. Ergo, DFID requires linear space.

3.5.4 Time complexity

While it pushes into the state space layer by layer, like BFS, in reality it revisits the older nodes again and again repeatedly. For every fresh layer it explores, it does the extra work of revisiting *all* the earlier layers completely. Where BFS would have inspected L nodes in the newest layer, DFID again explores, in addition, the entire I internal nodes again.

When BFS would have inspected L nodes, DFID inspects $(L + I)$ nodes.

Therefore,

$$N_{DFID}/N_{BFS} = ((L + I)/L)$$

Now, for a full tree with branching factor b, the following holds:

$$L = (b - 1)I + 1$$

This gives us

$$N_{DFID}/N_{BFS} = (bI + 1)/((b - 1)I + 1)$$
$$\cong (b/(b - 1)) \quad \text{for large } b$$

Thus, the extra work done by DFID is not significant for large b. When the search tree is a binary tree, with $b = 2$, we know from our data structures background that $L = I + 1$. That is, DFID would do double the work done by BFS. But as b increases, this ratio comes down. For $b = 11$, for example, DFID does only 10% more work than BFS. That is a price we would be willing to pay for the benefit of linear space complexity.

The fact that the extra work done by DFID is not too much is surprising for some people. But it should not be so. This is to be expected of exponential growth. The fact that as we go deeper into a tree, we multiply the nodes in the previous layer by the branching factor. *The work done on every new layer is much more than the work done on the entire tree before that!*

That is the nature of our adversary, CombEx!

3.5.5 Quality of solution

One can argue that this process will find the shortest path, because if a shorter path existed it would have been found in an earlier cycle. The reader would have noticed that in pushing forward layer by layer DFID is essentially a sequence of depth first searches masquerading as a breadth first search, which does find the shortest solution.

But this is not true for state spaces where there are multiple paths to a goal. The search tree that DFS generates precludes nodes visited once from being added to OPEN again. This can cause a problem, as illustrated in another small search problem[2] shown in Figure 3.14.

---

[2] This was pointed out by Siddharth Sagar, an undergraduate at IIT Dharwad in 2019, in a class I was teaching.

68 | Search Methods in Artificial Intelligence

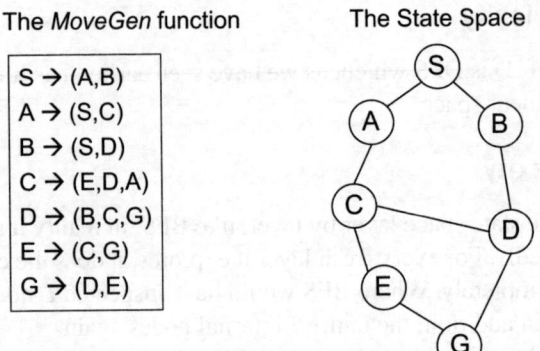

Figure 3.14 An example to highlight the need to add nodes in CLOSED to OPEN again, for DFID to find the shortest path from *S* to *G*.

The reader is encouraged to try out all the variations of DFS and BFS described in this chapter. We are interested in the variation in which only new nodes not already in OPEN or CLOSED are added to OPEN. The reader should ascertain that BFS finds the shortest path *S–D–G*. But, as illustrated in Figure 3.15, DFID does not.

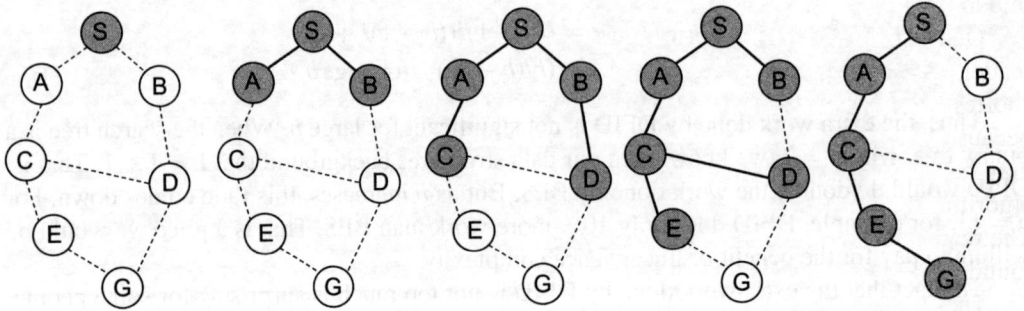

Figure 3.15 DFID explores paths of increasing length in successive cycles. When nodes in CLOSED are not added to OPEN, DFID does not explore the path from *B* to *D* when exploring *paths of length 3* because *D* is already on CLOSED as a child of *C*, and consequently fails to find the shortest path *S–B–D–G*.

The shortest path to the goal, *S–B–D–G*, is of length 3. However, node *D* on this path is not added to OPEN as a child of *B* because it has already been inspected as a child of *C* and is on CLOSED. It is left as an exercise for the reader to verify that not filtering nodes already on CLOSED does indeed work. We begin by modifying DBDFS function as follows. We do not filter out nodes already in CLOSED, and for the sake of simplicity, we add *all* new neighbours to OPEN after constructing *nodePairs* (now triples) with them. It would be natural to ask whether one should stop filtering out nodes already on OPEN as well. If the reader is convinced that the nodes on OPEN can be filtered out, she is encouraged to modify the algorithm to do so. This is left as something for the reader to ponder over (also see the exercises).

This, however, leads to another problem. If we allow nodes on CLOSED to be added to OPEN again, some nodes may be present in CLOSED more than once. In the above example, *D* would exist twice, once with *C* as the parent, and the other time with *B* as the parent. How do we make sure that the path reconstruction module picks the correct *nodePair* from CLOSED. One is to argue about the nature of the implementation and assert that the instance of *D* added later will be the one retrieved. This is true of our implementation. But what would happen if CLOSED were to be implemented as a hash table? Perhaps we can store depth information as well in each *nodePair* and choose the one with the lowest depth. We will do so anyway below. This may increase the complexity of the overall algorithm if we have to evaluate and choose between parents. In Chapter 6, when we look at storing only one parent pointer, the best one at any point in time, we will find an efficient solution to this problem. Another way is to store the entire path as prescribed in Exercise 1 at the end of this chapter.

3.5.6 Completeness

It is not yet time to celebrate though. The second problem is that of avoiding looping. Remember that we introduced CLOSED precisely to do this. If we do not filter nodes in CLOSED, which we *still maintain* to be able to reconstruct the path, how does one stop the program from going into loops? This question has two subparts to it.

One, if looping cannot be stopped, will we still find the shortest path? The interested reader should hand simulate search on the earlier examples and verify that indeed the shortest path will be found. There may be wasteful loops, but not infinite ones. This is because in every cycle the algorithm explores paths of up to a certain length, even if they may have cycles. For example, in the discussed problem when the depth bound is 3, the algorithm will also consider *S–A–S–B* as a candidate, before finding the path *S–B–D–G*. The reader should satisfy herself that when the algorithm picks the goal node it has indeed found the shortest path.

The other subpart needs more attention though. What if there is no path to the goal? This would happen if the goal node is not in the partition in which the start node is. Will the program go into incrementally longer and longer loops, and never terminate? That, indeed, is a danger. Imagine that in our tiny problem, *G*, or any other node, was not a goal node. Then it would continue looking at longer paths in an infinite loop. We still need a termination criterion when the solution does not exist. In our original DFS and BFS algorithms it was when OPEN became empty. Now we need a different one, since OPEN will never become empty.

One solution could be to count the number of *new nodes* visited by the algorithm in each call. While a node may be added to CLOSED more than once, because it can be visited more than once through different paths, it is counted only once.

The algorithm should now return two things. One is the count of the number of new nodes in the state space visited by DBDFS in a call. This count will depend upon the depth bound given to it. Initially it will be 1 when $b = 0$. When $b = 1$, the count is $b + 1$. And so on. If the goal node does not exist in the graph, then, after searching the entire graph, the count will be the same in two consecutive calls. The calling DFID algorithm will then be able to call it quits and report failure. The other parameter it returns is the path found, or the empty list if in that call a path was not found.

Earlier we had seen that pruning nodes already on CLOSED fails to find the shortest path on some graphs (see Figure 3.15). It turns out that there can be graphs where DFID fails to even find a path if we were to prune CLOSED and also use count as described above.[3] Consider the curious graph shown in Figure 3.16.

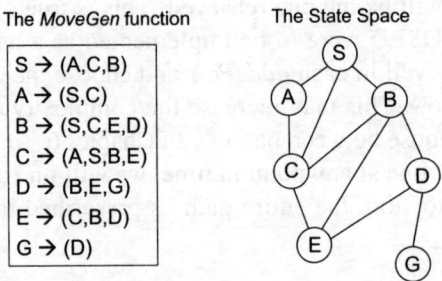

Figure 3.16 Another example to highlight the need to add nodes in CLOSED to OPEN again in DFID. Without that, *and* when the algorithm kept count of new nodes in every cycle, DFID would not even find a path from *S* to *G*.

Consider a version of DFID which filtered nodes already on CLOSED and also maintained a count of the number of nodes seen in each cycle. Then, on the above graph, when exploring paths of length 2 the algorithm would have visited the 6 nodes *S*, *A*, *C*, *B*, *E*, and *D*. In the next cycle it would again visit the same 6 nodes. In a manner reminiscent of Figure 3.15, DFID would visit *D* via *S–C–E–D* first, and later not extend *S–B* to *D*, thus missing the 3-step path to *G*. What is more, since the node count in the two cycles, 6, is the same, the algorithm will terminate and report failure.

This gives us another reason to not filter nodes already on CLOSED in DFID. What about filtering nodes already on OPEN? The reader is encouraged to find out.

3.6 Uninformed Search

Figure 3.17 illustrates the trajectory that BFS and DFS follow, irrespective of where the goal node may be.

Both DFS and BFS, and consequently also DFID, behave in an uninformed or blind fashion, dictated only by the order in which MoveGen generates the successors. Both are oblivious of the goal. It is as if their behaviour has genetic roots, with the environment or the problem context playing no role in shaping it. The next chapter will look at algorithms that have some sense of direction of which moves are likely to head towards the goal.

---

[3] As pointed out by my colleague S. Baskaran as we were formulating a quiz problem for the course on search methods.

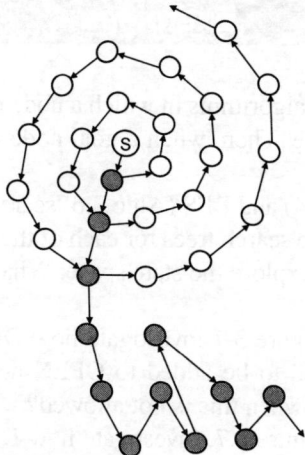

Figure 3.17 The trajectory of DFS and BFS in the state space. BFS gradually moves away from the start node *S*, while DFS, shown in grey nodes, dives headlong into the state space. The arrows depict the trajectory and not the edges in the state space. The goal node could be anywhere in the state space. Neither algorithm is cognizant of the goal node while searching and each traverses the space in a predetermined manner.

Summary

In this chapter we have built the basic machinery for search. Given a MOVEGEN or neighbourhood function, the search algorithm simultaneously generates and explores the search space. This happens until the GOALTEST function signals success, or for a finite search space it has exhausted all possibilities.

The search algorithm has its own perspective of the space being searched, and that manifests itself in the form of a search tree that it generates and explores. Every variation of the search algorithm has its own corresponding search tree that it explores.

We have studied three variations of our basic algorithm. DFS dives headlong into the search space. Its major advantage is that it requires only linear space. But it does not guarantee the shortest path, and for infinite spaces may not return any path even if there exists one. BFS conservatively pushes into the search space and guarantees the shortest path, even for infinite search spaces. DFID combines the best of both, with surprisingly low extra cost.

In terms of time complexity, all three are unable to combat CombEx, inspecting an exponential number of nodes as they go deeper. This is partly because they are all blind or uninformed. The way they search the given space is always the same in a given domain as defined by the MOVEGEN function, oblivious of where the goal node is. The GOALTEST function is only used to recognize a goal node when the search stumbles upon it. In the next chapter we study heuristic search, which exploits some extra knowledge to skew the exploration more towards the goal node.

Exercises

1. Implement the DFS and BFS algorithms in which a node in the search tree stores the entire path up to it in the state space. Then, when a goal node is found, the corresponding path would be readily available.
2. Investigate the DFS (Algo. 3.4) and BFS (Algo 3.6) search algorithms on the tiny problem shown in Figure 3.7. Draw the search trees for each of the two algorithms and list the order in which the two algorithms explore the state space. What is the path found by each of the two algorithms?
3. For the problem shown in Figure 3.7 investigate how DFS and BFS behave when nodes already in OPEN are allowed to be added to OPEN again. How does the performance compare with the algorithms when this is not allowed?
4. For the problem shown in Figure 3.7, investigate how DFS and BFS behave when nodes already in OPEN and CLOSED are allowed to be added to OPEN again. How does the performance compare with the algorithms when this is not allowed?
5. Repeat the above three paper and pencil simulations on the state space depicted in Figure 3.14.
6. [Baskaran] The following figure shows a map with several locations connected by 2-way edges (roads). The MOVEGEN function returns neighbours *in alphabetical order*. The start node is *S* and the goal node is *G*. Module REMOVESEEN removes neighbours already present in OPEN/CLOSED lists. List the nodes in the order visited by DFS and BFS. Draw the search trees generated by both, clearly identifying the nodes on CLOSED and on OPEN, and the path found.

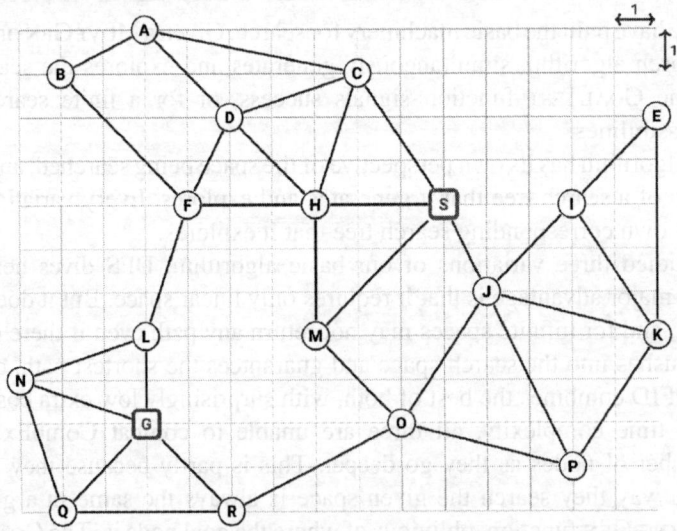

7. List the nodes in the order visited by DFID (when all new nodes are added to OPEN). What is the path found?
8. [Baskaran] The MOVEGEN function for the following graph generates the neighbours in alphabetical order. *S* is the start node and *G* is the goal node. List the nodes in the order in which DFS will visit the nodes till termination. What is the path found?

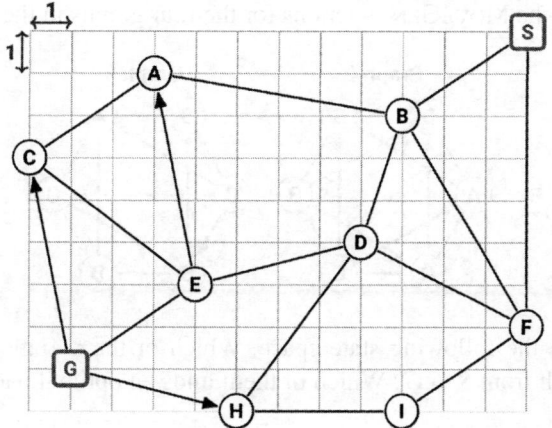

9. Will your answer be the same if the edge *GC* in the above graph were to be bidirectional?
10. Given the MOVEGEN function below, draw the state space. *S* is the start node and *G* is the goal node. Draw the search trees for each of the three variations for DFS and BFS and list the nodes in the order in which the six algorithms visit them.

```
S → (A B C D)
A → (E J B S)
B → (A F S)
C → (G H D S)
D → (C I S)
E → (K J A)
F → (J K B)
G → (L C)
H → (L M I C)
I → (H D)
J → (E F A)
K → (F E)
L → (M H G)
M → (H L)
```

11. Algorithm DFID (the version which adds all nodes generated by MOVEGEN without filtering any nodes) is traversing the state space generated by the above MOVEGEN function till termination. *S* is the start node and *G* is the goal node. List the nodes visited in each cycle up to depth 4 or till termination, whichever happens first.

12. [Baskaran] Write the MoveGen functions for the four graphs in the following figure.

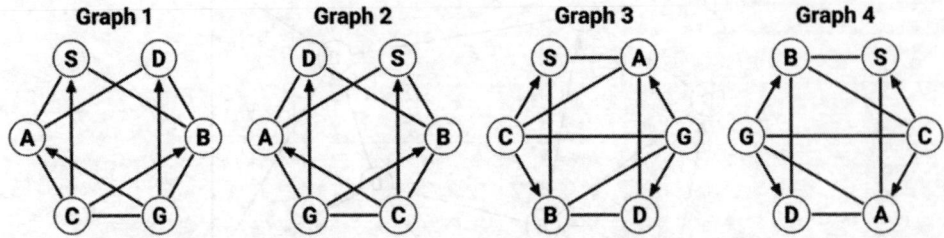

13. [Baskaran] Study the following state space. Which of the variations of DFS, BFS, and DFID, finds a path from *S* to *G*? Which of them finds an optimal path?

X	MoveGen(X)
S	[A, B]
A	[C, S]
B	[D, S]
C	[S, B, A, D]
D	[A, G, B, C]

14. Formulate the cryptarithmetic problem as a state space search problem. An example problem is shown here. Each letter needs to be assigned to a unique digit such that the arithmetic sum adds up correctly.

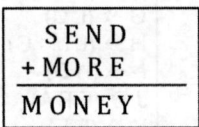

15. Modify the DFS algorithm and its ancillary functions with moves f_i and b_i as described in Section 3.2.4. Use this version to create a demo of the *N*-queens problem with one board on display and queens being placed and removed from the board as the search progresses.
16. Write down the modules RECONSTRUCTPATH and MAKEPAIRS along with the ancillary functions for implementing DB-DFS, given the fact that *nodePair* is now a triple after the depth information is added to it.
17. While developing DFID we argued in the chapter for not filtering out nodes already in CLOSED while adding new *nodePairs* to OPEN. Should one do the same for nodes already on OPEN? Hint: All nodes on OPEN are to the right of a given node in the search tree. Add an edge *A–B* to the problem in Figure 3.10 and try it out.

CHAPTER 4

Heuristic Search

Having introduced the machinery needed for search in the last chapter, we look at approaches to informed search. The algorithms introduced in the last chapter were blind, or uninformed, taking no cognizance at all of the actual problem instance to be solved and behaving in the same bureaucratic manner wherever the goal might be. In this chapter we introduce the idea of heuristic search, which uses domain specific knowledge to guide exploration. This is done by devising a heuristic function that estimates the distance to the goal for each candidate in *OPEN*.

When heuristic functions are not very accurate, search complexity is still exponential, as revealed by experiments. We then investigate local search methods that do not maintain an OPEN list, and study gradient based methods to optimize the heuristic value.

Knowledge is necessary for intelligence. Without knowledge, problem solving with search is blind. We saw this in the last chapter. In general, knowledge is that sword in the armoury of a problem solver that can cut through the complexity. Knowledge accrues over time, either distilled from our own experiences or assimilated from interaction with others – parents, teachers, authors, coaches, and friends. Knowledge is the outcome of learning and exists in diverse forms, varying from tacit to explicit. When we learn to ride a bicycle, we know it but are unable to articulate our knowledge. We are concerned with explicit knowledge. Most textbook knowledge is explicit, for example, knowing how to implement a leftist heap data structure.

In a well known incident from ancient Greece, it is said that Archimedes, considered by many to be the greatest scientist of the third century BC, ran naked onto the streets of Syracuse. King Hieron'II was suspicious that a goldsmith had cheated him by adulterating a bar of gold given to him for making a crown. He asked Archimedes to investigate without damaging the crown. Stepping into his bathtub Archimedes noticed the water spilling out, and realized in a flash that if the gold were to be adulterated with silver, then it would displace more water since silver was less dense. This was his epiphany moment when he discovered what we now know as the Archimedes principle. And he ran onto the streets shouting 'Eureka, eureka!' We now call such an enlightening moment a Eureka moment!

'Eureka' comes from the Ancient Greek word εὕρηκα (heúrēka), meaning 'I have found (it)'. We will interpret this assertion to mean 'I know (it)'. The word *heuristic* is etymologically related to it and refers to a similar state of knowing. In this chapter we focus on giving a sense of direction to our search algorithms which, hitherto, were blind, choosing the node from OPEN in an uninformed manner. We will now empower our search algorithm with some knowledge of which of the choices presenting themselves via OPEN is the best. Figure 4.1 illustrates the notion of an estimated distance to the goal node as embodied in a heuristic function. If there is more than one goal in the domain, then the heuristic value is the smallest of the estimated distances. As we will see, such heuristic knowledge is not infallible, but more often than not will favourably impact the time complexity of search.

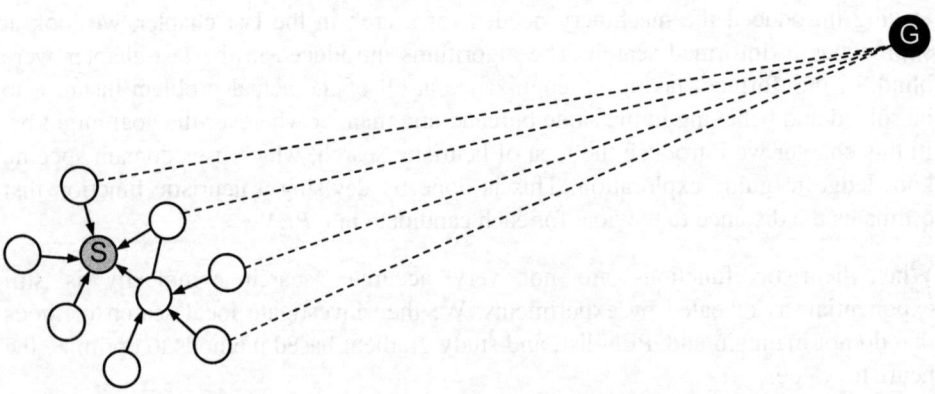

Figure 4.1 Node *S* is the start node and *G* the yet undiscovered goal node. Nodes in grey are on *CLOSED* and unfilled nodes are on OPEN. The arrows from nodes on OPEN point to the parent which added them first. The heuristic function *estimates* the distance of each node *N* to the goal. This estimate, $h(N)$, can help decide *which* node to pick from OPEN. The figure depicts the estimated distance only for four nodes.

4.1 Heuristic Functions

With every node *N* in our search space, we associate a function that estimates the distance from *N* to the goal node. Traditionally, we denote the heuristic function as $h(n)$ or $h(N)$. Observe that like the GOALTEST function, this does not take the goal explicitly as an argument. To begin with, we assign the responsibility of defining this function to the user. This is the third domain function we expect from the user, in addition to the MOVEGEN and GOALTEST functions. Later, in Chapter 10 on Planning, we will get a glimpse of how one can define domain independent heuristic functions.

For the moment, we will assume that the heuristic function is a static function. By this we mean that it is computed only by looking at the given node *N*, in the context of the goal node. Consequently, only by looking at a set of candidates we should be able to decide which of them appears to be closest to the goal, without doing any lookahead. The operative phrases here are 'estimated distance' and 'appears to be closest'. The heuristic function is only an estimate of distance, computed by an inexpensive procedure, and may not be accurate. If we were to search ahead from each of the candidates to decide which one to pick, that would defeat the purpose.

We look at heuristic functions for some example problems, including some from Chapter 2.

4.1.1 Map colouring

In the map colouring problem (Section 2.3.1), the heuristic function could be the number of colours already used up in a partial assignment of colours to regions. The idea is that the fewer the colours used so far, the more the likelihood of extending the partial assignment to other regions and completing the task. Using this heuristic, a set of regions conforming to a ring topology would be coloured with only two colours, without any need to backtrack. Algorithms that do not backtrack are also called greedy algorithms – algorithms that make the choice locally and are done with it, incurring low time complexity. They may or may not find a solution. One also says that one is employing a *greedy heuristic*. Later in this chapter we will look at some greedy approaches to search.

4.1.2 SAT

Defining a heuristic function for the Boolean satisfiability, or SAT, problem (Section 2.3.3) happens naturally. For a candidate solution, the value of $h(n)$ can be the number of clauses satisfied. For the example SAT problem,

$$(b \vee \neg c) \wedge (c \vee \neg d) \wedge (\neg b) \wedge (\neg a \vee \neg e) \wedge (e \vee \neg c) \wedge (\neg c \vee \neg d),$$

the candidate 10101 satisfies clauses 2, 3, 5, and 6. Thus $h(10101) = 4$. The reader would have noticed that this is not consistent with our idea of a distance function, which would have decreasing values as we satisfy more clauses, and presumably be closer to the goal. The observation is true. One simple way to convert this into such a function is to redefine it as follows:

$$h(n) = \text{Total clauses} - \text{the number of clauses satisfied}$$

This would have the property that the $h(goal) = 0$. Later in this chapter we will look at an algorithm that treats our search problem as a problem of minimization of $h(n)$. With the simpler definition of just counting the number of clauses satisfied, this would become a maximization problem instead. A variation of this function could be to consider the weighted sum of clauses satisfied, with larger weights for smaller clauses.

4.1.3 The 8-puzzle

The 8-puzzle (Section 2.4.1) has been the staple diet of artificial intelligence (AI) literature since work on heuristic search began (Nilsson, 1971). It is a planning problem well suited for experimentation, being easy to implement. At the same time, it is not trivial to solve, especially if one is looking for the shortest path. Figure 4.2 shows a typical situation. The start state is on the left, with two possible moves.

We consider two heuristic functions $h_{\text{Hamming}}(n)$ and $h_{\text{Manhattan}}(n)$. The first simply counts the number of tiles out of place. This assumes that each tile will take the same effort to get to its destination. The second adds up the distances for each from its destination. The distance is measured in terms of the number of moves either horizontally or vertically but does not consider the fact that the way has to be cleared for any tile to move.

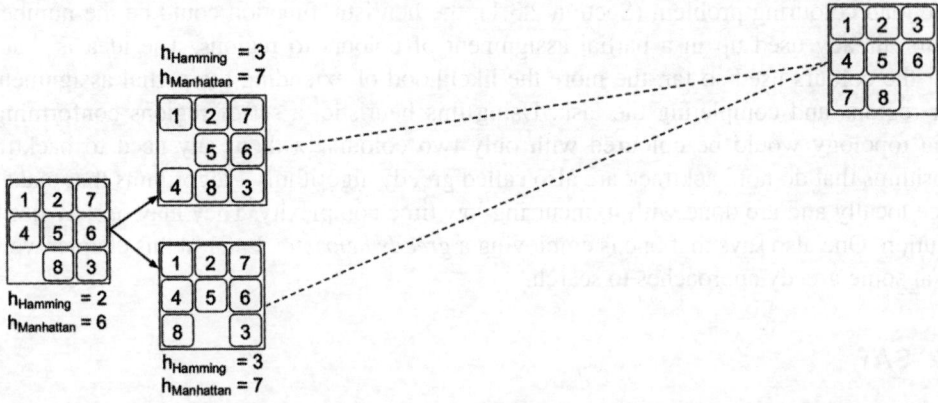

Figure 4.2 Two heuristic functions for the 8-puzzle. The Hamming distance counts the number of tiles out of place, and the Manhattan distance computes the Manhattan distance of each tile from its final location. Observe that neither is perfect, and both neighbours of the start node have a higher value.

Both the above heuristic functions vastly underestimate the actual number of moves that the solution will need. As we will see in Chapter 6 on finding optimal paths, it is desirable to underestimate the distance, but at the same time be as close to the actual distance as possible. Remember that the heuristic distance is only an estimate. If we had an oracle that would tell us the exact distance, there would be no need for search. The algorithm would directly head towards the goal.

The heuristic functions above are computed by inspecting the given state and the goal state. The more a given state looks like the goal state, the closer it is assumed to be. In that sense, a distance function is the inverse of a similarity function, though the latter normally has a range [0, 1]. However, similarity can be misleading. It could easily be the case of so near yet so far. One way to improve the heuristic function that researchers have tried is to add a certain value to the distance if two tiles are in the same row but in the wrong order. This kind of tuning has been common in the early research on heuristic search, until the time when machine learning made it possible to learn heuristic functions.

4.1.4 Route finding

Should I turn left, or right? Or go straight? As one emerges from the main gate of IIT Madras, one faces these choices at the signal. Of course, the answer depends upon where is it you want to go. An oracle would help. A city map can be converted into a graph with every junction a node, and every road segment between two junctions an edge. Figure 4.3 shows such a graph superimposed on a map of the South Chennai region. If you look carefully, you can also see a couple of routes to Anna University Alumni Club suggested to us by Google Maps.

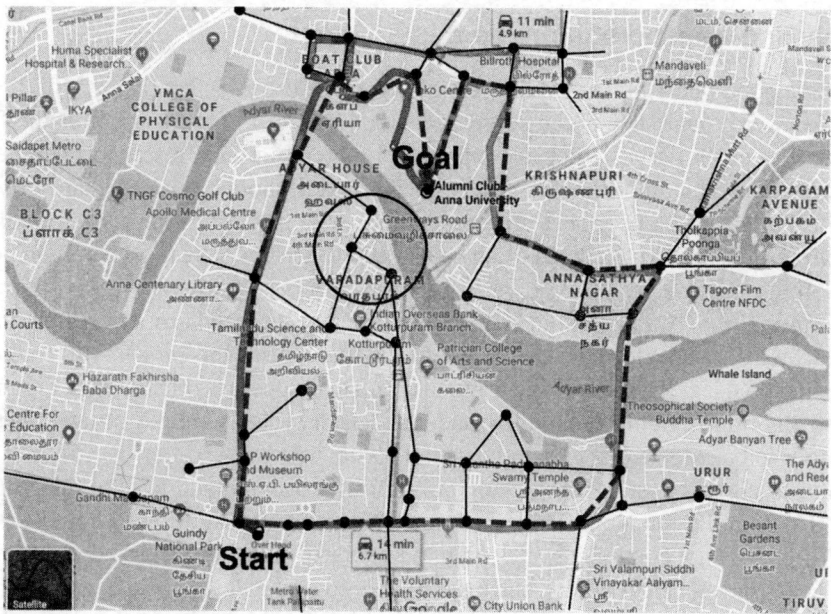

Figure 4.3 A city map can be represented as a graph, here shown superimposed on a map. Observe the two routes from IIT Madras to Anna Alumni Club suggested by Google Maps as shown with thick dashed edges. A heuristic function would first drive the search towards the nodes in the circle in the middle.

One expects the shortest route, or the fastest route, from a map finding algorithm, and that is what we generally get. What is underneath the hood of such applications? We will get some answers in Chapter 6, but for now let us focus on the use of a heuristic function to help us make the choices, as depicted in Figure 4.1.

The heuristic function $h(n)$ is an estimate of the distance from a given node to the goal node. While ideally we would like the road length, perhaps weighted by congestion, we assume we have access only to location information. Let the coordinates of the two points be (x_i, y_i) and (x_k, y_k). The following distance measures are commonly used in geometric spaces.

$$\text{Manhattan distance: } |x_i - x_k| + |y_i - y_k|$$

$$\text{Euclidean distance: } [(x_i - x_k)^2 + (y_i - y_k)^2]^{1/2}$$

The Minkowski norm, named after the German mathematician Hermann Minkowski, is a generalization of the Manhattan and Euclidean distance measures.

$$\text{Minkowski norm: } [(x_i - x_k)^n + (y_i - y_k)^n]^{1/n}$$

Observe that, whatever distance measure we use, the search algorithm will head to that point on the Adyar river that appears to be closest to the destination, before we end up seeking ways to cross the river. This is characteristic of heuristic functions, which are estimates based on incomplete information.

4.1.5 Travelling salesperson problem

What can be a good heuristic for solving the travelling salesperson problem (TSP)? Here we have used the word 'heuristic' in the sense of being a rule of thumb, and not as the function described earlier. Such usage is not uncommon. The literature also refers to meta-heuristics when talking of specific algorithms that may perform well, though not necessarily optimally. As far as the TSP goes, the heuristics are synonymous with the algorithms used for coming up with a solution. Constructive methods often use greedy heuristics. The most common being the following.

The NEARESTNEIGHBOUR algorithm begins at some city, proceeds to the nearest unvisited neighbour, which does not close the loop prematurely, and repeats this process till it has covered all cities. In the final leg, it returns to the original city. This process is illustrated in Figure 4.4 on the left. The algorithm extends the tour by one leg in each cycle. A simple variation is to allow the partial tour to be extended at either end of the partial tour.

The GREEDY algorithm starts by picking the shortest edge and adding it to the empty tour. Subsequently, from the remaining edges, it picks the next shortest allowed edge. An edge is allowed if (a) it does not complete cycles prematurely or (b) it is not a third edge for any given city. Recall that in a tour every city has exactly two edges. A few initial edges on the same problem are shown in Figure 4.4 or the right. This algorithm is similar to Kruskal's algorithm for constructing the minimum spanning tree. But while Kruskal's algorithm does guarantee the minimum spanning tree, the GREEDY algorithm for TSP does not. This is also an indication of the hardness of the TSP.

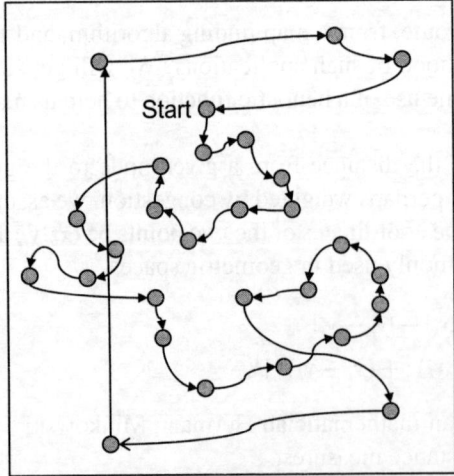 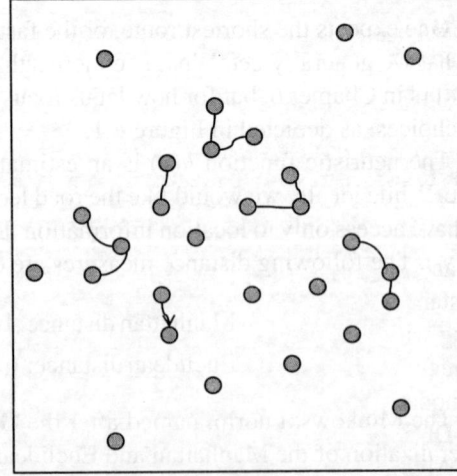

Figure 4.4 On the left, the tour constructed by the NEARESTNEIGHBOUR heuristic, which starts from a random node and goes to the nearest neighbour on the next hop. On the right, some initial edges added by the GREEDY heuristic, adding shortest edges first. Both algorithms have to carefully avoid constructing tours of smaller length.

Another popular algorithm is called the SAVINGS heuristic. The algorithm begins by choosing a pivot vertex and constructing $(N - 1)$ tours of length 2, anchored on the pivot. It then merges two tours by removing two edges containing the pivot and connecting the other ends of the two removed edges. The algorithm is illustrated in Figure 4.5. Let L_1 and L_2 be the lengths of the two edges removed. And let L_{new} be the length of the new edge added. The reduction in edge costs or the saving is $((L_1 + L_2) - L_{new})$. The two edges to be removed are chosen such that the saving is maximum, and that explains the name.

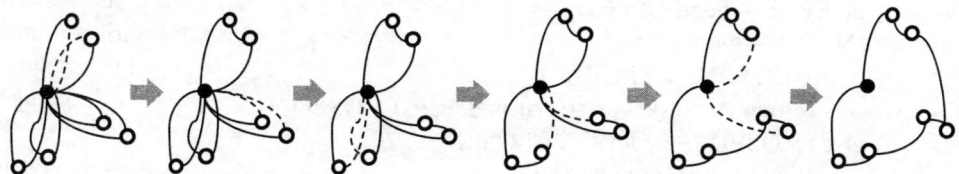

Figure 4.5 The SAVINGS heuristic begins with the construction of $(N - 1)$ tours of length 2 from a fulcrum node shown in black. Then in $(N - 2)$ cycles it progressively merges two tours by deleting one edge from each of the two and adding one edge to reconnect them.

The algorithms described above construct a solution in one shot. They do not search for alternatives. In Chapter 6 we will look at another constructive heuristic method for TSP that employs search, designed to guarantee the optimal tour. But search is more common in the solution space with perturbative methods, and we will explore them later in this chapter and the next. As far as search is concerned, the cost of a tour for TSP can itself be used as the estimate, with the understanding that the lower it is the better.

Heuristic functions add another dimension to the search space, along which the heuristic value for every node is specified. In a sense, a heuristic function defines the landscape that search operates on. The nature of the landscape is determined by the heuristic function, and the way we traverse it is determined by the neighbourhood or MOVEGEN function. We shall explore variations of these later. First, we look at modifying the search algorithm from Chapter 2 to incorporate the heuristic function.

4.2 Best First Search

Heuristic search is search guided by a heuristic function. As shown in Figure 4.1 every node has an estimate of goodness, which most of the time means closeness to the goal, expressed as a distance function. In this chapter we assume that the user supplies this heuristic function $h(n)$ along with the MOVEGEN and GOALTEST functions. When invoked, $h(n)$ returns the *heuristic value* for that node, often referred to as the *h-value*. We augment the representation of *nodePair*, the node in the search tree, to include the heuristic value. This is like the change we did for DB-DFS in Chapter 3 to store the depth parameter.

Algorithm 4.1, BESTFIRSTSEARCH, stores the heuristic value in *nodePair* which, despite its name,[1] is in fact now a triple. The only change from algorithm DFS and BFS is in Line 12, which sorts the OPEN on the heuristic value before picking the node at the head.

---

[1] As the great bard said, a rose by any other name would smell as sweet.

Algorithm 4.1. Algorithm BESTFIRSTSEARCH sorts the OPEN to bring the best node to the head. The third parameter in the nodePair stores the heuristic value of the node.

BESTFIRSTSEARCH(S)
1   OPEN ← (S, **null**, **h**(S)) : [ ]
2   CLOSED ← **empty list**
3   **while** OPEN **is not empty**
4   nodePair ← **head** OPEN
5   (N, _, _) ← nodePair
6   **if** GOALTEST(N) = TRUE
7   **return** RECONSTRUCTPATH(nodePair, CLOSED)
8   **else** CLOSED ← nodePair : CLOSED
9   children ← MOVEGEN(N)
10   newNodes ← REMOVESEEN (children, OPEN, CLOSED)
11   newPairs ← MAKEPAIRS(newNodes, N)
12   OPEN ← sort$_h$(newPairs ++ **tail** OPEN)
13 **return empty list**

We should point out that sorting the OPEN gives us an algorithm that is correct, although it may not be the most efficient. This is because sorting is expensive. The task is to pick the node with the minimum heuristic value from OPEN, and this is best done by maintaining OPEN as a priority queue. Using a similar argument, CLOSED should be maintained as a hash table, since the task is to retrieve a specific *nodePair* based on a node name.

The behaviour of BESTFIRSTSEARCH is illustrated in Figure 4.6 for an abstract search tree. The values in the nodes are heuristic values, and nodes in CLOSED are shaded. As can be seen in the illustration, search can jump around in the search tree and not follow a predetermined pattern like it did in DFS and BFS. Patrick Henry Winston (1943–2019), one of the pioneers of AI at the Massachusetts Institute of Technology (MIT), said that a *new branch may* sprout at any moment in the search tree. This can happen because the heuristic value does not always reflect the ground truth. Remember, the heuristic value is only an estimate, and hence it is perfectly possible that the heuristic function can go wrong and the focus of search may later shift to another part of the tree.

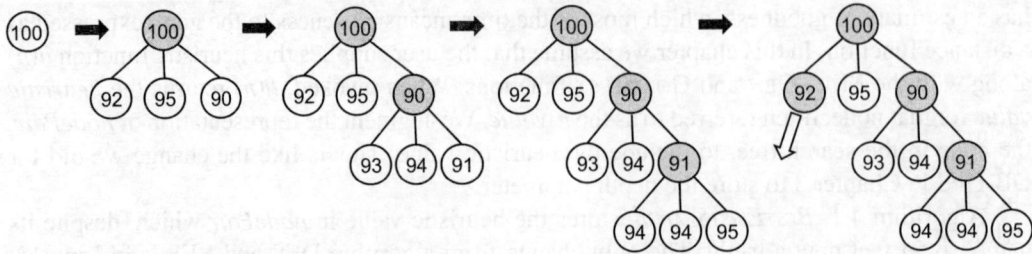

Figure 4.6 The nodes are labelled with their heuristic values. BESTFIRSTSEARCH expands the best node in each cycle, that is, the node with the lowest heuristic value. As Patrick Winston once said, a new branch may sprout in the tree any time.

Figure 4.7 depicts a small path finding problem that we shall test different algorithms on. The thick curve represents a river, not part of the problem description, which might explain the structure of the graph. As one can see, there are three places where one can cross the river. The nodes are placed on a grid of unit size 10 kilometres, so that their coordinates can be computed easily, as can be the heuristic function. The edge labels are the costs of traversing the edges. Node *I* is the start node and node *W* is the goal node.

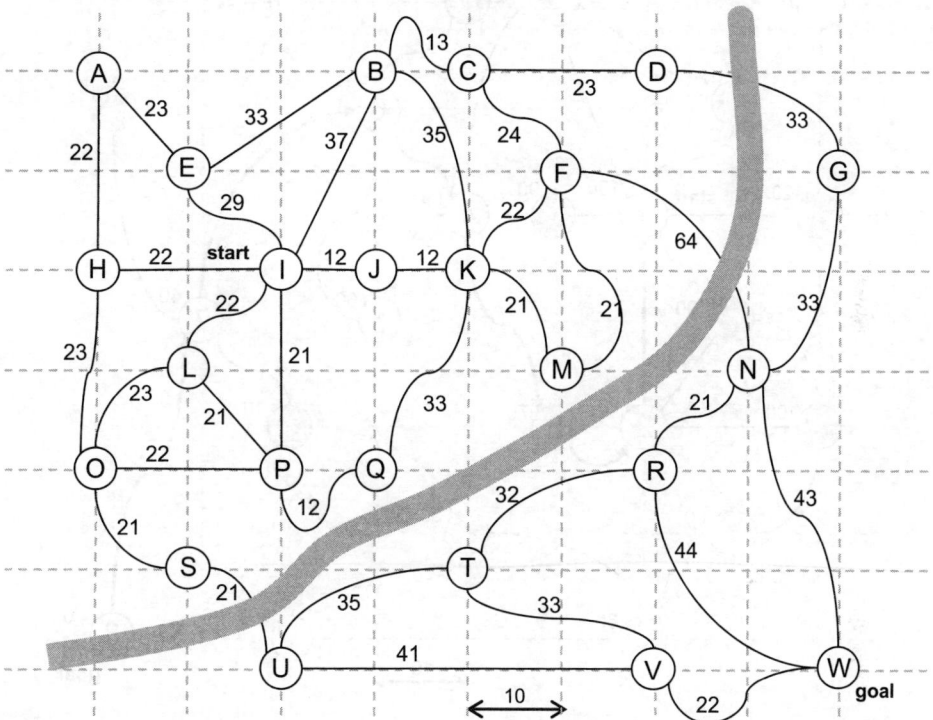

Figure 4.7 A tiny route finding problem. The nodes are placed on a grid where each edge is 10 kilometres. Each edge is labelled with the cost of traversing that edge. Node *I* is the start node and node *W* is the goal node. The thick grey curve is not a part of the graph and represents a river flowing with three bridges across it.

We will adopt the Manhattan distance function as the heuristic function for ease of computation. Thus $h(W) = 0$, being the goal node, and $h(R) = 40$ since it is four hops away. The start node has a value $h(I) = 100$, six steps horizontally and four vertically from the goal.

The node that BESTFIRSTSEARCH picks at each stage is determined completely by the heuristic function. In that sense, BESTFIRSTSEARCH is only a forward looking algorithm, focused on reaching the goal node. As we will illustrate, it does not guarantee the shortest path. In Chapter 6, we will look at the well known algorithm A* that guarantees the shortest path, while being guided by a heuristic function as well.

Figure 4.8 shows the progress of BESTFIRSTSEARCH on the problem given above. Since BESTFIRSTSEARCH ignores the edge costs, we have removed them from the figure. Instead, we have shown the *h*-value for each node, which is the Manhattan distance to the goal node.

The reader is encouraged to work out the order in which the algorithm inspects the nodes, by simulating the algorithm. This order is shown as labels of the nodes inspected. The shaded nodes are still on OPEN when the algorithm terminates. The directed edges show the parents pointers from the goal all the way back to the start node.

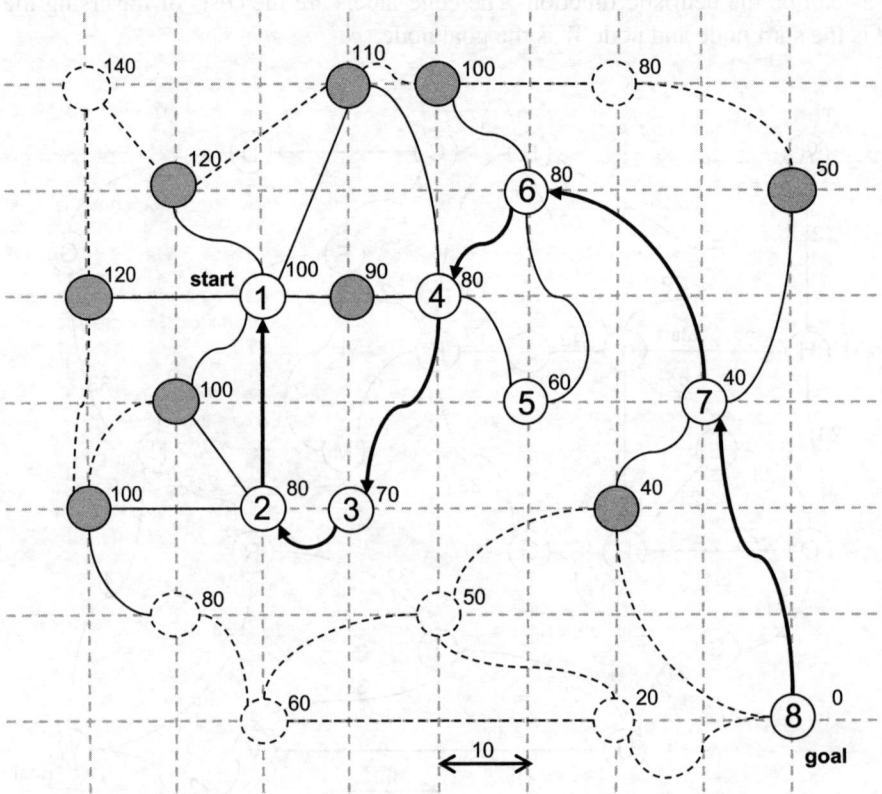

Figure 4.8 BESTFIRSTSEARCH is guided by heuristic function values shown next to each node. The numbers in the nodes show the order in which they are inspected. The path found is shown by the backward arrows. The shaded nodes are the nodes in OPEN when BESTFIRSTSEARCH terminates. The dashed edges are not generated by the algorithm.

The algorithm inspects eight nodes and finds the path <I P Q K F N W> with cost 195, as summed up from the edge costs in Figure 4.7. Observe that node M is visited after K, but is not on the path, because F was first generated as a child of K. Also observe that the path is not optimal.

Let us analyse the algorithm from the perspective of the four parameters we are looking at.

4.2.1 Quality of solution

The path found by BESTFIRSTSEARCH for the above problem is not optimal. The algorithm found a circuitous path of length 66 via Q to node K, because at each stage it picked nodes closest to the goal, and K was generated as a child of Q. The Manhattan distance of Q was 70, which is less than the 90 for J, which leads to the shorter path to K of length 24. Remember that in the version of BESTFIRSTSEARCH adopted from DFS, each node is generated and added to OPEN exactly once, and that determines, once and for all, who the parent of a node is.

When edge or move costs are equal, BESTFIRSTSEARCH can still find non-optimal paths. This happens because the heuristic function is not perfect. As described later, heuristic functions define terrains where the gradient embodies the direction the heuristic points to. As we will see in the blocks world domain later, these terrains define local optima which search gravitates towards. This is the case for problems like the 8-puzzle and also the Rubik's cube. BESTFIRSTSEARCH can still get around and find a solution, even though it may not be optimal. The local search algorithms we study next will in fact become incomplete.

4.2.2 Completeness

The BESTFIRSTSEARCH algorithm is complete for finite graphs. The argument is the same that we applied for BFS and DFS. In every cycle, the algorithm picks one node from OPEN and inspects it. There are a finite number of nodes in the connected component of the graph. And with filtering out of nodes already on OPEN or CLOSED, each node is added exactly once to OPEN. There are two ways the algorithm can terminate. One, when OPEN becomes empty. This means the goal node was not present in the graph. Or two, it finds the goal node.

When the graph is infinite, we cannot make the claim of completeness. While the heuristic function is meant to guide search towards the goal, if the function does not yield a good estimate, the search may wander off. If the heuristic function happens to be Machiavellian, it could even drive the search in the opposite direction.

4.2.3 Space complexity

With a good heuristic function, BESTFIRSTSEARCH has characteristics similar to DFS. So, we might hope that the space complexity might be linear. But it is possible that the algorithm may *change its mind* and sprout new branches in the search tree. This goes against the grain of linear space, and it has been empirically observed that the space required is often exponential. This is schematically depicted in Figure 4.9, which contrasts it with the OPEN list of BFS from the previous chapter as well.

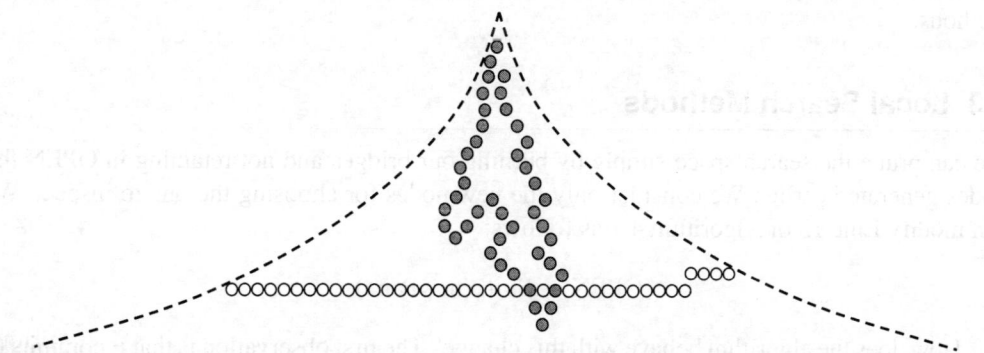

Figure 4.9 Given that the heuristic function is imperfect, the search frontiers for BESTFIRSTSEARCH shown in grey turn out to be exponential in practice too. Unshaded nodes show the frontier for BFS spanning the exponentially growing width of the search tree.

4.2.4 Time complexity

If the heuristic function were to be omniscient or perfect, it would drive the search directly towards the goal. Then the time complexity would be linear, as would be the space complexity. However, in practice, this is not seen to be the case and, more often than not, the time complexity is exponential.

In the early research on search methods reported in the pioneering books by Nils Nilsson (1971, 1980) and Patrick Henry Winston (1977), heuristic functions were compared by running extensive experiments on toy problems. The 8-puzzle was a favourite. Two parameters were devised to measure the effectiveness of heuristic functions. Let L be the length of path found and N the total number of nodes inspected by search. Then,

$$\text{Penetrance} = L/N.$$

If the heuristic function were to be perfect, then *penetrance* would tend to one.

The *effective branching factor*, B, is the number of successors generated by a 'typical' node for a given search problem. This is estimated by imagining a tree with branching factor B and depth L, with total N nodes. As we know from the last chapter, these three parameters satisfy the following constraint:

$$N = (B^{L+1} - 1)/(B - 1)$$

As one can see, the smaller N is, the lower B will be.

The bottom line is that *in practice* both time and space complexity of BESTFIRSTSEARCH are exponential in nature. This is also a consequence of the search being *global* in nature, retaining all nodes not yet inspected in OPEN. As we have discussed earlier, this is the basis of search being complete for finite spaces. Even if the heuristic function is poor, the algorithm will inspect all nodes before it terminates. In the real world, one is sometimes willing to trade completeness with complexity. This, for example, is the case with most TSP solvers. As we shall see, this is often also the case with SAT solvers.

In the following sections we look at search methods requiring low space. These are approaches that only look in the neighbourhood of the *current* node and are called *local search* methods.

4.3 Local Search Methods

We can prune the search space simply by burning our bridges and not retaining in OPEN the nodes generated earlier. We consider only the new nodes for choosing the one to inspect. We can modify Line 13 of Algorithm 4.1 as follows:

$$\text{OPEN} \leftarrow \text{Sort}_h (\text{newPairs})$$

How does the algorithm behave with this change? The first observation is that it commits to moving to one of the neighbours of the current node, the best one. Second, because we have not made any other change, and the filtering of nodes continues, the search will never turn back. It will terminate if it finds the goal node, or if no new neighbour can be generated.

4.3.1 An optimization problem

When new neighbours exist, the algorithm moves to the best one. Let us add one more criterion to the algorithm for making a move: That it moves to the next node *only* if it is better than the current node. That is, if it has a better heuristic value. We will revisit this criterion later on in the chapter. A consequence of this new criterion is that the algorithm will never look back, because the forward moves are made only to better nodes. Nodes in the past are always worse. And since the goal node is the one with the best heuristic value, zero when it is estimated distance, search cannot find a better neighbour at the goal node. This itself can be the termination criterion, without having to invoke the GOALTEST function. The following physical activity embodies the revised algorithm.

Imagine you are blindfolded and required to climb to the top of a hill from where you are. What strategy would you adopt? You could gingerly test the neighbourhood by stretching out one foot in different directions, and then take a step in the direction which appears to be highest. In other words, you would move in the direction of the *steepest gradient*. And you would stop when no neighbour is higher. This process of climbing the hill blindfolded is illustrated in Figure 4.10.

We have taken up the climbing problem for a reason which will become evident below. We could have equally chosen a problem of descending into a valley. The two problems are mirror

Figure 4.10 Algorithm HILLCLIMBING or steepest gradient ascent moves locally in the direction where improvement in the heuristic value is the highest. Figure from Khemani (2013).

images about the horizontal axis. In one, the highest point is the goal, and in the other, it is the lowest. Both are optimization problems. The first maximizes the height, the second minimizes it. Both choose the steepest gradient, and both terminate when the gradient becomes zero. The corresponding variation of our BESTFIRSTSEARCH algorithm is called HILLCLIMBING.

4.3.2 Hill climbing

This variation is a much simpler algorithm compared to BESTFIRSTSEARCH. We do not need CLOSED, except to reconstruct the path. We do not even need OPEN since we are not storing alternate candidates generated along the way. We do not need the GOALTEST function because the termination criterion has changed.

We have mentioned earlier that the heuristic function defines a landscape where the heuristic value is an added dimension in the search space. In our figures, we will assume a 1-dimensional search space, as in the Figure 4.10. The heuristic value will make up the second dimension. In general, of course, the terrain could be in multidimensional space.

The algorithm HILLCLIMBING is given below. It is a greedy algorithm, making only local choices. It begins with the start node being the best node, called *bestNode*. In every cycle, it looks at all the neighbours of the current node. If the best neighbour is better, it moves to it. Else it stops. And returns *bestNode*.

Algorithm 4.2. Algorithm HILLCLIMBING looks at the immediate neighbourhood and moves to the best neighbour nextNode, but only if it is better. Otherwise, it stops. At all times it keeps bestNode, the current best.

HILLCLIMBING(S)
1   bestNode ← S
2   nextNode ← **head sort$_h$** MOVEGEN(bestNode)   ▷ best to worst order
3   **while h** (nextNode) **is better than h**(bestNode)
4   bestNode ← nextNode
5   nextNode ← **head sort$_h$** MOVEGEN(bestNode)
6   **return** bestNode

The astute reader would have noticed that Algorithm 4.2 does unnecessary work. This is because it has been arrived at as an adaption of BESTFIRSTSEARCH. There is no need to sort the set of neighbours. Replacing Line 2 with the following would be more elegant:

$$2 \ \text{nextNode} \leftarrow \textbf{best} \ \text{MOVEGEN(bestNode)}$$

Here **best** is a function, *Max* or *Min*, that chooses the best neighbour. This can be done in one pass over the neighbours.

Algorithm 4.2 does not discuss the path reconstruction part. This is left as an exercise for the reader. Like the *SIMPLESEARCH* algorithm we started with, this is more suited to solving a configuration problem, where one has only to return the goal node. As mentioned before, the only termination criterion is when there is no better node in the neighbourhood. One implicitly assumes that that is the goal node. If wishes were horses...! When our climber stops, the situation is more likely to be the one in Figure 4.11.

We analyse the HILLCLIMBING algorithm on the four parameters we are looking at.

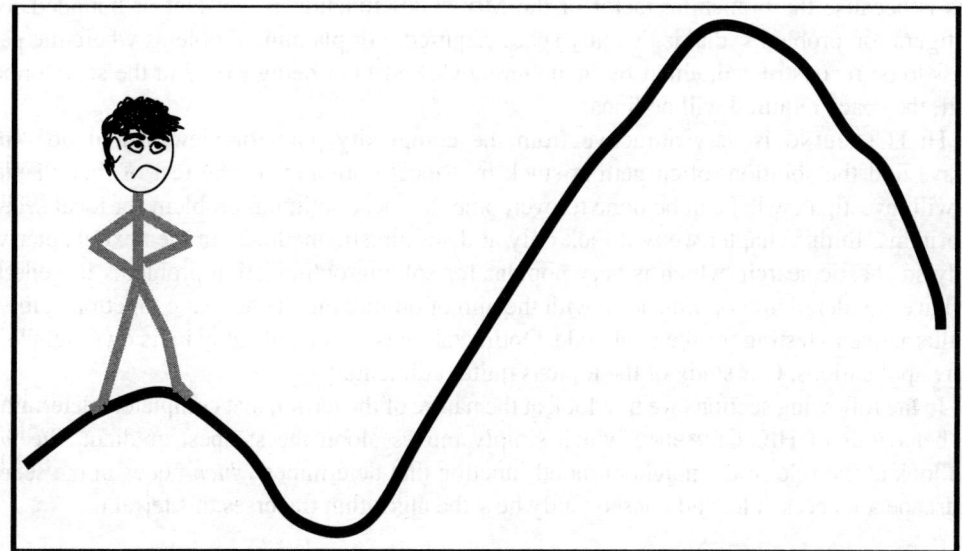

Figure 4.11 HILLCLIMBING may terminate at a local optimum. Figure from Khemani (2013).

4.3.3 Completeness

The algorithm terminates on all finite domains. It will also terminate on all infinite domains if the heuristic values are bounded. At some point, no neighbour will be better than the current node, and the algorithm will terminate.

However, the algorithm may *not* find a solution node, or a path to the solution node, because it may halt at a local optimum. It is not complete.

4.3.4 Quality of solution

For configuration problems, the global optimum corresponds to the solution. Problems like SAT or N-queens may have multiple solutions, and hence multiple global optima. There is no quality metric associated with such problems. Problems like TSP may have many local optima. These will correspond to non-optimal tours.

For planning problems too, a solution may not be found. This could depend upon the heuristic function. We will illustrate this with a blocks world example later. If we treat TSP as a planning problem, then the greedy construction methods described in Section 4.1.5 are instances of HILLCLIMBING. We have seen that they may not find optimal tours.

4.3.5 Time complexity

The algorithm is guided by the heuristic function to the nearest optimum. And then it stops. The time required is linear with depth, since it never looks back or sideways in its search.

4.3.6 Space complexity

The space required by HILLCLIMBING is its biggest selling point. Since it does not maintain an OPEN, and only needs enough space to generate the neighbours, the space needed is constant. This is because the branching factor of the MOVEGEN function is constant or bounded. For configuration problems, that is the only space required. For planning problems where the path needs to be reconstructed, either by maintaining CLOSED or being stored in the search node itself, the space required will be linear.

HILLCLIMBING is very attractive from the complexity point of view. But it does not always find the solution, often getting stuck in a local optimum. In the rest of this chapter, we will investigate what can be done to overcome the local optimum problem for local search algorithms. In this chapter we will look only at deterministic methods. In the next chapter we study stochastic search, which is very popular for solving optimization problems in general. We have wandered into optimization with the aim of optimizing the heuristic function value as an alternative to testing for the goal node. Optimization is an area of study in its own right with many applications. Our study of the topic is quite rudimentary.

In the following sections we first look at the nature of the terrain that completely determines the behaviour of HILLCLIMBING which simply moves along the steepest gradient. We will also look at the role of the neighbourhood function that determines *which* nodes in the search landscape are accessible, and consequently how the algorithm traverses this terrain.

4.4 Heuristic Search Terrains

Algorithm HILLCLIMBING evokes visions of an agent traversing a terrain along the steepest gradient. How is this terrain defined? As we have mentioned earlier, the heuristic function adds the 'height' dimension on the search space. In Figure 4.11 the search space is one dimensional. The agent can either move left or move right. The vertical axis is the height, representing the heuristic value. In higher dimensional spaces, the situation is similar, just harder to draw.

The point of interest for us is that the terrain is defined by the heuristic function. If the terrain were to be monotonic, with only one peak, like Mount Fuji in Japan, then HILLCLIMBING would always succeed. But if we were to have multiple peaks, like the Jagged Mountain in Colorado, then it would most likely fail, unless one started quite near the global maximum. Do look up pictures of the two mountains on the Internet.

We now describe a planning domain, the blocks world domain, widely used to illustrate planning. We also take up a simple problem in this domain and show how the choice of a heuristic function can make the difference between success and failure for HILLCLIMBING.

4.4.1 Hill climbing in the blocks world domain

One of the earliest domains conjured up by the planning community is the blocks world domain. In this domain, a collection of identical sized named blocks is on a table sufficiently large to accommodate as many blocks as needed. A block can be stacked on another block, and only one block can sit on another block. The stack has no limit on its height.

We will look at how the planning community formally defines planning domains in Chapter 11. There picking up and putting a block down are separate, named, moves. In this chapter we will adopt a simpler approach. A move is one where a one-armed robot can pick up one clear block, which has nothing on top of it, and place it on another clear block or on the table. The task is to rearrange the blocks in some desired way.

Figure 4.12 depicts a typical planning problem. This problem was used by Elaine Rich in her very popular book on AI that appeared in the 1980s (Rich, 1983). The example illustrates the efficacy of a good heuristic function. The figure depicts the start state on the left and the goal state on the right. The goal state is completely specified here. The planning community generally uses a partial goal description, which would correspond to a set of many states that satisfy the goal description.

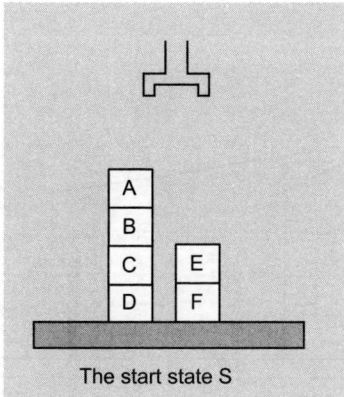
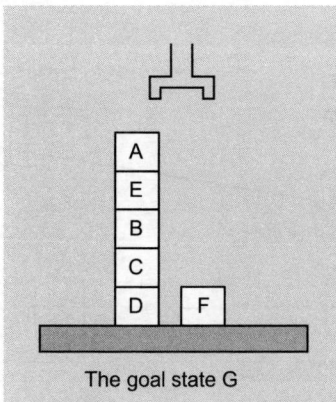

The start state S   The goal state G

Figure 4.12 The blocks world domain. A one-armed domain has the task of rearranging the blocks in the start state S, to achieve the goal state G. A move constitutes of picking up a block which has nothing on it and placing it on another block with nothing on it or on the table. Only one block can be placed on another block. The table is large enough to accommodate any number of the blocks.

4.4.2 Heuristic functions

Let us consider two heuristic functions defined for each state for the above planning problem.

The first heuristic function, $h_1(n)$, checks if each block is on the block it should be on in the goal state. If it is, then it increments the heuristic value by 1, else it decrements the heuristic value by 1. Adding up the values in alphabetical order for the blocks in Figure 4-12, we get:

$$h_1(S) = (-1) + 1 + 1 + 1 + (-1) + 1 = 2$$
$$h_1(G) = 1 + 1 + 1 + 1 + 1 + 1 = 6$$

The heuristic value of the goal is 6, signifying that all six blocks are where they should be. Observe that this makes our problem a maximizing problem, with higher values being better. Figure 4.13 depicts the progress of search on the state space. Remember that each move is reversible. It starts by generating the neighbours of the start state and computes their heuristic values.

$$h_1(P) = (-1) + 1 + 1 + 1 + (-1) + 1 = 2$$
$$h_1(Q) = 1 + 1 + 1 + 1 + (-1) + 1 \quad = 4$$
$$h_1(R) = (-1) + 1 + 1 + 1 + (-1) + 1 = 2$$
$$h_1(T) = (-1) + 1 + 1 + 1 + (-1) + 1 = 2$$

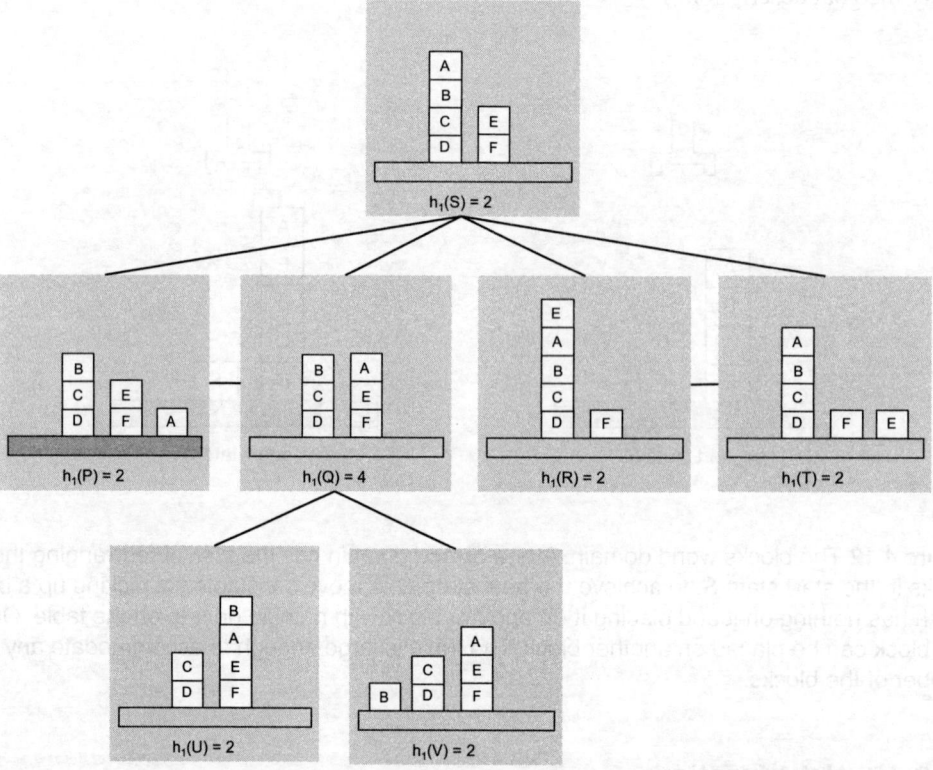

Figure 4.13 HILLCLIMBING begins by generating the neighbours of state S. All moves are reversible. Q is best neighbour with a value 4, as per $h_1(N)$, and it moves to it. But Q is also a local maximum as all four of its neighbours have a heuristic value 2. The algorithm terminates at Q without reaching the goal state.

The best neighbour is Q, in which the robot arm has moved block A on top of block E. HILLCLIMBING moves to node Q, which in turn has two new neighbours in which block B is moved, on top of A in node U, and on the table in node V. The other two neighbours of Q are S and P, when block A is moved, and they already exist in the state space.

$$h_1(U) = 1 + (-1) + 1 + 1 + (-1) + 1 = 2$$
$$h_1(V) = 1 + (-1) + 1 + 1 + (-1) + 1 = 2$$

All neighbours of Q are worse that Q, and it is a local maximum. HILLCLIMBING terminates here, without solving the planning problem.

The above heuristic function only checks whether a given block is perched on something it should be on. The second heuristic function, $h_2(n)$, is more perceptive. It checks whether the *entire tower* below it is as it should be. For every block that is sitting on a *correct tower* below, it adds the number of objects below it, including the table. That is, it adds the height of the block from the table. And for every block on a wrong tower, it subtracts the same. The following are the heuristic values of the start state and the goal state:

$$h_2(S) = (-4) + 3 + 2 + 1 + (-2) + 1 = 1$$
$$h_2(G) = 5 + 3 + 2 + 1 + 4 + 1 \qquad = 16$$

With the second heuristic function, HILLCLIMBING again begins by generating the four neighbours as shown in Figure 4.14. The h-values are shown below. It can be observed that $h_2(n)$ is more discriminating than $h_1(n)$ was, giving each of the four states different values. Also, it evaluates node P to be the best node, instead of node Q.

$$h_2(P) = (-1) + 3 + 2 + 1 + (-2) + 1 = 4$$
$$h_2(Q) = (-3) + 3 + 2 + 1 + (-2) + 1 = 2$$
$$h_2(R) = (-4) + 3 + 2 + 1 + (-5) + 1 = -2$$
$$h_2(T) = (-4) + 3 + 2 + 1 + (-1) + 1 = 2$$

The best neighbour is P, in which the robot picks up A and puts it on the table. Node P in turn has six new neighbours, three for moving block B and three for moving E. Figure 4.14 depicts only three of these, W, X, and Y, where block E is moved onto A, B, and the table respectively, ignoring the poorer moves for block B.

$$h_2(W) = (-1) + 3 + 2 + 1 + (-2) + 1 = 4$$
$$h_2(X) = (-1) + 3 + 2 + 1 + 4 + 1 = 10$$
$$h_2(Y) = (-1) + 3 + 2 + 1 + (-1) + 1 = 5$$

The reader is encouraged to verify that the three new neighbours in which block B is moved are all worse than P.

The best neighbour is node X where E has been moved onto block B. As one can see from the problem description in Figure 4.12, node X is quite close to the goal node. This is also reflected in its heuristic value $h_2(X) = 10$. Node X has five new successors. Three for moving block F, not shown in the figure, and two for block A. Of these, one is the goal node G with the best heuristic value. HILLCLIMBING moves to it, and terminates with a plan to reach the goal – *move A to table, move E onto B, move A onto E.*

The simple planning problem discussed above in some detail illustrates the effect of the choice of the heuristic function on local search. It also highlights the fact that the heuristic function defines a terrain on which search progresses.

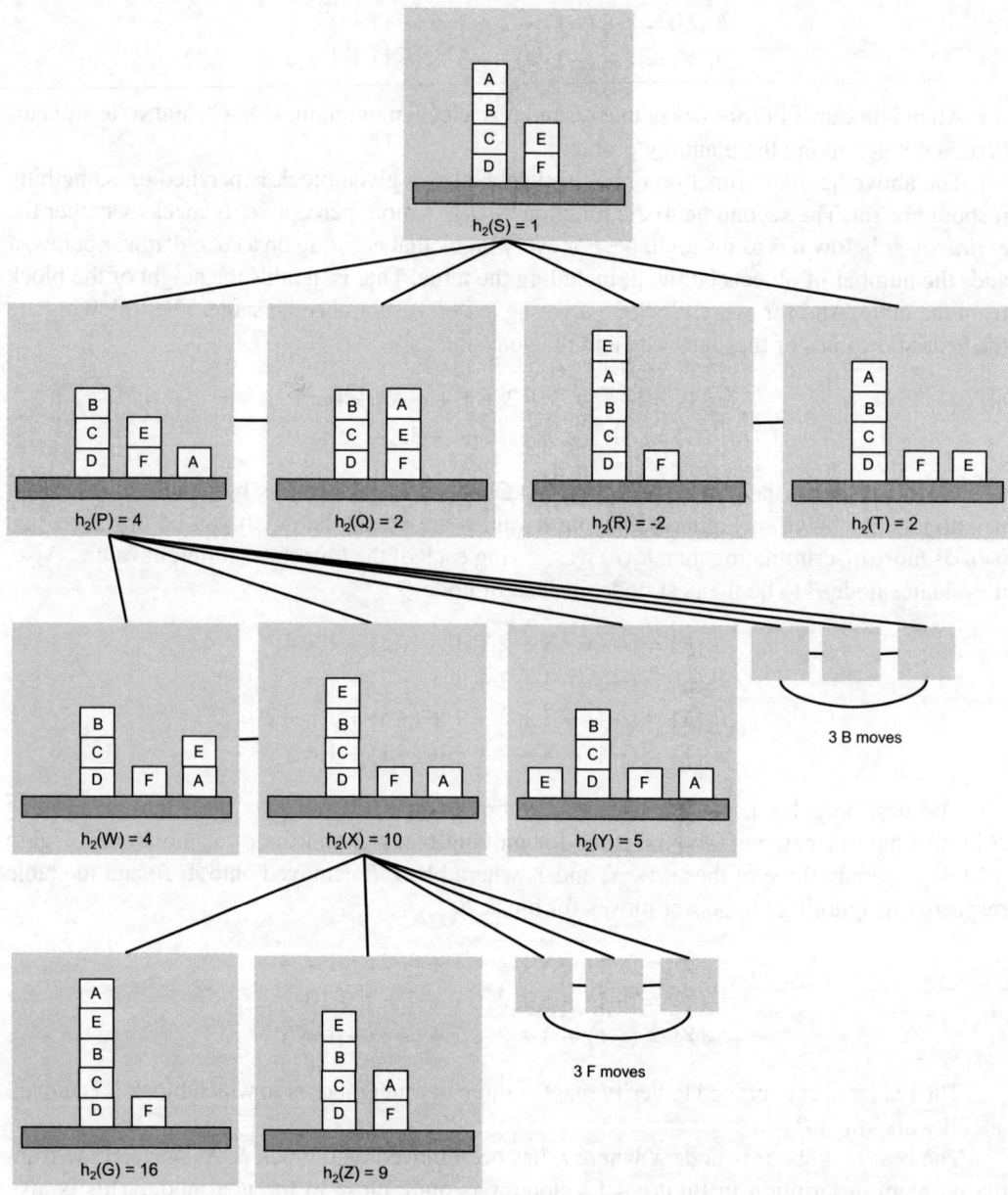

Figure 4.14 With the heuristic function h_2 again HILLCLIMBING starts with the same four neighbours of S. This time P is best with a value 4, as per $h_i(N)$, and it moves to it. P has six new neighbours, three moves each for blocks E and B. We have only drawn moves of E. Next, node X is best with a value 10 which leads to the goal G in the next cycle, and the algorithm terminates with a working plan.

4.4.3 The SAT landscape

The HILLCLIMBING algorithm follows the steepest gradient, and if there is a better node in the neighbourhood, it moves to it. We look at a tiny SAT problem to study the terrain defined by the heuristic function that counts the number of clauses a candidate satisfies. Consider the SAT problem on four variables $\{a, b, c, d\}$:

$$F = (a \vee b) \wedge (a \vee c) \wedge (c \vee d) \wedge (b \vee \neg d) \wedge (a \vee d) \wedge (b \vee c) \wedge (c \vee \neg d) \wedge (b \vee d) \wedge (\neg b \vee d)$$

The Boolean formula F has four variables and nine clauses. The number of candidates is 2^4. Let $h(n)$ be the number of clauses satisfied by a candidate, and let the neighbourhood function be flip-1-bit. The goal has a value 9 when all nine clauses are satisfied. With the flip-1-bit neighbourhood function, each node has four neighbours. The landscape is drawn in Figure 4.15.

There are four types of edges incident on the nodes, based on their heuristic values relative to their neighbours. Thick arrows lead to a unique best neighbour. Steepest gradient ascent

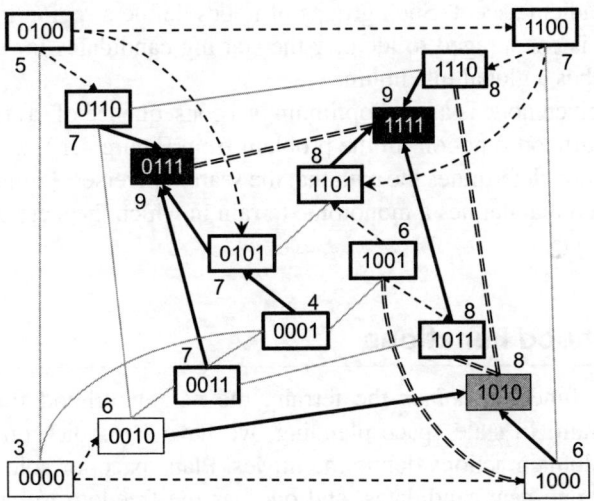

Figure 4.15 The heuristic terrain for a 4-variable SAT problem with nine clauses. The heuristic value is the number of clauses satisfied by a candidate's valuation. The darkest nodes are global maxima. There is one local maximum, 1010, shown in grey. Directed edges from any node indicate steepest gradient. Thick arrows are unambiguous steepest gradient edges and dashed arrows have other competing edges. Double lined edges are ridges. The remaining grey edges are never in contention.

has a clear path to follow. For example, the arrow from node 1000 to 1010. Dashed arrows represent the case when there is more than one best neighbour. HILLCLIMBING then has more than one option for making the move, and the tie break would depend upon the implementation. There are three such arrows emanating from 0100 with value 5 to nodes 1100, 0110, and 0101, each with value 7. The third kind of edge is shown with grey undirected edges, indicating that neither node is an option to move to from the other. Node 0010 is connected to two such nodes, 0110 and 0011 with value 7. All three nodes have a better option available elsewhere. Finally, a double lined undirected dashed edge represents a ridge, connecting to a node with the *same* value. HILLCLIMBING would not traverse such an edge. Node 1000 with value 6 is connected by such an edge to 1001. However, it has a better neighbour, 1010 with value 8, that it can move to. Node 1010 itself is connected by two ridges to nodes 1110 and 1011. Since these are the best neighbours, node 1010 becomes a maximum. In fact, it is a local maximum.

Observe that the nature of an edge depends only on what other nodes the given node is connected to. And this would be different for different neighbourhood functions.

There are two global maxima, 0111 and 1111, shown in dark rectangles with white text. There is one local maximum, 1010. These are nodes on which steepest gradient ascent halts. Observe that a local maximum like 1010 may be connected to nodes via ridges which are not conducive to gradient ascent. Such groups of nodes define a region in the terrain called a plateau. The reader is encouraged to identify the starting candidates for the SAT from which HILLCLIMBING reaches a global maximum.

The fact that a given node is a local optimum is a consequence of (a) the heuristic function and (b) the neighbourhood function. In the problem from Figure 4.12, we saw that the choice of the heuristic function determines the path that the search traverses. Figure 4.14 is an example of a heuristic function that defines a monotonic terrain in which the steepest gradient succeeds in finding the goal node.

4.5 Neighbourhood Functions

While the heuristic function *defines* the terrain, the neighbourhood function dictates *how* we traverse the terrain. In state space planning, we have no choice on the neighbourhood function, since the domain actions define the moves. Plan space, or solution space, relies on perturbation to generate new candidates, and one has the freedom to choose from different neighbourhood functions. We illustrate this choice with SAT and TSP. We also describe an algorithm VARIABLENEIGHBOURHOODDESCENT that capitalizes on this freedom.

4.5.1 Neighbourhood functions for the TSP

The simplest representation of a TSP candidate is known as the *path representation*. In the path representation, a tour is a permutation of the cities, indicating the order in which they are visited. The tour starts at the first city, proceeds to the last, and then culminates on the first city. Various perturbation operators have been explored to give us different neighbourhood functions. A few are given below.

In 2-city-exchange, some two cities are swapped in the tour, as shown in Figure 4.16 on the NEARESTNEIGHBOUR tour from Figure 4.4 as input. The shaded cities are the ones being swapped in the path representation. The tour on the left is the given tour and the one on the right is the result of the perturbation. The move is reversible.

If there are N cities, then the 2-city-exchange yields $^NC_2$ neighbours. One could devise similar, but denser, functions swapping more than two cities.

It is more common, however, to employ *edge exchange* neighbourhood functions, possibly because edges are the ones that contribute to the tour cost. In such operators, one removes a certain number of edges from a tour and inserts new edges to obtain a new tour. Figure 4.17 shows a 2-edge-exchange operator.

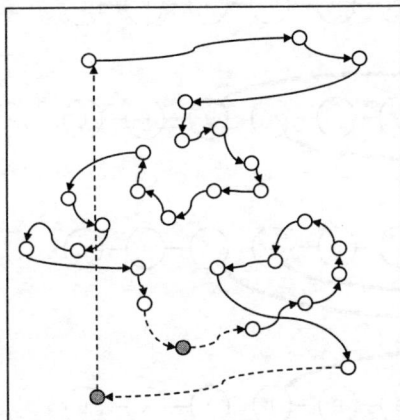 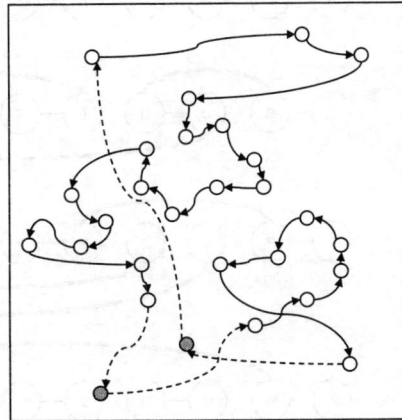

Figure 4.16 The perturbation operator 2-city-exchange swaps the position of two cities in the path representation of a tour. This results in four edges being deleted and four new edges being introduced.

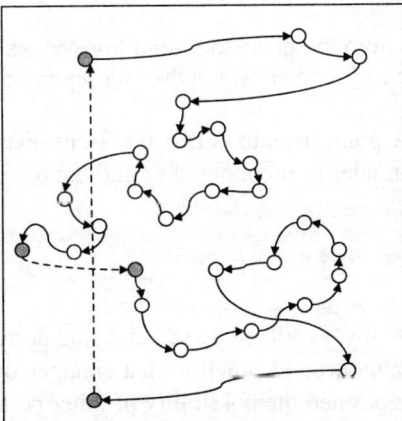 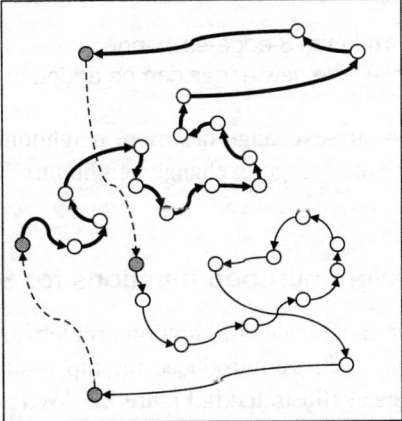

Figure 4.17 The perturbation operator 2-edge-exchange removes two dashed edges from tour on the left and adds two new dashed edges on the right. Observe that the direction of the arrows on the thick edges on the right has been reversed. In the path representation, 2-edge-exchange can be implemented by reversing a subtour of the original tour. In this illustration it is the subtour with these thick edges.

There is only one possible way of inserting two new edges and having a valid tour. Given that in an N city TSP there are N edges, the two edges to be removed can be chosen in $^NC_2$ ways, yielding as many neighbours. For the path representation, the 2-edge-exchange can be implemented by inverting a subsequence of the cities.

Likewise, in a 3-edge-exchange, three edges are removed from the tour, and three new ones are inserted to construct a complete tour. As shown in Figure 4.18, this can be done in four ways. The three edges can themselves be selected in $^NC_3$ ways.

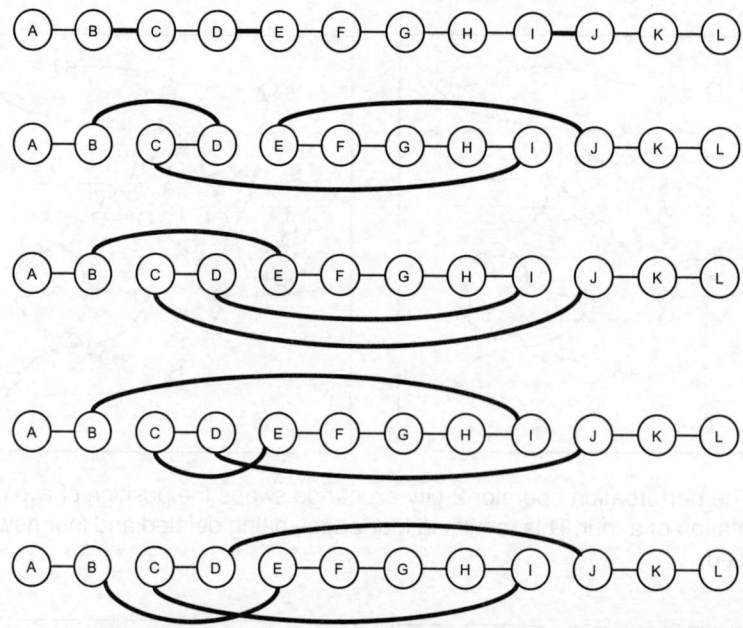

Figure 4.18 In the 3-edge-exchange, three edges from the given tour are removed, as shown on the top. Three new edges can be added in four ways, as shown in the four figures below.

The 4-city-exchange has more neighbours. A point to note is that the 2-city-exchange is just one of the 4-edge-exchange neighbours. The reader is encouraged to find the others.

4.5.2 Neighbourhood functions for SAT

SAT is the canonical configuration problem almost always solved by search using perturbation. In Section 2.5.2, we introduced the flip-1-bit neighbourhood function that changes one bit of a candidate as illustrated in Figure 2.8. We had mentioned the possibility of other perturbation methods.

Consider a SAT problem with five variables, $\{a, b, c, d, e\}$. The candidate solutions can be represented as 5-bit strings, one bit for the truth value of each variable. For example, 11111 represents the candidate $\{a = 1, b = 1, c = 1, d = 1, e = 1\}$. In *solution space search*, we define moves as perturbation in a given candidate. For the SAT problem, the perturbation could mean changing some k bits. For the above example, choosing $k = 1$ will yield five new candidates, as illustrated in Figure 2.8. They are 01111, 10111, 11011, 11101, and 11110. These five are the neighbours of 11111 in the search space. Let us call this neighbourhood function N_1. When $k = 2$, then the neighbourhood function is N_2. Each candidate would have ten new neighbours, since we can choose two bits in $^5C_2 = 10$ ways. For 11111, they are 00111, 01011, 01101, 01110, 10011, 11001, 11100, 11110, 11010, and 11100.

Figure 4.15 shows the heuristic terrain for a 4-variable SAT problem with the neighbourhood function N_1. The reader is encouraged to redraw the graph using the neighbourhood function N_2 which flips some two bits.

If there are N variables, then we could have a set of functions $\{N_1, N_2, ..., N_N\}$ which change a fixed number of bits. The kth neighbourhood function would have ^NC_k neighbours. We could also have neighbourhood functions like N_{12} that changes one *or* two bits. N_{12} would have fifteen neighbours. And then N_{123} and so on till $N_{1...N}$, the last one allowing one to change any number of bits. There are 2^N neighbours to choose from. Observe that the last one represents a fully connected graph.

Gradient based methods have traditionally been discussed in the context of optimization, which in general is a hard problem. Our interest in optimization arose from the objective of finding a node with the best heuristic value.

Maximization and minimization are two sides of a coin. A maximization problem can be converted to a minimization problem by prefixing a negation sign to the objective function (which in our case is the heuristic function). Thinking of it as a minimization problem does help us get some insight into the algorithm. Imagine rolling a ball down a slope with the intention of sending it to the lowest point of a valley. Keep aside physics for a moment and imagine that there is no momentum. Then the ball would roll down the steepest gradient, but only till the point where the gradient becomes zero. The moment it becomes zero, it would come to a halt.

4.5.3 Variable neighbourhood descent

As described earlier, we may have at our disposal a set of neighbourhood functions with a varying number of neighbours. A *sparse* neighbourhood function has fewer choices at each point, while a *dense* function would have more choices. But the denser neighbourhoods are also more expensive to inspect.

At the extreme, function $N_{1...N}$ can choose to flip *any* number of bits for the SAT problem. This would mean that *all* nodes in the search space are neighbours of a given node, and search would then reduce to inspecting *all* the candidates. When *all* the candidates are neighbours, the best among them is the only optimum. However, this reduces search to brute force, inspecting the entire space, which for SAT has a number of states which is exponential in the number of variables. Conversely, the more sparse the neighbourhood function, the more likelihood of there being local optima in the search space. The local optima arise because some nodes do not have a better neighbour. Better nodes exist in the search space, but the local optimum is not connected to any of them.

Neighbourhood functions that are sparse lead to quicker movement during search, because the algorithm must inspect fewer neighbours. But there is a greater probability of getting stuck on a local optimum. This probability becomes lower as neighbourhood functions become denser, but then search progress also slows down because the algorithm must inspect more neighbours before each move.

100 | Search Methods in Artificial Intelligence

Algorithm VARIABLENEIGHBOURHOODDESCENT tries to get the best of both worlds (Hansen and Mladenovic, 2002; Hoos and Stutzle, 2005). It starts searching with a sparse neighbourhood function. When it reaches an optimum, it switches to a denser function. The hope is that most of the movement would be done in the earlier rounds, and that the overall time performance will be better.

> **Algorithm 4.3.** Algorithm VARIABLENEIGHBOURHOODDESCENT begins with a sparse neighbourhood function and moves to a denser function on reaching an optimum. The algorithm assumes that the function MOVEGEN can be passed as a parameter. It assumes that there are n MOVEGEN functions sorted on the density of the neighbourhoods produced.
>
> VARIABLENEIGHBOURHOODDESCENT()
> 1 $node \leftarrow start$
> 2 **for** $i \leftarrow 1$ **to** n
> 3 $MoveGen \leftarrow MoveGen(i)$
> 4 $node \leftarrow $ HILLCLIMBING$(node, MoveGen)$
> 5 **return** $node$

Figure 4.19 revisits the SAT example with a neighbourhood function N_2. Each node has six neighbours, as illustrated for node 0000. We have not drawn all the forty-eight edges in the graph, but only those that participate in the hill climbing process.

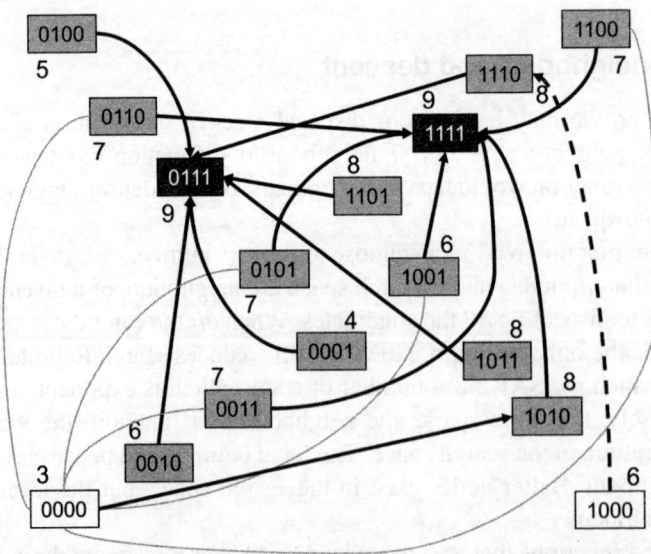

Figure 4.19 A subset of the edges on the four variable SAT problem when we use the neighbourhood function N_2 that flips two bits in a candidate. The nodes in grey are just one hop away from the solution. The remaining two nodes also have a gradient ascending to the solution.

As seen in the figure, the search terrain does not have any local maxima. The two nodes 0111 and 1111 are the solutions as before. Twelve of the remaining fourteen nodes, shown in grey, are just one step away from a solution. The remaining two nodes also have a steepest gradient path that leads to the solution.

4.5.4 Beam search

HILLCLIMBING puts all its eggs in one basket, the best neighbour. It often pays to have more than one string to your bow. Algorithm *BEAMSEARCH* does precisely that, keeping the best b options instead of just one, where b is known as the *beam width*. Given that memory is now abundantly available, this makes eminent sense. Figure 4.20 illustrates BEAMSEARCH with a beam width 2. Instead of keeping one best option, the algorithm maintains two best options at each level. With memory becoming abundant the algorithm has greater potential as beam widths can increase.

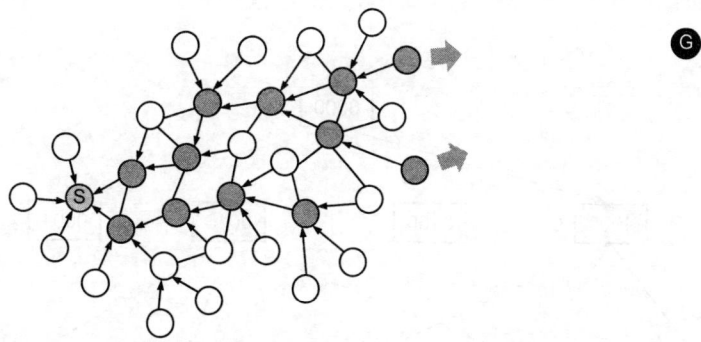

Figure 4.20 BEAMSEARCH picks the best b nodes from OPEN and expands them. The set of neighbours of these b nodes forms the new OPEN. At every level, it keeps b nodes from OPEN in contention. Hopefully, the path to the goal goes through one of b nodes. In this illustration $b = 2$.

BEAMSEARCH has been effectively used in speech recognition, where a set of phonemes need to be eventually combined into words, and then sentences. Eugene Charniak and Drew McDermott (1985) quote the following example where there may be ambiguity in speech understanding. If you think that someone from New York is telling you that 'everything in the city costs a nominal egg' they are more likely saying that 'everything in the city costs an arm and a leg'. Another touching example is that of a young child telling an acquaintance that she has 'sixty-five roses' when her diagnosis was 'cystic fibrosis'. Matt Payne (2021) says the following in his blog: 'First used for speech recognition in 1976, beam search is used often in models that have encoders and decoders with LSTM or Gated Recurrent Unit modules built in. To understand where this algorithm is used a little more let's take a look at how NLP models generate output, to see where Beam search comes into play.'

The algorithm that the speech community calls *VITERBI* search (Xie and Limin, 2004) maintains a short list of the most probable words at each time step, and only extends transitions from those words into the next time step. The algorithm is an implementation of BEAMSEARCH, keeping a few options in contention at each point of time as it processes the input sequence of phonemes.

Let us look at the search tree explored by BEAMSEARCH for the tiny SAT problem mentioned in Figure 4.15:

$$F = (a \vee b) \wedge (a \vee c) \wedge (c \vee d) \wedge (b \vee \neg d) \wedge (a \vee d) \wedge (b \vee c) \wedge (c \vee \neg d) \wedge (b \vee d) \wedge (\neg b \vee d)$$

The Boolean formula has four variables and nine clauses. Let $h(n)$ be the number of clauses satisfied by a candidate, and the neighbourhood function be flip-1-bit. The goal node would have a value 9 since there are nine clauses in the formula. Starting with the candidate 0000, the progress of BEAMSEARCH with beam width $b = 2$ is shown in Figure 4.21. The value alongside each node is the number of clauses satisfied. Observe that starting with a value 3 for 0000, both HILLCLIMBING and BEAMSEARCH can make a maximum of six moves, since the node at the next level needs to be better at each stage. Where there are more than two best nodes at any level, the algorithm selects the ones on the left. The nodes selected by BEAMSEARCH are shown in shaded rectangles in the figure. The solution found by the algorithm is 1111 after four moves.

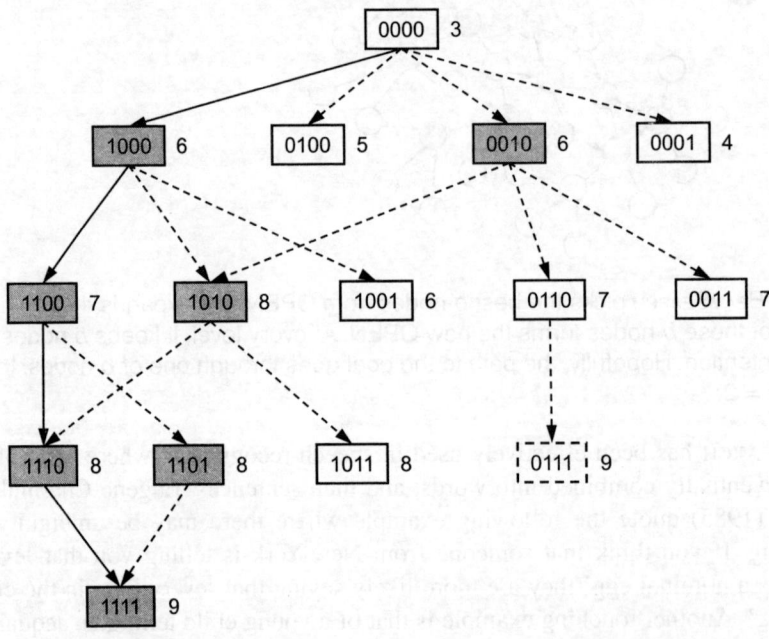

Figure 4.21 BEAMSEARCH with width $b = 2$ on the problem shown in Figure 4.15. If there are more than two best nodes in some level, then the leftmost two are selected. The algorithm finds the solution 1111 after four moves. Observe that HILLCLIMBING would not have found the solution because after 1000 it moves to 1010, which is a local maximum. Both do not reach the solution 0111.

For compactness, we have not drawn neighbours that exist at an earlier level, since they would have a lower heuristic value. For example, node 1000 would have 0000 too as a neighbour. Consequently, each candidate appears exactly once in the search tree. When there is a tie between nodes at some level, then we have broken it in favour of the moves on the left. This assumes that the neighbourhood functions flip bits from the left to right. Observe that HILLCLIMBING would have failed to find the solution under these conditions, because it would get stuck at node 1010, which is a local maximum. Since BEAMSEARCH keeps more than one option, our algorithm was able to find the solution.

The reader is encouraged to look at the search tree explored by BEAMSEARCH for the tiny SAT problem given below from Section 2.3.3. Choose 1111 as the starting node and beam width $b = 2$.

$$F = (b \vee \neg c) \wedge (c \vee \neg d) \wedge (\neg b) \wedge (\neg a \vee \neg e) \wedge (e \vee \neg c) \wedge (\neg c \vee \neg d)$$

This example shows us that even in small search spaces BEAMSEARCH can get trapped in local optima. However, the algorithm continues to be of considerable interest, not least because much larger beam widths are possible with memory becoming abundant.

4.6 Escaping Local Optima

The VARIABLENEIGHBOURHOODDESCENT algorithm described earlier incorporates a built-in mechanism for escaping from local optima. It switches to a denser neighbourhood function. But this luxury is only available for perturbation methods on configuration problems. For planning problems, one has only one MOVEGEN function, representing the moves one can do from any given state.

In the blocks world problem described earlier, we saw in Figure 4.14 that it is possible to devise heuristic functions that define monotonic surfaces amenable to steepest gradient ascent. However, this is not possible in many domains. For solving the Rubik's cube, for example, more of the same colour cubes on a face seems closer to the solution. But if you know how to solve it, you will recall the gyrations a function defined with such a criterion would have to undergo. The 8-puzzle displays similar behaviour. Figure 4.22 shows the progress of the Hamming distance and Manhattan distance heuristic functions (see Section 4.1.3) on the shortest solution path for a simple instance. Observe that the top row in the puzzle does not have to be disturbed in this short plan. Both functions still perceive a local minimum on the solution path.

In Chapter 10, we will look at an approach to planning, which reasons with subgoals. For example, someone solving the Rubik's cube by first principles could have subgoals like solving the top surface and then focusing on the adjacent surfaces, and so on. However, one soon realizes that working on later subgoals disrupts the ones achieved already. These are problems where subgoals are said to be non-serializable. They cannot be achieved independently in any serial order. The way we solve such problems is to accumulate knowledge about how to move from one subgoal to the next, despite temporarily disturbing goals achieved earlier, but knowing that they will be restored. Richard Korf called this approach a peak-to-peak heuristic and wrote programs to *learn* such *macro-operators* for his doctoral thesis (Korf, 1985).

104 | Search Methods in Artificial Intelligence

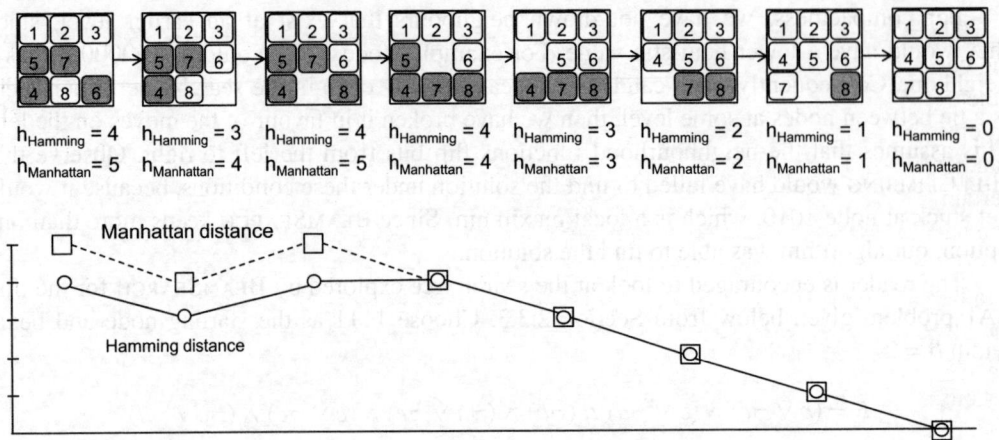

Figure 4.22 The values of two heuristic functions along the solution path for a simple 8-puzzle instance. The darker tiles are the ones out of place. The graph below plots the values of the two heuristic functions, the squares for the Manhattan distance and the circles for the Hamming distance. Observe that after the first move, the state is at a local minimum according to both heuristic functions.

We will, however, focus here on search, and turn our attention to mechanisms that enable local search to escape from local optima.

4.6.1 Exploration versus exploitation

A heuristic function guides search. It creates a landscape over which search follows the gradient. We say that the algorithm is *exploiting* the gradient. HILLCLIMBING and BEAMSEARCH do this faithfully and diligently, like horses on blinkers. However, this fixation with gradient leads to the possibility of being stranded on the first optimum that comes along. This can be lethal for local search.

Local search may need to stray a little from the path indicated by the steepest gradient. Or learn to go past a (local) optimum. For this, search needs another driver, *exploration,* the tendency to venture into newer areas. This is right up the alley of randomized algorithms, and we study them in the next chapter.

Here we look at one deterministic algorithm that goes happily sailing past optima.

4.6.2 Tabu search

The algorithm described here derives from the Tongan word *tapu* or Fijian *tabu*, which roughly translates to 'prohibited' or 'forbidden'. The algorithm allows search to proceed beyond a local optimum and prohibits it from going back to where it came from.

The main idea in *tabu search* is to augment the *exploitative* strategy of heuristic search with an *explorative* tendency that continues to look for new areas in the search space (Michalewicz and Fogel, 2004). That is, tabu search follows the diktat of the heuristic function only as long as better choices present themselves. But when there are no better choices, instead of terminating as HILLCLIMBING would, it gives in to its explorative tendency to continue searching.

This can be done simply by removing the criterion of the best neighbour being better than the current node. Once this restriction is removed, search, which is akin to HILLCLIMBING in other ways, merrily carries on endlessly. Remember that there is no way to determine whether an optimum is local or global. A different termination criterion has to be introduced again. This can be for a fixed number of cycles, or stopping when no significant improvement is made for a certain number of cycles. For problems where GOALTEST is available, like the SAT problem, one can have an additional exit condition if the goal is found, not least to avoid unnecessary work.

This simple modification would not work by itself though. The freedom from moving only to a better move allows the search to go beyond an optimum. But once past it, what is to stop the algorithm from coming right back in the next step? We prevent this by disallowing some moves at each stage, or making them prohibited, albeit temporarily. That is how the algorithm gets its name (Glover, 1986). The high level algorithm is described below.

Algorithm 4.4. TABUSEARCH moves to the best allowed neighbour till some termination criterion. Allowed here means not tabu or taboo.

TABUSEARCH(Start)
   N ← Start
   bestSeen ← N
   Until *some termination criterion*
        N ← best(allowed(MOVEGEN(N)))
        IF N *better than* bestSeen
            bestSeen ← N
   return bestSeen

Tabu search is different from hill climbing in two ways. One, the termination criterion is different. And two, it introduces a notion of *allowed* neighbours. By allowed, we mean moves that are not tabu, or taboo. The condition of being tabu is implemented in different ways for different problems. A simple way would be to maintain a small CLOSED as a circular list, disallowing the most recent states from being visited again. More often in literature, though, an embargo on some moves or perturbations itself is imposed in solution space search.

We illustrate the idea with our favourite flip-1-bit operator for a hypothetical 7-variable SAT problem. As search progresses, some bits cannot be changed for some time. The period for which a bit is tabu is set by a parameter called *tabu tenure* (tt). In our illustration, we use $tt = 2$. This means that once a bit is flipped, it is quarantined[2] for two cycles. One can keep track of which bits are allowed by keeping a memory array with a value for each bit initialized to zero.

$$M = [0\ 0\ 0\ 0\ 0\ 0\ 0]$$

The subsequent value for each bit could be the last time the bit was changed. The value could be compared with counter for the cycles to decide whether it is tabu or not. A simpler way is to maintain M as a timer, counting down to when that bit can be changed again.

---

[2] As I write this in the times of the Coronavirus, this seems to be the most appropriate word.

The moment it is changed, the value is initialized to *tt*, and decremented in every cycle. The process is illustrated in Figure 4.23.

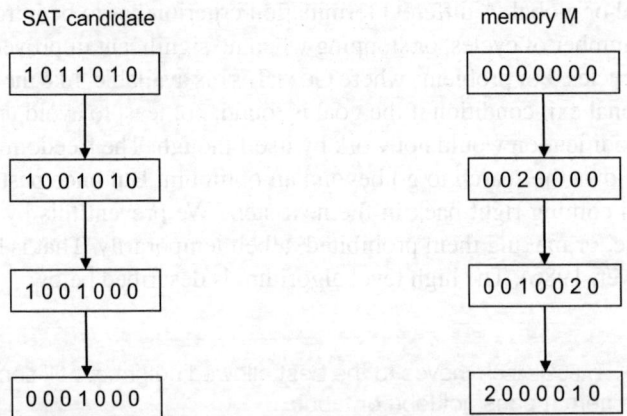

Figure 4.23 Illustrating TABUSEARCH with *tt* = 2. As bits are flipped on the left, the corresponding bit in *M* counts down to when they can be flipped again. The first move above changes the third bit as shown on the left. The shaded numbers on the right show the corresponding values in *M*. The third bit can only be flipped when it becomes 0 after two moves.

The algorithm works as follows. It flips all the *allowed* bits to generate neighbours and moves to the best neighbour, irrespective of whether it is better or not. The figure above depicts three cycles in the TABUSEARCH execution. In the first cycle, bit 3 is changed, and the corresponding value in *M* is set to 2, which is the tabu tenure. That value decrements to 1 in the next cycle, in which bit 6 is changed. When bit 1 is changed in the third cycle, bit 3 is available for flipping again.

How does tabu search perform on the SAT problem of Figure 4.15? Are there any states from which a goal node is not reached? Does it work when *tt* = 2? What about when *tt* = 1? This is left as an exercise for the reader.

Clearly, TABUSEARCH looks at a subset of the neighbours at each stage, ignoring the moves that are in quarantine. What if a tabu move leads to a very good candidate? Some implementations of tabu search include an *aspiration criterion*. This says that if one of the tabu neighbours has a value better than any seen in the entire run, then that tabu move should be allowed.

It may also happen that perturbations of some components may be happening much more often than others. One way of giving a boost to exploration is to drive the search to newer areas by devaluing the nodes generated by more frequent moves. This can be done by maintaining a frequency memory $F = [f_1, f_2, \ldots, f_N]$. Then, if the heuristic value of a node being generated by modifying (flipping for SAT) the *k*th component (a bit in the case of SAT) is $h(\text{node}_k)$, this can be attenuated as

$$h(\text{node}_k) \leftarrow h(\text{node}_k) - c \times f_k$$

where *c* is an appropriate constant.

A similar approach can be employed to solve TSP using tabu search. One way of marking tabu moves would be to maintain a 2-dimensional memory array showing which pairs of edges were removed in 2-edge-exchange or a 3-dimensional one for 3-edge exchange.

4.6.3 Iterated hill climbing

We take a first step towards randomized algorithms here, and will continue in the next chapter. Algorithms solving configuration problems are only interested in the goal state, and not *how* one arrives at it as it would be for a planning problem. In fact, they are not even interested in where the search *started* from. Only where it ends. This realization leads us to an effective algorithm for solving configuration problems.

The major drawback with gradient based local search is running into a local optimum. Whether it does so or not depends upon where it started from. There is no reason why HILLCLIMBING should start with the node 0000 in Figure 4.12 as BEAMSEARCH did in Figure 4.21. As observed earlier, HILLCLIMBING would have failed.

The *iterated hill climbing* (IHC) algorithm presented below randomly chooses different starting points and attempts HILLCLIMBING from each.

Algorithm 4.5. IHC does HILLCLIMBING *N* times starting each time from *N* different randomly selected starting points.

IHC(N)
1   bestNode ← **random candidate solution**
2   **repeat N times**
3   currentBest ← HILLCLIMBING(**new random candidate solution**)
4   **if h**(currentBest) **is better than h**(bestNode)
5   bestNode ← currentBest
6   **return** bestNode

The hope is that one of these instances of hill climbing will strike gold. The reader is encouraged to look at the SAT problem in Figure 4.15 and find out how many of the starting nodes in the solution space would lead to a solution.

Clearly, the performance of IHC depends upon the nature of the landscape defined by the heuristic function. The nodes from where hill climbing succeeds can be said to define the footprint of HILLCLIMBING. The boundary of the footprint is defined by the set of local minima surrounding the global maximum. If the algorithm were to start from any node in the footprint, it would have a smooth ascent to the summit. The larger the size of the footprint, the greater the probability of IHC succeeding. If the heuristic function were like the Jagged Mountain mentioned in Section 4.4, the footprint would be small, and many iterations would be required to have a reasonable chance of success. At the other extreme, if the landscape were to be like Mount Fuji, one iteration would be enough.

Even though both algorithms examine more than one path on the landscape, IHC is different from BEAMSEARCH. BEAMSEARCH has one starting point, and it spawns many paths from that, bounded by the beam width. IHC, on the other hand, begins from different randomly chosen starting points. It is the first algorithm we have seen that has an element of randomness. The next chapter is devoted entirely to randomized algorithms.

Summary

We arrived in this chapter in search of knowledge to guide search. Having developed the basic search machinery in the last chapter, we were looking for ways for search to be directed towards the goal. This was achieved by devising heuristic functions that somehow estimated the distance to goal, or conversely the similarity with goal. The more similar a state is to the goal state, the closer it should be. Heuristic functions estimate this closeness, and the BESTFIRSTSEARCH algorithm capitalizes on this knowledge to search in an informed manner.

The performance of heuristic search is only as good as the heuristic function. With a good function, both time and space complexity are low and may even be linear with very good ones. This is not the case in practice though, and the complexity is often exponential.

We then introduced the idea of local search, trading off completeness for lower complexity. Algorithm HILLCLIMBING views the problem as an optimization problem and chooses the steepest gradient path to reach the optimum. However, the problem of local optima then crops up, where the gradient is zero, and search gets trapped. That led our quest for algorithms to get around the local optima problem.

We introduced the notion of exploration in addition to exploitation. This chapter was devoted to deterministic search. We ended with IHC, which introduces an element of randomness. In the next chapter we build upon randomized algorithms and look at some popular stochastic methods employed by the optimization community.

Exercises

1. The BESTFIRSTSEARCH algorithm picks the node with the best heuristic value from OPEN. Compare the relative advantages of (a) appending the new nodes with the tail of OPEN and scanning for the best node, (b) appending the new nodes and then sorting OPEN, (c) sorting the new nodes and inserting them into a sorted OPEN, and (d) maintaining OPEN as a priority queue. How will the size of the search space influence your choice?
2. You are writing a program to solve the map colouring problem with regions A, B, C, D, E, F, G, and H using heuristic search. The input is defined as a graph in which the neighbours of a region are connected by edges. Each region has an allowed set of colours. What kind of heuristic functions (that look at neighbours only) can you think of (a) to choose which region to colour next and (b) what colour to pick from the set of available colours?

3. You are writing a program to solve an *N* variable SAT problem with clauses varying in size from 1 to *K*. Let us say you are trying a constructive method in which you pick a variable to try a value, *true* or *false*, one by one. Let us say you maintain a data structure in which you keep track of the sizes of clauses along how many variables in each clause have been assigned value in the partial assignment. What heuristic function would you choose the next variable to try a value for?

4. Let us say you are implementing best first search for path finding on a finite city map and are using the Euclidean distance as a heuristic function. Algorithm 4.1 removes the neighbours of a node returned by MOVEGEN that are already on OPEN or CLOSED before adding them to OPEN (Line 10). How would the performance of the algorithm be affected if this step was removed? Would it still be complete? What about complexity?

5. [Baskaran] The following figure shows a map with several locations on a grid where each tile is 1 unit by 1 unit in size, connected by 2-way edges (roads). The MOVEGEN function returns neighbours *in alphabetical order*. The start node is *S*, and the goal node is *G*. Module REMOVESEEN removes neighbours already present in OPEN/CLOSED lists. List the nodes in the order visited by BESTFIRSTSEARCH. Draw the search tree generated, clearly identifying the nodes on CLOSED and on OPEN, and the path found. How does HILLCLIMBING perform on this problem? What about BEAMSEARCH with *width*=2?

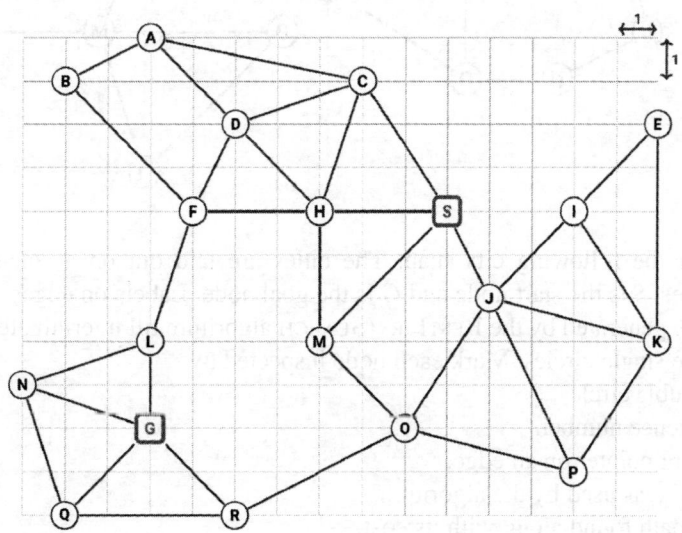

110 | Search Methods in Artificial Intelligence

6. [Baskaran] The following figure is another map, where the nodes are placed on a grid of 5 × 5 units. The MOVEGEN function returns neighbours *in alphabetical order*. The start node is S and the goal node is G. Module REMOVESEEN removes neighbours already present in OPEN/CLOSED lists. List the nodes in the order visited by BESTFIRSTSEARCH. Draw the search tree generated, clearly identifying the nodes on CLOSED and on OPEN, and the path found. How does HILLCLIMBING perform on this problem? What about BEAMSEARCH with *width=2*?

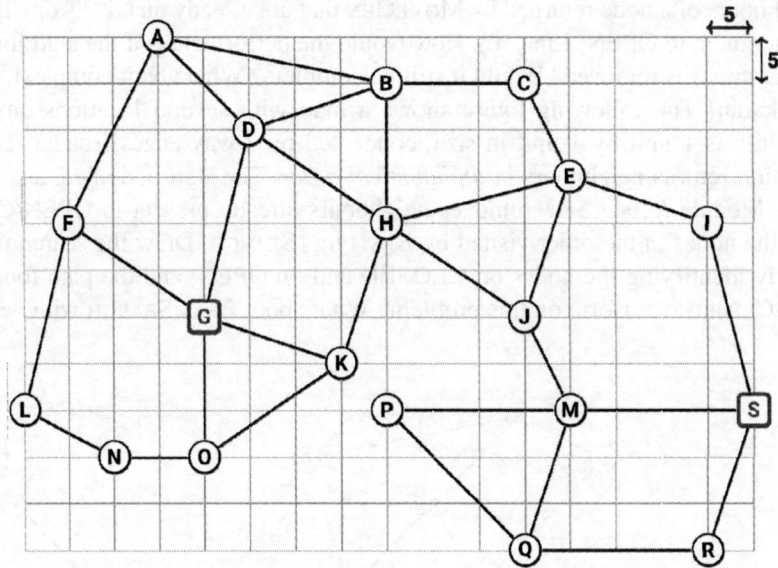

7. Consider the following city map. The cities are laid out on a square grid of side 10 kilometres. S is the start node and G is the goal node. Labels on edges are costs. Draw the *subgraph* generated by the BESTFIRSTSEARCH algorithm till it terminates. Let the nodes on OPEN be single circles. Mark each node inspected by
    * a double circle,
    * sequence number,
    * parent pointer on an edge,
    * its cost as used by the algorithm.
    List the path found along with its cost.

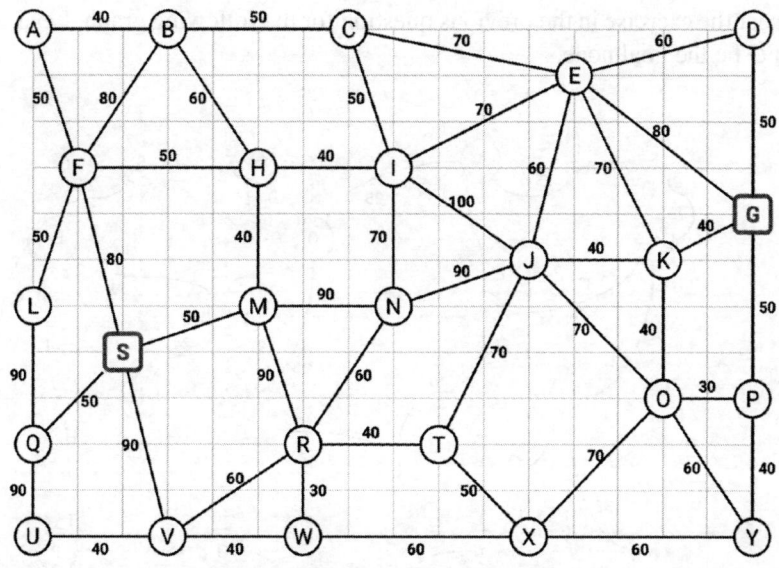

8. The following graphs represent a city map. The nodes are placed on a grid where each side is 10 units. Node *H* is the start node and node *T* is the goal node. Use Manhattan distance as the heuristic function where needed. The label on each edge represents the cost of the edge. Label the nodes with the heuristic values. List the nodes in the order inspected by (a) HILLCLIMBING and (b) BESTFIRSTSEARCH till it terminates. Does it find a path to the goal node? If yes, list the path found along with its cost.

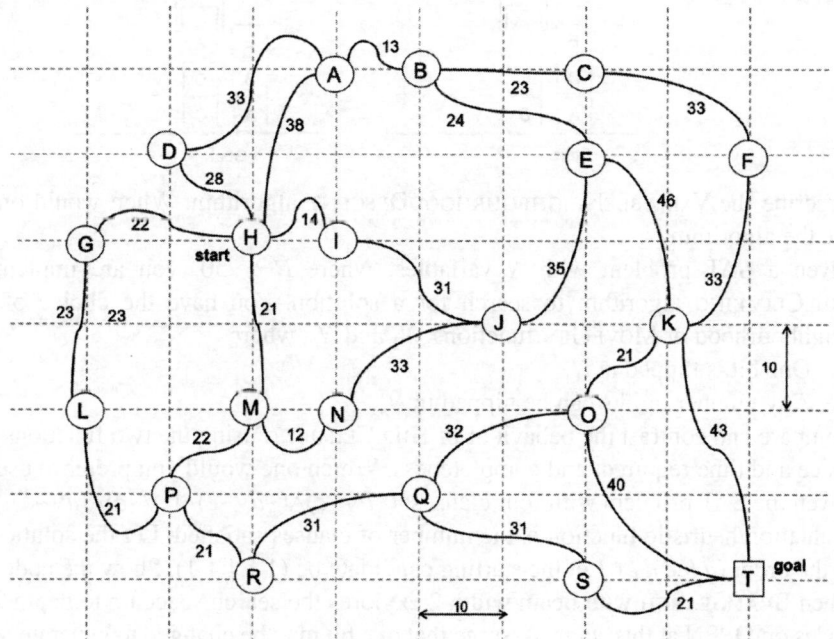

9. Repeat the exercise in the previous question for the following graph. Let S be the start node and G be the goal node.

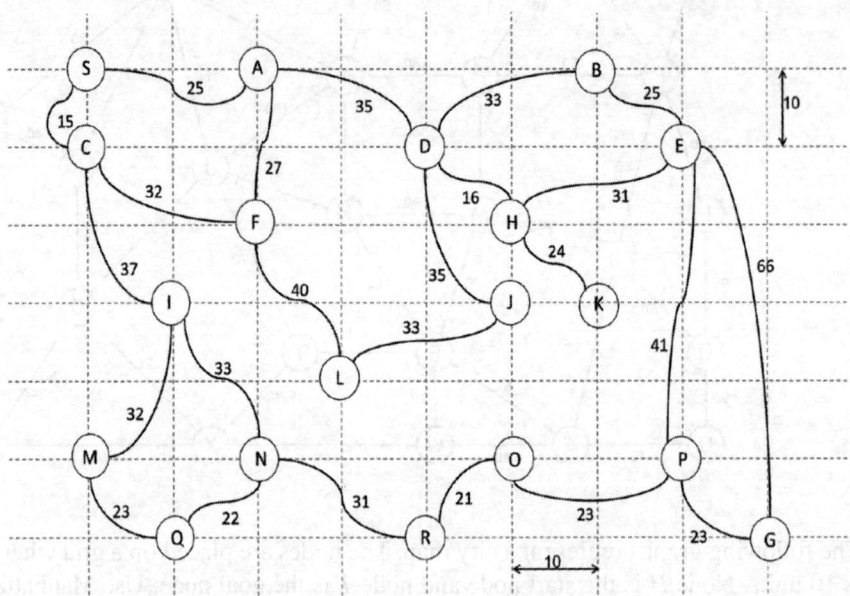

10. Try out the two heuristic functions in Section 4.4.2 for the blocks world domain on the problem defined in the following figure.

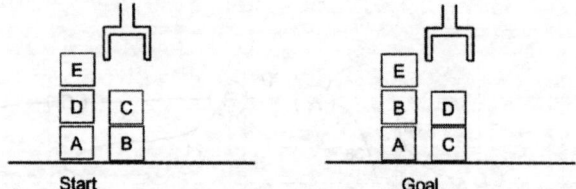

11. Describe the VARIABLENEIGHBOURHOODDESCENT algorithm. When would one prefer to use the algorithm?
12. Given a SAT problem with N variables, where $N > 50$, you are implementing the HILLCLIMBING algorithm to search for a solution. You have the choice of using two neighbourhood or MOVEGEN functions, N_1 and N_n, where
    A. One bit is flipped in N_1
    B. Any number of bits can be flipped in N_n
    Compare and contrast the behaviour of HILLCLIMBING using the two functions in terms of space and time required, and completeness. Which one would you prefer to use and why?
13. Given the SAT problem with four clauses: $(\sim b \lor \sim c) \land (\sim d \lor \sim e) \land (e \lor \sim c) \land (a \lor \sim d)$, assume the evaluation/heuristic function is the number of clauses satisfied. Let the solution vector be in the order $(a\ b\ c\ d\ e)$. Let the starting candidate be (1 1 1 1 1). Show the nodes generated when BEAMSEARCH with beam width 2 explores the search space up to depth 2. Mark the nodes on OPEN at this stage. Assume that one bit may be changed to generate a neighbour.

14. The following figure depicts the search terrain for the SAT problem defined below.
$$F = (a \vee b) \wedge (a \vee c) \wedge (c \vee d) \wedge (b \vee \neg d) \wedge (a \vee d) \wedge (b \vee c) \wedge (c \vee \neg d)$$
$$\wedge (b \vee d) \wedge (\neg b \vee d)$$
Study the figure and identify the nodes starting from which HILLCLIMBING reaches the solution node.

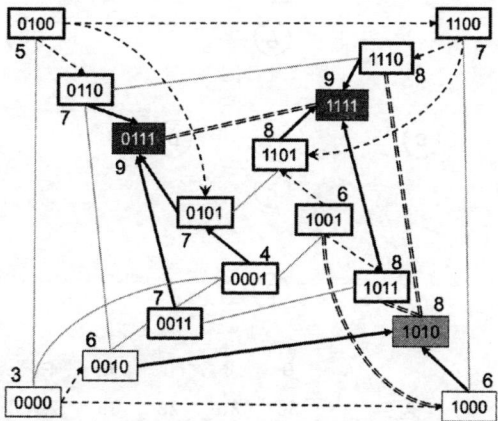

F = (a∨b) ∧ (a∨c) ∧ (c∨d) ∧ (b∨ d) ∧ (a∨d) ∧ (b∨c) ∧ (c∨¬d) ∧ (b∨ d) ∧ (¬b∨d)

15. Analyse the above SAT problem repeated below when the N_2 neighbourhood function is employed. What is the nature of the terrain?
$$F = (a \vee b) \wedge (a \vee c) \wedge (c \vee d) \wedge (b \vee \neg d) \wedge (a \vee d) \wedge (b \vee c) \wedge (c \vee \neg d)$$
$$\wedge (b \vee d) \wedge (\neg b \vee d)$$

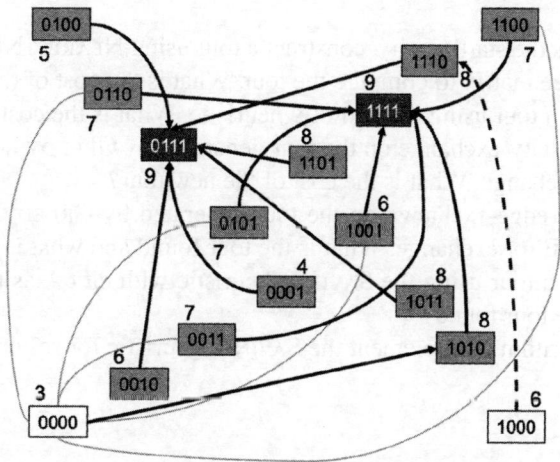

A subset of the edges on the 4-variable SAT problem when we use the neighbourhood function N_2 that flips two bits in a candidate. The nodes in grey are just one hop away from the solution. The node 0000 leads to 1010, and the next move is to a solution. Likewise for node 1000, where there are paths other than the one shown. Thus on this problem there is no local maximum when we use the neighbourhood function N_2.

16. How does TABUSEARCH perform on the SAT problem of Figure 4.15? Are there any states from which a goal node is not reached? Does it work when $tt = 2$? What about when $tt = 1$?

17. [Baskaran] A TSP problem on seven cities is shown in the accompanying table and city locations.

	A	B	C	D	E	F	G
A	-	50	36	28	30	72	50
B	50	-	82	36	58	41	71
C	36	82	-	50	32	92	42
D	28	36	50	-	22	45	36
E	30	58	32	22	-	61	20
F	72	41	92	45	61	-	61
G	50	71	42	36	20	61	-

   a. Using B as the starting city, construct a tour using NEARESTNEIGHBOUR heuristic. Use the distance matrix to compute the tour. What is the cost of the tour found?
   b. Construct a tour using the GREEDY heuristic. What is the cost of the tour found?
   c. Perform 2-city-exchange on the tour generated by GREEDY heuristic. Choose B and E for city exchange. What is the cost of the new tour?
   d. Perform 2-edge-exchange on the tour generated by GREEDY heuristic, use the edges BC and DE for exchange. What is the tour found and what is its cost?
   e. Construct a tour using the SAVINGS heuristic with city A as the fulcrum. What is the cost of the tour found?

18. Write the algorithm to implement the SAVINGS heuristic for solving a TSP.

CHAPTER 5
Stochastic Local Search

Search spaces can be huge. The number of choices faced by a search algorithm can grow exponentially. We have named this combinatorial explosion, the principal adversary of search, *CombEx*. In Chapter 4 we looked at one strategy to battle CombEx, the use of knowledge in the form of heuristic functions – knowledge that would point towards the goal node. Yet, for many problems, such heuristics are hard to acquire and often inadequate, and algorithms continue to demand exponential time.

In this chapter we introduce *stochastic* moves to add an element of randomness to search. *Exploiting* the gradient deterministically has its drawbacks when the heuristic functions are imperfect, as they often are. The steepest gradient can lead to the nearest optimum and end there. We add a tendency of *exploration*, which could drag search away from the path to local optima.

We also look at the power of many for problem solving, as opposed to a sole crusader. Population based methods have given a new dimension to solving optimization problems.

Douglas Hofstadter says that humans are not known to have a head for numbers (Hofstadter, 1996). For most of us, the numbers 3.2 billion and 5.3 million seem vaguely similar and big. A very popular book (Gamow, 1947) was titled *One, Two, Three ... Infinity*. The author, George Gamow, talks about the Hottentot tribes who had the only numbers *one*, *two*, and *three* in their vocabulary, and beyond that used the word *many*. Bill Gates is famously reputed to have said, 'Most people overestimate what they can do in one year and underestimate what they can do in ten years.'

So, how big is big? Why are computer scientists wary of combinatorial growth? In Table 2.1 we looked at the exponential function 2^N and the factorial $N!$, which are respectively the sizes of search spaces for SAT and TSP, with N variables or cities. How long will take it to inspect all the states when $N = 50$?

For a *SAT* problem with 50 variables, $2^{50} = 1,125,899,906,842,624$. How big is that? Let us say we can inspect a million or 10^6 nodes a second. We would then need $1,125,899,906.8$ seconds, which is about 35.7 years! There are $N! = 3.041409320 \times 10^{64}$ non-distinct tours (each distinct tour has $2N$ representations) of 50 cities. This would need more than a thousand trillion centuries to inspect. Surely you are not willing to wait that long, and brute force is not an option.

In Chapter 4 we introduced heuristic functions, and then we also looked at local search, which basically follows the diktat of the gradient.[1] The strategy is *exploitation* of the gradient function resulting in travelling along the steepest gradient (see Figure 4.11). We also observed that the HILLCLIMBING algorithm works when the 'hill' we are climbing is monotonous like Mount Fuji. The algorithm fails when there are local optima, as depicted in Figure 4.12.

5.1 Climbing Mount Improbable

In practice the surface defined by the heuristic function is more like the one depicted in Figure 5.1. We have borrowed the title of this section from the popular book written by Richard Dawkins (2006) in which he describes how the process of evolutionary adaptation can result in complex life forms. We will look at this process later in this chapter.

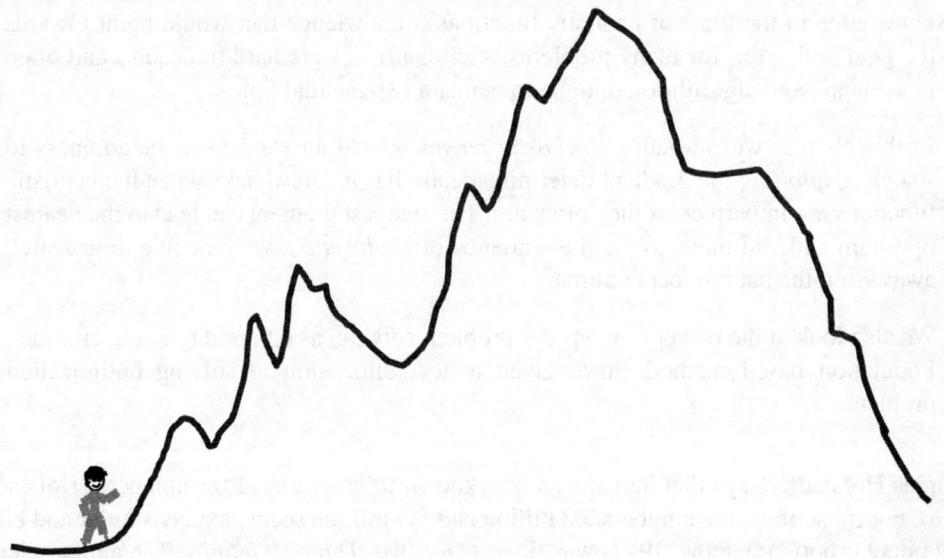

Figure 5.1 Most heuristic functions define a jagged mountain where HILLCLIMBING is doomed to fail.

The question is: if one has to reach the pinnacle in an optimization problem where the gradient is non-monotonic, what, other than brute force, would work?

Towards the end of the last chapter, we introduced the notions of *exploitation* versus *exploration*. Incidentally, while we did start off with the heuristic function defining the terrain for optimization, whose gradient we exploit, the techniques are more general. The optimization community often uses the phrase *objective function* or *evaluation function* for what is to be

---

[1] By gradient we will mean the difference between the values of the current node and the neighbour in question. By steepest gradient we refer to the neighbour for which this difference is the maximum.

optimized. The algorithms are of course name agnostic. A little later in this chapter we will also use the phrase *fitness function* in the context of evolutionary algorithms.

In the next section we inject a dose of *exploration* into the search process, allowing the local search to make some random moves, often against the yoke of the steepest gradient. As we will see, this facilitates the search to avoid getting stuck in a local optimum.

5.1.1 Random walk: pure exploration

Exploitation entails following the indicated path. Exploration is the exact opposite. Pure exploration is to move in a random direction at every step. Mathematicians have termed such sequences of moves a random walk. A random walk has no sense of direction. Of the available choices, we select one at random. In a 1-dimensional random walk on integers, one starts with a given value, and either adds one or subtracts one at each step.

For our optimization problem, the random walk will choose a move offered by the MOVEGEN function. We can implement this by randomly generating a neighbour and moving to it. We can also incorporate a ratchet mechanism by keeping track of the best node seen so far. In the context of solution space search, this means that the value of the *best* node seen can only improve with time. We will, however, need a new criterion for termination. We adopt the simple approach of doing a fixed number of iterations. The algorithm RANDOMWALK is described below.

Algorithm 5.1 Algorithm RANDOMWALK explores the search space in a random fashion. At each point, it moves to a random neighbour.

RANDOMWALK()
1 *node* ← random candidate solution or start
2 *bestNode* ← *node*
3 **for** i ← 1 **to** n
4   *node* ← RANDOMCHOOSE(MOVEGEN(*node*))
5   **if** *node* is better than *bestNode*
6   **then** *bestNode* ← *node*
7 **return** *bestNode*

If we assume that instead of selecting randomly from the choices offered by MOVEGEN, we can generate a random neighbour, then the time complexity of selecting a random neighbour will become constant, irrespective of how dense or sparse the neighbourhood function is.

Next, we look at a variation that is not purely random but selects a randomly generated neighbour with a probability that is dependent upon the gradient. In this framework, the RANDOMWALK can be seen as always accepting the random move with a probability half, irrespective of whether it leads to a better neighbour or worse.

5.1.2 Stochastic hill climbing: calibrated randomness

In the algorithm RANDOMWALK all moves are equal. But that means the algorithm is completely oblivious of the gradient. It has no sense of direction whatsoever, and no heuristic function to draw it towards the goal node. We do want our search to be attracted towards the goal. That is what HILLCLIMBING did single-mindedly, doing only exploitation. Now we inject an element of randomness into an algorithm that does prefer the gradient but is not bound by it. We call this algorithm STOCHASTICHILLCLIMBING.

Let C be the current node, and let N be the random neighbour being considered. Let EVAL(N) be the *objective function* for our optimization problem, the value that we want to optimize. This could well be the heuristic value $h(N)$ if we were doing heuristic guided search, as in HILLCLIMBING. We define the gradient ΔE as the difference between the values of N and C.

$$\Delta E = \text{EVAL}(N) - \text{EVAL}(C)$$

We move from C to N with some probability depending upon ΔE. For all neighbours N, this probability is non-zero. Even if the neighbour is worse than the current node. For a maximization problem, the larger the ΔE the better is node N, and the higher should be the probability of moving from node C to node N. A function that serves our purpose is the *sigmoid function*. The probability P of making the move is given by

$$P = 1/(1 + e^{-\Delta E/T})$$

where T is a parameter that determines the shape of the curve. Observe that the sigmoid function always evaluates to a value between 0 and 1. As ΔE tends to $-\infty$ the probability tends to 0. This means that for a very bad neighbour, the probability will be very low. Conversely, as ΔE tends to $+\infty$ the probability tends to 1. Thus, STOCHASTICHILLCLIMBING moves to better neighbours with a higher probability than to worse ones. The shape of the sigmoid function is shown in Figure 5.2.

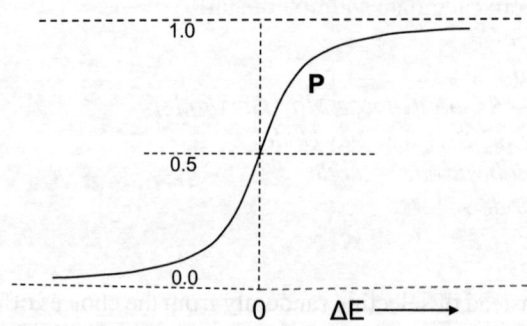

Figure 5.2 The sigmoid function is an increasing function, asymptotically approaching 0 and 1 at the two ends.

How does the parameter T influence the shape of the curve? When $T \to \infty$, then the probability P is 0.5 for all values of ΔE. This means that the probability does not depend on ΔE and is therefore like a random walk. Purely explorative. On the other hand, as $T \to 0$, the sigmoid function approaches a step function, with the probability being 0 for all ΔE less than 0, and 1 for all values greater than 0. Thus it is totally deterministic, like HILLCLIMBING. If the neighbour is better, it always moves to it, otherwise never. Purely exploitative.

For values of T between 0 and 1 the behaviour of stochastic hill climbing is a blend of exploitation and exploration. The higher the value of T, the more random the movement, and the lower the value, the more deterministic it is. The algorithm STOCHASTICHILLCLIMBING is described below.

Algorithm 5.2 STOCHASTICHILLCLIMBING generates a random neighbour and moves to it with a probability that depends upon the parameter T and the difference in the values of the two nodes. We use the function EVAL instead of h in the style used in optimization. Function RANDOM(0, 1) generates a random number in the range 0–1 with uniform probability.

STOCHASTICHILLCLIMBING(T, start)
1   node ← start
2   bestNode ← node
3   **while** some termination criteria   ▷ M cycles in a simple case
4   neighbour ← RANDOMNEIGHBOUR(node)
5   ΔE ← EVAL(neighbour) − EVAL(node)
6   **if** RANDOM(0, 1) < 1 / (1+$e^{-\Delta E/T}$)
7   node ← neighbour
8   **if** EVAL(node) > EVAL(bestNode)
9   bestNode ← node
10 **return** bestNode

The question now is: what should be the value of T?

We take a cue from techniques from material science and metallurgy. Producing materials with desired properties has been a goal since ancient times, as evidenced by the metal figurines produced by the tribes in Bastar, India. *Annealing* is a heat treatment process that changes the physical and chemical properties of a material to make it more workable. The idea is to heat the material and then cool it under controlled conditions so that the atoms settle down in a neat crystalline structure, which is characterized by low energy. It is a process of optimization, seeking the minimum energy state. The temperature to which the material is heated and the cooling rate determine the final properties.

We mimic this process in a popular algorithm called SIMULATEDANNEALING.

5.1.3 Simulated annealing: controlled randomness

The probability of moving to a random neighbour in STOCHASTICHILLCLIMBING is determined by two quantities. One, ΔE, is a measure of the gradient. The algorithm has a higher probability to move to better neighbours. The other, T, decides the degree of influence ΔE has on the probability. The higher the value of T, the more random is the behaviour of STOCHASTICHILLCLIMBING, favouring exploration. The lower the value of T, the more deterministic is the behaviour, exploiting the gradient. The algorithm SIMULATEDANNEALING is motivated by how annealing optimizes the energy levels in the material by the process of controlled cooling from a high temperature. We start with a high value of T, which we can now identify with temperature, allowing search to explore more initially, and gradually reduce T, allowing the gradient to have a greater say in the decision.

Figure 5.3 illustrates how the sigmoid function changes with decreasing values of T. On the top left, the probability is 0.5 when T tends to infinity. Then as we look at decreasing values of T, the sigmoid function emerges, becomes more pronounced, and ends on the bottom left as a step function when T approaches 0.

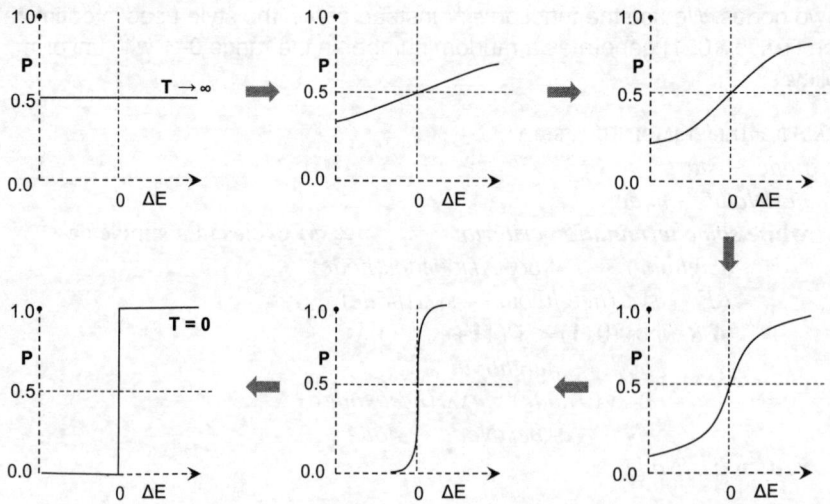

Figure 5.3 The probability curves (sigmoid function) for different values of T. When $T \to \infty$ the function has a value 0.5 irrespective of ΔE in the graph on the top left. As T decreases, the curve becomes more pronounced, and as $T \to 0$, the sigmoid function becomes a step function with value 1 for $\Delta E > 0$ and 0 otherwise, as shown in the graph on the bottom left.

The algorithm SIMULATEDANNEALING is essentially STOCHASTICHILLCLIMBING with the temperature T being gradually reduced. The *cooling rate* depends upon the problem at hand and is more often decided by a process of experimentation. The general idea is to let the algorithm explore more of the search space initially, and gradually settle down to moving along the gradient. The hope is, and this is validated experimentally, that the final phase of exploitation will happen closer to the global optimum. The algorithm is described below. It is common practice to let the STOCHASTICHILLCLIMBING run at a given temperature for a while, before reducing it.

Algorithm 5.3 Algorithm SIMULATEDANNEALING is essentially STOCHASTICHILLCLIMBING in which the temperature T is gradually reduced in stages.

SIMULATEDANNEALING(start, numberOfEpochs);
    1   *node* ← start
    2   *bestNode* ← *node*
    3   T ← some large value

```
4    for time ← 1 to numberOfEpochs
5      while some termination criterion do   ▷ M cycles in a simple case
7        neighbour ← RANDOMNEIGHBOUR(node)
8        ΔE ← EVAL(neighbour) − EVAL(node)
9        if RANDOM(0, 1) < 1 / (1+e^{−ΔE/T})
10         then node ← neighbour
11         if EVAL(node) > EVAL(bestNode)
12           then bestNode ← node
13       T ← COOLINGFUNCTION(T, time)
14   return bestNode
```

At each point in the inner loop, the algorithm generates a random neighbour and moves to it with a certain probability. Initially, when T is high, its moves are random, whether the neighbour is better or worse. But as T is reduced, the algorithm would prefer a better neighbour, with higher the probability the greater the gradient. It still can move to worse neighbours, but with lower and lower probability as T is reduced.

How does this stochastic search perform better than a random walk? The following intuition might help. Consider the situation in a maximization problem in which the algorithm is at a lower maximum A and must go down a valley (against the slope) and then climb up to a higher maximum B. The landscape is depicted in Figure 5.4. To travel from A to B, the search has to first arrive at L, after which the gradient will again be favourable. Likewise, if the algorithm has to travel from B to A, it has to also go via the low point L.

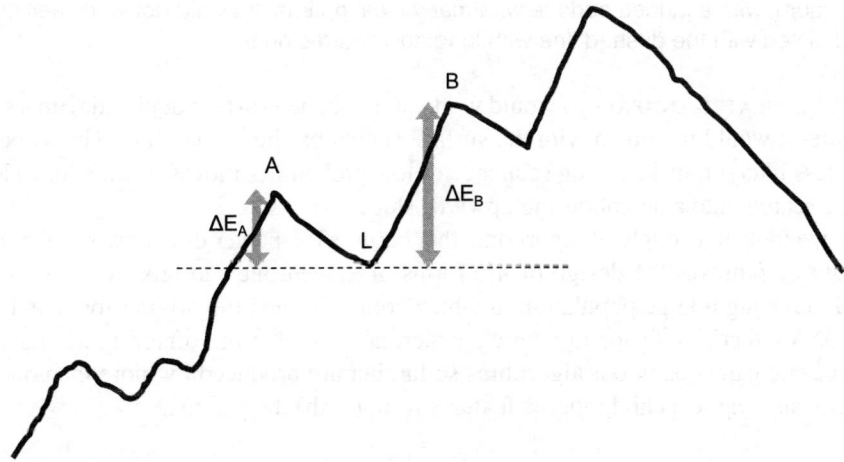

Figure 5.4 To move from peak A to peak B, the algorithm has to overcome a smaller energy difference than to move from peak B to peak A. Consequently, at high temperature, algorithm SIMULATEDANNEALING is more likely to move from A to B than vice versa.

To travel from A to L, the search must overcome the energy difference ΔE = EVAL(A) − EVAL(L), going against the gradient with a sequence of low probability moves. Likewise, if it has to travel from B to L, except that the difference here is larger. Over a period of time the probability of going from B to L is going to be lower than the probability of going from A to L because the energy difference is greater. Thus it is more likely that the search will end up at B rather than A. Even at a local optimum, the search carries on, stepping off it, like tabu search did.

When will SIMULATEDANNEALING perform well? Clearly when the magnitude of work against the gradient away from a local optimum is not too large. One can imagine a jagged surface with an overall upward trend (in the case of maximization) where search overcomes minor pitfalls on the way. This is illustrated in Figure 5.5.

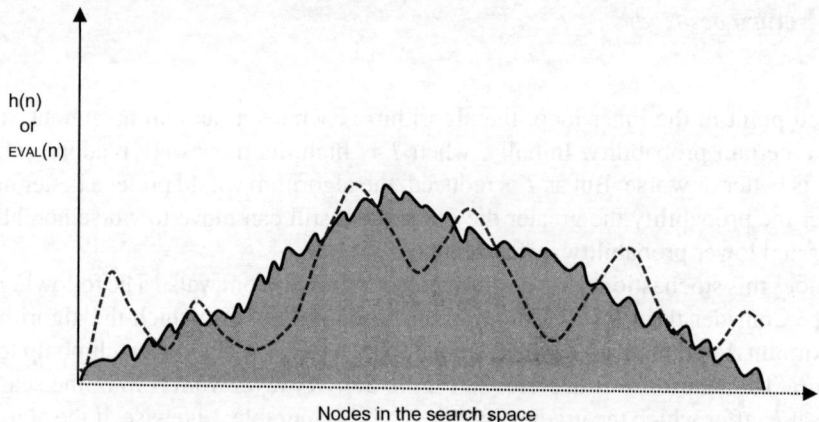

Figure 5.5 SIMULATEDANNEALING works well when there is an overall upward trend towards the global optimum, with a jagged surface with many local optima. It would not work well on the surface depicted with the dashed line with long downward slopes.

While SIMULATEDANNEALING would work on the shaded surface depicted with a solid line in the figure, it would not do so with the surface shown by the dashed line. This is because it would be less likely to make a long sequence of low probability moves down a long slope that would take search onto a neighbouring upwards slope.

Next, we look at a couple of algorithms that harness the power of teamwork. We first look at how nature optimizes the design of life forms, if indeed one can talk of nature as having agency. By creating a large population in which competition is the driving force and survival the mantra. Another key factor in nature's experiments is that the offspring are not a minor variation of one parent as in our algorithms so far, but are produced by more than one parent, usually two, and where a child inherits features from (both) the parents.

5.2 Evolution: Survival of the Fittest

What is life?

This question has been raised for millennia by humans pondering over their own existence, along with the wonder that is life on Earth. Many theories of creation of the Universe and life on Earth have been put forward by thinking humans. The main contenders are (a) Creationism, that everything was created by a God or a super being, versus (b) Darwinism, the theory of natural selection, named after Charles Darwin, who first published it. When we talk of creation, we often end up asking the question *why*? When we talk of natural selection, we ignore the why, and focus instead on the question *how?* We[2] dwell on the second question.

Let us, for argument's sake, say that nature, a name we give to everything around us, has a goal, and that it is to experiment with and create different life forms. Like a child with a Lego set.

The key idea behind evolution is that things may happen by chance, and nature is such that 'things that persist, persist; things that don't, don't' – a profound observation made by Steve Grand in his book *Creation* (Grand, 2001). In the cauldron of organic compounds that churned over aeons of time on Earth, certain molecules got together and had the happy habit of not being destroyed. Eventually these molecules that survived became large and sophisticated and acquired the ability to grow. The biological cell emerged as a stable structure – with a core and a cell body, and instructions stored in large molecules about its own structure. A single cell, however complex, by itself would be a drop in the sea of other forms getting assembled. When it learns to replicate, and create other similar *entities*, life begins to take root.

Then came the competition. Competition for matter and for energy to build *bodies* with. And here comes into play the phrase Darwin gave us, *survival of the fittest.* The fit are those that garner the resources. In fact, one can say that the fit are those that survive. If that would have been all, then some particular body form would have eventually dominated and the world have been homogenous and stable. But nature had more tricks up its sleeve. It allowed an occasional accident to happen during the reproduction process. Once in a while a small error crept in while copying the instruction molecules from parent to child. As accidents are prone to be, most accidents were disastrous. But once in an even rare while the random mutation would result in an improvement that would bequeath an advantage to the child in the survival game. An advantage that was evolutionary in nature, because the change would also be propagated to subsequent generations. And the ratchet mechanism comes into play. Once a superior life form is discovered, it dominates, thrives, and propagates.

Eventually, some better specimens would become markedly different, and a new *species* would be born.

Individuals need matter to build their bodies. Assembling bodies from 'raw' matter is a specialized task, something we can naturally think of plants as doing. Most life forms found it more expedient to *consume* other life forms, which gave them pre-processed matter – *food*. One species is the food for another. The prey and the predator. Giving rise to more competition. There was not only competition *between* the species, but, perhaps more importantly, competition *within* the species. Looking ahead in this narrative, the fastest foxes got the rabbits, and the fastest rabbits escaped and survived. Our next optimization method motivated by this phenomenon of survival of the fittest is presented in the following section.

The game of life gathered steam when nature hit upon the mechanism of bisexual reproduction. Instead of one parent creating an offspring, now two of them collaborated to produce one. The instruction set the child acquired was now an amalgamation of the instructions of the two parents, and there was always the possibility that it would be better than each of its parents. Combined with the thousands of mutation accidents that happened occasionally

---

[2] The thoughts presented here are the ruminations of the author and should by no means be treated as an authoritative source on the life sciences.

resulting in new species, a plethora of life forms have since evolved. About 540 million years ago this churning of genes in the Cambrian explosion 'filled the seas with an astonishing diversity of animals' (Fox, 2016). Almost all the creatures present in the world today – 'arthropods with legs and compound eyes, worms with feathery gills and swift predators that could crush prey in tooth-rimmed jaws' – can be traced back to the event.

One difference between plants and animals is that the latter are more sentient, though some people would dispute this. In any case, animals are mobile and have the freedom to move around. This becomes important since an individual needs to find a mate if it is to pass on its genes to an offspring. Nature has craftily injected the phenomenon of *attraction* into this fitness game. An individual seeks a mate with whom it would have a fitter offspring, and there is a plethora of ways in which individuals of a species seek to attract one, who in turn has similar genetically inherited reciprocal goals. As more couples produce better offspring, the species become fitter by natural selection. So much so, that in current times the *Homo sapiens* have become such a dominant species that we are threatening our own existence by the destruction of the ecosystem we live in. Paradoxical but true.

The natural ecosystem has emerged as an environment in which the same matter is recycled over and over to sustain the diverse life forms. Life inevitably leads to death. An individual does not hoard matter for ever and returns it upon its demise. Eventually every living creature's body is consumed by another life form, and the cycle of birth, growth, and death goes on incessantly. The individual dies but the species survives. Unless there is a catastrophic event that wipes out entire species. Remember the dinosaurs? This organization of the finite amount of matter into an unending life and death cycle is called the ecosystem, and the entire transformation is driven by energy, which to a large part comes directly or indirectly from the Sun. Figure 5.6 depicts a miniscule fragment of our ecosystem. The arrows depict a *qualitative positive influence* of the population of one species on another. For example, the more the number of grasshoppers in the environment, the more will be the number of robins who can feed on them.

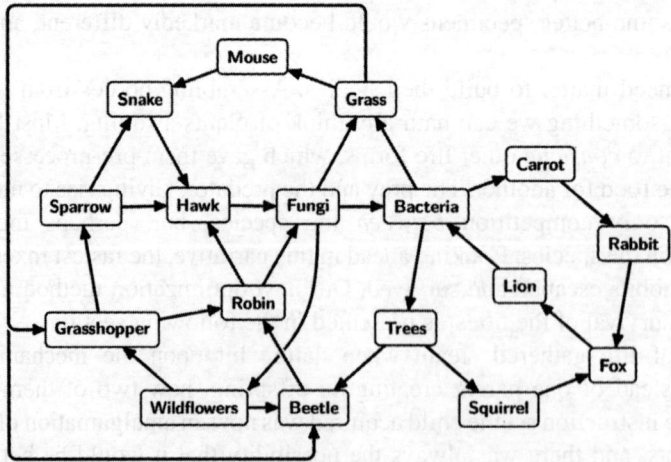

Figure 5.6 A small fragment of the vast ecosystem. The natural world contains millions of species interacting with each other. Arrows depict a positive influence of the population of one species on another.

Opponents of the evolution theory look at the clockwork-like movement of the Sun, the Moon, the planets and the stars, and the wonderful explosion of life on Earth and ask whether all this would have been possible without a maker. Richard Dawkins responds that nature is 'the blind watchmaker' which accepts and preserves any advancements that come about because of chance events (Dawkins, 1986). The *anthropic cosmological principle*[3] also comes out in support of natural selection. Just because it is so improbable that the Earth and life on it could have come about by a long sequence of improbable events *does not mean that that sequence of improbable events did not happen when it did*. If you toss a coin a hundred times and get a hundred heads, it does not mean that it could not have happened by chance.

Evolution by natural selection has two processes working in tandem. To quote the French poet Paul Valéry: 'It takes two to invent anything. The one makes up combinations; the other chooses, recognizes what he wishes and what is important to him in the mass of the things which the former has imparted to him' (Hadamard, 1945). All living creatures carry a blueprint of their design in their genes. The *genotype* is an organism's hereditary information encoded in the DNA. When a child is born of two parents, it receives a *combination* of genes inherited from both. This mixing up of genes is not deterministic but has an element of randomness. This is illustrated by the fact that human siblings can be very different from each other. The physical manifestation of the inherited genes is in the *phenotype*, for example, size and shape, metabolic activities, and patterns of movement.[4] The phenotype is the living being that is living out there in the world. The idea of survival of the fittest is embodied in the fact that it is the individuals of the species that compete in the real world. If they survive, find mates, and procreate, then their genotype is inherited by their offspring. In this fashion the species culls out weaker individuals, *selects* the better ones, and in the process becomes fitter. One can say that nature is continually improving upon its design of a species. It is this *evolutionary algorithm* that we seek to mimic when we devise GENETICALGORITHMS. John Holland is credited with inventing genetic algorithms along with his other work on complex adaptive systems (Holland, 1975).

5.2.1 Genetic algorithms: churning in a population of candidates

In the solution space search algorithms we have seen so far, new candidates are generated from old ones by perturbation of a parent candidate. The neighbourhood function makes a small change in a candidate, to produce a variant in the neighbourhood. This is possible when candidates and solutions are made up of components, which can be replaced by other components. The candidate can be thought of as a *chromosome*, and the components themselves as *genes*. GENETICALGORITHMS take a cue from nature and produce a new candidate from two parents. This is done by a process called *crossover*, which does this mixing up of genes.

We illustrate this process with the SAT problem using a *single point crossover*. Consider a 7 variable SAT problem. Given a candidate 1111111 and a perturbation function N1 which flips one bit, SIMULATEDANNEALING will generate a neighbour which is one of 0111111, 1011111, 1101111, 1110111, 1111011, 1111101, and 1111110. The GENETICALGORITHM generates a child from two parents. Let the two parents be 1111111 and 0000000. Then the single point

---

[3] https://www.thoughtco.com/what-is-the-anthropic-principle-2698848, accessed November 2021.

[4] https://plato.stanford.edu/entries/genotype-phenotype/, accessed November 2021.

crossover would choose a random point in the two chromosomes and create a child with the left part of one and the right part of the second. The second child would be a complement of the first. The first child would be one of 1000000, 1100000, 1110000, 1111000, 1111100, and 1111110. One can observe that the children produced by the crossover operation can be quite far in the solution space (in terms of Hamming distance). Figure 5.7 illustrates this process.

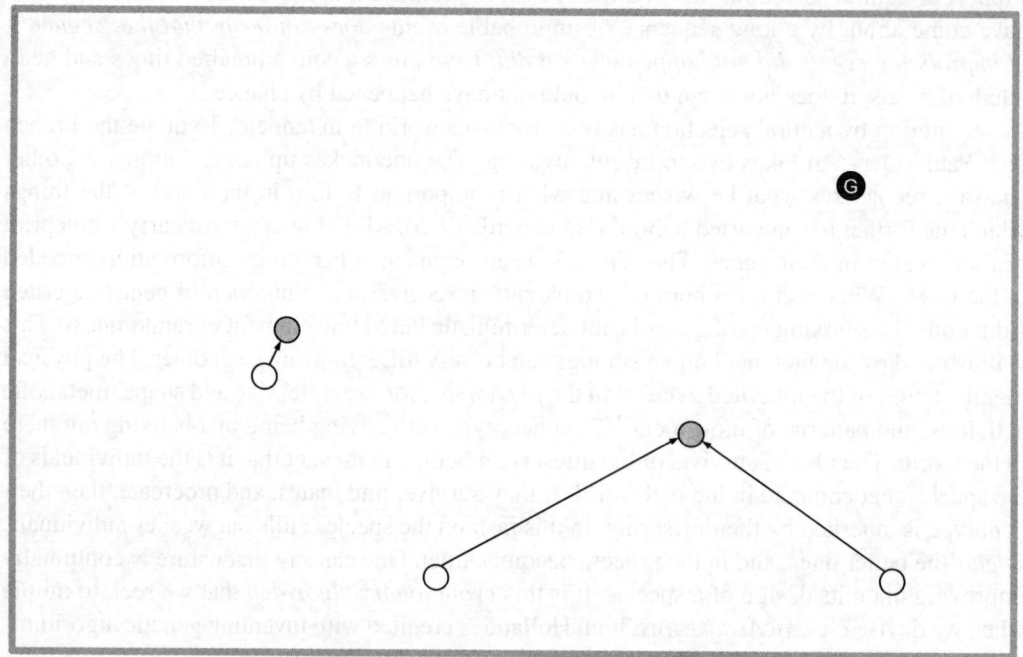

Figure 5.7 The pair of nodes on the left illustrate a move made by simulated annealing with a small perturbation on the current node. The three nodes on the right show how a new candidate is generated by a genetic algorithm by mixing up the genes from two parents. The resulting child may be far in the search space from both parents. *G* is the solution node.

Genetic algorithms, also known as *evolutionary algorithms*, work with a population of candidates, and are called population based methods. There are some differences between the way evolution happens in nature and the manner in which we implement genetic algorithms. While competition between individuals is the common theme, the way it comes about is different, as described below.

One, in nature the individual is out there living in the world. If it survives, we call it fit. In GENETICALGORITHMS, which are designed to solve specific optimization problems, we treat the objective function (or the evaluation function) as a measure of fitness. In fact, we call it a *fitness function*. For example, the cost of a candidate tour in TSP would be related to the fitness function, with low cost indicating higher fitness. In nature, survival in the real world defines fitness. In GENETICALGORITHMS, we know in advance what we are seeking to optimize. We seek to assemble a combination of genes that maximizes the fitness function.

Two, in nature, mating happens first and the offspring are produced after that. In GENETICALGORITHMS, the offspring are simply clones of the fittest parents, with fitter parents possibly getting to reproduce more than once. Given a population of N candidates $\{M_1, ..., M_N\}$, we produce a new population of N parents $\{P_1, ..., P_N\}$. This is done in N cycles, and in each cycle one member M_i is chosen with a probability in proportion to its fitness $f(M_i)$. Imagine a roulette wheel where each member gets a sector with arc length proportional to its fitness. The wheel is spun N times. After each spin, the winning member gets to reproduce, and the clone is added to the set of parents. This phase is called *Reproduction*.

Three, in nature bisexual reproduction happens when a male parent mates with a female parent. The male is the one who transfers some genetic material to the female, who nurtures it. The choice of the mate is exercised by the individuals concerned and is usually based on how attractive one finds a member of the opposite sex. In addition, physical proximity is a must, and that is why certain species thrive in specific geographic regions. In GENETICALGORITHMS this process of mating is egalitarian to the extreme. An individual mates with a randomly chosen individual. There are no conditions on gender or region. From the population $\{P_1, ..., P_N\}$ of parents produced in the *Reproduction* phase, pairs are randomly selected to mate. In the *Crossover* phase, each selected pair of parents produce a pair of children. In this way a new population $\{C_1, ..., C_N\}$ of N children is produced. The simplest idea is to replace the original population $\{M_1, ..., M_N\}$ of N members with these children, and begin the cycle again. Some researchers consider it prudent to retain some K of the fittest original members, or the K fittest parents, and only replace the remaining $(N-K)$ members with a the $(N-K)$ fittest children.

As one can imagine, the success of a GENETICALGORITHM would depend upon the abundant availability of the building blocks, or the genes or the components, in the population. This means that a large diverse population is more likely to throw up better candidates, as compared to a small or a less diverse one. This is especially true in a changing environment. It becomes imperative that a species retains genes that would be beneficial in a different environment. It is said that the cheetah evolved to be a perfect hunting creature for the grasslands. So much so that the genetic makeup in the species started becoming homogenous, with all individuals having similar genes. It is also speculated that for the same reason the cheetah is unable to adapt when there is a large scale destruction of its natural habitat, and is in danger of becoming extinct. On the other hand, Australians discovered in the recent weather upheavals that mice can adapt and proliferate, leading to huge losses for farmers. It is a fact that in current times, dominated by human activity, many species are becoming extinct at a rapid rate. The following example from the earliest book on genetic algorithms (Goldberg, 1989) illustrates this danger of losing out on a specific gene if the population is small.

Consider the problem of finding a 5-bit string over the alphabet $\{0,1\}$ where X is the number represented by the string and the fitness function is the value X^2. Let us say we start with four strings:

$$M_1 = 01101 \text{ with} f(M_1) = 169$$
$$M_2 = 11000 \text{ with} f(M_2) = 576$$
$$M_3 = 01000 \text{ with} f(M_3) = 64$$
$$M_4 = 10011 \text{ with} f(M_4) = 361$$

The total fitness of the population is 1170 and the average fitness is 293. The probabilities of the four candidates being reproduced are as follows:

$$\text{Prob}(M_1) = 169/1170 = 0.14$$
$$\text{Prob}(M_2) = 576/1170 = 0.49$$
$$\text{Prob}(M_3) = 64/1170 = 0.06$$
$$\text{Prob}(M_4) = 361/1170 = 0.31$$

As one can see, M_2 has the highest probability of being reproduced, followed by M_4, M_1, and M_3. We spin the roulette wheel four times and let us say that we get two copies of M_2, one of M_4, and one of M_1. The set of parents then is

$$P_1 = 01101$$
$$P_2 = 11000$$
$$P_3 = 11000$$
$$P_4 = 10011$$

Let us say that we (randomly) mate P_1 with P_2, and P_3 with P_4. We apply a single point crossover (randomly) after four bits for the first pair, and after two bits for the second. The set of children we now get along with their fitness values are

$$C_1 = 01100 \text{ with } f(C_1) = 144$$
$$C_2 = 11001 \text{ with } f(C_2) = 625$$
$$C_3 = 11011 \text{ with } f(C_3) = 729$$
$$C_4 = 10000 \text{ with } f(C_4) = 256$$

The total fitness of the population is 1754 and the average fitness is 439. The new population is a fitter population, with C_2 and C_3 being much fitter than the other two. It is quite likely that in the next cycle both C_2 and C_3 will get two copies each. But if that happens, the third gene, or bit, would have disappeared from the population, and however much we churn the genes after that, we can never generate the candidate 11111 which has the highest possible fitness.

To keep the possibility of breaking free from the confines of a restricted gene pool, genetic algorithms take another leaf out of nature's book. Every once in a while some gene is perturbed randomly. This, the third phase, is called *Mutation*. Now since a random move is more likely to be detrimental to the fitness of the candidate, this should be done very rarely. It does not promise a better candidate but keeps the possibility alive.

A GENETICALGORITHM is thus a process of starting with a population of N chromosomes, or candidates, and producing new candidates by trying different combinations of the genes, or components, inherited from two parents at a time. The following three steps are repeated until some termination criterion is reached.

1. *Reproduction*. Produce a new population in N cycles. In each cycle select one member with probability proportional to its fitness and add it to the new population.
2. *Crossover*. Randomly pair the resulting population, and for each pair do a random mixing up of genes, or components.
3. *Mutation*. Once in a while randomly replace a gene in some candidate with another one.

The algorithm is shown below.

> **Algorithm 5.4** A high level description of a genetic algorithm. Problem specific decisions to be made are – defining the fitness function, the population size, the crossover operation, the cooling rate, and the number of members to be carried forward.
>
> GENETIC-ALGORITHM()
> 1 P ← **create N candidate solutions**   ▷ initial population
> 2 **repeat**
> 3   **compute fitness value for each member of** P
> 4   S ← **with probability proportional to fitness value, randomly select** N **members from** P
> 5   offspring ← **partition** S **into two halves, and randomly mate and crossover members to generate** N **offsprings**
> 6   **with a low probability mutate some offsprings**
> 7   **replace** k **weakest members of** P **with** k **strongest offsprings**
> 8 **until some termination criteria**
> 9 **return the best member of** P

GENETICALGORITHMS have been popular because when carefully crafted they yield very good solutions. Starting with a random population of candidates, over a period of time the population becomes fitter, and the members gravitate towards the various maxima in the domain. This is illustrated in Figure 5.8.

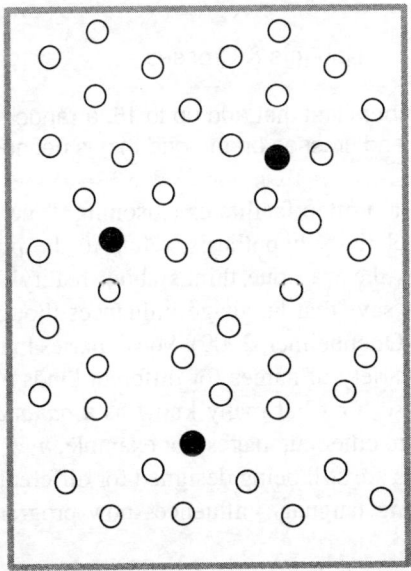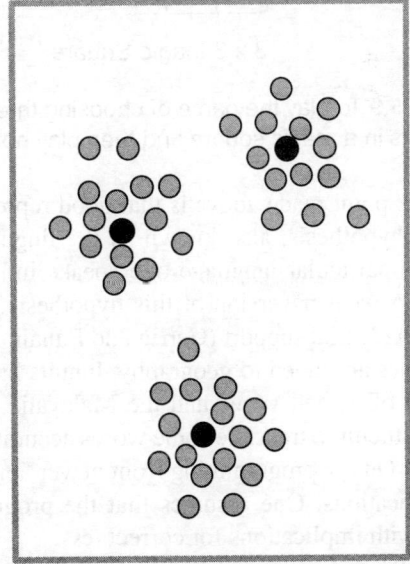

Figure 5.8 The initial population may be randomly distributed as shown on the left, but as GENETICALGORITHM is run the population has more members around the peaks, as shown on the right. The three dark circles represent three optima in the search space.

When we say that the GENETICALGORITHMS have to be carefully crafted, it pertains to the many choices that have to be made. What is the population size and how are the initial members generated? Are they generated randomly or are they the output of some other method, like *iterated hill climbing* or *simulated annealing*? What is the crossover operator that is employed? What is the nature of mutation and how often is it done? We look at the possibilities of crossover operators for the well-studied domain of TSP.

5.2.2 TSP: Representations and crossover operators

The representation we use often has a bearing on the way we can approach the problem. Consider, as a minor diversion, the following simple game. Two players make alternate moves choosing a number from the set {1 ... 9}. The first one to have three numbers that add up to 15 wins. One could play the game mentally adding up combinations of numbers one has chosen and looking out for what the opponent has chosen. Alternatively, one could transform the game into playing noughts and crosses (also called tic-tac-toe) by arranging the nine numbers as a magic square. A magic square is a 2-dimensional array filled with numbers such that each row, column, and diagonal adds up to the same number. Now, the winner in noughts and crosses is the player who gets a row, column, or diagonal. And playing nought and crosses is child's play, as most of us know. This transformation is shown in Figure 5.9.

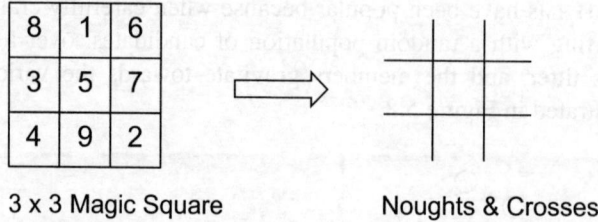

3 x 3 Magic Square   Noughts & Crosses

Figure 5.9 To play the game of choosing three numbers first that add up to 15, arrange the numbers in a magic square and then play noughts and crosses on the grid that is defined.

The point made above is that good representation often facilitates reasoning. 'The Sapir–Whorf hypothesis, also known as the linguistic relativity hypothesis, refers to the proposal that the particular language one speaks influences the way one thinks about reality' (Lucy, 2001). A weaker version of this hypothesis which says that language influences thought has found evidential support (Gerrig and Banaji, 1994; Ottenheimer, 2009). Vocabularies in natural languages are tuned to geography. Innuits have a variety of names for different kinds of what the rest of us call *snow*, and the Malayalis likewise for what many know as a *banana*. It is often difficult to translate some words accurately into other languages, for example, *hyggelig* in Danish. On the programming front newer languages are still being designed for different kinds of applications. One assumes that the programming language influences how programmers think, with implications for correctness.

TSP has sometimes been referred to as the holy grail of computer science and has attracted considerable attention. Gerhard Reinelt of Heidelberg University has a very informative webpage[5] called TSPLIB which has a collection of TSP problems and tools. We look at three different representations for a tour in TSP and the crossover operations that can be implemented on them (Larranaga et al., 1999).

In the following discussion, we will work with a TSP of 15 cities $\{A, B, ..., O\}$. Like in Chapter 4, we assume every city to be directly connected to every other city, even if some edge costs may be prohibitively high. This is to ensure that all possible tours are valid tours, even though some may have unacceptable cost.

5.2.2.1 TSP: path representation

The simplest representation of a tour is the path representation. As described in Chapter 4, a tour is described by the sequence in which the cities are visited. Thus, any permutation of the cities $\{A, B, ..., O\}$ is a valid tour. Observe that every tour has multiple representations. Let the string *AELNBIMODJFCKGH* represent the tour where the travelling salesperson starts from *A*, goes to *E*, then to *L*, and so on, till *H*, before returning back to *A*. Now if the starting city was changed to *M* then the string *MODJFCKGHAELNBI*, which is a rotation of the given string, represents the same tour, at least from the perspective of cost. There are 15 such rotations of any tour with 15 cities, each starting from a different city. Moreover, if we reverse any of these strings, we again get the same tour, albeit one in which the salesperson travels in the opposite direction.

Given N cities, every tour has $2N$ different path representations. Given that there are $N!$ permutations of N cities, we conclude that the search space with the path representation has $N!/2N = (N-1)!/2$ distinct tours that a search algorithm has to contend with. Furthermore, if the search algorithm does not distinguish between the $2N$ different representations of the same tour, its search space will have $N!$ candidates. As described in the beginning of this chapter, this is a very large number given that $N! = 3.041409320 \times 10^{64}$ for $N = 50$.

The *components* of the TSP candidate solutions are the edges, with each city having precisely two edges emanating from it. One with which the salesperson enters the city, and the other with which she departs. This is captured naturally in the path representation as the two cities adjacent to each city. And each city occurs exactly once in the path representation. Observe that the last property would be in danger of being disrupted if we employed the single point crossover. For example, given the two tours *AELNBIMODJFCKGH* and *ABCDEFGHIJKLMNO*, if we do a single point crossover after six cities, we would get *AELNBI + GHIJKLMNO* as one child in which cities *L*, *N*, and *I* occur twice and cities *C*, *D*, and *E* are absent. A similar problem occurs with the other child with *C*, *D*, and *E* being present twice and *L*, *N*, and *I* being absent.

Clearly, we need a more sophisticated crossover operator, one which would yield offspring that are valid tours. The following are some of the crossover operators for path representation that are well known in the literature.

The *cycle crossover* (CX) operator aims to make sure that every city gets a place in each of the two children. Consider the two parents P_1 and P_2 shown in Figure 5.10 from which we intend to produce two children C_1 and C_2 by copying cities in the same place from one of the

---

[5] http://comopt.ifi.uni-heidelberg.de/software/TSPLIB95/.

two parents. Let us say that we intend to copy the first city O in P_1 into C_1. Then we cannot copy H from P_2 in the same place in C_1. Therefore, we copy H too from P_1. This, in turn, leads to A being copied from P_1 as well. The corresponding city in P_2 is O, which we have already copied into C_1. This identifies a cycle, which we call Cycle 1 in both parents, with O, H, and A in P_1 and correspondingly H, A, and O in P_2. The identification of Cycle 1 is shown in the top half of Figure 5.10.

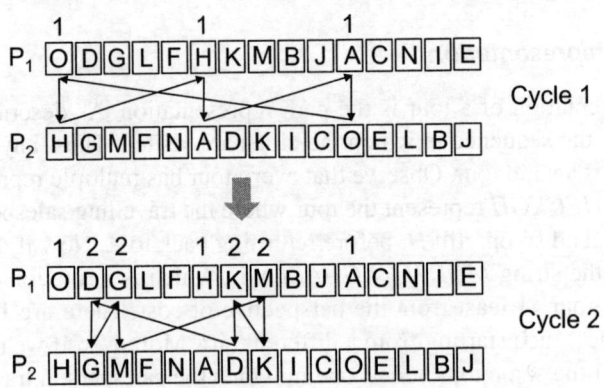

Figure 5.10 The top half shows the identification of Cycle 1 and the bottom half shows the identification of Cycle 2 in CX.

The bottom half shows the identification of Cycle 2. We start with city D in P_1 and follow through to G, K, and M before returning to D. In a similar fashion we identify Cycle 3 starting with city L in P_1, and then Cycle 4 and Cycle 5. The reader is encouraged to verify that the five cycles are as depicted in the top half of Figure 5.11. In the figure the odd numbered cycle cities in P_1 are shaded, as are the even numbered cycle cities in P_2.

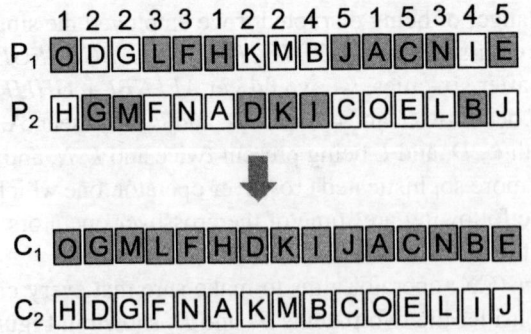

Figure 5.11 In the given example, five cycles are identified by CX. The first child C_1 gets the odd numbered cities from P_1 and the even numbered cities from P_2. These are shown in shaded boxes. The remaining cities go into forming C_2.

In CX we construct C_1 by copying all the odd numbered cycles from P_1 and the even numbered cycles from P_2. This is illustrated in the bottom half of Figure 5.11. All the shaded cities are in C_1 and we can observe where they have been copied from the two parents. The unshaded cities form the second child C_2, copied from the odd numbered cycles in P_2 and even numbered cycles in P_1.

The *partially mapped crossover* (PMX) also copies some cities from parents in the same position in the children, though with a slightly different procedure. One key difference is that it copies entire subtours from the two parents into the children, before filling in the rest. This potentially allows short subtours to survive from parents into the next generation.

Consider the parent $P_1 = ODGLFHKMBJACNIE$ from the previous example. In PMX a subtour is selected to be copied into the first child C_1. Let this subtour be $HKMBJ$. So far this is like the single point crossover which selects part of a parent chromosome. Remember that P_1 could have also been written as $HKMBJACNIEODGLF$ after rotation, and $HKMBJ$ is the left half here. The second half cannot be directly copied from P_2 because, as discussed earlier, this will include some cities already in C_1. To get around this duplication problem, PMX begins with a mapping of the chosen subtour with the corresponding cities in the other parent $P_2 = HGMFNADKICOELBJ$. This is illustrated in Figure 5.12.

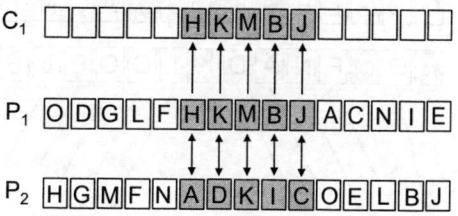

Figure 5.12 PMX begins by copying a subtour *HKMBJ* from P_1 into the first child C_1 and establishing a partial map of this subtour with P_2. One would like to copy the remaining cities from P_2 but the locations for cities *A*, *D*, *I*, and *C* in P_2 are already occupied by cities *H*, *K*, *B*, and *J* respectively. *K* has found a place in C_2.

Having selected a subtour from P_1, we need to copy the remaining cities from P_2, keeping in mind that we already have the subtour *HKMBJ*. City *A* has been displaced and we cannot copy *A* in place from P_2 because it is occupied by *H* in C_1. This is where the partial map comes into play. We look at the location of *H* in P_2. Its location is free in C_1 and serves as a refuge for *A*. We could have followed the partial map in the reverse direction as well, starting with *H* in P_2, which cannot be copied because it is already in C_1. Instead, we copy *A* which *H* maps to in the partial map.

Finding the destination for *D*, the second displaced city, requires a little more work. The location of *D* in P_2 is occupied by *K* in C_1. We follow the map and consider the location of *K* in P_2. That in turn is occupied by *M* in C_1. Finally, we look at the location of *M* in P_2 and find that the corresponding spot is available for *D*. This process of finding the destination for *D*, having found one for *A*, is illustrated in Figure 5.13.

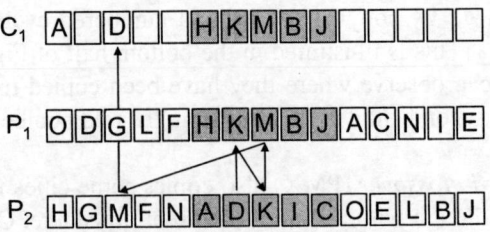

Figure 5.13 PMX copies the subtour *HKMBJ* from P_1 into C_1 and looks to copy the rest from P_2. After moving *A* to the position occupied by its image *H*, where should city *D* be in C_1? The answer: follow the partial map.

A similar process is followed for the cities *I* and *C*. City *K* from P_2 already has a place in C_1. After the five displaced cities *A, D, K, I,* and *C* have been accommodated, the remaining cities *G, F, O, N, E,* and *L* are directly copied from P_2 into C_1. The reader should verify that C_1 is *AGDFOHKMBJNELIC* and C_2, after a similar process, is *OMGLFADKICHJNBE*.

The *order crossover* (OX) also copies a subtour from P_1 into C_1. The remaining cities are then inserted in the order that they occur in the other parent P_2. This is illustrated in Figure 5.14.

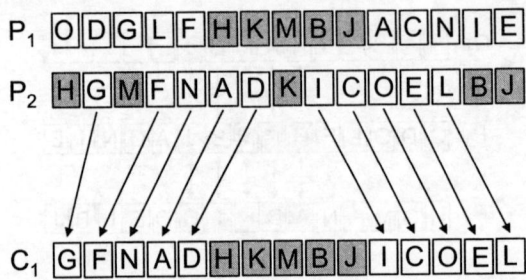

Figure 5.14 In OX we copy a subtour from P_1 into C_1 and insert the remaining cities in the order they occur in P_2.

As seen in the previous example, this crossover also has the property of preserving entire subtours from parents in the offspring. The reader should verify that the second child C_2 is *OGLFHADKICMBJNE*.

5.2.2.2 TSP: adjacency representation

The *adjacency representation* tells us from *where* one arrives to any given city in the tour. This is done by creating an index of the cities. City *A* is first in the index, the second one is *B*, and so on, as depicted in Table 5.1.

Table 5.1 An index of cities

Index	1	2	3	4	5	6	7	8	9	10	11	12	13	14	15
City	A	B	C	D	E	F	G	H	I	J	K	L	M	N	O

Consider the following tour in path representation:

ODGLFHKMBJACNIE

As discussed earlier this can be interpreted as a path *specification*:

$O \to D \to G \to L \to F \to H \to K \to M \to B \to J \to A \to C \to N \to I \to E \to O$

In the adjacency representation the above tour would be represented as *CJNGOHLKEAMFBID*. The way to interpret this is as follows. One arrives at C from A, the first city in the index, at city J from B, the second city, and so on. This is depicted in Table 5.2 with the index represented in the bottom row, and the corresponding city that one goes to from there in the top row. The main advantage of this representation is that one can quickly determine for any city the two options taken by the two parents.

Table 5.2 The Adjacency Representation of *ODGLFHKMBJACNIE*

Adjacency	C	J	N	G	O	H	L	K	E	A	M	F	B	I	D
From City	A	B	C	D	E	F	G	H	I	J	K	L	M	N	O

The first observation one can make is that for a tour being traversed in one direction, there is a unique representation. In the above example, the first city can only be C because one arrives at it from A, the first city in the index. Then, since we go from C to N, only N can be in the corresponding position in the representation with index C. And so on. If one were to traverse the tour in the opposite direction, we would have another unique representation in which the first entry can only be J, the other neighbour of A in the path representation. And A would be at the location indexed by C. Every tour then has exactly two representations, one for each direction of travel. This contrasts with the $2N$ representations any tour of N cities has in the path representation.

The second observation is that not every permutation of N cities in adjacency representation is a valid tour. For example, the permutation that begins with CAB cannot be a tour because this contains a cycle $A \to C \to B \to A$. A consequence of this is that the crossover operators must take care to avoid cycles of less than N cities.

One crossover operation that is popular with adjacency representation is the *alternating edges crossover* (AEX). Here one constructs the first child C_1 as follows. Given any starting city, choose the first successor from P_1 in the adjacency representation and then the next one from P_2, and so on. We illustrate this with the two parents *ODGLFHKMBJACNIE* and *HGMFNADKICOELBJ* we looked at earlier. First we must represent them in adjacency representation. The reader should verify that the representations are *CJNGOHLKEAMFBID* and *DJOKLNMGCHIBFAE* respectively. Observe that there is no identified starting point in the tour representation. For any city C_i Table 5.2 tells us what the next city C_j in the tour is. There is no starting point in the path representation too, where every rotation of a tour is the same tour.

The process of implementing the AEX is depicted in Figure 5.15, where the two parents are in the adjacency representation. The top half of the figure shows the two parents, and the bottom half shows the process of constructing the child C_1. Let us say we start constructing the tour starting at city F. Then, as is in P_1, we move to H. Then from H to G as in P_2, from G to L as in P_1, from L to B as in P_2, from B to J, and then from J to H!

136 | Search Methods in Artificial Intelligence

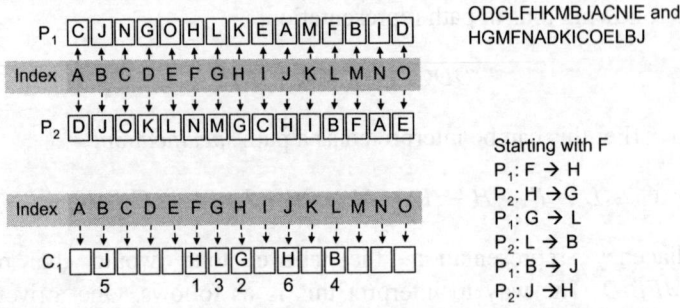

Figure 5.15 In AEX every step in the tour is dictated by alternating parents in the adjacency representation. If child tour begins at F, then the first move is $F \to H$ as in P_1, the second $H \to G$ as in P_2, the third $G \to L$ as in P_1, and so on. After six entries, H is repeated, and one may have to pick a different city.

As seen in this example, after six entries we are back to the city H, which is not allowed. This is typically resolved by choosing a city that is available instead of closing the loop. One could even optimistically choose the nearest available neighbour to form a shorter tour.

Choosing a shorter edge is the idea behind the *heuristic crossover* (HX) in which at every point the shorter of the two edges suggested by the two parents is selected, as long as it does not close the loop.

Both the above crossover operators produce one child from two parents. To generate N offspring one must go through N cycles of randomly selecting pairs.

One advantage of the adjacency representation and the crossover operators used is that one could produce offspring from multiple parents. We could choose, say, three parents for each child to be produced and select the moves made in the child by inheriting from them in a cyclic or heuristic manner. This would allow mixing up of the genes from more than two parents.

5.2.2.3 TSP: ordinal representation

The claim to fame of the *ordinal representation* is that with this encoding the single point crossover produces two valid tours. We describe the procedure for encoding a tour expressed in path representation into the ordinal representation. We begin with the index of cities introduced in the previous section repeated here.

Table 5.1 An index of cities

Index	1	2	3	4	5	6	7	8	9	10	11	12	13	14	15
City	A	B	C	D	E	F	G	H	I	J	K	L	M	N	O

Consider the following tour in path representation:

$$\text{Path-Tour} = ODGLFHKMBJACNIE$$

Let Ord-Tour be the tour we are constructing. We initialize Ord-Tour to be the empty list []. We process the cities in Path-Tour one by one and add the index of the city at the tail of the list. The first city is O, with index 15.

Ord-Tour = [15]

Next, we delete the inserted city from the index and shift the cities left. In case of city O, there is nothing to shift. The next city is D, with index 4.

Ord-Tour = [15, 4]

After deleting D from the index, we shift the remaining cities to the left. Observe that the index of cities beyond D in the index has decremented by 1, as shown in Table 5.3.

Table 5.3 Updated index used for ordinal representation

Index	1	2	3	4	5	6	7	8	9	10	11	12	13	14	15
City	A	B	C	E	F	G	H	I	J	K	L	M	N		

The next city is G and Ord-Tour = [15, 4, 6] and the updated index is shown in Table 5.4

Table 5.4 Updated index used for ordinal representation

Index	1	2	3	4	5	6	7	8	9	10	11	12	13	14	15
City	A	B	C	E	F	H	I	J	K	L	M	N			

The next city is L and Ord-Tour = [15, 4, 6, 10] with the updated index in Table 5.5.

Table 5.5 Updated index used for ordinal representation

Index	1	2	3	4	5	6	7	8	9	10	11	12	13	14	15
City	A	B	C	E	F	H	I	J	K	M	N				

We always use the updated index. The reader should verify that after all the cities have been processed the ordinal representation is

Ord-Tour = [15, 4, 6, 10, 5, 5, 7, 7, 2, 5, 1, 1, 3, 2, 1]

Given the above ordinal tour we can recreate the path by an inverse process. We begin again with the index of the cities in Table 5.1, reproduced here. We initialize the Path-Tour to and empty string (or list). Path-Tour = [].

Table 5.1 An index of cities

Index	1	2	3	4	5	6	7	8	9	10	11	12	13	14	15
City	A	B	C	D	E	F	G	H	I	J	K	L	M	N	O

We pick the first element from the Ord-Tour, which is 15, and add the 15th city, O, at the tail of Path-Tour.

Path-Tour = O

We delete the corresponding city in the index and shift the remaining cities to the left. At this point, there is nothing to shift. Next, we pick the second number from the Ord-Tour, which is 4, and append the corresponding city, D, at the end of Path-Tour.

$$\text{Path-Tour} = OD.$$

The updated index now is the same as Table 5.3.

Table 5.3 The updated index of cities

Index	1	2	3	4	5	6	7	8	9	10	11	12	13	14	15
City	A	B	C	E	F	G	H	I	J	K	L	M	N		

The next number is 6 and the corresponding city is G. Path-Tour $= ODG$. The update index is the same as Table 5.4 reproduced here.

Table 5.4 The updated index of cities

Index	1	2	3	4	5	6	7	8	9	10	11	12	13	14	15
City	A	B	C	E	F	H	I	J	K	L	M	N			

The next number is 10 and the corresponding city is L. Path-Tour $= ODGL$.

In the first four steps we have reconstructed the first four cities in the path representation. The reader is encouraged to complete the process and verify that this procedure decodes the ordinal representation back to the path representation.

5.3 Swarm Intelligence: The Power of Many

Evolution is a process facilitated by the churning of components by random processes. The idea behind genetic algorithms was to randomly inherit combinations of genes from two parents to produce new sets of specifications for building bodies. The bodies are themselves made up of organs which are the building blocks for the phenotype. We are ourselves colonies of a multitude of simple cells, organized into different organs. From this vast conglomeration somehow our notion of the self emerges.

When simple elements come together, a larger entity emerges. Our human minds are attuned to thinking about these *emergent* systems. We do not think of ourselves and our bodies as a collection of a large number of atoms somehow existing in a self-organized manner, like a flock of starlings. We think of ourselves as one person. In his book Stephen Grand talks about clouds that seem to hover on mountain passes, even though there is a steady breeze blowing there. It is just that the moisture condenses at the heights momentarily. The water molecules flow on, but leave an impression of a static cloud. Somewhat like our own bodies which we think of as the same, even as the molecules we are made up of are ever changing. It is said that a farmer once proudly claimed that he had Abraham Lincoln's axe (variations talk about George Washington) and when queried further admitting that 'the handle has been replaced 3 times and the head has been replaced twice'. John Holland compared cities to self-organized higher level entities, even though the people and their activities keep changing – 'Buyers, sellers, administrations, streets, bridges, and buildings are always changing, so that a city's coherence is somehow imposed on a perpetual flux of people and structures. Like the standing wave in front of a rock in a fast-moving stream, a city is a pattern in time.'

Emergent systems is a field which studies the emergence of complexity from the interaction and aggregation of many simple elements (Johnson, 2002).

'Emergence is what happens when a multitude of little things – neurons, bacteria, people – exhibit properties beyond the ability of any individual, simply through the act of making a few basic choices: Left or right? Attack or ignore? Buy or sell? The ant colony is the classic example, of course. This meta-organism possesses abilities and intelligence far greater than the sum of its parts: The colony knows when food is nearby, or when to take evasive action, or, amazingly, just how many ants need to leave the colony to forage for the day's food or ward off an attack' – (Ito and Howe, 2017).

Another field of study that looks at the interaction of a multitude of simple elements is *Chaos Theory* (Gleick, 1987; Holland, 1999). Chaos theory is an interdisciplinary branch that looks at compact mathematical models that describe patterns of behaviour that are highly sensitive to initial conditions. Even though the systems may be deterministic, minute differences in measurement of initial conditions can make prediction of future behaviour almost impossible. In the words of Edward Lorenz (1993), 'Chaos: When the present determines the future, but the approximate present does not approximately determine the future.' The book by Gleick describes concepts like self-similarity which can be used to model the structure of coastlines (which look the same at any level of magnification), trees, and other seemingly random structures in nature. Benoit Mandelbrot, a Polish-born French-American mathematician, devoted his research to finding the simple mathematical rules that give rise to the rough and irregular shapes of the real world. His work gave rise to the world of fractal geometry.[6] The fascinating images produced by the simple equations of the Mandelbrot sets, Julia sets, and the Sierpiński triangle all have unconstrained self-similar structures where any amount of zooming in presents the same picture. A fascinating documentary *The Secret Life of Chaos* by Jim Al-Khalili starts off with Turing and the study of chaos, and goes on to describe how genetic churning could have created complex life.

The key idea in emergent systems is that life forms have emerged through simple elements interacting with each other to give rise to complex structures. The British mathematician John Conway brought this to light in a spectacular 'game' that he invented called the *Game of Life*[7] or simply *Life*. It was publicized by Martin Gardner in his column on Mathematical Games in *Scientific American* (Gardner, 1970). The Game of Life is a cellular automaton made up of an infinite two dimensional grid of cells in principle. In computer implementations we generally stitch the left end of the screen with the right end, and the top with the bottom, defining an endless toroidal array of cells.

---

[6] https://users.math.yale.edu/public_html/People/frame/Fractals/, accessed November 2021.

[7] https://conwaylife.com/, accessed December 2021.

Each cell can be in two states, dead or alive, 0 or 1. The 'game' has no active players, and once initialized, takes a life of its own. At each time step, each cell obeys the following simple rules of birth, death, and survival based on the current states of its eight neighbours:

- If a cell is alive at time t and has two or three neighbours alive, then it stays alive at time $t+1$. Else it dies.
- If a cell is dead at time t it becomes alive at time $t+1$ if it has exactly three neighbours that are alive. Else it remains dead.

The rules are such that cells thrive unless there is overcrowding or, at the other end, loneliness. In the cellular automaton, time moves in discrete steps and the system evolves. The fascinating thing about the game is that depending upon the starting position the cells evolve in unpredictable ways. Three cells in a row left to themselves rotate around the middle cell, as two die of loneliness and two new ones are born in each cycle. There are static structures like four cells arranged in a square, since each has three live neighbours. But if another structure were to come and crash into it, the tranquillity would be broken when the number of alive neighbours cross the overcrowding threshold of three. Movement of patterns is possible, for example, the famous Gosper's *gliding gun* which 'emits objects or organisms' that tumble across the 2-dimensional space. This is illustrated in Figure 5.16. But there are more interesting patterns in which societies of cells grow and wane as time goes along. The reader is encouraged to implement the game and experiment with the rules.

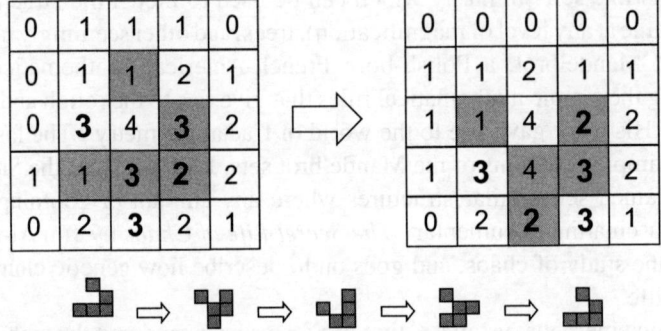

Figure 5.16 Gosper's glider gun emits this 5-cell 'organism' that creates an illusion of movement in the Game of Life. The detail on the top shows one time step. The shaded cells are alive. The number in each cell is the number of alive neighbours. The numbers in bold are the cells that will be alive in the *next* time step. The figure in the bottom shows four time steps ending in replicating the starting structure but shifted to the right and bottom.

What the game demonstrates is that simple elements operating with simple rules can lead to the emergence of higher level entities. We often see flocks of birds in flight coordinating their movements in tandem as if they were one entity. Collections of individual elements acting in a decentralized mode and self-organization principles is also referred to as *swarm intelligence*. Colonies of ants and termites, schools of fish, flocks of birds, and herds of land animals are some groups that characterize such behaviour.

Simple components interacting with simple rules result in the emergence of higher level patterns that persist. The phenomenon of reproduction emerged when nature figured out encoding body plans at the genomic level. These instructions coded in the genes orchestrate the diversification of cells into different kinds of tissues that make up a living creature, which consumes food which is matter to build bodies with and also energy to build them with. Competition for food and natural selection ensured that the best designs survived. Bisexual reproduction accelerated the production of better designs when individuals competed for mates in addition to food. The world we live in with all its diversity is the culmination of this mindless application of simple rules over millions of years, including the evolution of creatures with minds that could contemplate and ponder over this very phenomenon.

In Chapter 10 of his book *Gödel, Escher, Bach*, Hofstadter (1979) describes how composite systems can be viewed. In an accompanying writeup, *Ant Fugue*, he describes how a colony of ants can be treated as a more complex organism behaving in a purposeful manner, even creating living bridges with their own bodies. The human brain too is a complex system made up of large numbers of simple elements, the neurons. Neither the ant in a colony nor a neuron in a neural network has the big picture which the composite entity has.

We look at an optimization algorithm that is inspired by ant colonies.

5.3.1 Ant colony optimization

Populations of chromosomes in genetic algorithms form a large pool of *candidate solutions* from which new combinations are churned out and presented to the selection phase. Ant colonies on the other hand are *populations of agents* going about their simple, often repetitive, business but which gives the colony a higher level of purposefulness and agency not witnessed in individual ants.

Anyone who has left some sugary food out in the open in Chennai is soon confronted with a caravan of ants diligently ferrying it away. How does this happen?

The secret behind the success of an ant colony is communication. Bees are known to communicate information about the location of a meal by a dance that is performed on one of the honeycomb walls (Schirber, 2005). Ants rely on the phenomenon of *biosemiotics*, in which chemical deposits act as symbols.

The chemical used by ants is *pheromone*, which acts as a marker for other ants to follow. Wherever an ant goes, it leaves behind a trail of pheromone. An individual ant acting alone would have hardly any chance of discovering your lump of sugar, but an army of ants in search of food will find it almost without fail. This happens because the ant which chances upon your food dutifully carries it back to its nest. Like the ball of red thread that Ariadne gave Theseus in Greek mythology, which he unrolled as he penetrated the Labyrinth and was able to find his way back, the ant can follow its own pheromone trail back to its nest. While doing so, it deposits even more pheromone on the trail, out of sheer habit. This has the happy effect of strengthening the pheromone trail. Now ants are addicted to pheromone, and other ants that emerge from the nest instinctively follow the strongest trail, further strengthening it as they do so. Eventually this positive feedback cycle results in all ants marching along in a single column. This saga of finding food is illustrated in Figure 5.17.

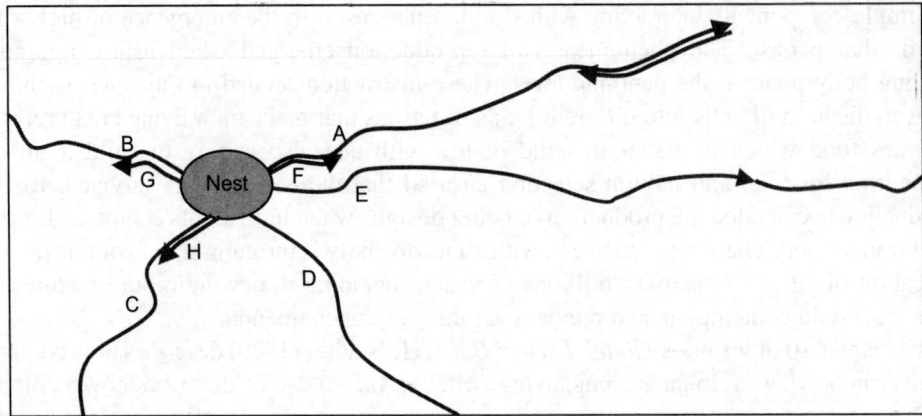

Figure 5.17 Initially ants A, B, C, D, and E set out in search of food. When ant A find some food, it grabs it and dutifully heads back. It follows its own pheromone trail back, and strengthens it. Ants F, G, and H have set out meanwhile. Ant F followed the trails left by ant A, and will come back with food too, further strengthening the trail.

Motivated by the behaviour of the ant colony, Marco Dorigo and his associates devised an optimization algorithm in which simple problem solving agents cooperate with each other, finding better and better solutions by following the cues given out by other agents (Colorni, Dorigo, and Maniezzo, 1991; Dorigo, 2004). The algorithm is called *ant colony optimization* (ACO). The main idea here is that an army of simple agents repeatedly solves a problem individually, and in each new cycle is influenced by the solutions synthesized by all agents in the previous cycle. A little bit like how a cell in the Game of Life is influenced by the states of its neighbours at the previous time step. The algorithm is best illustrated with the TSP problem.

Let us say we have an N city TSP to solve. We place M ants randomly on the nodes in the graph. Then, in each cycle, each ant constructs a tour independently using a simple stochastic greedy algorithm as described below.

In each cycle beginning at time t the ants start constructing a tour, completing it at time $t + N$. At each choice point at $City_i$ an ant chooses a city to move to in a probabilistic manner. The probability of moving to $City_j$ is influenced by two factors. One, η_{ij}, called *visibility*, which is inversely proportional to the cost of the edge between $City_i$ and $City_j$. The other, $\tau_{ij}(t)$, is the total pheromone on $edge_{ij}$ at the beginning of the cycle at time t. The two factors are moderated by parameters α and β. The probability of the kth ant moving from $City_i$ to $City_j$ in the cycle beginning at time t is given by

$$P_{ij}^k(t) = \begin{cases} \dfrac{[\tau_{ij}(t)]^\alpha * [\eta_{ij}]^\beta}{\sum_{h \in allowed_k(t)} ([\tau_{ih}(t)]^\alpha * [\eta_{ih}]^\beta)} & \text{if } j \in allowed_k(t) \text{ the cities } ant_k \text{ is allowed to move to} \\ 0 & \text{otherwise} \end{cases}$$

In the real world, an ant constantly deposits pheromone as it ambles along. If the path that it travels is a short one, it returns quickly with food and deposits more pheromone on the way back. In this way shorter paths have more pheromone, and an ant colony finds the shortest path to the lump of food. The goal in TSP is to construct the shortest tour. Shorter tours should have more pheromone. This is achieved by assuming that an ant deposits the same amount of pheromone on all the edges of the tour it has found. The amount of pheromone deposited is inversely proportional to the length L_k of the tour constructed by the kth ant. After constructing a tour in N time steps, each ant k deposits an amount of pheromone Q/L_k on all the edges it traversed, where Q is a parameter the user can control.

Being a chemical substance, pheromone evaporates in real life. In ACO, this phenomenon is incorporated by a parameter ρ, which is the rate of evaporation. We assume that a fraction proportional to ρ evaporates in every cycle. The total pheromone on edge$_{ij}$ after the cycle is over is

$$\tau_{ij}(t+n) = (1-\rho) * \tau_{ij}(t) + \Delta\tau_{ij}(t, t+n)$$

where $\Delta\tau_{ij}(t, t+n)$ is the pheromone deposited by all the ants in the cycle beginning at time t and ending at time $t+N$.

In this manner, M ants repeatedly construct tours in a greedy stochastic manner, choosing the next city heuristically at each point in time, and not looking back. Each ant deposits an amount of pheromone inversely proportional to the cost of the tour it finds. Shorter tours get more pheromone on the edges, and more ants follow those edges, further strengthening the levels of pheromone. The termination criterion can either be a fixed number of cycles or a threshold between the best tour cost in two consecutive cycles. The algorithm is given below.

Algorithm 5.5 The ant colony algorithm for TSP (TSP-ACO) employs M ants to construct tours independently in a greedy stochastic manner. In every cycle each ant is drawn towards edges on shorter tours found by other ants.

TSP-ACO()
1 bestTour ← **nil**
2 **repeat**
3   randomly place M ants on N cities
4   **for each ant** a   ▷ construct tour
5   **for** n ← 1 **to** N
6   ant a selects an edge from the distribution P_n^a.
7   update bestTour
8   **for each ant** a   ▷ update pheromone
9   **for each edge** (u, v) **in the ant's tour**
10   deposit pheromone ∝ 1/tour-length on edge (u, v)
11 **until** some termination criteria
12 **return** bestTour

The reader is encouraged to implement this simple algorithm, all the better with a graphical user interface depicting the pheromone with thickness of the edges.

Summary

In this chapter we studied randomized algorithms in the context of stochastic local search for optimization. All the algorithms we looked at were motivated by real world phenomenon. The first, SIMULATEDANNEALING, is based on a transition from random search to gradient descent (or ascent) in a quest to head towards the global optimum while avoiding the local ones. This is based on a technique used in moulding materials into desired low energy states.

Then we moved on to drawing inspiration from natural phenomena. The world around us has emerged from random mixing and matching combined with the survival of the persistent. As simple elements combine to give rise to higher level phenomena, competition for resources leads to selection of the fittest elements, and the rise of different species that evolve. Selection provides a ratchet mechanism that preserves better designs, as and when the random processes stumble upon them. This led us to GENETICALGORITHMS which mimic the process of evolution. Next we looked at how populations of ants coordinating with each other with biosemiotics provide us with a cue for distributed optimization techniques in the form of the ACO algorithm.

Emergence refers to complex entities emerging from the interaction of many simple ones, like in a colony of ants. The human brain is one such complex entity that is made up of millions of simple elements called neurons. We defer the discussion on neural networks, which are information processing architectures that learn, to a later chapter when we discuss learning and its relation to search.

In the next chapter we come back to deterministic approaches to finding optimal solutions. We had ventured into optimization, motivated by the desire to find the node with the best heuristic value.

The heuristic function was devised to speed up search. We return to problem solving where we have a goal node, with the objective of finding the shortest path. The need for employing a heuristic function will come in again. We seek not only to find the optimal path to a goal, but also to do so quickly in an informed manner.

Exercises

1. Devise an algorithm to take a value P between 0 and 1, and return a 1 with probability P, and 0 otherwise.
2. Extend the above algorithm to accept fitness values of a set of candidates $\{P_1, P_2, ..., P_N\}$ and produce a new population where each candidate is cloned with a probability proportional to its fitness. A candidate may have more than one cloned child, and some candidates may have none.
3. Write a program to generate random instances of the TSP problem, given the number of cities N as an input. This entails generating the random coordinates of the N cities. Assume Euclidean distance as the cost of each edge. Implement and compare the performance of the following three sets of algorithms.
   a. Choose different perturbation operators for the path representation. Choose a starting temperature T, a cooling function, and a random starting tour. Implement the SIMULATEDANNEALING algorithm and plot the temperature and the tour cost with time.

b. Accept the population size P as an input, experiment with different crossover operators for solving the TSP with a GENETICALGORITHM. Plot the best and average tour costs with time.
   c. Accept the number of ants M as an input, and experiment with the three parameters α, β, and ρ used in the ACO algorithm. Plot the best and average tour costs with time.
4. A tour is shown in the figure below; the edges are bidirectional. Use A, B, C, D, E, F, G, H, I, and J as reference sequence (index sequence) for preparing ordinal and adjacency representations. Use this tour to answer the following questions:

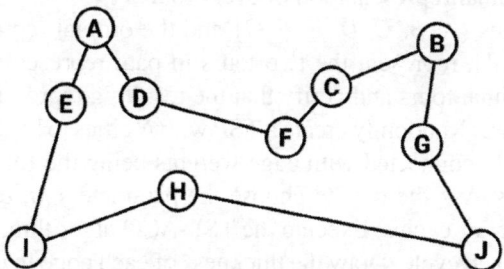

   a. Give two different path representations of the above tour.
   b. How many adjacency representations are there for the given tour? List all of them.
   c. For your answer to (a), what are the corresponding ordinal representations?
5. Consider 10 cities {A, B, C, D, E, F, G, H, I, J} and two parent tours (P_1 and P_2) in the path representation:

$$P_1: BIGCHEFJAD$$
$$P_2: FEDGABICHJ$$

   a. What is the ordinal representation for tour P_2? Take the alphabetical ordering of the cities as the reference sequence.
   b. While doing the CX with the two parents P_1 and P_2, identify the first cycle starting from city B in P_1.
   c. Identify the second cycle starting from city G in P_1.
   d. Construct a child C_1 from P_1 and P_2 using CX; select ODD cycles from P_1.
   e. Construct a child C_2 from P_1 and P_2 using CX; select EVEN cycles from P_1.
6. Convert the two parents *ODGLFHKMBJACNIE* and *HGMFNADKICOELBJ* into ordinal representation. Choose an arbitrary point between 1 and 15, and do a single point crossover on the two parents. Decode the resulting children back into path representation and verify that the two offspring are valid tours.
7. Convert the two parents *ODGLFHKMBJACNIE* and *HGMFNADKICOELBJ* into ordinal representation. Choose two arbitrary points between 1 and 15, and do a 2-point crossover on the two parents. In a 2-point crossover, the middle segment is exchanged between the parents to produce the two children. Decode the resulting children back into path representation and verify that the two offspring are valid tours.
8. Given the set of cities {A, B, C, D, E, F, G}, is there a tour whose ordinal representation is [1 1 1 1 1 1 1]? If yes, then what is the path representation of the tour? If no, explain why.

9. Given the set of cities {A, B, C, D, E, F, G}, is there a tour whose ordinal representation is [2 2 2 2 2 2 2]? If yes, then what is the path representation of the tour? If no, explain why.
10. Given the set of cities {A, B, C, D, E, F, G}, is there a tour whose ordinal representation is [7 6 5 4 3 2 1]? If yes, then what is the path representation of the tour? If no, explain why.
11. Given the set of cities {A, B, C, D, E, F, G}, is there a tour whose ordinal representation is [2 2 2 2 2 2 1]? If yes, then what is the path representation of the tour? If no, explain why.
12. Given the set of cities {A, B, C, D, E, F, G}, is there a tour whose ordinal representation is [2 1 2 1 2 1 1]? If yes, then what is the path representation of the tour? If no, explain why.
13. Is there a unique ordinal representation of every tour?
14. Given the set of cities {A, B, C, D, E, F, G} and the ordinal representations [2 1 2 1 2 1 1] and [7 6 5 4 3 2 1], represent the two tours in path representation. Do a single point crossover on the ordinal tours and verify that the resulting representations are valid tours.
15. Programming exercise: Randomly create a TSP with N cities to be displayed on the monitor. Let the graph be fully connected with edge weights being the Euclidean distance. Do not display all the edges. Ask the user to choose the parameters M, α, β, and ρ. Ask the user to choose the number of cycles. Execute the TSP-ACO algorithm, displaying the best tour found at the end of each cycle. Draw the thickness of each edge in proportion to the amount of pheromone on the edge. If there are any edges that have a higher amount of pheromone than the best ones on the tour, draw them too.
16. Trace the evolution of the following three patterns in Conway's Game of Life.

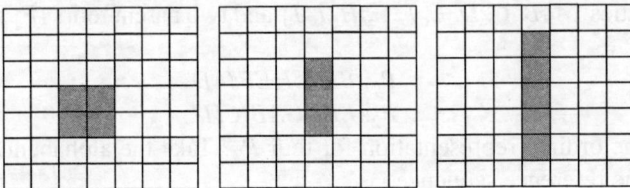

CHAPTER 6

Algorithm A* and Variations

Finding a solution is one aspect of problem solving. Executing it is another. In certain applications the cost of executing the solution is important. For example, maintaining supplies to the International Space Station, a repetitive task, or sending a rocket to Jupiter, an infrequent activity. Coming down to Earth, the manufacturing industry needs to manage its supplies, inventory, scheduling, and shipping of products. At home, juggling the morning activity of cooking, sending off kids to school, and heading for office after grabbing a coffee and a bite could do with optimized processes.

In this chapter we look at the algorithm *A** for finding optimal solutions. It is a heuristic search algorithm that guarantees an optimal solution. It does so by combining the goal seeking of *best first search* with a tendency to keep as close to the source as possible. We begin by looking at the algorithm *branch & bound* that focuses only on the latter, before incorporating the heuristic function.

We revert to graph search for the study of algorithms that guarantee optimal solutions. The task is to find a *shortest* path in a graph from a start node to a goal node. We have already studied algorithms BFS and DFID in Chapter 3. The key idea there was to extend that partial path which was the shortest. We begin with the same strategy. Except that now we add weights to edges in the graph. Without edge weights, the optimal or shortest path has the least number of edges in the path. With edge weights added, we modify this notion to the *sum of the weights* on the edges.

The common theme continuing in our search algorithms is as follows:

Pick the best node from *OPEN* and extend it, till you pick the goal node.

The question that remains is the definition of 'best'. In DFS, the deepest node is the best node. In BESTFIRSTSEARCH, the node that appears to be closest to the goal is the best. In BFS, the node closest to the start node is the best. We begin by extending the idea behind *breadth first search*.

We can generalize our common theme as follows. With every node *N* on OPEN, we associate a number that stands for the *estimated* cost of the *final* solution. For BESTFIRSTSEARCH, this is the estimated distance from *N* to the goal node, ignoring the cost up to node *N*. For BFS, it is, implicitly, the depth in the search tree. For the branch and bound algorithm described below, it is the sum of the edge costs from the start node to *N*. Here we ignore the cost beyond node *N*.

6.1 Branch & Bound

The branch & bound (B&B) algorithm extends the partial path that is the cheapest so far. We devise a new search space so that for every node *N* in the state space we represent the partial path from the start node *S* to *N*, along with the cost of that path. A search node is a pair *pathPair* of the form ([*N B A S*], 30) where the first element is a reversed list of the partial path from the start *S* to the node *N*, and the second element is the known cost of that partial path. B&B treats this known cost as the estimated cost of the solution, quite optimistically ignoring future costs. Like BFS it extends the cheapest partial path. The first cut of the algorithm is given below.

Algorithm 6.1. Algorithm B&B searches in the space of partial paths. It extends the cheapest partial path till it finds one to the goal.

B&B-FIRSTCUT(S)
1   OPEN ← ([S], 0) : [ ]
2   **while** OPEN **is not empty**
4   pathPair ← **head** OPEN
5   (path, cost) ← pathPair
6   N ← **head** path
6   **if** GOALTEST(N) = true
7   **then return reverse**(path)
8   **else**
9   children ← MOVEGEN(N)
10   newPaths ← MAKEPATHS(children, pathPair)
11   OPEN ← sort$_{cost}$(newPaths ++ **tail** OPEN)
12 **return empty list**

MAKEPATHS(children, pathPair)
1   **if** children **is empty**
2   **then return empty list**
3   **else**
4   (path, cost) ← pathPair
5   M ← **head** children
6   N ← **head** path
7   **return** ([M : path], cost + k(N, M)) : MAKEPATHS(**tail** children, pathPair)

The function *MakePaths* generates all children of the node N and extends the path $[N \ldots S]$ by adding each child M in turn to the head of the path, along with the cost of the path $[M\ N \ldots S]$. The new search nodes of the form $([M\ N \ldots S],\ \text{cost})$ are added to OPEN in Line 11. OPEN is maintained as a sorted list, or more efficiently, as a priority queue. We look at the performance of this algorithm on the tiny search problem depicted in Figure 6.1.

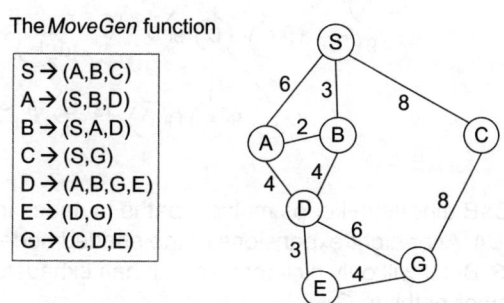

Figure 6.1 A tiny search space. S is the start node and G is the goal node. The labels on the edges are edge costs. Note that the placement of nodes in the diagram does not reflect edges costs.

The algorithm begins by adding the *pathPair* $([S], 0)$ to OPEN. This is the only node on OPEN. It is removed and the neighbours of S are added in a sorted order.

$$\text{OPEN} = [([B\ S], 3), ([A\ S], 6), ([C\ S], 8)]$$

The path SB with cost 3 is removed, and the neighbours of B are added.

$$\text{OPEN} = [([A\ B\ S], 5), ([A\ S], 6), ([D\ B\ S], 7), ([C\ S], 8)]$$

Observe that the algorithm has found two paths to the node A, one with cost 5 and the other with cost 6. It will next pick $([A\ B\ S], 5)$ for expansion. At this point one can argue that it should abandon and delete the other path $([A\ S], 6)$. We will investigate this a little later, when we will keep only one copy of a *node* instead of all the *paths* that lead to it. Meanwhile, we can observe that the simple strategy of extending the cheapest path at all times will lead to an unfettered explosion of the search space as depicted in Figure 6.2.

150 Search Methods in Artificial Intelligence

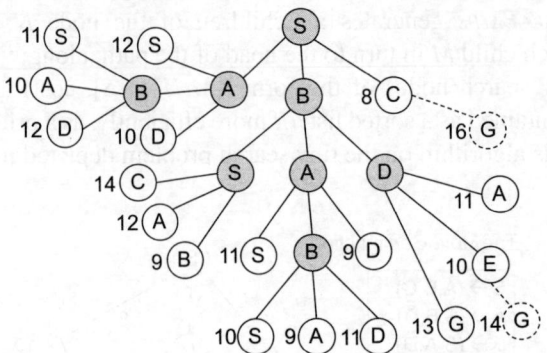

Figure 6.2 The first cut B&B algorithm keeps multiple paths to nodes in its search space including loops like *SBABA*. After eight expansions of the shaded nodes, it has found one path to the goal *G* with cost 13. But it will only pick that when it has exhausted all cheaper paths, in the process generating other paths to *G*.

When there is a collection of short edges in the state space, then the algorithm goes into loops because of its strategy of always extending the cheapest path. In our example, the nodes *S*, *A*, *B*, and *D* have short edges connecting them. Apart from the wasted work, there is a larger danger lurking here. While the algorithm is guaranteed to find the shortest path if there is one, if there is no path to the goal it will keep looping endlessly searching for one. We saw this danger lurking in algorithm DFID in Chapter 3. One way to mitigate this problem is to stop the search from going into loops. This can be done by checking if a new child already exists on the partial path being extended. Algorithm 6.2 is a version of B&B which does that.

Algorithm 6.2. A version of B&B that avoids getting into loops. If a child of *N* is already present in the path, then that is discarded.

B&B(S)
```
 1   OPEN ← ([S], 0) : [ ]
 2   while OPEN is not empty
 4           pathPair ← head OPEN
 5           (path, cost) ← pathPair
 6           N ← head path
 6           if GOALTEST(N) = true
 7               then return reverse(path)
 8           else
 9               children ← MOVEGEN(N)
10               noloops ← REMOVESEEN(children, path)
11               newPaths ← MAKEPATHS(children, pathPair)
```

```
12                    OPEN ← sort_cost(newPaths ++ tail OPEN)
13    return empty list

REMOVESEEN(children, path)
  1  if children is empty then return empty list
  3  else
  4    M ← head children
  5    if OCCURSIN(M, path) then return REMOVESEEN(tail children, path)
  6    else return M : REMOVESEEN(tail children, path)

OCCURSIN(node, list)
  1. if list is empty then return False
  2. else if node == head list then return True
  3.    else return OCCURSIN(node, tail list)
```

The function *REMOVESEEN*, similar to the one we saw in Chapter 3, removes any nodes that are looping back to the path. It does so by calling *OCCURSIN* which checks if a node is already in a given list. Figure 6.3 is the same search space as in Figure 6.2 except that loops have been removed by the modified B&B algorithm.

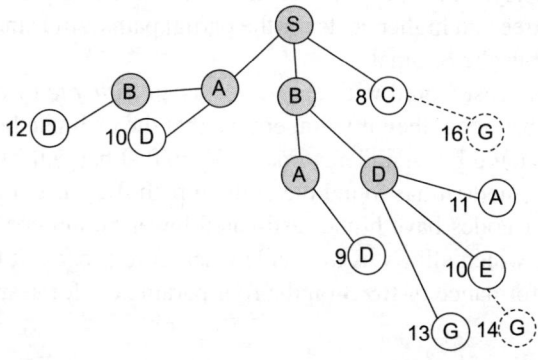

Figure 6.3 The pruned search space from Figure 6.2 generated by B&B with loop checking. Values next to nodes on OPEN are known path costs.

With the above modification, the B&B algorithm will not enter an infinite loop. It will still explore all distinct paths without loops. Even this can be wasteful. For example, the algorithm will first extend the path *SBAD* which has cost 9 to *G* with cost 15, before it eventually picks the node *G* on *SBDG* with cost 13. This is because of the implicit assumption that the known cost of *SBAD*, viz. 9, is the estimated cost of the final solution, which is less than 13. This happens because it does not know that the node *D* in the two options is the same. If we store the graph itself, keeping exactly one copy of any node, then it would circumvent this problem. Figure 6.4 shows the search space at the moment when B&B is about to finally pick the goal node from OPEN with cost 13. As can be seen, it has found the optimal path.

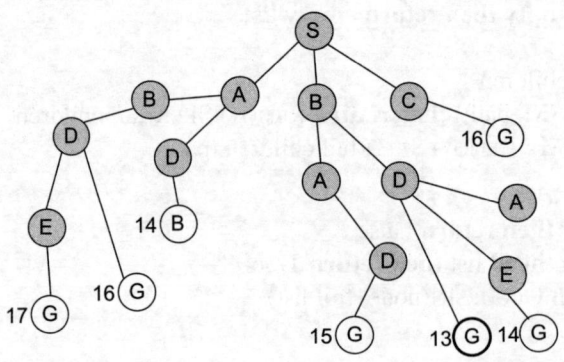

Figure 6.4 The moment when B&B is about to pick the goal node with path cost 13. Observe that all other nodes on OPEN have a higher cost.

One observation is key. At the point where the B&B algorithm picked the goal G, all other branches in the search tree had higher costs for the partial paths. And since those costs can only go up, the path found must be optimal.

If one is to treat the cost stored with each node as an *estimate* of the cost of the optimal solution leading from that node, then it is imperative that the estimate be a lower bound on the actual cost. That is, it should be an underestimate. When that happens, then at the moment the algorithm picks the goal node it has found the optimal path. Because it is the lowest cost node on OPEN, and all other nodes have higher estimated lower bound costs. In B&B, we have a *bound* on every *branch* which allows us to decide whether to process it further.

We look at the performance of B&B on the four parameters for search algorithms.

6.1.1 Performance of B&B

1. Completeness: The algorithm is complete. It will even terminate on infinite graphs if there is a path from the start node to the goal. This is because it explores paths in increasing order of cost and will eventually find a path to a goal node.
2. Quality of solution: For the same reason, it always finds the optimal path to the goal node. This is because it picks the least cost path to goal first.
3. Space complexity: The algorithm needs exponential space since it keeps all partial paths in a tree at all times. In each cycle, it extends the cheapest partial path.
4. Time complexity: Time complexity is exponential in depth. This is because it combinatorially explores all partial paths of cost less than the optimal path to the goal.

We will return to graph search a little later and prove that underestimation leads to optimal solutions. Before that we look at an application of B&B to the travelling salesperson problem (TSP).

6.1.2 B&B on the TSP

The TSP problem is an optimizing configuration problem we have studied earlier. We now look at a constructive approach in the solution space. We begin with a search node representing all possible solutions. The move at any point is to pick a node in our search space, an edge E in the problem graph, and partition the node into two, with one child containing the set of all tours that include the edge E, and the other the set of all tours that exclude E. This process is called refinement. Eventually a node will be fully refined to contain a single tour.

The key question is *which* node to pick for refinement. The answer is to pick that node which has the lowest cost estimate. To this end, we associate a cost with each node, which is an estimate of the cost of the best tour in that node. For reasons discussed earlier, it is necessary that the estimated cost is a lower bound. When this is done, the following high level algorithm finds the optimal tour.

While the best node is not a singleton set, choose the *cheapest available edge* and split the node into two

Observe that in addition to using the B&B strategy of extending the cheapest partial solution, we have also incorporated a heuristic that the *cheapest available edge* should be selected. This introduces a forward looking heuristic and anticipates the algorithm A* discussed below. There are two choices our algorithm makes. One, the node to refine. Here we follow the backward looking least cost principle of B&B. Two, the edge to add to the selected node. Here we use the heuristic that the shortest edge, as used on the GREEDY algorithm discussed in Section 4.1.5, is the best.

We illustrate this algorithm with a tiny TSP defined by the following distance matrix (Table 6.1). We have not included the zero cost from any city to itself in the matrix.

Table 6.1 A tiny TSP

Cities	A	B	C	D	E
A		2	6	100	110
B	2		4	80	90
C	6	4		60	70
D	100	80	60		10
6E	110	90	70	10	

We begin with a search node S containing all tours. The value associated with this node is an estimate of cost of the best tour in S. One way to estimate the absolute lower bound for the cost of any TSP is to associate each city with the two shortest edges emanating from it. This follows from the fact that in a valid tour each city is connected to exactly two cities. This will clearly be a lower bound.

The following are the two cheapest edges and the costs associated with each city in our tiny problem:

A: COST(AB) + COST(AC)   = 2 + 6   = 8
B: COST(AB) + COST(BC)   = 2 + 4   = 6
C: COST(AC) + COST(BC)   = 6 + 4   = 10
D: COST(DE) + COST(DC)   = 10 + 60   = 70
E: COST(DE) + COST(EC)   = 10 + 70   = 80

COST(S) = (8 + 6 + 10 + 70 + 80) / 2 = 174/2 = 87

Following the heuristic, we select the cheapest edge AB and partition S into two. One, called S_{AB}, includes the edge AB, and the other, $S_{\overline{AB}}$, excludes it. The estimated cost of S_{AB} is same as cost of S since edge AB is counted in S. For $S_{\overline{AB}}$ the edge $\overline{AB}$ must be excluded. The following are now the two cheapest edges and the costs associated with each city:

A: COST(AD) + COST(AC)   = 100 + 6   = 106
B: COST(DB) + COST(BC)   = 80 + 4   = 84
C: COST(AC) + COST(BC)   = 6 + 4   = 10
D: COST(DE) + COST(DC)   = 10 + 60   = 70
E: COST(DE) + COST(EC)   = 10 + 70   = 80

COST($S_{\overline{AB}}$) = (104 + 86 + 10 + 70 + 80) / 2 = 350/2 = 175

Clearly S_{AB} has a lower estimated cost. We repeat the process by selecting the next available cheapest edge BC. This gives us two sets: $S_{AB,BC}$ that includes both AB and BC, and $S_{AB,\overline{BC}}$ that includes AB but excludes BC. The cost of $S_{AB,BC}$ remains the same.

$$S_{AB,BC} = S_{AB} = S$$

For $S_{AB,\overline{BC}}$, we need to exclude edge BC.

A: COST(AB) + COST(AC)   = 2 + 6   = 8
B: COST(AB) + COST(BD)   = 2 + 80   = 82
C: COST(AC) + COST(DC)   = 6 + 60   = 66
D: COST(DE) + COST(DC)   = 10 + 60   = 70
E: COST(DE) + COST(EC)   = 10 + 70   = 80

COST($S_{AB,\overline{BC}}$) = (6 + 8 + 10 + 70 + 80) / 2 = 306/2 = 153

The search space at this stage is shown in Figure 6.5.

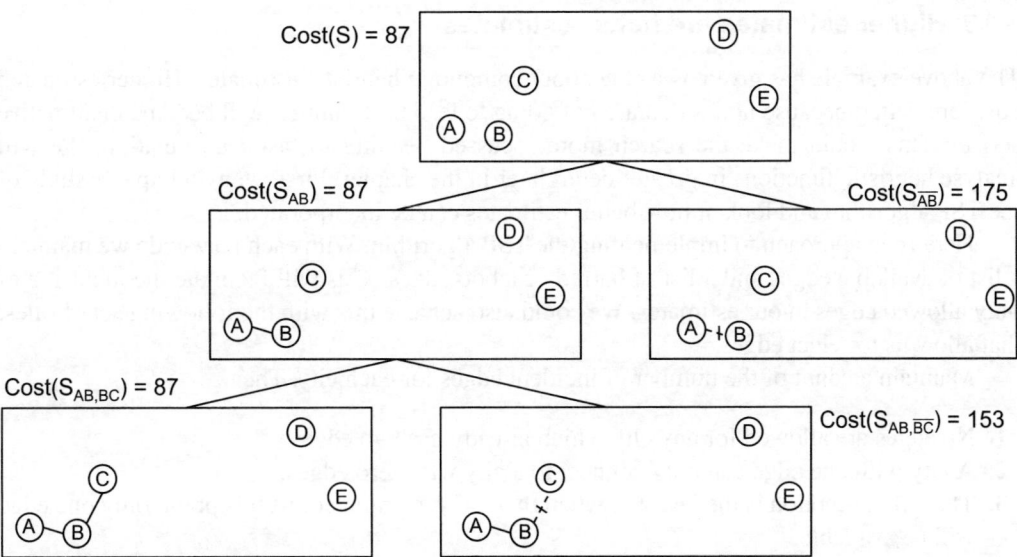

Figure 6.5 The search space in B&B TSP after two expansions.

It can be seen that the best node is $S_{AB,BC}$ with estimated cost 87, and our search will proceed to expand that. It cannot add the next shortest edge AC because that would result in a premature loop. It will instead select the edge DE and continue.

The reader is encouraged to verify that after DE the next edge chosen will be CD, and then the only option would be to select AE. When the last edge is chosen, the cost of the tour will shoot up, because AE is the most expensive edge, and search will have to abandon this node, and shift to the next cheapest node.

One can see that the estimate for $S_{AB,BC}$ was overly optimistic. This is because the two edges being considered for A in this step are AB and AC with costs 2 and 6 respectively. Looking at Figure 6.5 one can see that the edge AC cannot be part of a tour that already has AB and BC, because together they form a subtour. Thus, for $S_{AB,BC}$ we need to exclude edge AC.

A: $\text{Cost}(AB) + \text{Cost}(AD)$ = 2 + 100 = 102
B: $\text{Cost}(AB) + \text{Cost}(BC)$ = 2 + 4 = 6
C: $\text{Cost}(CD) + \text{Cost}(BC)$ = 60 + 4 = 64
D: $\text{Cost}(DE) + \text{Cost}(DC)$ = 10 + 60 = 70
E: $\text{Cost}(DE) + \text{Cost}(EC)$ = 10 + 70 = 80

Revised $\text{Cost}(S_{AB,BC})$ = (8 + 6 + 10 + 70 + 80)/2 = 322/2 = 161

Now that the cost of $S_{AB,BC}$ has been revised upwards to 161, the algorithm will instead shift its attention to $S_{AB,\overline{BC}}$ with cost 153. The reader is encouraged to verify that this will indeed lead to the optimal solution.

6.1.3 Higher estimates are better estimates

The above example has given us a clue about computing heuristic estimates. Higher estimated costs are better, because nodes that are not going to lead to a solution will become unattractive sooner. This would make the search more focussed, leading to faster termination. We will analyse heuristic functions in greater detail later in the chapter. First, we wind up our study of the TSP algorithm and look at how better estimates can be incorporated.

Here is an approach to implementing the TSP algorithm. With each new node we maintain a list of available edges and a list of barred or taboo edges. This will facilitate the inclusion of only allowed edges in our estimates. We could also achieve this with the following set of rules, that allow us to select edges.

Maintain a count of the number of incident edges for each city. Then,

1. No edges are allowed for any city which already has two edges.
2. A city with one edge can only connect to a city with zero edges.
3. The only exception is the last step when the tour is completed. At this point, only one edge will be available.

The above rules apply to selection of edges both for refinement and for estimation. Why are higher estimates better? By excluding cheaper disallowed edges in the estimates, we get a more accurate estimate, which is higher. That branch will not make an unrealistic claim for being picked from OPEN.

On the other hand, computing higher estimates involves more work, where one has to do additional reasoning. But the more accurate the heuristic function, the lower is the amount of search one has to do, because it has a better sense of direction. Choosing between the amount of effort on the heuristic function versus the effort doing search is often a delicate balance.

6.2 Algorithm A*

Algorithm B&B has no sense of direction. It proceeds to the next nearest candidate and checks whether it is the goal node. While correct, this can be extremely inefficient. Consider the problem of finding the route from IIT Madras in Chennai to the Shore Temple in neighbouring Mahabalipuram. B&B would exhaust every nook and corner of Chennai city before heading out and south. It is restrained by the desire to stay as close to source as possible. What one also needs is a pull towards the goal, like in BESTFIRSTSEARCH (Chapter 4).

6.2.1 Dijkstra's algorithm

Another source of inefficiency in B&B is the fact that each partial path was a node in the search space, leading to a proliferation of paths to any node. This is tackled much better in Dijkstra's algorithm (Dijkstra, 1959; Cormen et al., 2001) which, though, is an algorithm to find the shortest path to all nodes from a single source. One could easily modify it to terminate when a goal node is found. We describe the algorithm informally here. The input to the algorithm is the complete graph at the very outset.

- Dijkstra's algorithm begins by assigning infinite cost estimates to all nodes except the start node, which has cost zero.
- It assigns the colour white to all the nodes initially.
- It picks the cheapest white node and colours it black.
- Relaxation: Inspect all neighbours of the new black node and check if a cheaper path has been found to any of them. If yes, then update the cost of that node, and mark its new parent.

Dijkstra's algorithm reverts to maintaining the graph. Instead of keeping a *nodePair* or a *nodePath*, it marks the parent of each node visited by a pointer *parent*(node). This is first set to the parent that generated this node, but may be reassigned if a cheaper path is found later from another parent. Since it picks the *cheapest* 'white' node (read node on OPEN), it has found the optimal path to that node. Figure 6.6 illustrates this for our tiny search problem (Figure 6.1).

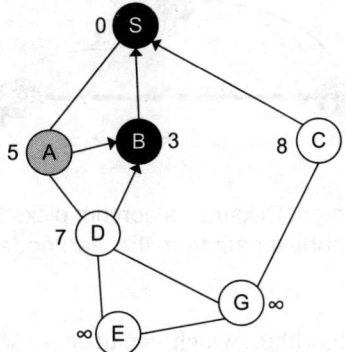

Figure 6.6 Dijkstra's algorithm first generates node *A* as a child of *S* and marks *S* as a parent. When it colours and relaxes node *B* black (adds to CLOSED), it also finds a cheaper path to *A* and becomes the parent of *A*. It will next colour node *A* which has the lowest cost on OPEN.

Dijkstra's algorithm was devised to find shortest paths to all nodes given a graph as input. In our search task we have a goal node or goal specification in mind, and a move generation function to generate the graph on demand. The problem graph could in principle have infinite nodes, as long as there is a goal at a finite distance.

Figure 6.7 shows the problem graph from Figure 4.7 at the moment Dijkstra's algorithm picks the goal node. Observe that the algorithm has ended up searching the entire graph, but found the optimal path with cost 148, where BESTFIRSTSEARCH had found a path with cost 195 after inspecting only eight nodes.

158 | Search Methods in Artificial Intelligence

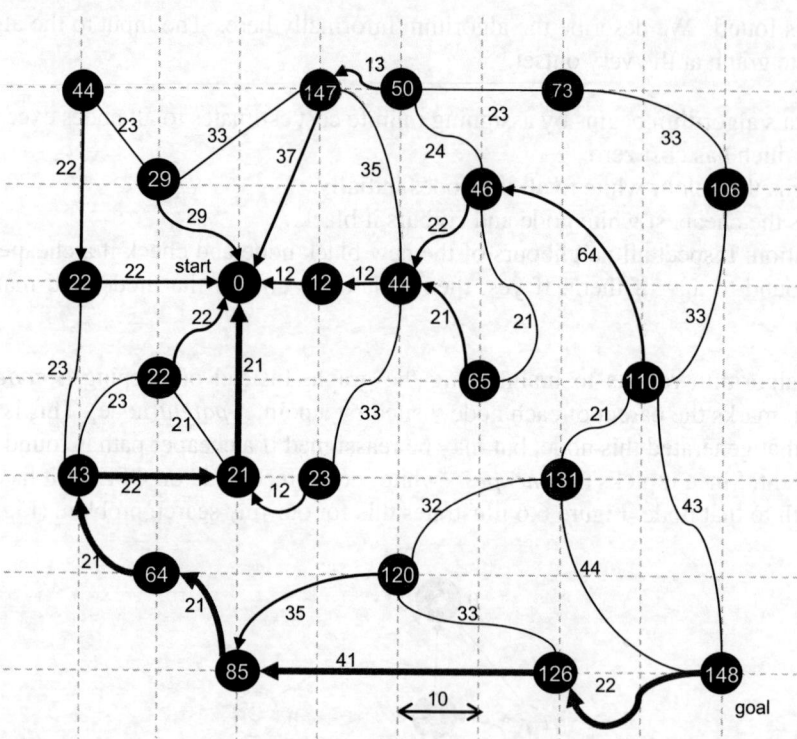

Figure 6.7 The graph at the moment Dijkstra's algorithm picks the goal node. The values in the nodes are the costs of the shortest path from the start node. Observe that it has visited all 23 nodes in this graph.

We now introduce the A* algorithm, which imparts a sense of direction to search. It is one of the most important algorithms we study in this book.

6.2.2 A*

The algorithm A*, first described by Hart, Nilsson, and Raphael (Hart, Nilsson, and Raphael, 1968; Nilsson, 1971, 1980) is a heuristic search algorithm that has found wide-ranging applications.

Algorithm A* combines several features of the algorithms we have studied so far.

1. Like BESTFIRSTSEARCH it employs a heuristic function $h(n)$ that is an estimate of the distance to the nearest goal. This gives search a sense of direction, so that it is focussed on the goal. For simplicity, we will assume there is one goal node in our description.
2. Like B&B it keeps track of the known distance from the start node. This is done by a function named $g(n)$. Preferring low g-values results in a tendency to stay close to the start node. This is necessary since we are interested in shortest paths.
3. Like Dijkstra's algorithm it works with a graph representation, keeping exactly one copy of every node. Associated with each node is a function $parent(n)$ that points to the

parent node. Like Dijkstra's algorithm it marks the parent which is on the shortest path to n. The parent pointer is instrumental in reconstructing the path.

Every node n in A* has an estimated cost $f(n)$ of the path from the start node to the goal node passing through n.

$$f(n) = g(n) + h(n)$$

That is, $f(n)$ is the estimated cost of the solution containing node n. The algorithm always refines or expands the node with the lowest f-value. The search space for A* is depicted in Figure 6.8.

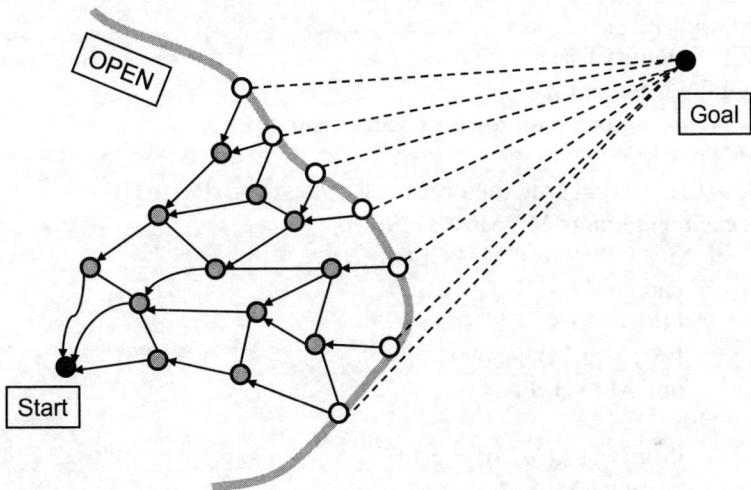

Figure 6.8 The search space for A* is the state space itself. The estimated cost $f(n)$ for every node on OPEN has two components. One, $g(n)$, is the *known* cost from the start to n. The second, $h(n)$, is an *estimated* distance to the goal.

Like our earlier algorithms A* picks the best node n from OPEN and checks whether it is the goal node. If it is, then it follows the pointers back to reconstruct the path. If it is not, it adds the node to CLOSED and generates its neighbours.

For each neighbour m of n it does the following:

1. If m is a new node, it adds it to OPEN with parent n and $g(m) = g(n) + k(n, m)$ where $k(n, m)$ is the cost of the edge connecting m and n.
2. If m is on OPEN with parent n', then it checks if a better path to m has been found. If yes, then it updates $g(m)$ and sets its parent to n.
3. If m is on CLOSED with parent n', then it checks if a better path to m has been found. If yes, then it updates $g(m)$ and sets its parent to n. This possibility exists since $h(n)$ is an estimate and may be imperfect, choosing n' before n. The algorithm will also need to propagate this improved cost to other descendants of m.

The algorithm A* is described below. We have been deliberately vague about the lists OPEN and CLOSED. For small problems, one could implement them as lists. For larger problems, it would be prudent to implement OPEN as a priority queue and CLOSED as a hash table.

Algorithm 6.3. Algorithm A* maintains one copy of each node either in OPEN or in CLOSED.

A*(S)
1. parent(S) ← **null**
2. g(S) ← 0
3. f(S) ← g(S) + h(S)
4. OPEN ← S : []
5. CLOSED ← **empty list**
6. **while** OPEN **is not empty**
7.   N ← **remove node with lowest f-value from OPEN**
8.   add N to CLOSED
9.   **if** GoalTest(N) = True **then return** ReconstructPath(N)
10.   **for each neighbour** M ∈ MoveGen(N)
11.   **if** (M ∉ OPEN **and** M ∉ CLOSED)
12.   parent(M) ← N
13.   g(M) ← g(N) + k(N,M)
14.   f(M) ← g(M) + h(M)
15.   **add** M **to** OPEN
16.   **else**
17.   **if** (g(N) + k(N,M)) < g(M)
18.   parent(M) ← N
19.   g(M) ← g(N) + k(N,M)
20.   f(M) ← g(M) + h(M)
21.   **if** M ∈ CLOSED
22.   PropagateImprovement(M)
23.   **return empty list**

PropagateImprovement(M)
1. **for each neighbour** X ∈ MoveGen(M)
2.   **if** g(M) + k(M,X) < g(X)
3.   parent(X) ← M
4.   g(X) ← g(M) + k(M,X)
5.   f(X) ← g(X) + h(X)
6.   **if** X ∈ CLOSED
7.   PropagateImprovement(X)

We have used the *ReconstructPath* function from Chapter 3, though it may need minor tweaks given the different representation. While implementing the algorithm one may want to also return the cost of the solution found.

Figure 6.9 illustrates the algorithm on the problem graph from Figure 4.7. As done with BestFirstSearch we use the Manhattan distance heuristic function as the heuristic function $h(n)$. We do this for ease of manual computation (especially in exam papers). In an implementation, one would use the Euclidean distance.

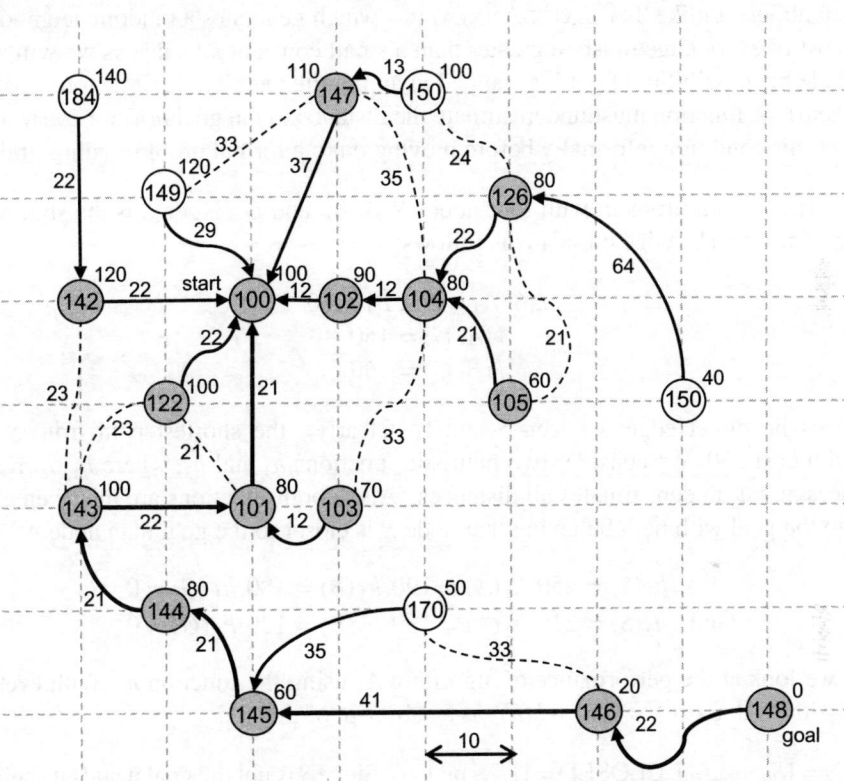

Figure 6.9 The graph generated by algorithm A* when it picks the goal node. Shaded nodes are on CLOSED. The values in the nodes are the *f*-values, and the values outside the nodes are the Manhattan distance $t - l$ values. A* has inspected 14 nodes and found the optimal path with cost 148.

As can be observed, A* inspects much fewer nodes than Dijkstra's algorithm, but still finds the same optimal path. It did inspect more nodes than BestFirstSearch but found a much better path. What is the secret of its success?

6.2.3 A* is admissible

The star in the name A* is indicative of the fact that the algorithm is admissible. This means that A* always finds the optimal path when a path to the goal exists. This is true even if the graph is infinite under the following conditions:

1. The MoveGen or neighbourhood function has a finite branching factor. Clearly, with infinite branching, it would not be able to even generate the neighbours. It does generate all neighbours, unlike SimulatedAnnealing which generates a random neighbour.
2. The cost of every edge must be greater than a small constant ε.[1] This, as we will see, is to preclude the possibility of getting trapped in an infinite path with a finite total cost.
3. The heuristic function must underestimate the distance to the goal $h(n)$ for every node. We look at this condition informally before moving on to a formal proof of admissibility.

Consider a tiny search problem with four nodes S, A, B, and G. Node S is the start node and G is the goal node. The edge costs are as follows:

$$k(S, A) = k(S, B) = 80$$
$$k(A, G) = 150$$
$$k(B, G) = 160$$

There is no direct edge between S and G. Clearly, the shortest path from S to G is S–A–G with cost 230. We consider two heuristic functions h_1 and h_2, where h_1 overestimates all distances and h_2 underestimates all distances. Also, both functions are mistaken about the distance to the goal with both believing that node B is closer to the goal than node A is. Let

$$h_1(S) = 250, h_1(A) = 190, h_1(B) = 180, h_1(G) = 0$$
$$\text{and} \quad h_2(S) = 210, h_2(A) = 130, h_2(B) = 120, h_2(G) = 0$$

First we look at the performance of algorithm A_1 using the function h_1. With every node, we display its f-value ($f(n) = g(n) + h(n)$) as a subscript of the node.

1. OPEN = $[S_{0+250=250}]$, CLOSED = []. A_1 picks S. Since S is not the goal it adds its neighbours A and B to OPEN. S itself is added to the CLOSED.
2. OPEN = $[A_{80+190=270}, B_{80+180=260}]$, CLOSED = $[S_{250}]$. A_1 picks node B with $f(B) = 260$. It adds its neighbour G with $g(G) = 80+160 = 240$ and $f(G) = 240 + 0 = 240$. Parent(G) = B.
3. OPEN = $[G_{240+0=240}, A_{80+190=270}]$, CLOSED = $[S_{250}, B_{260}]$. A_1 picks node G with $f(G) = 240$ and terminates.

A_1 has failed to find the optimal path and is therefore undeserving of being decorated with a star. Next we look at the performance of A_2* using the function h_2.

1. OPEN = $[S_{0+210=210}]$, CLOSED = []. A_2* picks S.

---

[1] This was observed by an alert student, Arvind Narayanan, during my class in the mid-1990s. Traditional wisdom then was that the edge cost just be greater than zero.

2. OPEN = $[A_{80+130=210}, B_{80+120=200}]$, CLOSED = $[S_{210}]$. A_2* picks node B with $f(B) = 200$. It adds its neighbour G with $g(G) = 80+160 = 240$ and $f(G) = 240 + 0 = 240$. $Parent(G) = B$.
3. OPEN = $[A_{80+130=210}, G_{240+0=240}]$, CLOSED = $[S_{210}, B_{200}]$. A_2* picks node A and finds a better path to G with $g(G) = 80+150 = 230$ and $f(G) = 230 + 0 = 230$. It updates the value of G and resets the parent of G to A.
4. OPEN = $[G_{230+0=230}]$, CLOSED = $[S_{210}, B_{200}, A_{210}]$. A_2* picks node G with cost 230 and terminates.

As one can observe, both versions of the algorithm were misinformed and wrongly picked the node B and found a longer path to G first. For A_1 the cost 240 of S–B–G appeared better than the overestimated f-value 270 of node A and it terminated. For A_2* the same cost 240 of S–B–G appeared worse than the underestimated value 210 of node A and it expanded node A to find the optimal path S–A–G with cost 230. The differentiating factor was the underestimation done by h_2.

The fact that A_2* needed four steps to terminate against the three needed by A_1 is directly related to the fact that the estimate h_2 is consistently lower than the estimate h_1. This is similar to what we saw in the TSP example in the previous section where the version with a *higher* estimate terminated faster. The difference between this example and the TSP example is that the latter found the optimal tour in both cases. This leads us to the conjecture that higher estimates are better but only up to a certain point. That point is the threshold set by the actual optimal cost. Within this threshold, the higher the estimate the better. We formalize this notion in the next section.

6.2.4 Proof of admissibility

We present a proof of admissibility of A* as described in (Rich and Knight, 1991; Nilsson, 1998). We begin by introducing some additional terminology.

Let g*(n) be the optimal path cost from the start node S to node n. Observe that this is not known in general. What we do know is $g(n)$ which is the cost of the path found by the algorithm. Observe that g*$(n) \leq g(n)$ because the algorithm may not have found the optimal path.

Let h*(n) be the optimal path cost from the node n to a goal node G. Again, this is not a known quantity. Then f*$(n) = g$*$(n) + h$*(n) is the optimal cost of a path from S to G via node n. This value is the same for any node on an optimal path. In particular, f*$(S) = h$*(S) stands for the cost of the optimal path.

In addition, we have the conditions stated above for admissibility, repeated here.

1. For all n: |MOVEGEN(n)| < b for some value of b.   - C1
2. For every edge <n, m>: $k(n,m) > \varepsilon$ - C2
3. For all n: $h(n) \leq h$*(n) - C3

Given the above, we prove a series of lemmas leading to the proof of admissibility of A*.

L1: The algorithm always terminates for finite graphs.

Proof: The algorithm keeps exactly one copy of every node generated, either in OPEN or in CLOSED. In every cycle, it picks one node from OPEN and moves it to CLOSED if it is not a goal node. It also adds some previously not seen nodes to OPEN. Since the total number of nodes is finite, there will be eventually none left to add to OPEN. If the goal node is not in the graph, OPEN will become empty and the algorithm will terminate.

L2: If a path exists from the start node to the goal node, then at all times before termination OPEN always has a node n' on an optimal path. Furthermore, the *f*-value of this node is optimal or less.

Proof: Let $(S, n_1, n_2, ..., G)$ be such an optimal path. The algorithm begins by adding S to OPEN. When S is removed from OPEN, its successor n_1 is added to OPEN, and when that is removed then n_2 is added, and so on. If G is removed, then the algorithm has terminated. Else let n' be the node on OPEN.

Then,

$$\begin{aligned} f(n') &= g(n') + h(n') \\ &= g^*(n') + h(n') \quad \text{because } n' \text{ is on the optimal path } g(n') = g^*(n') \\ &\leq g^*(n') + h^*(n') \quad \text{because } h(n') \leq h^*(n') \text{ by C3} \\ &\leq f^*(n') \\ &\leq f^*(S) \text{ because } n' \text{ is } on \text{ the optimal path } f^*(n') = f^*(S) \end{aligned}$$

$$\therefore f(n') \leq f^*(S)$$

L3: If there exists a path from the start node to the goal, A* finds a path. This is true even if the graph is infinite.

Proof: A* always picks a node n with the lowest *f*-value and extends the path to one or more neighbour m of n. From C1 the number of neighbours is finite. From C2 each extended path cost increases at least by ε. This means that each of the finite number of paths gets extended by a non-infinitesimal amount and will eventually become greater than $f^*(S)$. Since $f(G) \leq f^*(S)$, the algorithm will eventually pick the goal G and terminate.

L4: A* finds the least cost path to the goal.

Proof (by contradiction): Let A* terminate with node G' with cost $g(G') > f^*(S)$.

This could not happen. Because by L2 there is always a cheaper node n' in OPEN that is on the optimal path. It would instead pick n'. Therefore, A* terminates by finding the optimal cost path.

L5: For every node n expanded by A*, $f(n) \leq f^*(S)$

Proof: A* picked node n in preference to node n'. Therefore,

$$f(n) \leq f(n') \leq f^*(S)$$

6.2.5 Higher is better

We have seen anecdotal evidence from two examples that higher heuristic estimates result in a more focussed search and faster execution. We look at a formal argument to support this fact. In this section we restrict our heuristic function to underestimating actual costs, maintaining the guarantee of optimal solutions. Later we will explore the impact of still higher estimates.

L6: Let A_1^* and A_2^* be two admissible versions of A* and let $h^*(n) > h_2(n) > h_1(n)$ for all n. We say h_2 is *more informed* than h_1, because it is closer to the h^* value. Then the search space explored by A_2^* is contained in the space explored by A_1^*. Every node visited by A_2^* is also visited by A_1^*. Whenever a node is visited it is also added to CLOSED.

Proof (by induction): To show that if $n \in \text{CLOSED}_2$, then $n \in \text{CLOSED}_1$.

Base step: if $S \in \text{CLOSED}_2$, then $S \in \text{CLOSED}_1$.

Induction hypothesis: Let the statement be true for all nodes up to depth d. If $n \in \text{CLOSED}_2$, then $n \in \text{CLOSED}_1$.

Induction step (by contradiction):

Assumption: Let L be a node at depth $(d+1)$ such that $L \in \text{CLOSED}_2$ and let A_1^* terminate without inspecting L.

Since A_2^* has picked node L,

$$f_2(L) \leq f^*(S) \quad \text{from L5}$$

That is $\quad g_2(L) + h_2(L) \leq f^*(S)$

or $\quad h_2(L) \leq f^*(S) - g_2(L)$

Now, since A_1^* terminates without picking node L,

$f^*(S) \leq f_1(L)$ because otherwise A_1^* would have picked L

or $\quad f^*(S) \leq g_1(L) + h_1(L)$

or $\quad f^*(S) \leq g_2(L) + h_1(L)$

because $g_1(L) \leq g_2(L)$ since A_1^* has seen all nodes up to depth d seen by A_2^*, and would have found an equal or better cost path to L.

We can rewrite the last inequality as

$$f^*(S) - g_2(L) \leq h_1(L)$$

We already have $h_2(L) \leq f^*(S) - g_2(L)$.

Combining the above two, we get

$$h_2(L) \leq h_1(L)$$

which contradicts the given fact that $h_2(n) > h_1(n)$ for all nodes.

The assumption that A_2^* terminates without expanding L must be false, and therefore A_2^* must expand L. Since L was an arbitrary node picked at depth $d+1$, the following is true at depth $d+1$ as well:

If $n \in \text{CLOSED}_2$, then $n \in \text{CLOSED}_1$

Hence by induction, it is true at all levels.

6.2.6 The monotone condition

When Dijkstra's algorithm picks a node from OPEN, it has already found the optimal path to that node. In contrast, when A* picks a node and adds it to CLOSED, it may yet find another shorter path to it. That is why the algorithm has the steps to update g-values of nodes on CLOSED as well. The reason why A* may not have found the optimal cost to a node is that it picks nodes based on f-values which have an h-value that could be inaccurate in its estimates. The condition that $h(n)$ underestimate the distance to the goal is sufficient for admissibility, and it will find an optimal path to the goal. To find an optimal path to every node *en route*, it needs a stricter condition. Let n be a successor of m on an optimal path to the goal. The monotone or consistency condition is

$$h(m) - h(n) \leq k(m,n)$$

Remember that the heuristic function is an estimate of the distance to the goal. As one moves towards the goal, the h-value is expected to decrease. The monotone condition says that this decline cannot be greater than the edge cost. One can think of $h(m) - h(n)$ as the cost of the m–n edge as estimated by the heuristic function. This too must be an underestimate.

Rearranging the above inequality, we get

$$h(m) \leq h(n) + k(m, n)$$

Adding $g(m)$ to both sides,

$$h(m) + g(m) \leq h(n) + k(m, n) + g(m)$$
$$\text{or } h(m) + g(m) \leq h(n) + g(n)$$
$$\text{or } f(m) \leq f(n)$$

That is, as we move towards the goal, the estimated cost of the solution increases, becoming closer to the optimal cost. A more significant consequence of the monotone condition is that when A* picks a node n from OPEN, it has already found the optimal path from the start node S to n. That is, $g(n) = g^*(n)$. We look at the proof.

Let A* be about to pick node n with a value $g(n)$. Let there be an optimal path P from S to n which is yet unexplored fully. On this path P let n_L be the last node on CLOSED and let n_{L+1} be its successor on OPEN. Given the monotone condition,

$$h(n_L) + g(n_L) \leq h(n_{L+1}) + g(n_{L+1})$$
$$\text{or } h(n_L) + g^*(n_L) \leq h(n_{L+1}) + g^*(n_{L+1})$$

because both are on the optimal path. By transitivity, this inequality extends to node n.

$$h(n_{L+1}) + g^*(n_{L+1}) \leq h(n) + g^*(n)$$

Algorithm A* is about to pick node n when n_{L+1} is on OPEN. Hence,

$$f(n) \leq f(n_{L+1})$$
$$\text{or } h(n) + g(n) \leq h(n_{L+1}) + g^*(n_{L+1})$$

This gives us

$$h(n) + g(n) \leq h(n) + g^*(n)$$
$$\text{or } g(n) \leq g^*(n)$$

But $g(n)$ cannot be less than $g^*(n)$ which is the optimal cost from S to n. Therefore,

$$g(n) = g^*(n)$$

A direct consequence of this is that A* does not have to update $g(n)$ for nodes of CLOSED. This allows us to implement some space saving versions of A* which prune CLOSED drastically. We look at space saving versions later in this chapter.

6.2.7 Performance of A*

1. Completeness: The algorithm is complete. As shown earlier, it will even terminate on infinite graphs if there is a path from the start node to the goal.
2. Quality of solution: As shown earlier, A* is admissible. Given an underestimating heuristic function, it always finds the optimal path to the goal node.
3. Space complexity: The space required depends upon how good the heuristic function is. Given a perfect heuristic function, the algorithm heads straight for the goal, and the OPEN list will grow linearly. However, in practice, perfect heuristic functions are hard to come by, and it has been experimentally observed that OPEN tends to grow exponentially.
4. Time complexity: The time complexity is also dependent on how good the heuristic function is. With a perfect function, it would be linear. In practice, however, it does a fair bit of exploration, reflected by the size of CLOSED, and generally needs exponential time as well.

We now explore variations of A*. First we look at a variation that compromises on the quality of the solution to gain on time and space complexity.

6.2.8 Weighted A*

We have observed that A* explores more of the space than BestFirstSearch, but finds an optimal solution. We look at a variation of A* that allows us to choose the trade-off between quality and complexity, of both time and space which go hand in hand for A*. We have also observed earlier that heuristic functions with higher estimates result in more focussed search. We have so far imposed a condition that the heuristic function must underestimate the distance to the goal. We now explore how the algorithm behaves when we relax that condition.

Consider a weighted version of the estimated cost $f(n) = \alpha g(n) + \beta h(n)$. What would be the behaviour of A* with different values of α and β? When $\beta = 0$ we have the uninformed search algorithms BFS and B&B. We can even model DFS by defining $g(n) = 1/\text{depth}$. If $\alpha = 0$ we have BestFirstSearch.

When $\alpha = 1$ and $\beta = 1$ we have A*. Choosing a value of $\beta < 1$ would push the algorithm towards B&B, without any material advantage. Choosing $\beta > 1$ gives us the weighted A* algorithm. Traditionally in the literature the algorithm is known as wA* where w is the weight in $f(n) = g(n) + w \times h(n)$. As the value of w increases, the algorithm becomes more and more like BESTFIRSTSEARCH, finding the solution faster but possibly one with a higher cost.

Figure 6.10 shows wA* with $w = 2$ on our example graph of Figure 4.7 on which Dijkstra's algorithm and A* found solutions with cost 148. As can be seen, wA* expands fewer nodes than A* but finds a more expensive solution with cost 153, which though is better than the one found by BESTFIRSTSEARCH with cost 195.

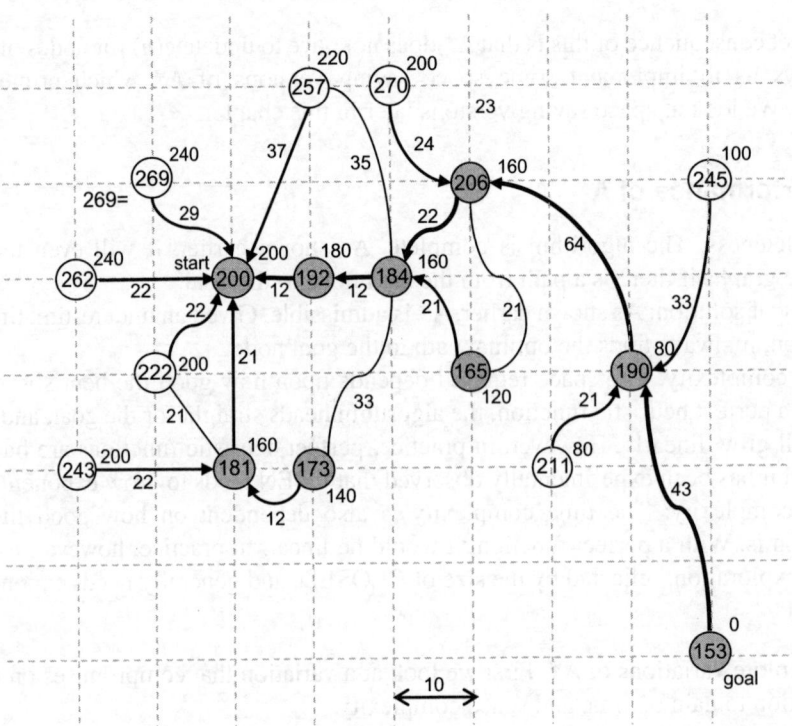

Figure 6.10 Weighted A* with $w = 2$ finds a path to the goal after inspecting nine nodes. But it finds a more expensive path to the goal than A* with cost = 153. The values outside the nodes are $2h(n)$ values.

To summarize, there is a spectrum of search algorithms guided by an estimated cost that varies from only $g(n)$ to only $h(n)$ with various combinations in between. An equal contribution from both results in an optimal algorithm that finds the shortest path fastest. Increasing the contribution of $h(n)$ beyond that may speed up the algorithm, but at the expense of optimality.

As machines become bigger and faster we try and solve larger problems. With both space and time growing exponentially, it is sometimes meaningful to look at space saving versions of A*, even at the cost of increased time complexity. We take a brief look at some space savings versions of A* next, without spelling out the detailed algorithms. We hope that the interested reader will be able fill in the details without difficulty.

6.3 Space Saving Versions of A*

In this section we discuss space saving variations of A* that have been developed over a period of time. We begin with variations that attempt to reduce overall space requirements to linear. Then we look at ways to prune CLOSED in A*, followed by variations that target OPEN. The common theme running through these algorithms is trading off time for space.

6.3.1 Iterative deepening A*

Algorithm iterative deepening $A*$ (IDA*) (Korf, 1985a) replicates the strategy used by the algorithm DFID (Section 3.6), which is to mimic a BFS using a series of DFSs of increasing depth. DFID pushes deeper into the search space in each cycle of DFS. Thus the first time it picks the goal it has found the shortest path, but being DFS it uses linear space.

IDA* is to A* as DFID is to BFS, finding the optimal path but using linear space. The main difference is that edges have a non-uniform cost, and the depth of a node is not a measure of the cost. Instead we use the f-values to determine the bound till which DFS searches in each cycle. The algorithm DB-DFS-2 (Algorithm 3.8 in Section 3.4) is adapted as follows:

1. Instead of *depth* the algorithm uses the f-value of a node to check that it is within the bound to continue searching. This bound is initially set to $f(S) = h(S)$, which we know is a lower bound on optimal cost.
2. When the neighbours of a node are generated, one needs to check that their f-value is within the bound before adding it to OPEN.
3. One has to keep track of the lowest value among the nodes that were not added to OPEN. This lowest value will become the bound in the next cycle.

The above changes are left as an exercise for the reader. The high level IDA* algorithm is given below.

Algorithm 6.4. IDA* does a series of DB-DFSs with increasing depth bounds.

IDA*(*start: S*)
    1   *depthBound* ← f(S)
    2   **while** *True*
    3   DB-DFS(*start, depthBound*)
    4   *depthBound* ← *f(N) of cheapest unexpanded node on OPEN*

The above algorithm suffers from the same drawback we observed in DFID. The algorithm may loop forever even for a finite graph which does not contain the goal node. The reader is encouraged to modify it to count the number of new nodes generated by IDA* and terminate with failure, like we did for DFID, if the count does not increase in some cycle. If the graph is infinite and there is no goal node, there is no way of reporting failure.

One problem with using IDA* without checking for CLOSED is that in many problems the number of paths to any node grows combinatorially. The algorithm will end up exploring all these paths. Another problem with IDA* is that of thrashing. When the *f*-values of most nodes are distinct, extending the bound to the lowest one will only include that node in the next cycle. This is a waste of effort. DFID worked because one expected the next layer to have more nodes than all the visited ones put together. One can ameliorate this problem a little by compromising on quality. This could be done by increasing the bound by a fixed increment δ, which then becomes the tolerance for the drop in quality.

6.3.2 Recursive best first search

IDA* lacks a sense of direction, being essentially DFS inside. The next algorithm attempts a different tack that exploits the heuristic function. Richard Korf is again the author of *recursive best first search* (RBFS) which also requires linear space (Korf 1993). Like HILLCLIMBING algorithm, RBFS follows the heuristic function, locally choosing the best successor at each stage. Unlike the HILLCLIMBING algorithm, it does not insist on a better node at each step, since it works with *f*-values which tend to increase. It reserves the right to try a different path. It keeps track of the second best node, and rolls back when all successors of the current node are worse than that. Observe that rolling back is different from simple backtracking. When it rolls back a path, it also updates its estimate of the cost of going down the node it rolls back to. Figure 6.11 illustrates the roll back process.

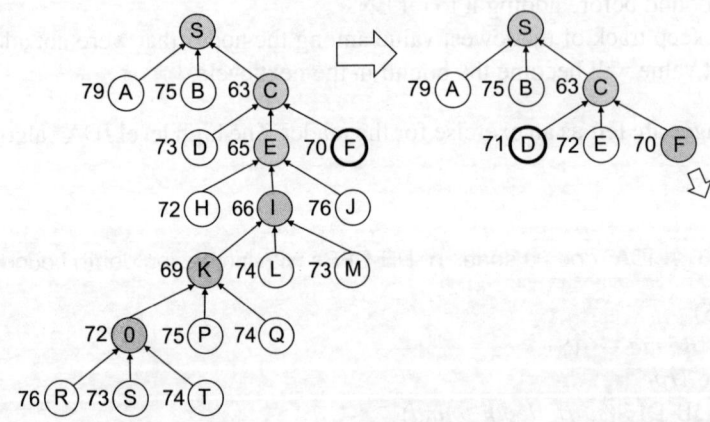

Figure 6.11 RBFS maintains a linear number of nodes down one path indicated by the heuristic function. When the path shown on the left looks less promising it rolls back to the next good looking node, updating the *f*-values on the path as it rolls back.

In Figure 6.11, RBFS expanded node *C* at level one with the value 63, marking *B* as the second best. It moved on to *E* and then marked *F* as the second best. It moved forward and had finished inspecting node O with the value 72 while F remained the second best. None of the successors of O is better than F. It starts the rollback, updating the f-value of each node to the best value of its children. O gets the value 73 from S, which gets transmitted to K, and then on to I. The parent E at this level gets its value 72 from H. At this point RBFS shifts to node F and marks D as the second best node.

As one can imagine, going down the path from F could soon look less appealing and it might roll back and try D. Experimentally this kind of frequent rolling back, known as thrashing, has been observed in RBFS, and is a major negative property. One could alleviate this problem by setting a level of tolerance, like we suggested in IDA*. That is, roll back from the current path if it is worse than second best by a given margin. For example, if the margin was 10 in the previous example, then RBFS would continue searching beyond O as long as the value was better than 80, which is $f(F) + 10$.

Next we look at a problem in which CLOSED grows much faster than OPEN and is a candidate for pruning.

6.3.3 Sequence alignment

The sequence alignment problem is one of the fundamental problems of biological sciences, aimed at finding the similarity of two amino-acid sequences. The task is to find a best match alignment between two or more sequences of characters. Saul B. Needleman and Christian D. Wunsch published a dynamic programming algorithm to solve the problem (Needleman and Wunsch, 1970). Since then, the use of heuristic methods have made improvements in the time and space complexity.

Deoxyribonucleic acid (DNA) and ribonucleic acid (RNA) molecules are described using sequences of letters (G, A, C, T and G, A, C, U respectively) representing their chemical constituents. In DNA the letters are A, C, G, and T, which stand for the four nucleotide bases – adenine, cytosine, guanine, and thymine.

Given two sequences composed of the characters C, A, G and T, the task of sequence alignment is to list the two alongside with the option of inserting a gap in either sequence. The objective is to maximize the similarity between the resulting two sequences with gaps possibility inserted between two characters. The similarity can be quantified by associating a positive cost of alignment and a negative or zero cost of misalignment. An example similarity matrix is shown in Table 6.2.

Table 6.2 Match Scores

Char	A	G	C	T
A	10	−1	−3	−4
G	−1	7	−5	−3
C	−3	−5	9	0
T	−4	−3	0	8

One may also insert a gap in one of the two sequences being aligned if the resulting match score improves. Then we impose an *indel* cost to the alignment. If the indel penalty is −5, for example, then the following alignment yields a total match score of 1.

AGACTAGTTAC
CGA___GACGT

The task of sequence alignment is to find the alignment with the highest total score. The highest possible score is when the two sequences are identical. Posed as graph search the two sequences are laid out on two axes, and a grid created with diagonal edges added as shows in Figure 6.12.

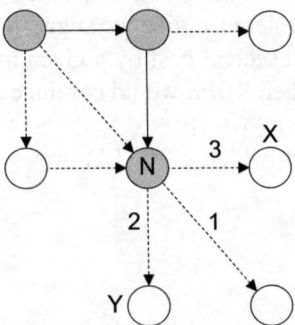

Figure 6.12 Graph search for sequence alignment has arrived at node *N* and is looking at two letters *X* and *Y* to match next. It could align them (move 1), or insert a blank before *X* (move 2), or insert a blank before *Y* (move 3).

Let us say the next two characters to consider are *X* in the horizontal sequence and *Y* in the vertical sequence. A diagonal move means the *X* is aligned with *Y*, along with the match score. This is marked by the edge labelled 1. A horizontal move, edge marked 3, means that a blank has been inserted before *Y*. This means *X* aligns with the blank, and the next character in the sequence will come into play. A vertical move, marked 2, likewise means a blank has been inserted before *X*.

The search space for an example problem is shown in Figure 6.13. The sequence along the horizontal axis is GCATGCA and the one along the vertical direction is GATTACA. Observe that they are of unequal length, which means that at least one blank will be inserted in the former. As can be seen from the figure, the state space grows quadratically with depth. But the number of distinct paths grows combinatorically. Consider two strings of length *N* and *M* being aligned. The grid size then is $(N + 1) \times (M + 1)$.

The number of ways that gaps can be inserted (*moving only horizontally or vertically*) is $(N + M)! / (N! \times M!)$

Taking *diagonal moves also* into account the number of paths is

$$\Sigma (M + N - R)! / (M - R)! \times (N - R)! \times R!$$

where *R* varies from 0 to min(*M*,*N*) and stands for the number of diagonal moves in the path

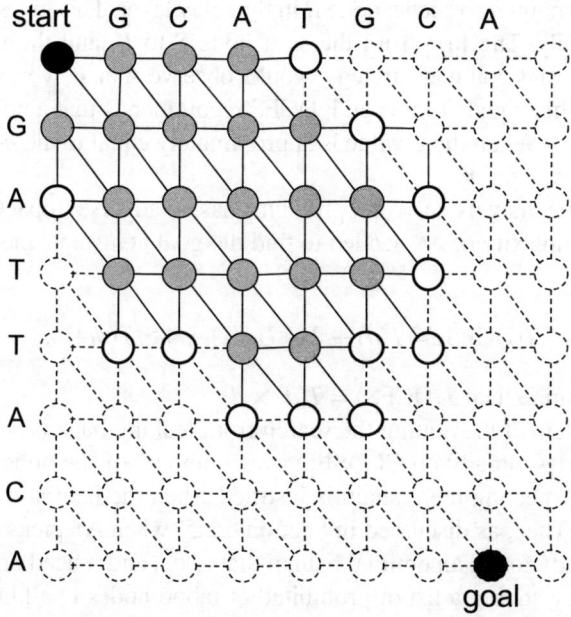

Figure 6.13 The search space in sequence alignment. The shaded nodes are on CLOSED. The unshaded nodes are on OPEN. The dashed nodes are yet to be generated. Observe that the two sequences are of unequal length.

As can be observed in the figure, the size of OPEN grows linearly, while CLOSED grows as a quadratic. In biology the sequences to be aligned may have *hundreds of thousands* of characters. Quadratic is then a formidable growth rate. This gives us a motive to prune CLOSED. But CLOSED serves the following two vital functions:

1. It prevents the search from 'leaking back' and going into loops.
2. It serves as a means for reconstructing the path. In the A* version of search we do this by maintaining parent pointers, and the parents are in CLOSED.

If one can address these two requirements, then one could prune the CLOSED and, as a consequence, solve much larger sequence alignment problems. The next section describes two variations that achieve that.

6.3.4 Pruning CLOSED

We look at two variations of an algorithm that prune the set of nodes in CLOSED.

6.3.4.1 Divide and conquer frontier search

The algorithm *divide and conquer frontier search* (DCFS) was the first algorithm to prune CLOSED (Korf and Zhang, 2000).

The problem of reconstructing the path is tackled as follows. Instead of maintaining parent pointers one maintains pointers to nodes on a judiciously chosen *relay layer*. When the goal node is found, it has a pointer to such a node R in the relay layer. This gives us two subproblems to be solved recursively. The first from the start node S to R, and the second from R to G. Proponents of divide and conquer strategy would observe that it works best when the two subproblems are roughly equal. To this end, DCFS recommends that a node R be made a relay node when $g(R) \cong h(R)$. When the g-value is approximately equal to the h-value, search should be roughly at the halfway mark.

The fact that we recursively solve subproblems has an adverse impact on time complexity. If $T(d)$ is the time complexity of A* needed to find the goal at d steps, then time complexity of DCFS is

$$T(\text{DCFS}) = T(d) + 2 \times T(d/2) + 4 \times T(d/4) \ldots$$

If $T(d)$ is exponential, then $T(\text{DCFS}) = T(d) \times d$.

The second problem of preventing the search from leaking back is addressed quite simply by not generating nodes already on CLOSED as neighbours of the node picked from OPEN. The strategy of not inspecting them again is justified when the heuristic function satisfies the monotone condition. Then, as discussed in Section 6.2.5, when A* picks a node and adds it to CLOSED, it has already found an optimal path to that node, and it need not be generated again. DCFS achieves this by adding a list of prohibited or taboo nodes to all nodes on OPEN. With these two modifications one does not need to store CLOSED, thus achieving the space saving. This is illustrated in Figure 6.14.

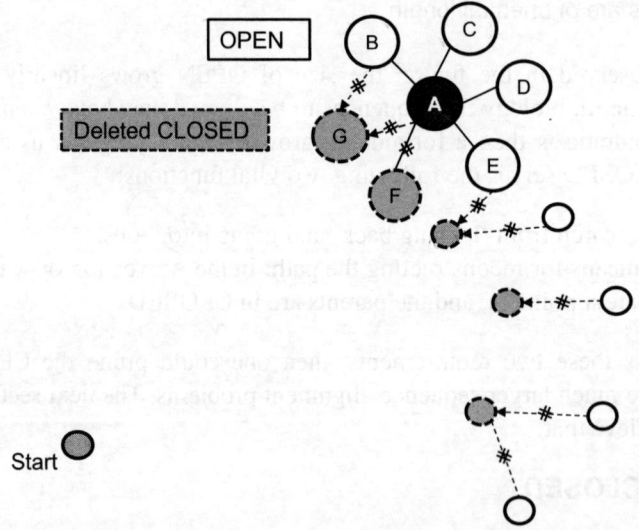

Figure 6.14 DCFS is about to pick and expand node A. When A was generated as a child of G and F, they were added as taboo neighbours. Only nodes B, C, D and E are generated. Of these B and E are already on OPEN. C and D are new nodes.

In the figure, nodes on CLOSED have been deleted. The shaded nodes shown in dotted circles only depict the nodes which are taboo for the corresponding nodes on OPEN. Algorithm DCFS is about to pick node A from OPEN. Node A has six neighbours, B, C, D, E, F, and G, which along with A have been magnified in the figure. Of the neighbours, F and G are on CLOSED, G being the parent of A. Both had generated A earlier, and both were placed on the taboo list of A, as depicted by crosses on the edges. They will not be generated now. Nodes B and E are generated and already on OPEN, and A has not found a better path to them, so their parent pointers will remain as they are. Nodes C and D are new nodes that will be added to OPEN, with A as their parent. Now node A gets added to the taboo list of nodes B, C, D, and E, so that when they are picked and expanded, they will not generate A as neighbour.

In this manner DCFS pushes into the search space with only the OPEN layer of nodes, modified as described above to avoid leaking back. Each node on OPEN has a pointer to the start node. Around the halfway mark, when $g(R) \cong h(R)$ for node R, it is stored as a relay node. Beyond the relay layer, all children will carry a pointer to their corresponding relay node, and nodes that would have gone into CLOSED are pruned. This goes on till the goal node is picked, and two recursive subproblems created.

6.3.4.2 Smart memory graph search

Smart memory graph search (SMGS) by Zhou and Hansen is a variation on DCFS that makes a more informed decision on pruning (Zhou and Hansen, 2003).

SMGS adopts a different approach to preventing the search from leaking back. It distinguishes the nodes on the boundary of CLOSED from the nodes deeper inside CLOSED. Nodes in CLOSED that have at least one neighbour on OPEN form the *BOUNDARY* layer. The remaining nodes, which have all neighbours in CLOSED, are identified as the *KERNEL*. It is only the nodes on the BOUNDARY that are needed to prevent search from going into loops and must be always retained. The search space of SMGS is shown in Figure 6.15.

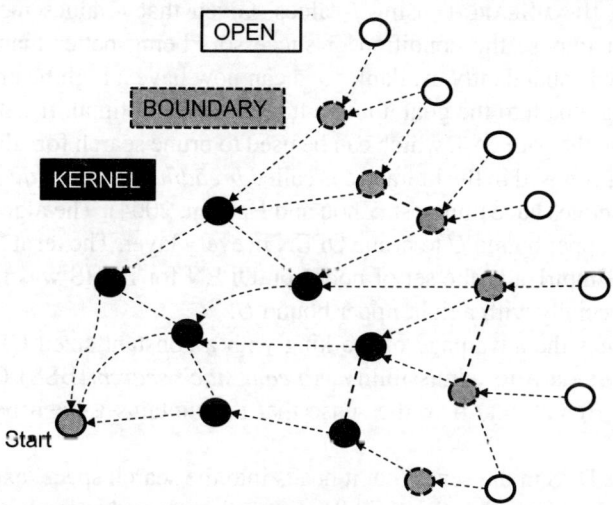

Figure 6.15 SMGS keeps the nodes on CLOSED while it can. Nodes on CLOSED with at least one neighbour on OPEN are BOUNDARY nodes. The rest form the KERNEL and are expendable. When SMGS senses memory shortage it deletes the KERNEL and converts the BOUNDARY into a RELAY layer. It continues searching, creating a new KERNEL and BOUNDARY as OPEN pushes forward.

When the time comes to prune nodes, SMGS does the following. It deletes all nodes in the KERNEL, and it converts the BOUNDARY layer into a RELAY layer, with pointers to the previous RELAY layer, or the start node. On the one hand, if the problem being solved is small enough to have been solved by A*, SMGS does not prune CLOSED at all. When it finds the goal node, it simply traces the path back. This path is known as the dense path.

On the other hand, when the problem is really big, SMGS may prune CLOSED more than once, creating multiple RELAY layers. It does this by being aware of the memory available to it, and smartly deciding to prune nodes when it is running out of memory. When it finds the goal node, it may have several relay layers left behind, creating a sparse path with back pointers to the previous relay layer, and needs to find the path between them recursively.

Observe that the recursive calls in SMGS are not likely to be nested deep. This is because the problem size may now be small enough to solve without further subdividing it. A corollary of this awareness of available memory is that it may prune CLOSED less often when the memory is abundant.

Pruning CLOSED is beneficial in special situations like the sequence alignment problem. In the more general case, it is the OPEN that grows faster, in an exponentially growing search tree. We next look at approaches to prune OPEN.

6.3.5 Beam stack search

In Chapter 4 we moved from BESTFIRSTSEARCH to HILLCLIMBING and BEAMSEARCH to save on space. Both these algorithms did local search, and the price we had to pay for space saving was incompleteness. The two algorithms do not guarantee a solution. We now try the same with A*, but with the requirement of both completeness and admissibility.

We begin with BEAMSEARCH, using f-values. Given that f-values increase as we move forward, we do not impose the condition of successors being better than the current node. Also, since memory is abundantly available, one can now have a high beam width and greater likelihood of finding a path to the goal, though it may not be optimal. It will, however, give us an upper bound U on the path cost, which can be used to prune search for admissible variations.

One algorithm proposed in the literature is called *breadth first heuristic search* (BFHS) that is admissible when edges have unit cost (Zhou and Hansen, 2004). The algorithm is essentially BFS, but it uses the upper bound U to prune OPEN in every layer. The term 'heuristic' in BFHS comes from there. Empirically the set of nodes on OPEN for BFHS was found to be smaller than that of A*, especially with a tight upper bound U.

BEAMSEARCH has the advantage of working with a constant sized OPEN. An algorithm that extends BEAMSEARCH to admissibility is *beam stack search* (BSS) (Zhou and Hansen, 2005). BSS is like BEAMSEARCH in the sense that it maintains a fixed beam width b while searching the space.

BSS is also like DFS in the sense that it heads into the search space, except that it chooses the nodes with best f-values rather than blindly in a predetermined order. It backtracks when the f-values become greater than the upper bound U.

BSS is different from DFS in that it does not keep the entire OPEN in a stack. Remember that in DFS backtracking happened naturally as the next node to be visited pops out of the stack.

In BSS backtracking has to be done explicitly by going back to the parent and regenerating the next set of nodes that could not be accommodated in the beam earlier. To make this process systematic BSS maintains a *beam stack* which stores a pair of f-values $[f_{min}, f_{max})$ at each level. The lower value f_{min} is the value of the cheapest node *in* the beam, and the higher value f_{max} is the lowest value *not* in the beam. If one imagines that the search tree is sorted on increasing f-values from left to right at each level, then one can imagine the search sweeping from left to right keeping a constant number of nodes at each level. The role of the values in the *beam stack* is depicted in Figure 6.16.

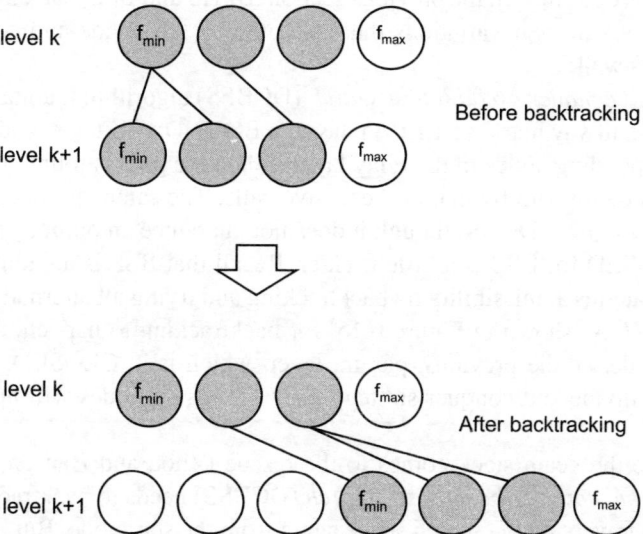

Figure 6.16 The *beam stack* stores pairs of values $[f_{min}, f_{max})$ at each level where, f_{min} is *lowest value in the beam,* and f_{max} is the lowest value *not* in the beam. When BSS backtracks to level k, it knows which of the neighbours to add next to the beam at level $k + 1$. The old f_{max} becomes the new f_{min} at level $k + 1$.

Figure 6.16 shows two layers of the beam, with shaded nodes. In the figure the nodes are arranged from left to right with increasing f-values. The beam width is 3. The figure on top shows two layers at levels k and $k + 1$ before backtracking happens. The leftmost node has the value f_{min}, and the lowest value node excluded from the beam has value f_{max}.

When BSS needs to backtrack from level $k + 1$, it goes up and generates the nodes with the next higher values. These are shown as shaded nodes in the bottom part of the figure. The old f_{max} has become the new f_{min}, and a new f_{max} value is identified at level $k + 1$. One must remember that the upper bound U is the largest f-value one is willing to explore, beyond which the beam will not be extended.

Algorithm BSS systematically explores the search space within the bound U and will guarantee an optimal path. In fact, as and when it finds a new path to the goal with a f-value lower that U, it updates the upper bound, and the space to be searched shrinks further.

Let the beam width be b. The size of OPEN is b. Overall, searching at depth d algorithm BSS stores $O(bd)$ nodes. It also keeps d pairs of [f_{min}, f_{max}] values in the *beam stack*. Thus, its overall space requirement is linear.

We next briefly discuss the divide and conquer versions of the algorithms studied in this section.

6.3.6 Pruning OPEN and CLOSED

The two algorithms described in the previous section, BFHS and BSS, had emphasis on pruning OPEN. We now look at their variations that also employ the divide and conquer strategy to prune CLOSED as well.

The *divide and conquer breadth first search* (DCBFS) algorithm maintains three layers of nodes beyond the halfway mark – a RELAY layer, a BOUNDARY layer, and OPEN with back pointers to corresponding nodes in the relay layer. When the goal is found, it must convert the sparse path into a dense path by making recursive calls. The same approach applies to *divide and conquer beam search* (DCBS) though it does not guarantee an optimal path.

Pruning CLOSED for BSS is a little trickier. Recall that BSS is an admissible version of BEAMSEARCH. It attains admissibility by backtracking and trying all alternatives that are within the upper bound U. As shown in Figure 6.15, for backtracking to happen, nodes at any layer need to access nodes in the previous, parent, layer, which is in CLOSED. If CLOSED is to be pruned by the divide and conquer strategy, then the parent nodes will not be available for backtracking.

This is where the beam stack comes to the rescue (Zhou and Hansen, 2005). When the algorithm *divide and conquer beam stack search* (DCBSS) needs to backtrack from level $k + 1$ to level k, it starts traversing the search space again from the start node. But it does not visit the nodes visited earlier again. At each level, the *beam stack* identifies the b nodes that would have existed in the beam, where b is the beam width. The first node is identified by its f-value f_{min}, and there are b nodes on the beam. At each level, all the children of the b nodes are generated and of those b identified by f_{min} are considered. When it arrives at level $k + 1$, then it selects b nodes starting from f_{max}, and in this way achieves backtracking in a roundabout way.

Like BSS the algorithm backtracks when the f-value has crossed the upper bound U. Whenever the algorithm finds a better path, it updates U and continues in quest of a still better path. Both BSS and DCBSS can be used as an anytime algorithm, asking for a solution when one needs it. Given that the algorithm explores the lowest f-value nodes first, the solution at any stage is expected to be a good one, especially if the heuristic function is well informed.

Algorithm DCBSS stores three layers of constant width nodes. One for OPEN, one for the BOUNDARY, and one for the RELAY needed to reconstruct the path. A snapshot of its search space is depicted in Figure 6.17.

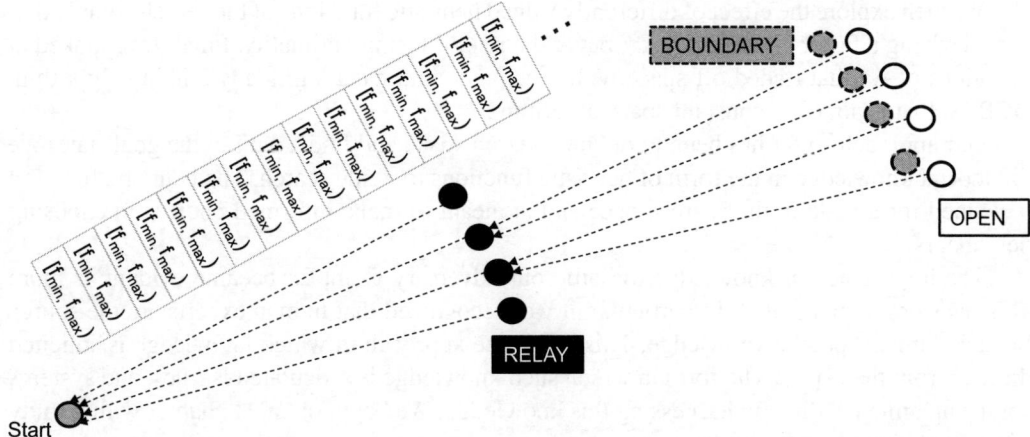

Figure 6.17 DCBSS maintains a constant number of nodes in three layers. The BOUNDARY nodes prevent search leaking back, and the RELAY nodes are used to reconstruct the path when the goal node is found. It also maintains a *beam stack* marking f_{min} and f_{max} at each layer. The stack keeps track of the edge nodes in the beam and is instrumental in facilitating backtracking.

Algorithm DCBSS is as close to a constant space version of A* as one can be. Strictly speaking, it is not constant space because, though it stores a constant number of nodes, it does have to maintain the *beam stack* which grows linearly with depth. The price that it must pay for this space economy is twofold. One, like BSS, it has to backtrack and explore the unseen part of the space that may yet yield a better solution. Second, in DCBSS each backtracking move requires it to regenerate the entire beam from the start node. This is extra work. But given that memory is becoming abundant, one hopes that with a large beam width, the need for backtracking would be reduced.

Summary

In this chapter we have focussed exclusively on finding optimal solutions. We postulate the problem as a graph search problem when edge weights are arbitrary. The goal is to find the shortest or least cost path. This goal is different from that of Chapter 4 in which we employed heuristic functions to find the goal node *quickly*. Finding the optimal path is important in situations when costs are significant, and also when the solution found has to be executed multiple times.

Starting with the uninformed B&B algorithm we added a heuristic function to guide search and proved that the resulting algorithm, A*, is admissible given certain conditions. An important one is that the heuristic function underestimates the distance to the goal node.

We then explore the effect of differently valued heuristic functions of the search complexity, even devising a faster algorithm wA*, but at the risk of losing optimality. Finally we looked at variations of A* that traded off space with time, and culminated our study with an algorithm, DCBSS, that is almost a constant space algorithm.

Our approach so far has been to define a search space and find a path to the goal state. We did look at knowledge in the form of heuristic functions to guide search. The heuristic function is defined for a node in the search space and is meant to make informed choices in choosing neighbours.

The importance of knowledge to battle our adversary CombEx became evident as more systems were implemented. In particular, it was recognized that human experts are the source of such domain specific knowledge. This led to the approach in which knowledge is solicited directly from the expert. The form in which such knowledge is articulated is rules, and systems sprang up aimed at directly harnessing this knowledge. We begin out next chapter with a study of such rule based systems.

Exercises

1. Recall the method used for computing the lower bound estimate of nodes in the B&B search for an optimal solution for the TSP. What is the lower bound estimate for the following TSP problem? Assume that the edges that have not been drawn have infinite cost.

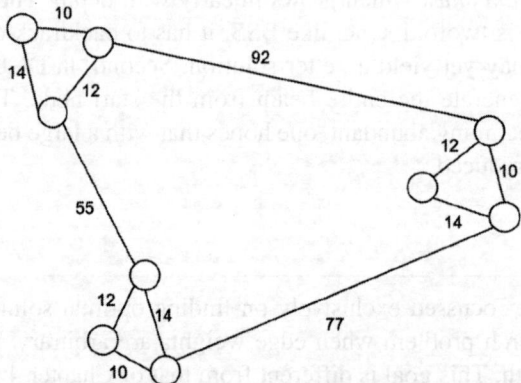

2. [Baskaran] A TSP problem on seven cities is shown on the accompanying table and city locations. Simulate the B&B algorithm on the problem.

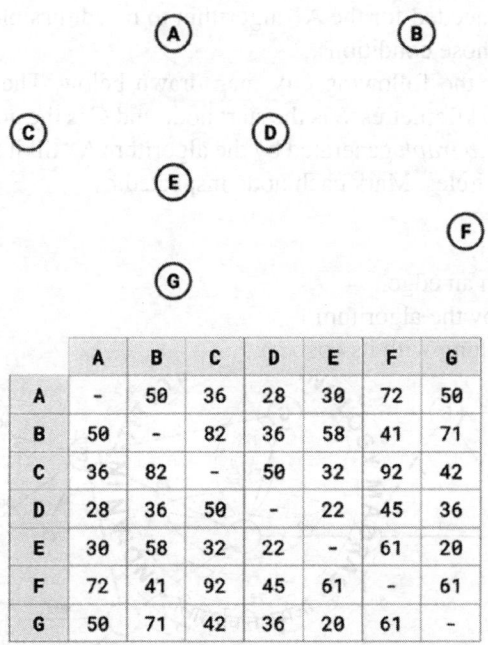

3. Algorithm A* is about to expand the node *N* in the graph below. Thick rectangles are on CLOSED and dotted rectangles are on OPEN. One node is about to be added to OPEN. Labels on edges are cost of moves. These are only shown for some edges where they cannot be computed from the parent. Show the graph after A* has finished expanding the node *N*. Clearly show the parent pointers and mark where the *g*-values have been updated.

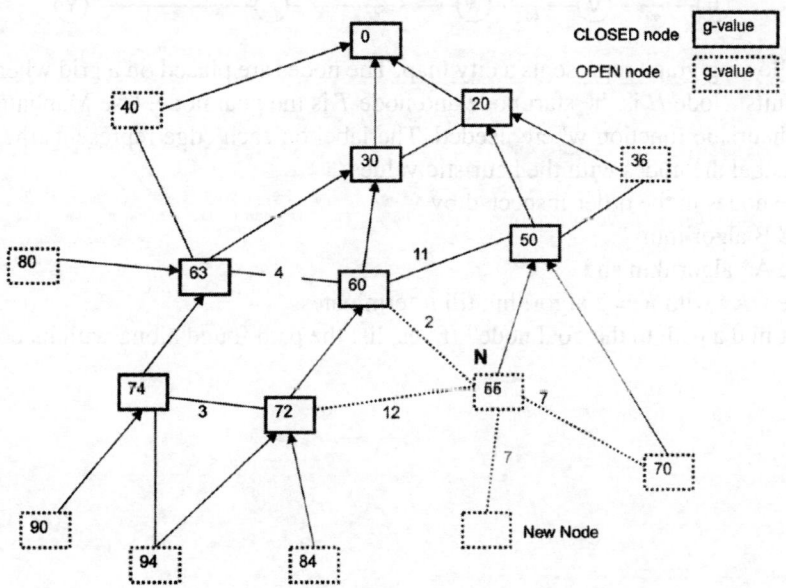

4. State the conditions needed for the A* algorithm to be admissible, and give a proof of its admissibility under those conditions.
5. [Baskaran] Consider the following city map drawn below. The cities are laid out on a square grid of side 10 kilometres. S is the start node and G is the goal node. Labels on edges are costs. Draw the *subgraph* generated by the algorithm A* till it terminates. Let the nodes on OPEN be single circles. Mark each node inspected by
   a. a double circle,
   b. sequence number,
   c. parent pointer on an edge,
   d. its cost as used by the algorithm.
   List the path found along with its cost.

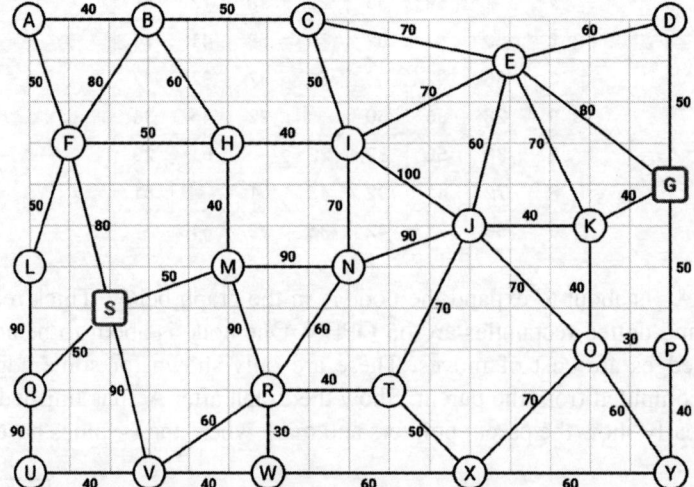

6. The following graph represents a city map. The nodes are placed on a grid where each side is 10 units. Node H is the start node and node T is the goal node. Use Manhattan distance as the heuristic function where needed. The label on each edge represents the cost of the edge. Label the nodes with the heuristic values.
   List the nodes in the order inspected by
   a. B&B algorithm
   b. the A* algorithm and
   c. the wA* with $w = 2$ algorithm till it terminates.
   Does it find a path to the goal node? If yes, list the path found along with its cost.

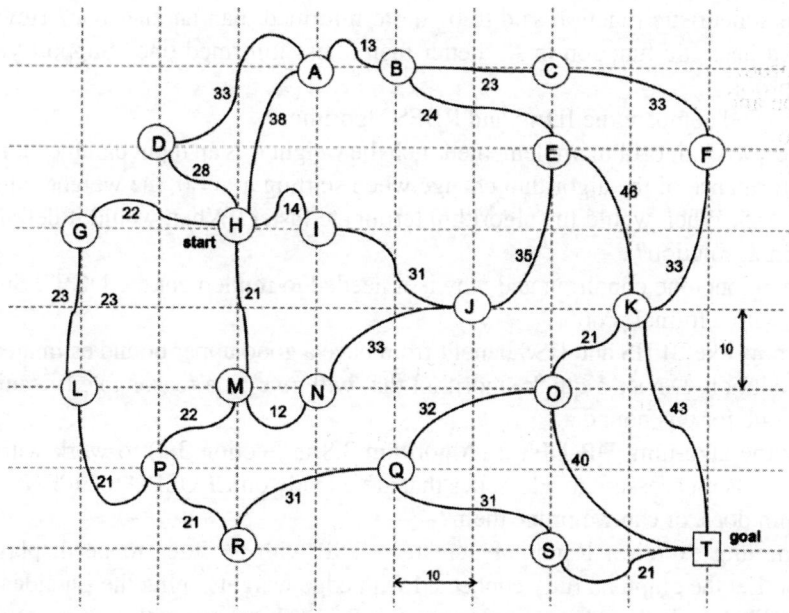

Draw the sub-graphs when the algorithms A* and wA* (with $w = 2$) terminate. Clearly show the nodes that are on OPEN and those that are on CLOSED. Show the f-values and the parent pointers of all nodes.

7. Repeat the exercise in the previous question for the following graph. Let S be the start node and G be the goal node.

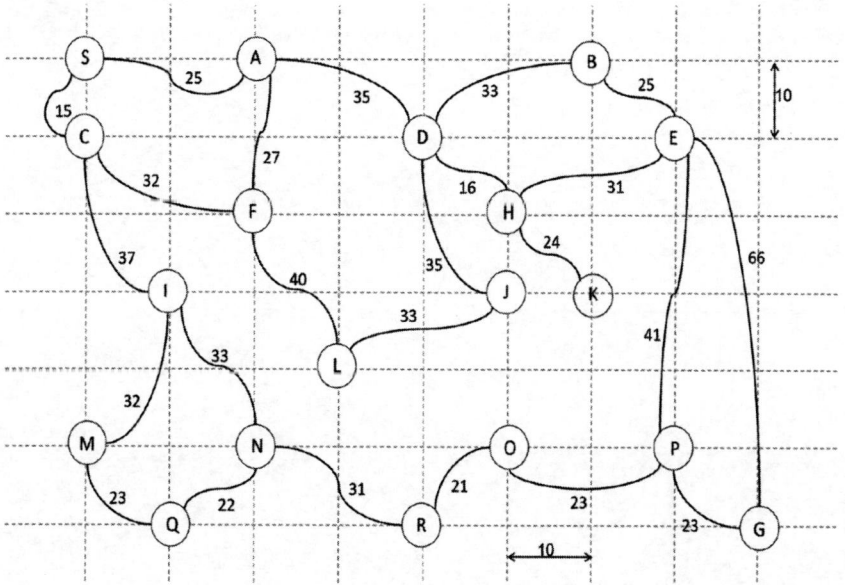

8. When is a heuristic function said to be more informed than another one? How is a more informed heuristic function in A* better than a less informed one? Support your answer with a proof.
9. Describe and compare the IDA* and RBFS algorithms.
10. Imagine a wA* algorithm implemented such the weight w is an input parameter. How would the performance of the algorithm change when starting at $w=0$, the weight is increased in steps of 0.5. When would the algorithm terminate faster? When would it definitely return the optimal solution?
11. State the monotone condition and why is it needed to implement the DCFS. Support your answer with a formal proof.
12. Algorithms like BFHS and BSS benefit from have a good upper bound estimate of the cost of the solution. Devise a quick algorithm that will give us a reasonable (meaning not too high) value for this bound.
13. Modify the algorithm DB-DFS-2 (Algorithm 3.8 in Section 3.4) to work with the IDA* algorithm. Is it necessary to add nodes that are already on CLOSED to OPEN again as this algorithm does, or can we prune them?
14. Programming exercise: Randomly create a TSP with N cities to be displayed on the monitor. Let the graph be fully connected with edge weights being the Euclidean distance. Do not display all the edges. Implement the B&B algorithm with the heuristic *cheapest available edge first*. Maintain a list of allowed edges and taboo edges for estimation as well as refinement. Display the best tour (node in the search space) along with the estimated cost at the click in each cycle.
15. Adapt the RECONSTRUCTPATH algorithm (Algorithm 3.2) from Chapter 3 to work with the node representation and a separate parent pointer. Modify this algorithm to return the cost of the path found as well.

CHAPTER 7
Problem Decomposition

So far our approach to solving problems has been characterized by state space search. We are in a given state, and we have a desired or goal state. We have a set of moves available to us which allow us to navigate from one state to another. We search through the possible moves, and we employ a heuristic function to explore the space in an informed manner. In this chapter we study two different approaches to problem solving.

One, with emphasis on knowledge that we can acquire from domain experts. We look at mechanisms to harness and exploit such knowledge. In the last century in the 1980s, an approach to express knowledge in the form of if–then *rules* gained momentum, and many systems were developed under the umbrella of *expert systems*. Although only a few lived up to expert level expectations, the technology matured into an approach to allow human users to impart their knowledge into systems. The key to this approach was the *Rete algorithm* that allowed an *inference engine* to efficiently match rules with data.

The other looks at problem solving from a *teleological* perspective. That is, we look at a goal based approach which investigates what needs to be done to achieve a goal. In that sense, it is reasoning backwards from the goal. We look at how problems can be formulated as *goal trees*, and an algorithm *AO** to solve them.

The search algorithms we have studied so far take a holistic view of a state representing the given situation. In practice, states are represented in some language in which the different constituents are described. The state description is essentially a set of statements. As the importance of knowledge for problem solving became evident, using rules to spot patterns in the description and proposing actions emerged as a problem solving strategy.

7.1 Pattern Directed Inference Systems

An approach to problem solving that was developed in the mid-1970s was called *pattern directed inference systems* (Waterman and Hayes-Roth, 1978). The basic idea is that patterns in a given state are associated with actions. These pattern–action pairs would contribute to the *MoveGen* function. These pattern–action pairs can also be represented as rules or productions and used directly. Rule based systems are programs that facilitate their use directly. We begin with a few example domains.

Example 1: Consider the blocks world planning problem discussed in Figure 4.13. The planning community has devised languages of varying expressivity to describe planning domains. These languages are a series called planning domain definition languages (PDDL). The simplest of these languages, called PDDL 1.0, is used to describe the blocks world domain using the following set of predicates:

On(X,Y)	: Block X is on block Y
OnTable(X)	: Block X is on the table
ArmEmpty	: The robot arm is empty
Holding(X)	: The robot arm is holding block X
Clear(X)	: There is no block on block X

The domain has a table of unbounded extent, and a set of named cuboid blocks, all of equal size. Exactly one block can be placed on another block, and the height of a tower of blocks is unbounded. Given the above predicates, the start state in Figure 4.13 can be described by a set of sentences or predicates: {OnTable(D), OnTable(F), On(C,D), On(B,C), On(A,B), On(E,F), Clear(A), Clear(E), ArmEmpty}. A subset of the domain description is a *pattern* that can trigger an action. Such pattern–action combinations define the moves, or operators as they are called by the planning community. For example, an operator called Unstack(X,Y) is *applicable* if the robot arm is empty, X is on Y, and Y is clear. An action is a ground instance of an operator. In our example problem, two actions are possible in the start state – Unstack(A,B) and Unstack(E,F).

In general, a collection of such applicable actions together define the MoveGen functions from the search perspective. MoveGen is then not a uniformly described function, but depends on the patterns that exist in the domain description. Each state has its own sets of applicable moves. The setup is depicted in Figure 7.1.

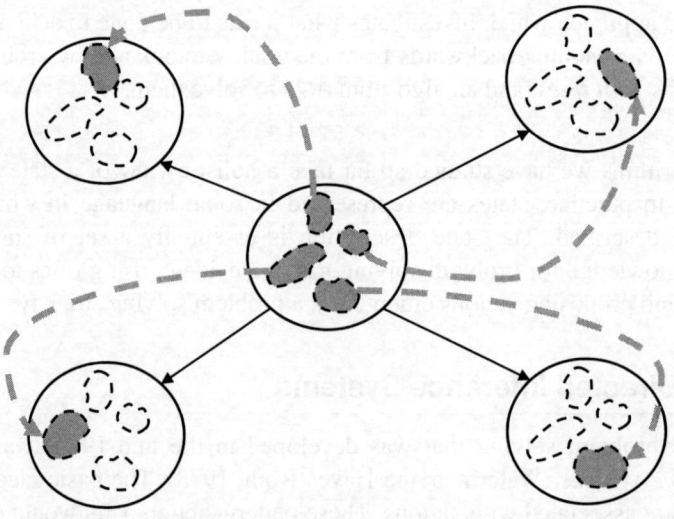

Figure 7.1 Beneath the hood of the MoveGen function, there are typically a set of pattern–action associations known as rules. MoveGen is simply a collection of all applicable actions.

We shall look at planning in Chapter 10 where we will describe the planning operators and the algorithms they are used in. In this chapter we will discuss how to compute the MOVEGEN function efficiently as moves are made and the state changes as a consequence.

Example 2: Consider the game of tic-tac-toe or noughts and crosses, commonly played by children. The game is played on a grid of nine squares typically drawn by two vertical and two horizontal lines. The two players play alternately, one marking a X (or cross) and the other marking a O (or nought). The first one to place three in a row, column, or diagonal wins. Figure 7.2 shows the moves that Cross can make on her turn.

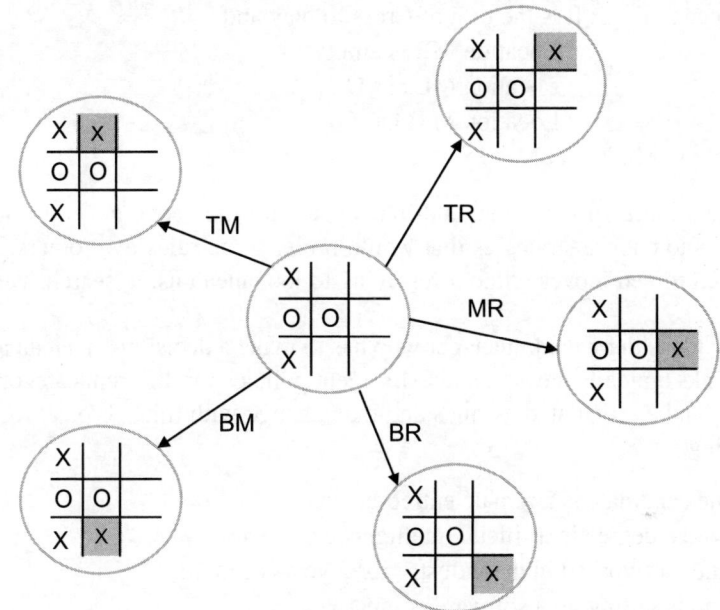

Figure 7.2 Given the tic-tac-toe board position in the centre, there are five moves where a X can be played. The moves are identified by their location TM (top middle), TR (top right), MR (middle right), BM (bottom middle), and BR (bottom right).

We will look at a standard game playing algorithm in Chapter 8. Here we focus on the representation and move generation. Of the several representations that are possible we choose a simple one amenable to move generation using rules. We identify each square by its coordinates with T (top), M (middle), and B (bottom) being the three values along the vertical axis, and L (left), M (middle), and R (right) on the horizontal axis. In Figure 7.2, the move TM says that Cross can mark a X on the top-middle square which is empty. The MOVEGEN function, in the style of planning operators, starts with a move name, followed by its preconditions, followed by its actions.

Move Play-Cross(L)
        Preconditions: It is the turn of Cross to play and
                Location L is empty
        Action: Mark X at L

Here we have treated L as a variable whose value is every location that Cross can play on. In our example above, the moves are Play-Cross(TM), Play-Cross(TR), Play-Cross(MR), Play-Cross(BM), and Play-Cross(BR). We could then hand the moves over to the game playing program which would search the game tree and choose the best move.

The interesting thing about describing the moves as rules is that one can have additional rules that capture our expert knowledge as well. For example, one could have a rule that defines a forced move, as in the board position of Figure 7.2.

> Move Forced(MR)
>   Preconditions: It is the turn of Cross to play and
>   Location MR is empty
>   Location ML has O
>   Location MM has O
>   Action: Mark X at MR

Now this is a more specific rule, and may exist with the above general rule. Rule based systems allow us to choose strategies that would prefer some rules over others, and we could then choose such forced moves without resolving to lookahead using search. We will develop this idea next.

Example 3: Consider a bank manager who has to make a decision on a loan application by a candidate. Banks typically have loan disbursement policies for different categories of people. These policies can be articulated as rules and may change with time. Typical rules could look like the following.

> If   the candidate is a female entrepreneur
>   with a degree in artificial intelligence
>   and has worked in the industry for 2 years or more
>   and is setting up a software company
> then   the candidate is eligible for a loan

> Or,

> if   the candidate is a farmer
>   who has completed class 12
>   and has access to land for farming
>   and has no outstanding loans
> then   the candidate is eligible for a loan

There could be a multitude of such rules, many of them with exceptions, that could define the policy the bank uses for deciding upon loans. The management would benefit if all they had to do was to add and delete such rules, and the application software could classify a candidate as eligible or not eligible.

We describe below the mechanism which makes such *business rule management systems* possible, and they have applications in many areas.

Example 4: The Indian Railways has a plethora of rules for concessional ticket prices, wherein various categories of people get different levels of concession.

Example 5: In these Covid times, governments of various countries made and revised rules depending upon the status of the pandemic. For example, the Indian government first allowed frontline health workers to get vaccinated, followed by senior citizens, other adults, and finally teenagers and children. At any time, various exceptions like comorbidities and other health conditions could influence the decisions. Allowing travel between different regions and such rules could have been efficiently handled by a rule management system.

Such rule based systems are very well suited to do classification tasks, in which the knowledge is articulated by human users. The inputs to the rules are not uniform, and different rules may look at different sets of preconditions. Many such application areas are not amenable to machine learning approaches to classification, not least because there could be heavy costs for misclassification, there may be insufficient training data, and the human users may demand explanations which are possible only with articulated symbolic rules.

7.2 Rule Based Production Systems

Rule based systems, or production systems as they are also known, embody the idea of expressing problem solving knowledge independently from the algorithms that use this knowledge. This is also the idea of declarative programming, in which the user has to only express this knowledge and an inference engine takes care of control in the program. This idea is also embodied in the notion of logic programming, which we will study briefly in Chapter 9.

The idea of production systems emerged primarily from research at Carnegie Mellon University (CMU) and draws upon the study by Herbert Simon and Alan Newell on how humans solve problems (Newell and Simon, 1963; Newell, 1973). This evolved into a programming language, OPS5 (Forgy, 1981; Brownston et al., 1985), we will look at below, and later into a cognitive architecture for problem solving called SOAR (Laird et al., 1987) which is still evolving. Rule based systems have three components.

1. A working memory (WM). The WM is a set of statements that define a state of the problem to be solved. These statements are known as working memory elements (WMEs). As the world changes, some of these WMEs may get deleted and new ones may get added. Thus the WM is transient knowledge of the world at the moment and is the short term memory (STM) of the problem solver. In rule based systems, the WM is initialized in the beginning and is modified by the application of rules.
2. Rules. The problem solving knowledge of the agent is captured in the form of rules, which are pattern–action combinations. Rules reside in the long term memory (LTM) of the problem solver and are not specific to a particular problem being solved. The rules have a left hand side (LHS), which is a set of patterns. Patterns can have slots for variables which make them general. Each pattern in the LHS must match a WME for the rule to be triggered. If a rule is executed or fired, then it may add or delete WMEs among other possible actions.
3. The inference engine (IE). The IE is the workhorse of the rule based system. It matches the rules with the WMEs, decides upon which rule–data combination is to be selected, and then executes the rule. It does this repeatedly till some termination criterion.

In the following sections we look at rule based systems in a little more detail. To make the discussion more concrete, we ground the description of rule based systems in the language called OPS5 mentioned earlier. OPS5 is said to expand to Official Production System language version 5 and was specifically devised for implementing forward chaining with rules (Brownston et al., 1985).

7.2.1 The working memory in OPS5

The basic data structure in the OPS5 language is made up of a class name, and a set of attribute names along with data. The attribute names are identified by a marker, which in OPS5 is ^. Thus the data structure has the form

    (class-name
        ^attribute1 value1
        ^attribute2 value2
        .
        .
        attributeN valueN)

The order in which the attributes are written is not important. The following two examples represent the same WME describing a student:

    (student ^name Shreya ^rollNo 1111 ^major CS ^year 3 ^age 20)
    (student ^rollNo 1111 ^age 20 ^name Shreya ^major CS ^year 3)

As we will see, the match algorithm does not require the attributes to be in a specific order. Observe that white spaces are ignored. Also note that there is no notion of explicit types for values, though implicitly the language recognizes numbers.

The WM is a collection of WMEs. The WMEs are *indexed* by a *time stamp*, indicating the order in which they were added to the WM. The following is an example of what the data could look like:

1. (next ^rank 1)
2. (student ^name Shreya ^rollNo 1111 ^major CS ^year 3 ^age 20)
3. (student ^name Aditi ^rollNo 1112 ^major CS ^year 3 ^age 20)
4. (student ^name Garv ^rollNo 1113 ^major CS ^year 3 ^age 20)
5. (student ^name Atish ^rollNo 1114 ^major CS ^year 3 ^age 20)
6. (marks ^subject AI ^rollNo 1111 ^midSem 41 ^endSem 35 ^total nil ^rank nil)
7. (marks ^subject AI ^rollNo 1112 ^midSem 40 ^endSem 40 ^total nil ^rank nil)
8. (marks ^subject AI ^rollNo 1113 ^midSem 43 ^endSem 36 ^total nil ^rank nil)
9. (marks ^subject AI ^rollNo 1114 ^midSem 42 ^endSem 35 ^total nil ^rank nil)

Note that there is no structure in the data except for the time stamp in the WM. If one wants to implement an array, then one must represent it as a set of records or WMEs and explicitly add an attribute for the index. WMEs with these index values will continue to be organized by the time stamp. If we want to sort the records, then one may need to swap index values, rather than keeping the index sorted and swapping the records as would traditionally be done.

7.2.2 Patterns in rules

Rules have an LHS and a right hand side (RHS) separated by an arrow. Rules or *productions* have a name as well, marked by the symbol *p*.

    (p rule-name
        LHS
    →
        RHS)

The LHS is a set of patterns and the RHS is a set of actions.

    (p rule-name
        pattern1
        pattern2
        .
        .
        – patternK
    →
        action1
        action2
        .
        .
        actionM)

The patterns can be identified by their sequence number. Alternatively, each pattern can be prefixed by a label which serves as an identifier. Each pattern must match some WME, unless it is preceded by a negation symbol, in which case there must be no WME that matches it. In OPS5 negated patterns are written after the positive ones.

Each pattern conforms to some WME. It begins with a class name and is followed by a sequence of attribute names followed by a *test* for the attribute *value*. Each of the tests specified in a pattern must be satisfied. The attributes being tested need only be a subset of the attributes the corresponding WME has. The pattern has a set of attributes to be tested and ignores other attributes in the WME. The tests on values of attributes are as follows.

If the attribute value in the rule is a constant, then it must match an identical constant in the WME. For example, '^name Shreya' in the pattern matches '^name Shreya' in the WME.

If the attribute in the rule is a variable, then it can match any constant in the WME. For example, '^name <n>' in the pattern matches '^name Shreya' as well as '^name Aditi'. Variables are enclosed in angular brackets. In addition, if the same variable name occurs more than once in the LHS of a rule, then all occurrences must match the same constant.

If an attribute in the rule contains a Boolean test, then the value in the WME must satisfy that test. The following are examples:

^name <y> = Shreya: the value is a variable <y> and must be equal to Shreya
^name <y> = <x>: the value is a variable <y> which must be equal to <x> also in the LHS
^name <y> <> Shreya: the value is a variable <y> and must not be equal to Shreya
^name <y> <=> Shreya: the value is a variable <y> and must be the same *type* as Shreya

If the value in the WME is numerical, then these additional Boolean tests apply: > (greater than), >= (greater than or equal to), < (less than), and <= (less than or equal to). Some examples are

^age >= 19: the value in the WME can be any number greater than or equal to 19
^age <y> > 19: the value is a variable <y> and in the WME must be a number greater than 19
^age <y> > <x>: a variable <y> and in the WME must be greater than <x> from the LHS

In addition, one can combine two or more tests using logical connectives. Curly brackets stand for the AND connective and double angular brackets stand for the OR connective. All the test inside { } must be satisfied, and at least one test inside << >> must be satisfied. For example,

^age {> 12 < 20} says that the matching value must be greater than 12 and less than 20
^day << Wednesday Friday>> says that the value must be Wednesday or Friday

7.2.3 Actions in rules

The RHS of a rule is a set of actions. The main actions we are interested in are *Make* and *Remove*. The action Make adds a new element into the WM, with the latest time stamp. For example, (Make (Winner ^Round 1 ^Name Ayesha)). The Make action can access variables from the LHS and has access to mathematical operators like plus and multiplication. One can say, for example, (Make (Price ^today (<x> + 12))) where <x> could be a variable in the LHS.

The Remove action deletes a WME referenced in the LHS of the rule. This could be done by using the pattern number. For example, (Remove 2) says that remove the WME that matched the second pattern in this rule. One could also reference the pattern by its label if there is one.

The action Modify is combination of Remove and Make. It assigns a new time stamp to the element it adds. The following rules illustrate these actions:

(p sumTotal
 (marks ^subject <S> ^midSem <m> <> nil ^endSem <e> <> nil ^total nil)
 →
 (Modify 1 ^total <m> + <e>))

(p ranking
 (next ^rank <r> ^subject <S>)

(marks ^subject <S> ^student <s> ^total <m> ^rank nil)
–(marks ^subject <S> ^total > <m> ^rank nil)

→

(Modify 1 ^rank (<r> + 1))
(Modify 2 ^rank <r>))

We could express the same rules in English as follows:

Rule sumTotal
  IF
    there is a WME marks for a given ^subject with ^midSem and ^endSem attribute values non-nil and whose ^total attribute has value nil
  THEN
    modify the ^total attribute to the sum of the ^midSem and ^EndSem values

Rule ranking
  IF
    the next rank to assigned in subject <S> is <r>
    and there is an entry for some student in <S> with total <m> and rank nil
    and there is no entry for any student in <S> with total > <m> and rank nil
  THEN
    modify the next rank to be assigned as <r> + 1
    and modify the rank of the student to <r>

Between the two rules, they add up the midsemester and end-semester marks for every student and assign ranks to the students. The third pattern in rule *ranking* says that there must be no matching WME which has a total greater than <m> and rank nil.

Observe that in the second rule, next ^rank is updated in the first action and the rank is assigned to the student in the second action. These actions are not sequential. If the next rank to be assigned was 4 to begin with, then that would get modified to 5 for the *next* round, while the student in *this* rule firing would still get rank 4. One could also write the same rule as follows:

(p ranking
    {<rank> (next ^rank <r>)}
    {<total> (marks ^student <s> ^total <m> ^rank nil)}
    – (marks ^total > <m> ^rank nil)

→

    (Modify <total> ^rank <r>)
    (Modify <rank> ^rank (<r> + 1)))

Observe that the order of the actions is different. The reader would no doubt have started viewing OPS5 as a programming language, which indeed it is. It is in fact a Turing complete language, which means that whatever you can do in any other language you can also do with OPS5. The other actions we have not mentioned are the ones needed to make it a complete programming language. Actions like Read, Print, Load a file, and so on.

But having given up the flow of control to an inference engine (see the next section), one still needs to be careful in how one writes the rules. The above two rules are a program meant to first sum up the marks of the students and *then* assign ranks to them based on the total. But this sequencing of tasks may not happen in practice. Clearly rank assignment must be done only after completing the adding up of the total for all students. Initially, for our four students, four instances of the sumTotal rule would match. One of them would get executed. As soon as that happens, an instance of the ranking rule for that student would now also match. What is to prevent that student from being assigned rank 1? In that sense, the 2-rule program has a bug. One way to address that problem is to introduce a *context*. For example, to start with, we can add the WME (current ^context totalling) and add that as a pattern in the sumTotal rule. Likewise, one could add a pattern (current ^context ranking) to the ranking rule. No ranking rule will match till that appears in the WM. Then we can have a rule to switch contexts as follows. If the context is *totalling* and there is no WME with ^total nil, then modify the context to *ranking*. In general, rule based programs may have many rules that match a context, and thus they can all be bunched up to execute together. Within a context we still need a strategy to select one from the set of matching rules. We will look at some strategies after describing the inference engine, the third component in rule based systems.

7.2.4 The inference engine

The rules embody the problem solving knowledge of the agent, and the WM has a description of the current problem. The task of selecting and applying rules is assigned to the inference engine (IE). The IE is a program that executes the following three modules in a loop.

1. Match. The task of the match phase is to determine the set of rules that are matching along with the WMEs they are matching. Each pattern in each rule must be checked against each element in the WM. Each rule may have multiple matching instances, and every WME may match multiple rules. Thus the number of tests that have to be done is the product of the number of patterns in all the rules, the number of attributes being tested in each rule, and the number of WMEs. The output of the match phase is a set of tuples of the form <rule-name $t1$ $t2$ ... tK> which specifies the rule name and the K timestamps of the K WMEs matching the K patterns.
2. Resolve. The output of the match phase is a set called the conflict set, so named because one can imagine a set of matching rules vying to be selected. The resolve phase selects one such rule along with its matching data. This is called conflict resolution, since it resolves the conflict between the matching rules. This selection is done by means of a conflict resolution strategy. We describe the common conflict resolution strategies in a separate section below.
3. Execute. The actions in the RHS of the selected rule are then executed. The execution is parallel in principle, even though in practice the actions are executed sequentially. Our interest is in the actions Make, Remove, and Modify. These make changes in the WM, adding and deleting elements, and ostensibly requiring one to do the Match phase all over again. As Charles Forgy showed, that is not necessary and we discuss his approach later.

The above cycle concludes either when there is no matching rule or one of the rules executes an action called Halt. Figure 7.3 shows the Match–Resolve–Execute cycle of a forwarding chaining rule based IE. It is forward chaining, or data driven, because rules match data and add (or delete) data to the WM, and again match rules going forward in this fashion. In contrast, one can have a goal driven backward chaining approach that selects rules that match a goal to be achieved. We look at this in Chapter 9 on logical reasoning.

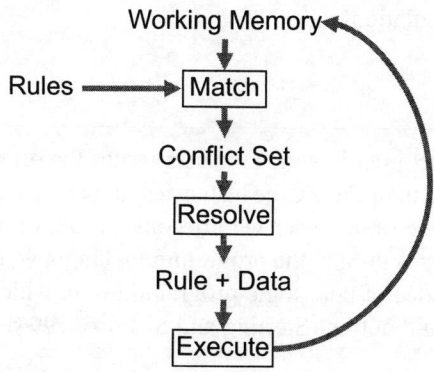

Figure 7.3 The match–resolve–execute cycle of a forward chaining rule based IE.

7.2.5 Conflict resolution strategies

The key to intelligent problem solving is to select the right moves. Algorithms like A* employ a heuristic function to guide search towards a goal. Rule based systems evolved to harness expert knowledge in specific domains (Jackson, 1986). Some examples are MYCIN, an expert system for medical diagnosis (Shortliffe, 1976; Buchanan and Shortliffe, 1984), and PROSPECTOR a system for mineral exploration (Campbell et al., 1982). The system R1 (internally called XCON, for eXpert CONfigurer) was a production rule based system written in OPS5 to assist in placing orders for DEC's VAX computer systems by automatically selecting components based on the customer's requirements (McDermott, 1980a, 1980b). Each expert system was designed to solve a particular kind of problem. The common approach is that expert knowledge is captured in the form of rules, and an IE chooses and applies rules to solve specific problem instances. The choice of which of the matching rule instances to execute is instrumental in finding a good solution. The following conflict resolution strategies have been proposed.

Refractoriness

A rule–data combination may fire only once. If a rule along with a set of matching WMEs has been selected to execute once, then it cannot be executed again. This is particularly relevant when the selected rule does not modify the WMEs matching its precondition. Given that the rule–data combination was selected to fire, it is possible that it would be selected again. Selecting it would not contribute anything new, and could even result in the system going into a loop. Observe that the same rule can still fire with different data.

The idea of refractoriness comes from the way neurons fire in animal brains. When a neuron has received excitation that crosses its threshold, it fires once, and then waits for new signals to come in. As defined in the *Merriam-Webster* dictionary, refractoriness is 'the insensitivity to further immediate stimulation that develops in irritable and especially nervous tissue as a result of intense or prolonged stimulation'.

Lexical Order

By lexical order we mean the order in which the user writes the rules. And if a rule has multiple instances with different data, then choose the instance that matches the earlier data. This strategy places the onus of this choice on the user who becomes more of a programmer, deciding the flow of control. This strategy is used in the programming language Prolog which, for efficiency reasons, deviates from the idea of pure logic programming in which the user would only state the relation between input and output (Sterling and Shapiro, 1994).

Specificity

This says that of all the rules that have matching instances, choose the instance of the rule that is *most specific*. Specificity can be measured in terms of the *number of tests* that patterns in rules need.

The intuition is that the more specific the conditions of a rule, the more appropriate the rule is likely to be in the given situation. Remember that the working memory models the STM of the problem solver and rules constitute the problem solver's knowledge and reside in the quasistatic LTM.

Specificity can facilitate default reasoning. When one has only a little information then one can make some default inferences (Reiter, 1980; Lifschitz, 1994). But if you have more information then you might make a different inference. The most popular example is as follows. If one knows that Tweety is a bird, then one can make an inference that Tweety can fly. But if in addition one knows that Tweety is also a penguin, then we infer that it cannot fly.

The following is another example of a default reasoning. If one were to be writing a program to play contract bridge (Khemani, 1989, 1994), then the bidding module might have many rules that cater to different hand holdings. Two examples are as follows:

- Default rule: If it is your turn to bid, then pass.
- 1NT rule: If it is your turn to bid, and no one else has bid before you, and you have 15–17 high card points and a balanced hand, then make an opening bid of one no trump.

Such default reasoning is common in many situations and can easily be implemented in rule based systems.

Recency

Looking at a pattern and adding a new element to the working memory is akin to making an inference. In logical reasoning one builds an argument step by step, from one lemma to the next. The difference between the mechanism of a rule based system and deriving proofs in logic is that the rules have to be sound for the proof to be valid. Rule based systems do more than logical reasoning and can address tasks like planning and classification as well. Nevertheless, the last lemma or WME added could provide the cue for the next step. This is behind the conflict resolution strategy called *recency*.

Recency aims to maintain a *chain of thought* in reasoning, with rules matching newer WMEs gaining preference over others. Of all the rules that have matching instances, choose the instance that has the most recent WME. Recency can be implemented by looking at the time stamps of the matching WMEs. The intuition is that when a problem solver adds a new element to the WM, then any rule that matches that WME should get priority. Recency can be implemented by maintaining the conflict set as a priority queue.

Means–Ends Analysis

In their pioneering work on human problem solving, Newell and Simon proposed a strategy of breaking down problems into subproblems, addressing the most significant subproblem first, and then attempting the rest (Newell and Simon, 1963; Newell, 1973). Each subproblem reduces some difference between the given state and the desired state, and is solved by *analysing* the *means* to achieve the *ends*. They named this strategy as *means–ends analysis* (MEA). We describe MEA a little later in this chapter.

OPS5 has the option of choosing MEA as a conflict resolution strategy. The idea is to partition the set of rules based on the *context* and focus on one partition at a time. One can think of each partition as solving a specific subgoal or reducing a specific difference. The context is set by the *first pattern* in a rule. All rules in the same partition have the same first pattern. The MEA strategy applies *recency* to the first pattern in each rule, and *specificity* for the remaining patterns.

Next, we look at how the IE can be implemented efficiently.

7.2.6 The Rete net

Consider a rule based system with a few hundred rules and a few thousand WMEs. Each rule has a few patterns, each of which has to be matched with each of the WMEs in the match phase. And this would have to be done in every cycle. Except, as Charles Forgy showed, it does not.

In the simplest algorithm, each pattern in each rule is compared with each WME in each cycle. Forgy reversed this idea in his RETEALGORITHM (Forgy, 1982). Instead, the patterns extracted from all the rules are arranged in a network which awaits each element that is added to the WM as a token. The patterns shared by different rules, even partially, are conflated in the network. What is more, once a token is drawn into the network, it is directed to one or more matching rules, and it stays there. This happens with each new WME that is created and added as a token. The key point is that WMEs are matched only once when they are created, and not in each match–resolve–execute cycle.

There are two kinds of tokens that are generated in each cycle. One <+WME> when a WME is added, and the other <–WME> when one is deleted. The positive token will go and reside in the network. The negative token will go and neutralize the existing positive token.

The origin of the name *rete* is based on the middle English word *riet*, itself derived from the Latin word *rete* meaning[1] 'an anatomical mesh or network, as of veins, arteries, or nerves'.

The Rete network is composed of two parts. The top half is a *discrimination tree*, in which at each node a property is tested and the token sent down different paths. A bit like a decision tree or a generalization of a binary search tree. The nodes in this part are called *alpha* nodes and are the locations of the tokens that travel down the discrimination tree. Alpha nodes have exactly one parent. Sitting below the alpha network is the beta network which assimilates the different tokens that match the different patterns in the rules. Beta nodes may have more than one parent, in which case the WMEs are joined based on equality of the values of variables. Rules themselves are attached to beta nodes which have collected all the WMEs that match their patterns. Figure 7.4 is a schematic of the Rete net.

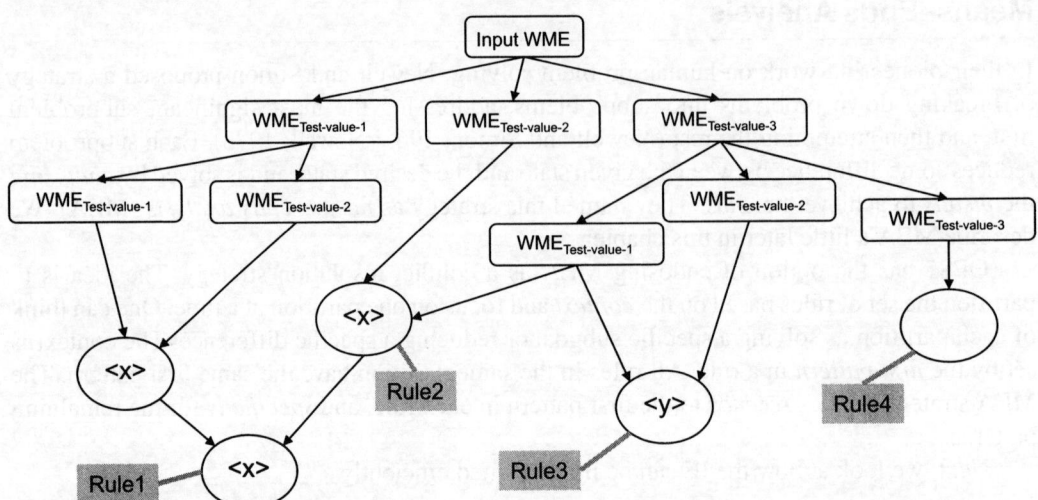

Figure 7.4 A schematic of a Rete network. The top half is a discriminatory network of alpha nodes shown in rectangles. The bottom half is made of beta nodes drawn as ovals and assimilate WMEs that match the patterns of a rule. The joins are typically made on values of variables. Rules themselves are attached to beta nodes, where WMEs matching all the patterns arrive. Observe that Rule2 is the default version of Rule1.

In Figure 7.4, Rule1 needs four matching WMEs which arrive from the four paths from the root to the corresponding beta node. Above Rule1 resides Rule2 which needs only two of those four WMEs, and thus can be seen as a default version of Rule1.

---

[1] https://www.yourdictionary.com/rete, accessed November 2022.

The Rete network is a directed acyclic graph that is a compilation of the rules that it embodies. The tokens generated by rule firing are inserted at the root and follow a distinct path for each class name that begins a pattern. Subsequently, they are routed based on the values of tests that each pattern makes. The bottom half is the beta network that knows the rules. The WMEs needed by different patterns in each rule are identified. Tokens may need to have the same values of a variable in different patterns for a join to happen. We look at an example to illustrate this relation between a set of rules and the corresponding Rete net.

Consider a database in which there are a set of named figures, along with a set of properties associated with each figure. We consider polygons with three or four sides. The properties are the number of sides the figure has, the number of equal sides the figure has, the number of parallel sides the figure has, and the number of right angles the figure has. These are defined by four class names. The task is to classify the figure and identify what geometrical shape that figure is. The following are the rules for the different shapes. The first rule simply states that if a polygon has three sides, it is a triangle. This is a default 3-sided figure. More specific rules identify different kinds of triangles.

(p triangle
      (sides ^in <name> ^ are 3)
→
      (make (polygon <name> ^instance-of triangle))

(p isosceles-triangle
      (sides ^in <name> ^ are 3)
      (equal-side ^in <name> ^are 2)
→
      (make (polygon <name> ^instance-of isosceles-triangle))

(p right-triangle
      (sides ^in <name> ^ are 3)
      (right-angles ^in <name> ^count 1)
→
      (make (polygon <name> ^instance-of right-triangle))

(p right-isosceles-triangle
      (sides ^in <name> ^ are 3)
      (equal-side ^in <name> ^are 2)
      (right-angles ^in <name> ^count 1)
→
      (make (polygon <name> ^instance-of right-isosceles-triangle))

(p trapezium
      (sides ^in <name> ^ are 4)
      (parallel-sides ^in <name> ^pairs 1)

→   (make (polygon <name> ^instance-of trapezium))

(p rhombus
    (sides ^in <name> ^ are 4)
    (equal-side ^in <name> ^are 4)
→   (make (polygon <name> ^instance-of rhombus))

Figure 7.5 shows the Rete net for the rules given above, along with some rules for other shapes that are left as an exercise for the reader. The root node tests for the class name in the input WME and sends the token down the corresponding path, where other alpha nodes apply tests for some attributes. In this scheme of things, the tokens reside in the first beta node they encounter and can be accessed by lower nodes.

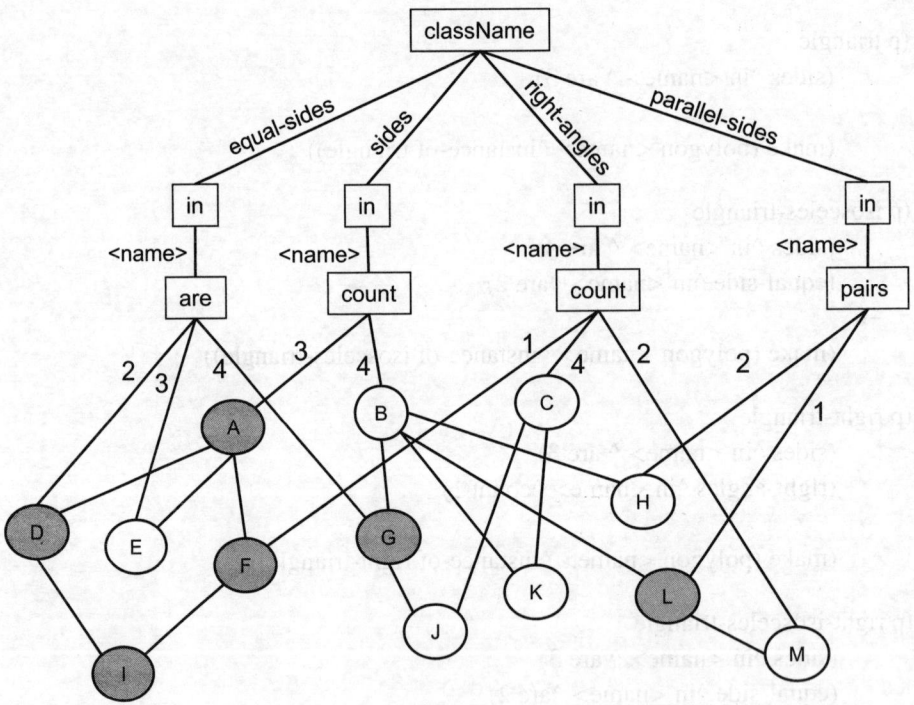

Figure 7.5 A Rete net for geometric shapes. The patterns describe the number of sides, number of equal sides, number of parallel sides, and the number of right angles. Rectangles are alpha nodes. Circles are beta nodes, and joins are on <name> variable. The shaded beta nodes have the rules described in the text attached.

If you only know that a shape has three sides, you can conclude that it is a triangle. The corresponding beta node has label *A* in Figure 7.5. In addition, if we know that there is a right angle in the figure, then we can classify it as right angled (node *F*). Instead, if we know that it has two equal sides, it is an isosceles triangle (node *D*). If it is both right angled and isosceles, then it is right isosceles (node *I*). Nodes *D, E, F*, and *I* are special cases of triangles. Node *G* hosts the rule for a rhombus, and node *L* defines a trapezoid. The reader is encouraged to write the rules for other figures in the Rete net.

7.2.7 The Rete algorithm

With the Rete net the IE has transmogrified away from the explicit match in the match–resolve–execute cycle, into a scheme where the WMEs find their place in the network. As discussed earlier, the RETEALGORITHM begins by compiling the rules into the Rete network. Then to begin with, the WMEs describing the initial state are inserted into the net.

When a WME is inserted into the Rete net it is routed by the tests at every node *en route*. It then resides in the node following the last successful test, till the point when it is deleted by a Remove action in the RHS of some rule. Till that time, it is available to join with other WMEs that may arrive in the future. Consider the following sample data describing some geometric figures. The WMEs are like the proverbial features of the elephant that the blindfolded men are trying to recognize, and are identified by their time stamp.

1. (sides ^in shape1 ^count 3)
2. (sides ^in shape2 ^count 4)
3. (sides ^in shape3 ^count 4)
4. (sides ^in shape4 ^count 3)
5. (equal-sides ^in shape1 ^are 2)
6. (equal-sides ^in shape2 ^are 3)
7. (equal-sides ^in shape3 ^are 4)
8. (right-angles ^in shape1 ^are 1)
9. (right-angles ^in shape3 ^are 4)
10. (parallel-sides ^in shape4 ^pairs 0)
11. (parallel-sides ^in shape2 ^pairs 1)

The above WMEs are converted into positive tokens and inserted one by one into the Rete net. We show their locations in the alternate, but equivalent, depiction of the net in Figure 7.6. This diagram views each alpha node as a placeholder with a test. Initially the token is placed in the root node named *token-in*. The alpha nodes below that include tests that a WME must satisfy to be accepted. This is different from the depiction in Figure 7.5 where each alpha node was labelled with the attribute being tested, and the edges emanating below were labelled with the accepted value of the test. Observe that, irrespective of the diagram schema, a token may be replicated and may go down more than one path if more than one test is satisfied. Remember that the tests come from patterns in different rules. For example, one rule may test for a value being greater than 5 while another may test for a value being greater than 11. If an incoming WME has a value 16 for that attribute, it will satisfy both tests.

202 | Search Methods in Artificial Intelligence

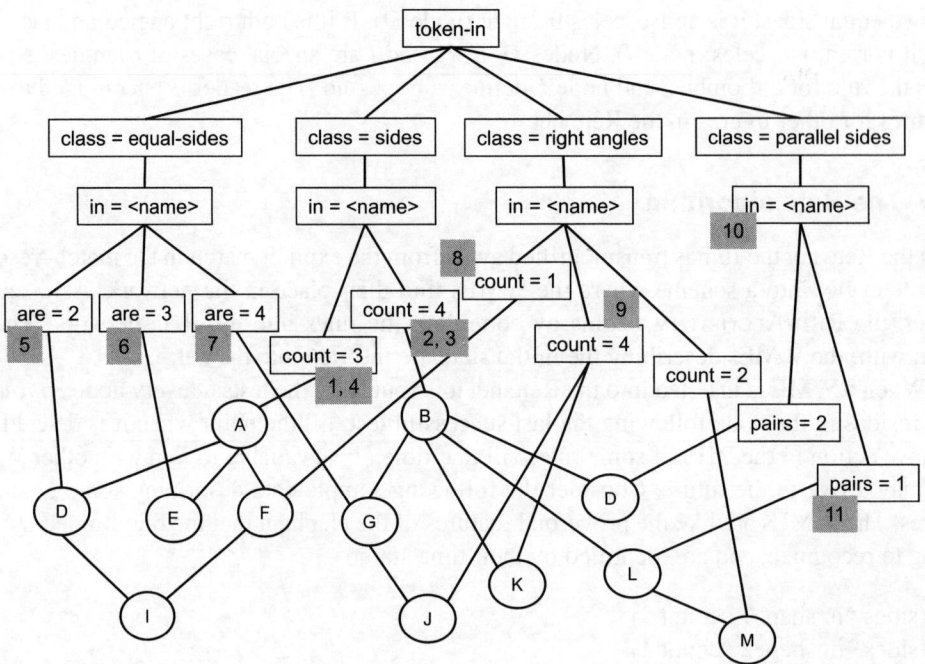

Figure 7.6 An alternate depiction of the Rete net from Figure 7.5 in which each alpha node contains the test that must be satisfied to accept a WME. The numbers in the shaded boxes are the time stamp values for the eleven WMEs inserted into the net.

Each token travels down the network as long as it satisfies the tests. Figure 7.6 depicts the location of the eleven WMEs described above. In this version of the figure, we assume that the tokens reside in the alpha node whose test they satisfy. In the figure they are shown as numbers in shaded boxes representing time stamps of the WMEs for brevity. In practice they should have a positive sign as well, for example, <+ 1>, <+ 2>, and so on. In our simple rule base there are no Remove actions, but if there were, then negative tokens like <–15> could have been inserted. A negative token would follow the same path as its positive version. When it collides with the positive token, both get destroyed. Any rules matching the positive WME must be retracted from the conflict set as well.

The beta nodes below the alpha nodes accept tokens from one or more parents. Receiving exactly one WME is rare. It typically happens with a rule with only one pattern, for example, the triangle rule attached to node *A* which adds two instances <triangle 1> and <triangle 4> for shape1 and shape4, the names of the figures, to the conflict set. Later, when the WME 5 arrives, the beta node *D* adds the instance <isosceles-triangle 1 5> to the conflict set for shape1. Observe that token 1 is still residing in the same alpha node, even though it has triggered two rules for shape1. In fact, when token 8, also for shape1, arrives, two new rule instances with data are triggered, <right-triangle 1 8> and <right-isosceles-triangle 1 5 8>. Thus, there are four rules waiting to classify shape1.

Observe that when the token for WME 10 arrives which says that shape4 has zero parallel sides, it stays in the alpha node as shown and does not trigger any rule. Tokens 2 and 3 talk of

4-sided figures and trigger quadrilateral rules <quadrilateral 2> and <quadrilateral 3> in node B for shape2 and shape3. When token 7 for shape3 arrives then <rhombus 3 7> is added to the conflict set, and when token 9 arrives <square 3 7 9> is also added. Finally, when token 11 arrives for shape2 <trapezium 2 11> is added.

The initial data is uploaded into the WM before any rule is selected. In our example, we have nine rules that enter the conflict set. Which rule to select for execution is determined by the conflict resolution strategy. We quickly review how the RETEALGORITHM implements the strategies described in Section 7.2.5.

Refractoriness

This happens naturally in the RETEALGORITHM. This is because the match happens when the WMEs are inserted into the net, and that happens only once. Let us say three WMEs with time stamps 10, 15, and 20 match a rule *someRule*. When 10 is inserted, it goes and sits in its place, waiting for its partners that could trigger any rule. Likewise for WME 15. Now when WME 20 arrives <someRule 10 15 20> is added to the conflict set. Since WMEs 10, 15, and 20 have already been generated and inserted, this rule–data combination can fire only once. When it does fire, it is removed from the conflict set.

Lexical Order

Forward chaining rule based systems in general and OPS5 in particular ignores the lexical order of rules that would be present in a text file. This is because the Rete net obliterates that order. Hence lexical order cannot be realized.

Specificity

This says that of all the rules that have matching instances, choose the instance of the rule that is *most specific*. Specificity can be measured in terms of the *number of tests* that patterns in rules need. In the Rete net this can be measured by summing up the lengths of the paths for all matching patterns.

Recency

Recency gives preferences to the rules that match the most recent WMEs. This can be implemented by maintaining a priority queue. When a new WME goes in and triggers a set of rules, then all these rules will go to the front of the priority queue.

Note that this implementation only takes care of one pattern in the rule, the one that matches the most recent WME. Breaking ties by choosing a rule with the second most recent WME is a harder task, and is left as an exercise for the reader to ponder over. Perhaps one could add the time stamp values, but what when one rule has more patterns than another?

Means–Ends Analysis

The MEA strategy looks at the recency of the first pattern in every matching rule. Ties can be broken by specificity.

When rule firing begins on the Rete net with WME tokens depicted in Figure 7.6 then, in some order, the following inferences are made:

- shape1 is a triangle
- shape2 is a quadrilateral
- shape3 is a quadrilateral
- shape4 is a triangle
- shape1 is an isosceles triangle
- shape3 is a rhombus
- shape1 is a right triangle
- shape1 is a right isosceles triangle
- shape3 is a square
- shape2 is a trapezium

The reader is encouraged to explore the order in which the above conclusions are asserted with the different conflict resolution strategies. In this example, all the ten instances of rule–data fire, and then the program terminates.

Different conclusions are drawn about the different figures that are represented in the WM. As an exercise, modify the rules so that exactly one conclusion is drawn by each figure. Hint: you can include Remove actions in the RHS. What conclusion will be drawn for each of the four figures – shape1, shape2, shape3, shape4 – under the different conflict resolution strategies?

Finally, we look again at how the match–resolve–execute cycle is handled by the RETEALGORITHM. In the brute force approach, the MATCHALGORITHM compares the patterns in the rules with WMEs and produces the conflict set. In the RETEALGORITHM, the rules are compiled into a network that accepts changes in the WM and produces changes in the conflict set, as shown in Figure 7.7.

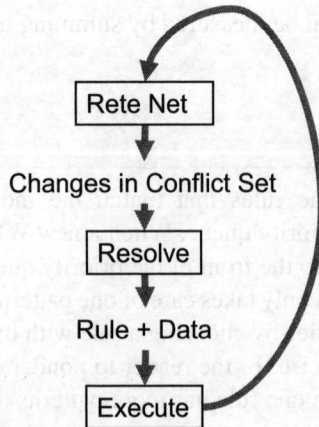

Figure 7.7 The Rete net is a compilation of the given rules and also hosts the WMEs. The tokens generated when a rule is fired are the new input, resulting in changes in the conflict set.

This selective processing of the incoming data makes the rule based system an order of magnitude faster. Even though the original goal of building expert systems was on the wane, the business community adopted the technology with great enthusiasm, since it allowed them to focus on the rules. Adopted as a business rule management system technology, the Rete algorithm was refined many times, resulting in more speedups.

Carole-Ann Berlioz of Sparkling Logic wrote the following in her blog[2] in 2011.

> The best usage of the Rete network I have seen in a business environment was likely Alarm Correlation and Monitoring. This implies a 'stateful' kind of execution where alarms are received over time as they occur. When you consider that the actual Telecom network is made of thousands of pieces of equipment to keep in 'mind' while processing the alarms, there is no wonder that Rete outperforms brute force. When one alarm is raised on one Router, the Rete network does not have to reprocess all past events to realize that we reached the threshold of 5 major alerts on the same piece of equipment and trigger the proper treatment, eventually providing a probable diagnostic. Hours of processing time turned into seconds in Network Management Systems. Fabulous.

At some point Charles Forgy implemented a version called Rete-NT™ that was another order of magnitude faster. He, however, decided not to publish the algorithm and kept it as a trade secret.

7.3 Problem Decomposition with And–Or Graphs

The idea of rule based expert systems emerged as a response to the difficulty of formulating heuristic functions to guide search. Rule based systems encapsulate knowledge elicited from human experts in the form of rules, and use that to drive reasoning with the IE. Search received less attention. The hope was that one could accrue sufficient knowledge to eliminate the need for search. But acquiring large amounts of knowledge has proven to be difficult and is called the *knowledge acquisition bottleneck* (Hayes-Roth, Waterman, and Lenat, 1983). Most of the systems that were built were hand-crafted and consequently were also narrow in their coverage. For the same reason, they were also brittle, failing to perform outside their regions of competency. Modern machine learning methods have attempted to get around this bottleneck by learning from data, even making remarkable progress in classification tasks. Looking at images in medical diagnosis where they can exploit the data gleaned from thousands of human radiologists has been a huge success. But again, such learning is task specific.

One cannot escape search if one aims for completeness. Rules can be acquired or learnt to look at patterns and predict class labels, but many problems require a combination of search and knowledge, for example, in the system DENDRAL described later. A different kind of search space is generated in which problems are broken down into smaller problems. This space is an And–Or (AO) graph in which some of the choices emanating from a node may be hyperarcs that link the problem to its subproblems.

---

[2] https://www.sparklinglogic.com/rete-algorithm-demystified-part-1/, accessed November 2022.

The AO graph is a directed graph rooted at a node representing the problem to be solved. Figure 7.8 shows a contrived AO graph in which the root node G is the goal to be solved. The And edges are shown by connecting the related edges with an arc. As can be seen, there are three ways the goal G can be transformed. One can solve G by solving A, or solving B, or solving both C and D. Whichever choice one makes, one can transform the resulting problem into one or more subproblems. The shaded nodes in the graph are primitive problems or *solved* nodes. Each *solved* node may have an associated cost, and each edge transforming a problem may have an associated cost too.

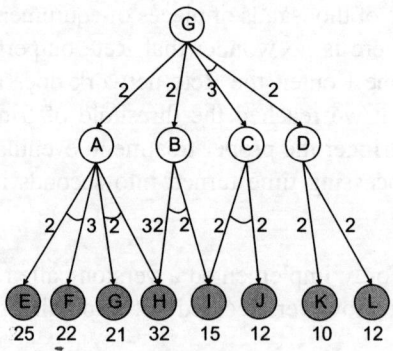

Figure 7.8 An And–Or graph decomposes a problem into subproblems. Here goal G can be transformed into problem A, or problem B, or problem $C + D$. Edges are labelled with transformation costs. Shaded nodes are primitive problems whose solutions are known along with their cost.

The reader should verify that the combinations of *solved* nodes for solving G are $\{E, F\}$, $\{G, H\}$, $\{H, I\}$, $\{I, J, K\}$, and $\{I, J, L\}$. Which combination has the lowest cost?

We first look at programs from literature that employ AO graphs, and then we look at an algorithm to solve And–Or graphs.

7.3.1 DENDRAL

DENDRAL (for DENDRitic ALgorithm) was an early expert system, developed beginning in 1965 by Edward Feigenbaum and the geneticist Joshua Lederberg, at Stanford University (Lindsay et al., 1980, 1993). It was designed to help chemists in identifying unknown organic molecules, by analysing their mass spectra and using knowledge of chemistry. DENDRAL is often considered the first expert system, though the authors of MYCIN would have contested that.

DENDRAL was a program designed to assist a chemist in the task of deciphering the chemical structure of a compound whose molecular formula was known and some clues about its structure were available from mass spectroscopy. The problem is important because the physical and chemical properties of a substance depend upon how the atoms are arranged. Figure 7.9 shows some ways in which the atoms of $C_6H_{13}NO_2$ could be arranged. The reader might recall that hydrogen atoms have valency 1, carbon has 4, oxygen 2 and nitrogen 3. This is reflected as edges in the graph representation. The hydrogen atoms are shown only in the structure on the right for brevity.

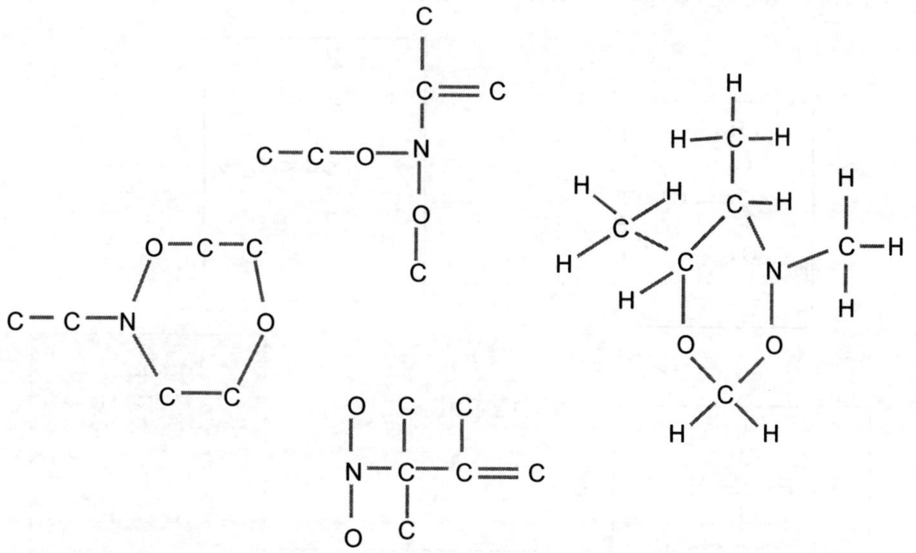

Figure 7.9 DENDRAL was designed to assist chemists in finding the structural formula of compounds. The above are some candidate structures for $C_6H_{13}NO_2$. The hydrogen atoms are shown only for the arrangement on the right. Figure adapted from [Buchanan 82].

Earlier versions of DENDRAL could handle compounds up to a hundred atoms, but even there millions of arrangements are possible, making the chemist's task quite hard. The problem has been addressed by handling substructures, called 'superatoms', in the search process. To discover the structure, chemists use a mass spectrometer that breaks up the chemical into some constituents and produces a spectrogram in which peaks correspond to the mass of the constituent. Chemists have some knowledge of how fragments could be created when bombarded by atoms, and the kind of spectrogram that might be produced. But looking at the spectrogram and determining the structure is harder, because there are combinatorially increasing possibilities. What spectroscopy reveals is the arrangement in some substructures, which become constraints in exploring the possible structures. The DENDRAL program had a planner that could decide which constraints to impose in the search for possible candidates. It applied knowledge of spectral processes to infer constraints from instrument data. DENDRAL extends the generate-and-test approach of search to plan-generate-and-search. The planner strives to generate only meaningful candidates to narrow down search, while not ignoring solutions.

A constrained generator (CONGEN) accepts constraints and generates complete chemical structures. Some of the symbols manipulated by the structure generation algorithm stand for 'superatoms', which are collections of atoms arranged in a particular way, along with their collective valency. The program has to expand these to get the final structure. The chemist may specify the superatoms to be used. For example, such a substructure may have four carbon atoms, two of which may have free valencies of 1 and 2. These will be linked by CONGEN to other structures. Figure 7.10 illustrates the kind of search space DENDRAL explores. It is an And–Or graph, with hyperedges connected by arcs.

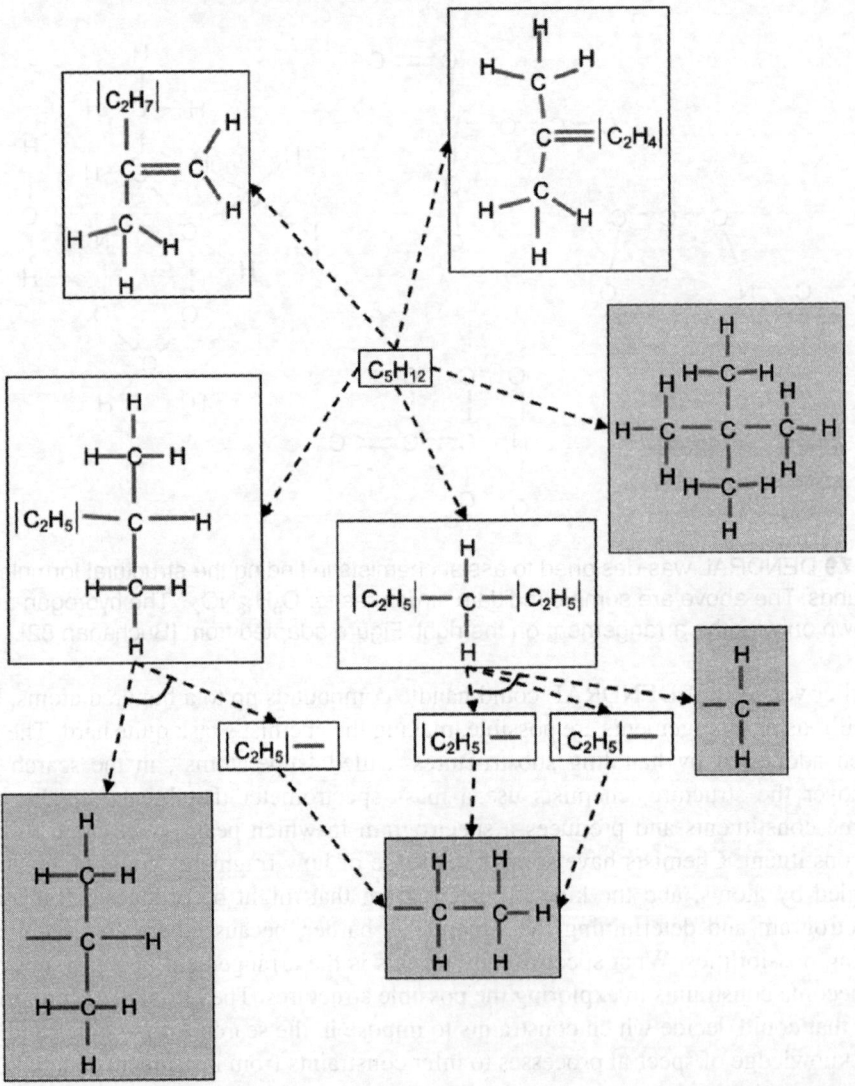

Figure 7.10 The program DENDRAL explores and And–Or graph. It generates candidate structures and a synthetic spectrogram. Formulas in vertical lines are superatoms. The shaded nodes represent completely specified structures or substructures. Figure from Mainzer (2003: ch. 6).

The following is a record of a session with CONGEN (Lindsay et al., 1993). The constraints listed below illustrate the flexibility and power of the program to use information about chemical structure that may be inferred (manually or automatically) from a variety of analytical techniques.

C1. The empirical formula is $C_{12}H_{14}O$.
C2. The compound contains a keto group in a 5-membered ring.
C3. There are three protons (H's) alpha (adjacent) to the carbonyl group.
C4. There are two vinyl groups (–C=C–) and four vinyl protons.
C5. There is no conjugation (alternation of double and single bonds).
C6. There are no diallylic protons (hydrogens at the middle carbon of a diallylic structure: –C=C–C–C=C–), nor protons alpha to both a vinyl and the keto group.
C7. There are only two quaternary carbons, one in the keto group and one in one of the vinyl groups.
C8. There are no additional multiple bonds.
C9. It is assumed there are no 3- or 4-membered rings.
C10. There are no methyl groups.

DENDRAL is not a single program, but a collection of programs. The project consisted of research on two main programs, Heuristic DENDRAL and Meta-DENDRAL, and several subprograms. The first uses knowledge of chemistry and mass spectroscopy data to hypothesize possible structures. The second accepts known mass spectrum/structure pairs as input and attempts to infer the specific knowledge of mass spectrometry that can be used by Heuristic DENDRAL (Lindsay et al., 1993).

7.3.2 Symbolic integration

Mathematics is often symbolic, involving manipulation of symbolic expressions into simpler ones, or closed form representations more amenable to finding answers. One such activity that has drawn computer scientists is symbolic integration. Many of us have struggled with solving integral equations. One of the earliest programs for symbolic integration was called SAINT (Symbolic Automatic INTegrator) written by James Slagle as part of his PhD thesis (Slagle, 1961, 1963). This prompted Joel Moses to implement SIN (Symbolic Integration). Moses felt that 'one should endow computers with knowledge of the problem domain so that searches could be largely eliminated, at least much of the time' (Moses, 1967, 2008). His work was subsequently integrated into the *Macsyma* (Project MAC's SYmbolic Manipulator) project. Symbolic integration is also a part of WolframAlpha's Mathematica.[3]

Integral calculus, as students of mathematics are well aware, is replete with techniques to transform problems into other problems by substitution. Such substitutions may involve choices (Or edges) or may involve problem decomposition (And nodes). The goal is to simplify and reduce integration problems to a set of primitive problems whose solutions are available in a table. Figure 7.11 shows the search space as an And–Or tree in an example from (Nilsson, 1971). Observe that many of the transformations are reversible, being substitutions.

---

[3] https://www.wolfram.com/mathematica/, https://mathworld.wolfram.com/, accessed November 2022.

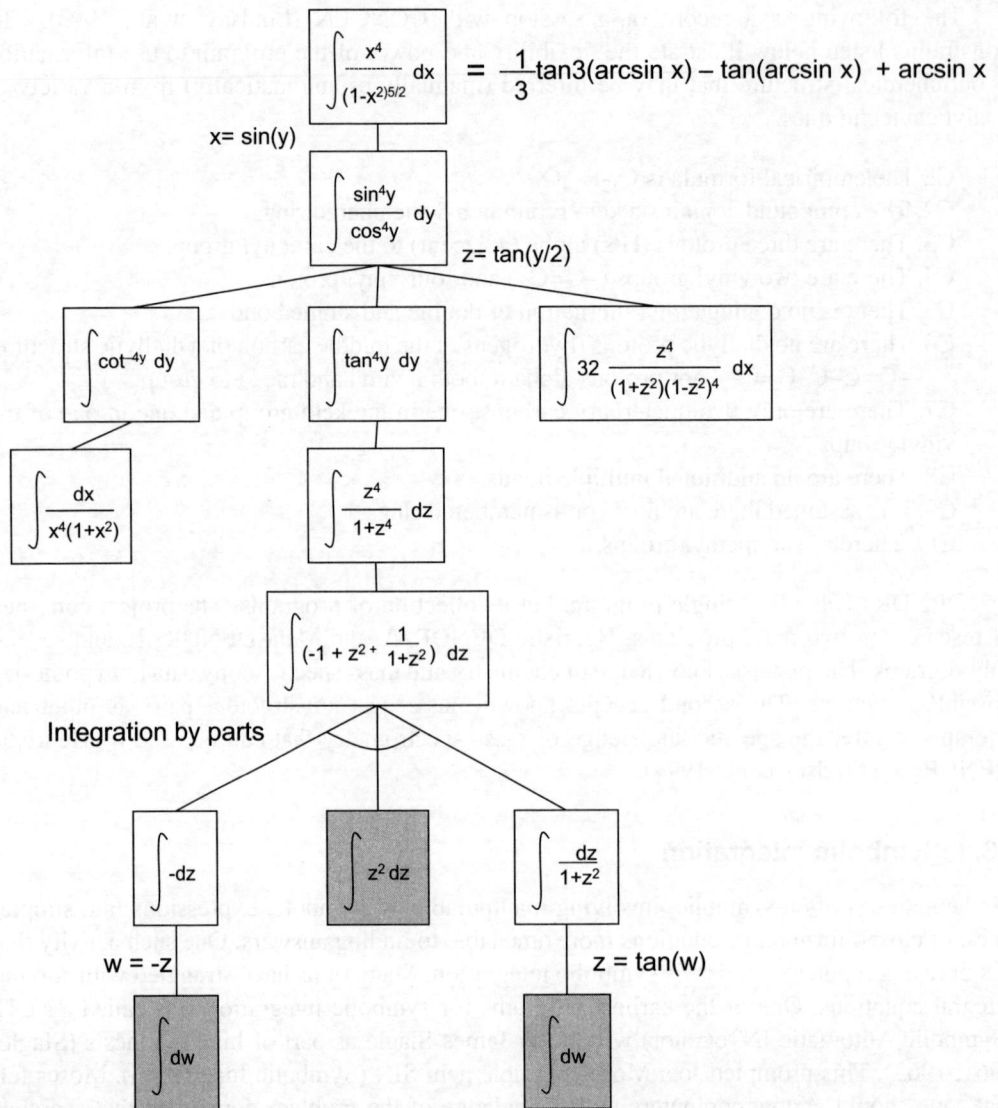

Figure 7.11 Symbolic integration involves searching through a space of substitutions and decompositions in the quest for primitive integration problems. The figure shows a part of the search tree, along with a subtree that has three primitive problems. Figure from Mainzer (2003: ch. 6).

Problem decomposition is an integral part of problem solving. It represents a goal directed or teleological view of problem solving where the thought process begins with a goal that represents a problem to be solved. Planning a dinner, for example, may involve planning the different courses, which could be addressed independently. Unless, of course, there is a theme for the dinner in which case the dessert may depend upon what is the main course. Designing a house likewise could be broken down into subtasks of the foundation, the walls, the doors and windows, the kitchen, the roof, and so on. For each of the subproblems, there may be options to choose from, and thus the search space is an And–Or tree (or a graph if solution components can be reused). Such search spaces are also called goal trees, because they transform goals into subgoals. More recently And–Or trees have been used in graphical models (Dechter and Mateescu, 2007).

We make an assumption here that the subgoals can be solved independently. Sometimes this may not be true, as when planning a thematic dinner. In Chapter 10 on planning, we will encounter this problem of not independent subgoals again, even for simple planning problems. We say that the subgoals are non-serializable. This means that one cannot solve the subgoals independently in a serial order and arrive at the whole solution. In Chapter 9 when we study goal directed reasoning, we will encounter this dependency between subgoals. The solution of a previous subgoal will constrain the solution of the current one.

Next, we study the well-known algorithm AO*, which as the name suggests, is an admissible algorithm. This means that under certain conditions, including an underestimating heuristic function, the algorithm finds an optimal solution. The assumption is that the subgoals can be solved independently.

7.3.3 Algorithm AO*

And–Or graphs are also known as *goal trees* because they represent the process of breaking down a complex goal into simpler subgoals. The leaf nodes are simple problems whose solution is trivial or known. They are labelled *solved*.

The AO* algorithm for solving goal trees begins with a single node that represents the goal to be achieved (Martelli and Montanari, 1978; Nilsson, 1980). As the search progresses, it builds the graph piece by piece. At every choice point in the graph, an *Or* node, it marks the best choice which has the lowest estimated cost. The algorithm alternates between two phases.

In the forward phase, it follows the marked path and chooses an unexpanded node. If there is more than one candidate for refinement, it means that they are all linked by *And* hyperedges, and any one can be picked in principle. In practice choosing the most expensive one may allow one to switch to a different candidate early. Expanding or refining a node entails adding edges to the selected node. The edges point to transformation and decomposition options. It employs a heuristic function that estimates the cost of the solution for each node. The algorithm looks at the options and chooses the best option, and marks it as the best option, and also updates the estimated cost of the node just refined. Figure 7.12 illustrates the forward phase for an example graph, in which every edge has a cost 10 associated with it. The algorithm begins from the root G, follows the markers, and arrives at node J with estimated cost, $h(J)$, 170. Refinement produces two options, $\{L, M\}$ and $\{N, O\}$. The heuristic values are $h(L) = 100$, $h(M) = 80$, $h(N) = 120$, and $h(O) = 120$. The cost of solving J using $\{M, N\}$ is $200 = h(L) + h(M) + 10 + 10$. The cost of solving J using $\{N, O\}$ is $260 = h(N) + h(O) + 10 + 10$. The former is cheaper and AO* marks it as the better choice, and updates $h(J)$ to 200. One of the children may be a *solved* node, in which case it cannot be refined further. If it is the best option for its parent, then the parent node (J in this example) is labelled *solved*. In the case of *And* hyperedges, *all* children below must be labelled *solved* for the parent to be labelled *solved*. All other nodes in the graph which are not *solved* are assigned the status *live*.

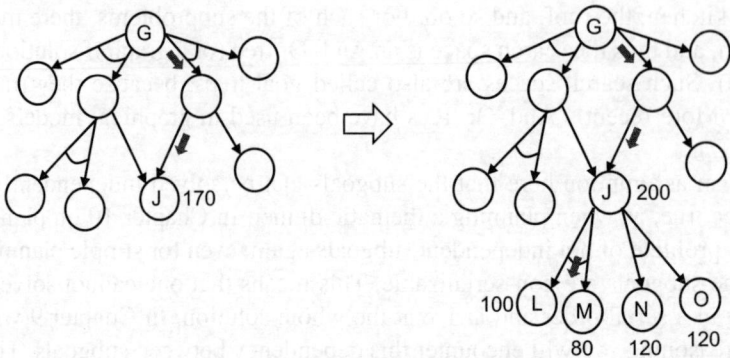

Figure 7.12 In the forward phase, algorithm AO* follows the markers and picks a node (here node J with $h(J) = 170$) for refinement. It expands J into two options, {L, M} and {N, O}, and marks the cheaper option {L, M}. Each edge has a cost 10. The estimate of node J changes to 200 which is $h(L) + h(M) + 10 + 10$.

The last step when $h(J)$ was updated triggers the second phase, which is the backward phase. In this phase the updated estimates are propagated upwards. In the above example, when node J changes, the change must be propagated to all its parents.

The following is the outline of the algorithm AO*. It uses a value *futility* to set the limit of an acceptable cost of the solution (Rich, 1983).

WHILE the root is labelled *live* AND $h(root) < futility$
1. Forward: Follow the marked path from the root to a set of live nodes L. Pick some node N from L.
2. Refine N to get a set of options $O_1 \ldots O_P$.
3. Compute the estimate for each option O_i. This may be the sum of all constituents if it is an *And* hyperedge, plus the cost of each edge. For each node in O_i which is a primitive problem, label it *solved*.
4. Choose the best option O_{best} and mark it. Update $h(N)$ with cost of O_{best}. If every node in O_{best} is labelled *solved*, assign label *solved* to N.
5. If N has changed, initialize the set of changed nodes C with {N}.
6. Backward: WHILE C is not empty
   a. Pick a node N from C and remove it.
   b. Propagate change in $h(N)$ to *each* parent P in the graph.
   c. If P has changed, add P to C.

IF root is labelled *solved* return the subgraph of marked nodes
ELSE return *fail*

It should be emphasized that when a node changes in the backward propagation phase, all its parents must consider the change, and not just the parent figured in the forward phase. This is because the other parents may also get affected, and it might have an impact on the solution found. Figure 7.13 illustrates the propagation in the backward phase.

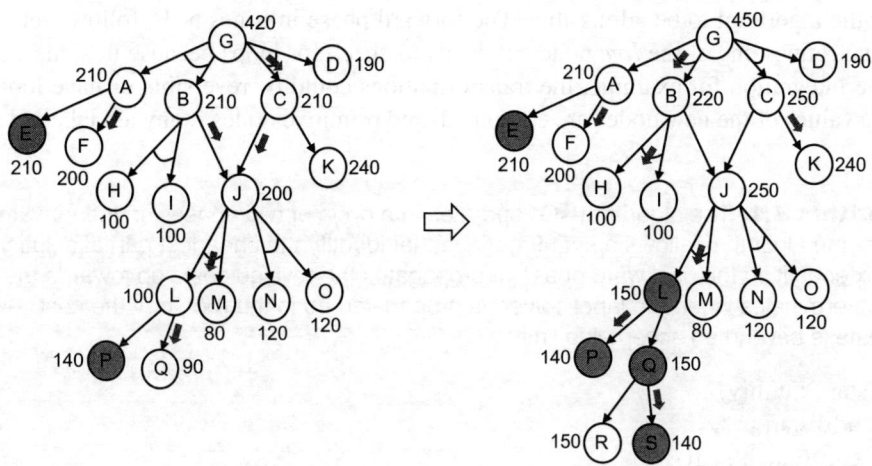

Figure 7.13 In the backward phase, algorithm AO* propagates any changes seen back towards the root. In this example, after node Q is expanded, it reaches a *solved* node S which is cheaper than the *live* node R. Consequently, Q gets labelled *solved* as well, but its sibling P gets the marker as it is cheaper. Node L is now *solved* too, but the hike in cost results in AO* shifting its attention to the {A, B} option. Notice that the estimate of G has escalated as well.

Each edge in the figure has a cost 10. The shaded nodes in the figure are *solved* nodes and have an associated known cost. In the forward phase, AO* lands up on the *live* nodes M and Q, with estimates 80 and 90 respectively. It picks Q, which is an *Or* node, and refines it. Of the two children R and S of Q, the latter is cheaper, and so it gets the marker from Q, whose estimate goes up to 150 (= 140 + 10). S is also a *solved* node and, since it is the marked option, Q is labelled *solved* as well. But its sibling P is cheaper and so gets the marker from L, which is now labelled *solved* as well. Its parent J now has a revised estimate of 250. This is propagated up to *both* the parents B and C, even though C is the partial solution being currently refined. Notice that both B and C abandon J and shift to other options. With the other option K, the cost of C has gone up to 250, so that the solution containing {C, D} now has an estimated cost 460. The option {A, B} offers a solution with a lower cost 450, and AO* will go down that road in the next forward move.

The cycle of forward and backward moves continues till the root is labelled *solved*, or till the estimated cost of the root becomes greater than an acceptable limit, *futility*. When it finds a solution, it must be the optimal solution, *provided* the heuristic function underestimates the cost of each node. The argument is similar to the one made in the B&B for TSP and A*. The markers serve the purpose of identifying the cheapest partial solution to be refined. If the estimate of another option is higher, then it can be safely excluded because the actual cost will be still higher. Thus, the AO* algorithm is an instance of the common theme described earlier.

Refine the candidate solution with the lowest estimated cost,
till the lowest cost candidate is fully refined.

Algorithm 7.1 presents the details of AO*. *Futility* is a large number that sets the limit for the acceptable cost. Line 1 begins by adding the goal as the start node. A heuristic function $h(N)$ returns an estimate for solving N. As discussed earlier, this must underestimate the actual cost for the algorithm to be admissible. The forward phase in Lines 5–16 follows the marked nodes and refines one of the *live* nodes. It needs to check for loops because in some domains, symbolic integration for example, the transformations could be reversible or have loops. The heuristic values of the new nodes are computed, and primitive nodes if any are labelled *solved*.

Algorithm 7.1. The algorithm AO* operates in a cycle of two phases. In the forward phase, the algorithm follows a set of markers that identify the cheapest partial solution and extends it. In the backward phase, it propagates the revised costs up towards the root. It terminates when the label *solved* is propagated up to the root, or if the cost estimate is beyond an acceptable limit.

AO*(start, Futility)
1   **add** start **to** G
2   **compute** h(start)
3   solved(start) ← FALSE
4   **while** solved(start) = FALSE **and** h(start) ≤ Futility
          ▷ FORWARD PHASE
5   U ← **trace marked paths in** G **to a set of unexpanded nodes**
6   N ← **select a node from** U
7   children ← SUCCESSORS(N)
8   **if** children **is empty**
9   h(N) ← Futility
10   **else check for looping in the members of** children
11   **remove any looping members from** children
12   **for each** S ϵ children
13   **add** S **to** G
14   **compute** h(S)
15   **if** S **is primitive**
16   solved(S) ← TRUE
          ▷ PROPAGATE BACK
17   M ← {N}   ▷ set of modified nodes
18   **while** M **is not empty**
19   D ← **remove deepest node from** M
20   **compute best cost of** D **from its children**
21   **mark best option at** D **as** MARKED
22   **if all nodes connected through marked arcs are solved**
23   solved(D) ← TRUE
24   **if** D **has changed**
25   **add all parents of** D **to** M
26 **if** solved(start) = TRUE
27   **return the marked subgraph from** start **node**
28 **else return null**

The backward phase begins by adding the newly refined node to a set of modified nodes M. Choosing the lowest node D from M, one computes its estimated cost and marks the subtree from where it is propagated. If the subtree is labelled *solved*, D is labelled *solved* as well, and its cost becomes the actual cost. If D has changed, then *all* its parents are added to M, and they will be updated in turn.

If the root is labelled *solved*, then the algorithm returns the subtree containing the marked edges and the corresponding nodes.

As described above, the solution returned by AO* is a subtree or a subgraph of the problem graph. The leaves of the solution are the primitive *solved* nodes and represent the solution parts that together solve the original problem. In this way, solving And–Or graphs is different from path finding. The fact that the solution is a subgraph has bearing also on the contribution that heuristic value of an individual node makes during the search process. In A* the heuristic value $h(N)$ of a node directly represents an estimate of the distance to the goal, and $f(N)$ the estimated total cost. In AO* an individual node may only be a part of the solution, and the total estimated cost is determined by other constituents of the solution as well. This explains the need for the backward phase where the estimated cost of the entire solution is aggregated, and the best options marked, which guide the forward movement in each cycle.

Figure 7.14 shows a solution on a plausible problem graph on which the progress of AO* is shown in Figure 7.13. The labelled nodes are as explored by the algorithm, and the unlabelled nodes are yet to be explored. The shaded nodes are the primitive nodes. The solution is identified by solid edges, while the rest of the problem graph is drawn with dashed edges.

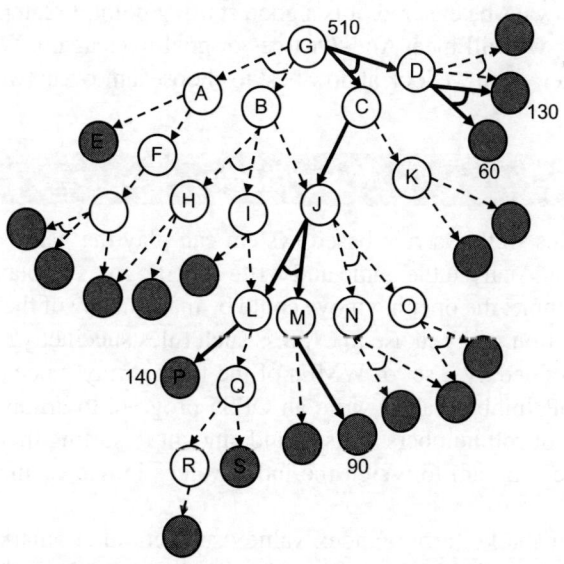

Figure 7.14 The partial solution of Figure 7.13 extended to a plausible complete problem graph. The unlabelled nodes are yet to be explored. A solution is a subtree with solved nodes as leaves. The subtree of solid edges is one solution with total cost 510. This is the sum of the cost of the four leaves that are part of the solution, along with a cost 10 for every edge.

As can be seen, the solution shown here is a refinement of the option $\{C, D\}$. There are four nodes in the solution with costs 140, 90, 60, and 130, of which only node P has been explored in Figure 7.13. Along with the costs of the edges in the solution, the total cost is 510. The question is whether this is the optimal solution. For that we need to know the costs of all the primitive nodes. We pose this question again in the exercises where the complete graph is given.

Summary

In this chapter we have taken a different view of problems and problem solving, in which we focus on subproblems while searching for a solution. We have looked at two approaches, each requiring its own representation scheme.

The first is a continuation of the forward chaining approach which moves from the start state in search of achieving the goal. The difference is that, instead of a monolithic state transition system, we look at how patterns in the state description can trigger moves or rules. The idea behind this modular rule or production representation arose from the desire to elicit problem solving knowledge from human experts and use that to drive the problem solving process. The evolution of this approach led to the development of a formalism that could not only capture expert heuristic knowledge, but also serve as a complete programming language.

The second approach is the goal-directed backward reasoning approach which starts with the goal to be achieved and explores means of breaking it down into parts that can be solved independently. This becomes possible by capturing the choices as well as the part–whole relations in an And–Or graph, which the algorithm AO* explores. While the independence of subproblems cannot always be ensured, it is a good starting point. In later chapters on planning and logical deduction, we will meet And–Or trees or goal trees again. We will also encounter them in the next chapter when we look at how best to choose a move in two player board games.

Exercises

1. Write a set of rules so that a rule based system can play the game of tic-tac-toe without resorting to search. Analyse the game and write specific rules to play a move in a specific situation. For example, the opening move could be made in any of the corner squares. What kind of representation will you use to express such rules succinctly?
2. Given an array defined by a set of WMEs of the form (array ^index value ^subject value ^marks value ^rollNumber value) write an OPS5 program to arrange the index values in increasing values of roll numbers. This would amount to sorting the array on roll number. Modify the above program to assign the index values in decreasing values of the marks attribute.
3. Given an array of marks (array ^index value ^subject value ^marks value ^rollNumber value) and a set of cutoffs for a set of grades $\{A, B, C, D, E, F\}$ for each subject, write a program to assign grades to each student for each course. How would you choose to represent the cutoffs, and how many rules would you need for three subjects – economics, history, and political science?

4. Six rules for geometric shapes have been defined in the chapter, and these are attached with the shaded beta nodes in Figure 7.5. Define the rules for the unshaded beta nodes and give the shapes their commonly accepted name. Does node *C* have a common name? [Hint: the number of sides is not known.]
5. The six rules and the eleven WMEs together result in ten conclusions about shape 1, shape 2, shape 3, and shape 4, as given in Section 7.26. Simulate the execution of these ten rule–data instances using the specificity, recency, and MEA strategies.
6. Assume the following schema for WMEs:
   a. (Person ^name ^age ^gender) gender: *M/F/N*
   b. (Work ^person ^nature) nature: self-employed/government/corporate/NGO
   c. (Habits ^person ^activity) activity: smoking/trekking/cricket
   d. (Education ^person ^completed) completed: highSchool/bachelors
   e. (Eligible ^person ^loan) loan: yes/no

   A bank uses the following rules for deciding whether a person is eligible for a loan or not:
   a. If the person has finished high school and smokes, then s/he is not eligible
   b. If the person is in a corporate or government job and does not smoke, s/he is eligible
   c. If the person's name is Vijay or Nirav, he is eligible
   d. If the person is a female graduate, then she is eligible
   e. If the person is a self-employed female, she is eligible

   Express the above rules in an OPS5-like language, and construct and draw a Rete net for the above rules.
7. For the above problem, given the following WM, list the conflict set for the above set of rules:
   1. (Person ^name Sunil ^age 37 ^gender *M*)
   2. (Person ^name Jill ^age 22 ^gender *F*)
   3. (Person ^name Sneha ^age 27 ^gender *F*)
   4. (Habits ^person Sneha ^activity cricket)
   5. (Habits ^person Sunil ^activity smoking)
   6. (Education ^person Sneha ^completed bachelors)
   7. (Education ^person Sunil ^completed bachelors)
   8. (Education ^person Jill ^completed highSchool)
   9. (Work ^person Jill ^nature corporate)
   10. (Work ^person Sunil ^nature self)
   11. (Work ^person Sneha ^nature NGO)

   Which element of the conflict set would be selected if the conflict resolution strategy is specificity?

   Which element of the conflict set would be selected if the conflict resolution strategy is recency?

   Who is/are eligible for a loan?
8. The rule based program as given in the text produces all possible category labels for the input figures. Modify the program to assign only one category to each figure and simulate the different conflict resolution strategies. [Hint: you can include *Remove* actions in the RHS.] What conclusion will be drawn for each of the four figures – shape1, shape 2, shape 3, shape 4 – under the different conflict resolution strategies?

218 | Search Methods in Artificial Intelligence

The graph below is a problem graph similar to the one used in Figure 7.14.

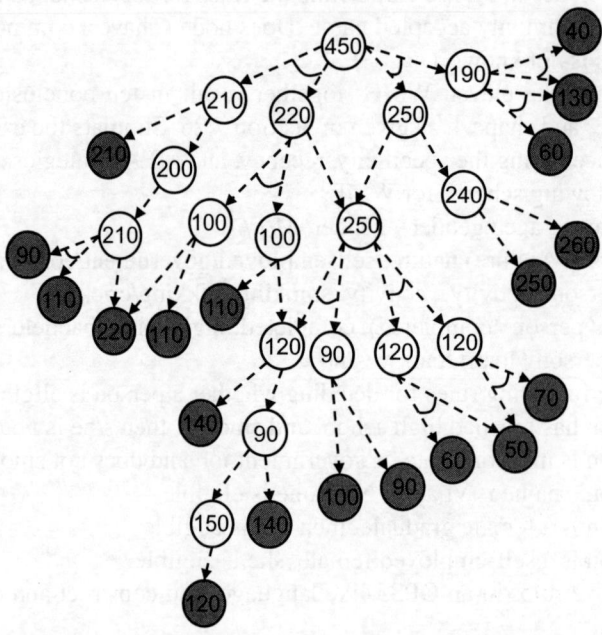

9. Propagate the costs in the *solved* nodes upwards in the above graph and replace the heuristic estimates with actual costs. What is the cost of the optimal solution? Does the heuristic function underestimate the actual costs?
10. Hand simulate the AO* algorithm on the above graph and draw the solution found. What is the cost of the solution found?
11. Divide the heuristic estimates of the internal nodes by 2 and repeat the hand simulation. What is the impact of the lower heuristic estimates?
12. Multiply the heuristic estimates of the internal nodes by 2 and repeat the hand simulation. What is the impact of the lower heuristic estimates?
13. [Baskaran] Show how the algorithm AO* will explore the following And–Or graph. Assume that each edge has a cost 1. Draw the graph after each move. Repeat the process after assuming that each edge has cost 10.

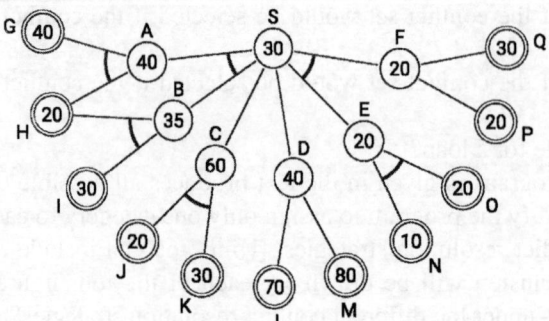

14. [Baskaran] Show how the algorithm AO* will explore the following And–Or graph. Assume that each edge has a cost 2. Draw the graph after each move. Repeat the process after assuming that each edge has cost 10.

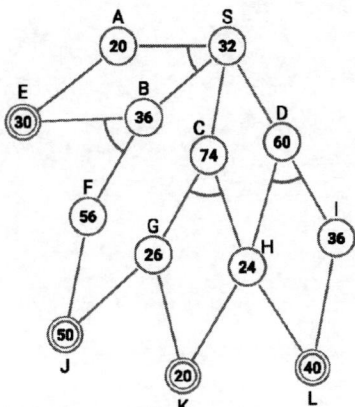

15. Multiplying a sequence of matrices can be posed as an And–Or graph, though it is usually seen as a dynamic programming problem because the costs are known. Given four matrices $A[5 \times 15]$, $B[15 \times 2]$, $C[2 \times 20]$, and $D[20 \times 1]$ we need to find the order in which to pick two matrices to multiply, so that the total number of multiplications is minimized. Pose this problem as an And–Or search problem and identify the optimal solution.

CHAPTER 8

Chess and Other Games

Acting rationally in a multi-agent scenario has long been studied under the umbrella of *games*. Game theory is a study of decision making in the face of other players, usually adversaries of the given player or agent. Economists study games to understand the behaviour of governments and corporates when everyone has the goal of maximizing their *own* payoffs. A stark example is the choice of NATO countries refusing to act directly against the Russian invasion of Ukraine given the threat of nuclear escalation.

In this chapter we turn our attention to the simplified situation in which the agent has one adversary. Board games like chess exemplify this scenario and have received considerable attention in the world of computing. In such games each player makes a move on her turn, and the information is complete since both players can see the board, and where the outcome is a win for one and a loss for the other. We look at the most popular algorithms for playing board games.

Chess has long fascinated humankind as a game of strategy and skill. It was probably invented in India in the sixth century in the Gupta empire when it was known as *chaturanga*. A comprehensive account of its history was penned in 1913 by H.J.R. Murray (2015). The name refers to the four divisions an army may have. The infantry includes the pawns, the knights make up the cavalry, the rooks correspond to the chariotry, and the bishops the elephantry (though the Hindi word for the piece calls it a camel). In Persia the name was shortened to *chatrang*. This in turn transformed to *shatranj* as exemplified in the 1924 story by Munshi Premchand (2020) and the film of the same name by Satyajit Ray, *Shatranj Ke Khiladi* (The Chess Players). It became customary to warn the king by uttering *shāh* (the Persian word for king) which became *che*ck, and the word *mate* came from *māt* which means defeated. *Checkmate* is derived from *shāh māt* which says that the king has been vanquished.

Table 8.1 lists the names of the chess pieces in Sanskrit, Persian, Arabic, and English (Murray, 2015). In Hindi users often say *oont* (camel) for bishop and *haathi* (elephant) for rook.

From India the game spread to Persia, and then to Russia, Europe, and East Asia around the ninth century. It was introduced to southwestern Europe by the Moors and finds mention in the *Libro de los Juegos* (The Book of Games)[1] commissioned by King Alphonso in the thirteenth

---

[1] https://www.ancientgames.org/alfonso-xs-book-games-libro-de-los-juegos/, accessed November 2022.

Table 8.1 The names of chess pieces

Sanskrit	Persian	Arabic	English
Raja	Shah	Shah	King
Mantri	Vazir	Wazir/Firzān	Queen
Gajah	Pil	Al-Fil	Bishop
Ashva	Asp	Fars/Hisan	Knight
Ratha	Rukh	Rukh	Rook
Padati	Piadeh	Baidaq	Pawn

century. Buddhist pilgrims and traders on the Silk Road spread the game to the Far East, where it sprouted newer variations like Chinese chess and *shogi*. Another game that was played on a board was *go* which is still very popular today.

The moves that the pieces could make also changed over time as the game spread. The rules used currently in tournaments correspond to *international chess* as compared to the Indo-Arabic chess which had different rules. Castling, for example, was a new addition, as well as the ability of the pawn to move either one step or two.

The game of chess is played on an 8 × 8 board of alternating black and white squares, and each player has sixteen pieces at the start of the game. These are two rooks, two knights, two bishops, a king, and a queen, along with eight pawns. Each piece has well defined rules of movement and thus embodies different tactical abilities. The pieces of the two players are initially lined up on the respective ends of the board, like the armies of yore, and then each player in turn makes a move. A move can result in the *capture* of an opponent piece on the destination square, which then goes out of play. The objective of the game is to threaten to capture the opponent's king, and the aggressor is obliged to utter the word *check* while doing so as a warning. If there is no escape for the king being attacked, then the word used is *checkmate*, and the game ends at that point. This game with simple rules on a small terrain gives rise to virtually countless possibilities, and tomes have been written, and read, on the strategies for playing the game. So much so that the discourse is split into three parts. The first is the well documented *opening game*, with strategies like Ruy Lopez and the Queen's Gambit. After a few moves when the board opens up and pieces become more mobile is the *middle game*. This is not as well documented because of the humungous possibilities, and is driven by heuristics like retaining control of the centre, maintaining connected pawn structures, lining up doubled rooks, and so on. The real battle of wits is fought here. Then, as pieces are captured and removed from the board, the game becomes more tractable, and seasoned chess players recognize patterns with known outcomes well in advance. This phase is the *end game*, and more easily documented. Beginners, for example, are often taught how to checkmate a lone opponent king with two rooks.

The earliest pioneers of computing including Alan Turing and John von Neumann were deeply interested in chess which, though being a game of well defined rules and outcomes, is difficult to master for most of us, and they treated it as an alibi for intelligent behaviour (Larson, 2021). In 1950 Claude Shannon published a paper on computers and chess (Shannon, 1950). At the Dartmouth conference in 1956, Alex Bernstein presented a chess program developed at IBM, but the limelight was stolen by Arthur Samuel's checkers playing program which learnt from experience and is reputed to have beaten its creator. Chess was still considered to be a difficult game, and in 1968 the British grandmaster David Levy wagered a bet that no machine could beat him in the next ten years. He did win his bet, but in 1997 the program *Deep Blue* developed at IBM beat the reigning world champion Garry Kasparov in a legendary six game match (Campbell, Hoane, and Hsu, 2002). Chess machines continued their march over humans over the years. David Levy himself turned an artificial intelligence (AI) proponent and went so far as to predict that robots will provide the romantic companionship that humans crave (Levy, 2008).

From a game theoretic perspective, the objective would be to determine the outcome of the game when both players are perfect, but that is not yet computable. Instead, beating the best humans has been set as the benchmark. After conquering chess, attention shifted to the oriental game of *go* which was considered to be much harder. The game is played with black and white coins called *stones*, with each player free to place one on any grid location. Go is played on a 19×19 grid, thus presenting a much larger set of choices. The first move can be played in 361 ways, the second in 360, and so on. In all, there are 10^{170} possible board positions. This is much larger than the estimated 10^{120} chess games that are possible. Both these numbers are practically incomprehensible for most of us. The reader is encouraged to do a back of the envelope calculation to estimate how long it would take to inspect the 10^{120} chess games even if every one of the estimated 10^{75} fundamental particles in the universe were to be a supercomputer inspecting billions of games a second. Nevertheless, in 2016 the program *AlphaGo* developed by DeepMind, then a company in the United Kingdom, beat the reigning *go* champion Lee Sedol 4–1 in a much publicized match in Seoul (Silver et al., 2016). A gripping account of the match, and a film made on it, can be seen on DeepMind's website[2] which says —

> The game earned AlphaGo a 9 dan professional ranking, the highest certification. This was the first time a computer Go player had ever received the accolade. During the games, AlphaGo played several inventive winning moves, several of which – including move 37 in game two – were so surprising that they upended hundreds of years of wisdom.

We will briefly comment on AlphaGo later in the chapter.

As we will see below, board games like checkers, chess, and *go* can be abstracted away into layered trees known as game trees which can then be analysed and searched. It is just that the trees are too large to be analysed completely, unlike the smaller tree of the game tic-tac-toe, or noughts and crosses, which even children figure out will end in a draw unless one of the players bungles. A point to note is that while the best humans have been defeated in chess and *go*, we still do not know what the outcome of the game would be if both the players were to be perfect. That is the kind of result that the people working in the field of game theory would be interested in.

8.1 Game Theory

Game theory is a field of study that studies the outcomes of the interaction between rational agents each trying to maximize their self-interest. John von Neumann and Oskar Morgenstern

---

[2] https://www.deepmind.com/research/highlighted-research/alphago, accessed November 2022.

(1944) are generally credited with formalizing the idea of game theory. Game theory has also been defined[3] as 'the study of the ways in which *interacting choices* of *economic agents* produce *outcomes* with respect to the *preferences* (or *utilities*) of those agents, where the outcomes in question might have been intended by none of the agents'.

When we say rational agents we mean selfish agents, whose only goal is to maximize their own reward. This could mean that the outcome of the actions of all agents might not be the best for all concerned. A stark example is the concerted actions required to mitigate the effects of climate change, which are imperative if one is to save the world, but do not happen because individual nations have their own short term goals, most often in energy requirements. Many nations are critically dependent on fossil fuels, the very culprit behind detrimental climate change. Energy requirements are also behind the conundrums that many nations found themselves entangled in during the recent Russian invasion of Ukraine. The fact is that self-preservation was the motive behind the reluctance of most powerful nations to overtly intervene to prevent the massive destruction of people and property in Ukraine. When peoples and nations act in 'rational' self interest, they can still push the entire world towards a catastrophe. That is what we mean when we say that rational refers to being selfish, even though it may be short sighted. This is illustrated by the well known dilemma when two suspects are being interrogated independently.

8.1.1 The prisoner's dilemma

Rational self interest often leads to players not cooperating with each other in the interest of the larger good. This was starkly demonstrated by the *prisoner's dilemma,* named so by Albert Tucker, and originally framed by Merrill Flood and Melvin in 1950 (Poundstone, 1993). The problem faced by two prisoners accused of a crime is as follows. They are being interrogated separately by the police who have no concrete evidence but are hoping for a confession. They offer each a deal in which they would be let off with a light sentence and their accomplice would be given a harsher punishment if they confessed. Game theory shows that they will both end up confessing if they act in rational self interest, even though the opposite action would benefit them both.

Each prisoner has the option of cooperating with his accomplice by refusing to confess, or betraying him by confessing. The outcome of the separate interrogations is captured in Table 8.2 where A and B are the two prisoners. The outcome for each is as follows. If both cooperate, then each gets a payoff R (reward). If both betray the other, they get a payoff P (punishment). If only one betrays the other, then he gets a payoff T (temptation) and the other gets a payoff S (sucker punch). The payoff matrix is shown in Table 8.2. Acting alone without any communication with the accomplice, what should each prisoner do?

Table 8.2 The payoff matrix for prisoner's dilemma

(A, B) payoff	B cooperates	B betrays
A cooperates	(R, R)	(S, T)
A betrays	(T, S)	(P, P)

---

[3] https://plato.stanford.edu/entries/game-theory/, accessed November 2022.

It has been shown that as long as $T > R > P > S$ then it is rational for both *players* to betray each other, irrespective of what the other *player* does.[4] Consider the case when $T = 0$, meaning that the prisoner is let off, and the other payoffs are $R = -1$, $P = -2$, and $S = -3$. Then each prisoner reasons as follows.

Case 1: The other cooperates and does not confess (Column 1 for A, and Row 1 for B). If I cooperate then I get $R = -1$, and if I betray him then I get $T = 0$. I am better off betraying him.

Case 2: The other confesses and betrays me (Column 2 for A, and Row 2 for B). If I confess and betray him I get $P = -2$, and if I do not confess then I get $S = -3$. I am better off betraying him.

In both cases, the best action for each prisoner is to confess and betray the other. Clearly, rational self-interest results in each player getting a payoff $P = -2$ which is lower than the payoff $R = -1$ if both had cooperated. Therein lies the dilemma. This decision point is known as the *Nash Equilibrium* after the mathematician John Nash. Nash Equilibrium is the outcome from the choices of all the players such that if *any* player were to make a different choice the player would lose. As shown by the above example, this is not necessarily the best possible outcome.

The above analysis holds for a one time game for the two rational players, and not for seasoned criminals who may have developed trust in their partners in crime over a period of time.

It is not always necessary that a stable decision point will be reached. Consider the task of dividing the proceeds of a bank robbery between three robbers. Let the amount be 10 units, and let the decision be made by majority vote. Then, if A and B propose a 5–5 division among themselves, C can offer 4–6 to, say, A. Now B can make a counteroffer to one of them, and in this manner a decision will never be arrived at. This example is of a game between three players.

8.1.2 Types of games

In general there can be many scenarios that can be modelled as different kinds of games. Games can be classified based on the following criteria:

Number of Moves

Many games have one decision or one move to be made, like in prisoner's dilemma. We will be interested in games where each player makes a sequence of moves, and where the payoff is received after the game ends. As we will see, these can also be analysed to arrive at one complex decision, which we will call a *strategy*. But in games like chess, it is not feasible to compute the winning strategy, and we end up making decisions afresh at every turn.

---

[4] https://plato.stanford.edu/entries/prisoner-dilemma/, accessed November 2022.

Payoff

Games need not always be adversarial. They can be a basis of cooperation as well. The nature of the game is characterized by the sum of the payoffs received by each player. Games can be classified as follows:

- Zero sum games: Here the total payoff is zero. Some players may gain while others lose. Board games like chess are zero sum games. One player's win is the other player's loss.
- Positive sum games: Here the total payoff is positive, and most or all players gain. Such games are the basis for cooperation. For example, two researchers working jointly on a project, two students studying together preparing for an exam, two nations indulging in free trade benefiting the economies of both, and people of different religions participating in each other's festivals. Cartels between manufacturers result in increased profits for them, though at the expense of the consumer.
- Negative sum games: Here the total sum is negative and most or all players lose. A price war between two companies will result in reduced profits for both, though if we include the consumer in the game then it will become zero sum. Malevolent neighbours disrupting each other's festivities results in negative payoffs for all. War is the ultimate negative sum game, even when we count the profits made by arms manufacturers and dealers. Heads of arms manufacturing countries are known to promote sales during official visits to other countries.

Number of Players

The number of agents involved is another characteristic of the game.

- Two player games: Many games are treated as two player games, for example, a price war between two companies is a negative sum two player game. Games like noughts and crosses, checkers, and chess are two person games.
- Multiplayer games have more than two players. Price wars become a zero sum game when the consumer is included.
- Team games: Multiplayer games are sometimes modelled as competition between two teams – often zero sum. Each team may have team members collaborating with each other. Examples are contract bridge, football, armies on a battlefield, teams of lawyers in a courtroom, and members of a species in an ecosystem where survival of the fittest is the theme.

Uncertainty

Most real world situations in which agents act are fraught with uncertainty. Uncertainty arises from two sources. One, due to incomplete information. A bridge player does not know what cards an opponent, or the partner, is holding. The executives of a company do not know what their competitors are planning. Nations likewise do not know what other nations are planning. In both cases, espionage is a move employed by the people in power. Authoritative governments are known to spy upon their own citizens. Most recently, the use of spyware, for example, Pegasus, is testimony to that. The other source of uncertainty is due to one's actions not being deterministic.

- Incomplete information games lead to uncertainty. In card games like contract bridge and poker one cannot see other players' cards. In the corporate world we are not aware of what others are planning, and hence corporate espionage and lobbying with governments. In war, what the enemy is up to is not known, and we have spies reporting from across the line. We may even have to contend with misinformation when double agents are active. Generals often resort to deception, for example, Operation Fortitude in World War II in Normandy.
- Stochasticity in the domain leads to uncertainty. The throw of the dice in backgammon or in snakes and ladders cannot be predicted, as cannot be the draw of cards in poker or rummy. Except for the best in the field, most of us can only have the intent of shooting a basket on the basketball court.

As we can observe, the study of games involves a multitude of scenarios, with the common theme being the choice of actions designed to maximize one's payoff. We will confine our attention to two-person zero-sum complete-information alternate-move games like chess. In these games there is not just one move but many. We will call these board games, glossing over for the moment the fact that there are games played on a board like backgammon that involve the throwing of dice, and games like Chinese checkers and snakes and ladders that additionally may have more than two players. The games we intend to write algorithms for can all be abstracted into a game tree and have well studied algorithms to play them.

8.2 Board Games

We focus on two player games played on a board visible to both players. We assume that the players make alternate moves. In some games, for example, Othello, a player may not have a legal move and the opponent gets to play again. One can handle this case by modelling those two consecutive moves as a compound move, as can also be done in checkers while making multiple captures.

In the game of chess the two players are traditionally called White and Black. White always makes the first move. We will call the first player *Max* and the opponent *Min*. The rules of the game prescribe a set of allowed moves for Max, and then for each resulting board position a set of moves for Min. Max moves again after Min, and the game continues with players making moves alternately. The rules also determine when the game ends, at which point the outcome of that game is known. The outcome is either a win for one of the players or a draw.

All board games can be abstracted into a game tree. The game tree is a layered tree in which the two players make moves at alternate levels. The leaves of the tree are labelled with the following: W (a win for Max), D (a draw), and L (a win for Min, and thus a loss for Max). Figure 8.1 shows a small game tree.

228 | Search Methods in Artificial Intelligence

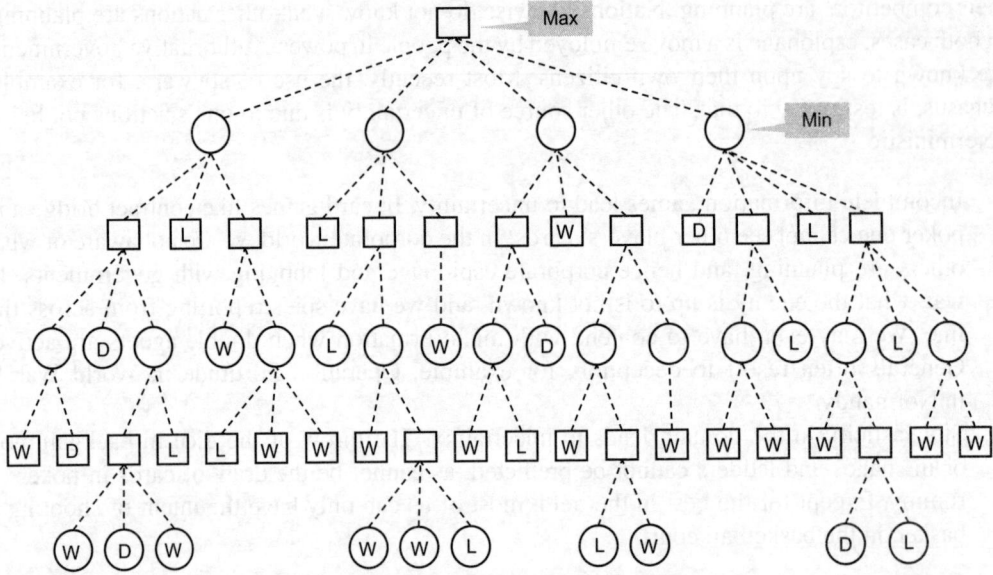

Figure 8.1 A generic game tree is a layered game tree with the two players choosing a move at alternate levels. Max plays first and can choose one of the four moves at the root. Each path represents a possible game, and the games end at leaf nodes. The leaf nodes are labelled with the outcome – win, draw, or loss – from Max's perspective.

Max chooses a move at the root, and then Min chooses one at the next level. This continues till the game ends. Every path in the tree represents a game, with the leaf labelled with the outcome. Observe that some games in the tree are shorter than others.

Associated with every game tree is a value that is the outcome when both players play perfectly. This is known as the minimax value of the game, and is the Nash Equilibrium. The minimax value can be computed by backing up the values from the leaves to the root in a bottom fashion. The procedure is as follows.

Pick a node whose children are all already labelled with W, D, or L. If the node is a Max node, then back up a W if some child is labelled W, else D if some child is labelled D, else L. This reflects perfect play for Max. The outcome can also be labelled with values 1, 0, and −1 respectively for W, D, and L. With these numeric labels one can see that Max prefers the highest value, and hence the name Max. Min is the opposite, preferring the lowest value from its children. Min's first preference is L, because it is a loss for Max and a win for Min. In this fashion we back up the values from the leaves to the root, choosing the maximum and minimum of the children at alternate levels. The value of the root is the minimax value of the game. It is the maximum value Max can pick from the Min child, which in turn chooses the minimum of the values offered by its Max children.

The reader is encouraged to compute the minimax value of the above game tree and verify that the game is a draw. Observe that this is so even though most leaves are labelled with a W. The reason is that at critical points it is Min that prefers another outcome over W.

8.2.1 Strategies

The game tree represents a game in which multiple moves are made by both players, and the outcome is known only at the end of the game. Chess players, for example, ponder over every move, as do children playing noughts and crosses. How can one relate such multiple move games with the single choice games like the prisoner's dilemma seen in the last section?

We can do that by defining the notion of a *strategy* for each player. A strategy is a statement of intent, once and for all, by a player. In other words, a strategy *for Max* freezes the choices for Max, while catering to all possible choices for Min. A strategy is a subtree of the game tree, where for Max we choose one move and choose all moves for Min. A strategy for Min likewise freezes the choices for Min, while catering to all subsequent moves by Max. Figure 8.2 shows two strategies for Max for the above game tree. One is the subtree on the left with bold edges, and the other on the right with bold-dashed edges. The leaves in the two strategies are shaded. We have used labels +1, 0, and −1 instead of W, D, and L.

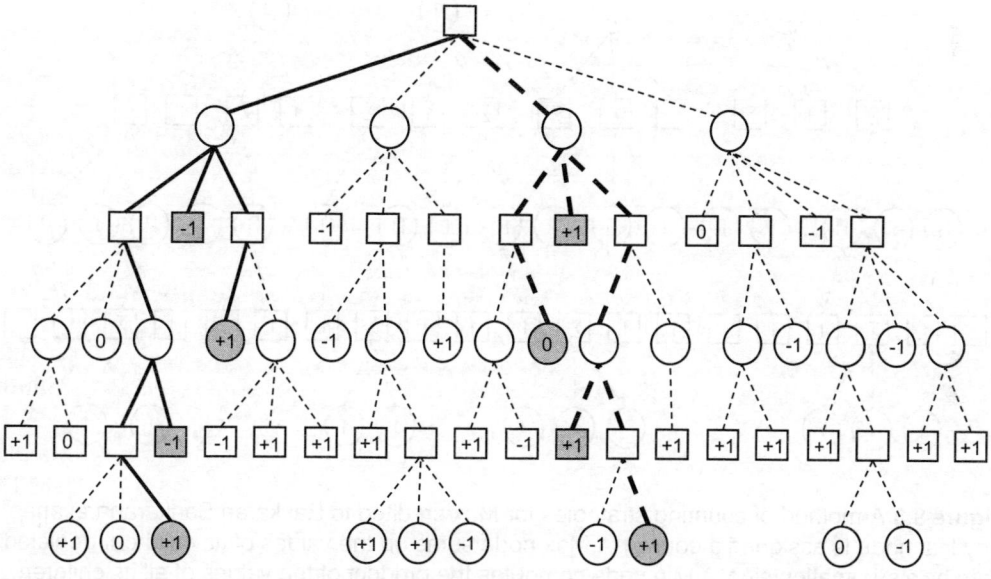

Figure 8.2 Two strategies for Max. Each strategy chooses one move for Max and all moves for Min. The shaded leaves are the leaves in the two strategies. The labels on the leaves are +1, 0, and −1, instead of W, D, and L.

Each strategy represents Max's decisions over the entire game. The strategy on the left has four leaves with labels +1, −1, −1, and +1. Since Max has frozen her moves, the game played, which is a path in that subtree, will be determined by Min. Clearly, any rational Min will drive the game towards a leaf with the lowest value, in this case −1. This is the case in general, and we can make the following observation:

> The value of a strategy for Max is the minimum of the values of the leaves in the strategy. That is the best that Min can do, given the frozen choices for Max. If the value of the strategy is +1, then it is called a *winning strategy* for Max.

The reader should verify that the value of the strategy with dashed thick edges on the right is 0. Clearly, it is the better of the two strategies that Max can choose. In general, a rational Max would choose the strategy with the highest value. That will also be the minimax value of the game, since it represents the best choices both the players can make.

How many strategies does Max have? Figure 8.3 depicts an approach to counting strategies given by Baskaran Sankaranarayanan in 2020.[5] Like computing the minimax value, we count the strategies from the leaf nodes up to the root. We begin with 1 for every leaf node. At the Max level, we sum the values from its children, since each branch represents a choice for Max. At the Min level, we compute the product of the children, since Max has to account for all combinations of choices that Min has.

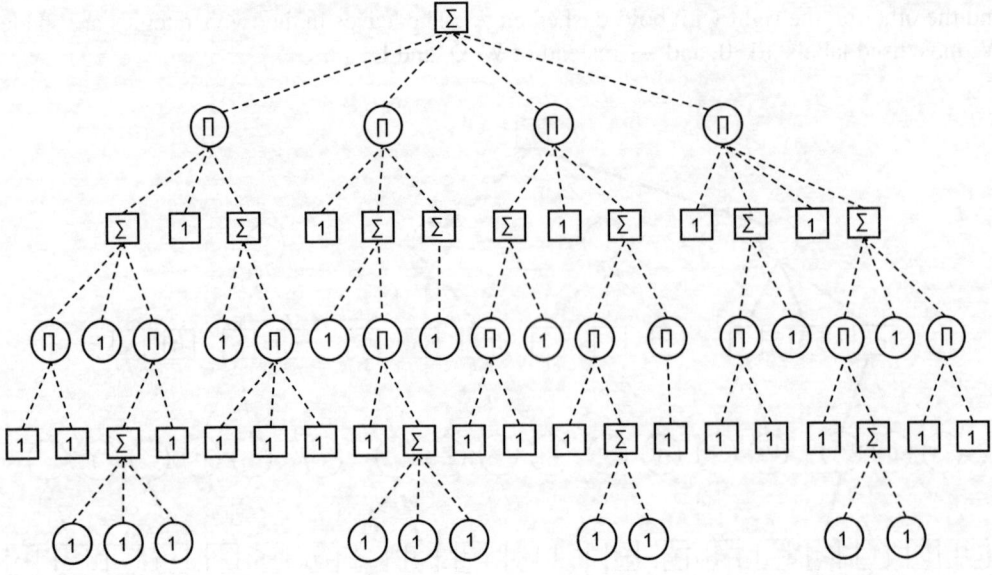

Figure 8.3 A method of counting strategies for Max credited to Baskaran Sankaranarayanan. Any leaf node is assigned a count 1. A Max node sums up the values of its children, depicted here by a summation sign. A Min node computes the product of the values of all its children.

The minimax value then represents the outcome of the game when both players choose their best strategies. This can be computed by analysing the entire game tree. For games like noughts and crosses, this can easily be computed and, as most children realize over a period of time, one cannot win the game against an opponent who makes no mistakes. And it was only in this century that the game of checkers was analysed (Schaeffer et al., 2007). Perfect play by both sides leads to a draw. Games like chess and *go* are another matter. The game trees are just too large to be analysed completely, and we still do not know whether White, the player who moves first, always wins in chess or not. The best we can do is to play *better* than our opponent. Which is why chess still fascinates us.

---

[5] Personal communication with author.

8.2.2 Limited lookahead with the evaluation function

On their first move in chess, each player has 20 options, represented by 20 edges in the game tree. Then, as the board opens up and pieces become mobile, the branching factor increases. In the end game, the branching factor declines again as pieces are captured and removed. It is estimated that the average branching factor in the game of chess is about 35, and if we assume that a game is typically 80 moves long, though estimates vary, the size of the chess tree would be $35^{80} = 10^{120}$ games or leaves in the tree. If we consider a lower estimate of 40 moves in a game, we still have a game tree of size $35^{40} = 5 \times 10^{62}$. In either case, the tree is too large to be traversed completely. If a machine could inspect a trillion (10^{12}) games in a second, it would need 5×10^{50} seconds for the smaller 40 move tree. This is about 1.5×10^{43} years. Contrast this with the age of the universe estimated[6] to be 13.8×10^9 years. Clearly inspecting the entire chess tree is out of the question and we have to resort to alternative methods.

One alternate method is to try and *judge* which of the moves leads to the *best* board position. This can be done by defining an *evaluation* function EVAL(J) which looks at a board J and returns a value from the perspective of Max. Given that we associate $+1$ with a win and -1 for a loss, the range of the evaluation function should be $[-1.0, +1.0]$ with the two extremes representing a loss and a win, with 0.0 being a draw, and values in between indicating how good or bad the position is for Max. In practice one works with integer values and the range may be defined as $[-\text{Large}, +\text{Large}]$, where Large is an appropriately sized integer. Figure 8.4 shows how Max could choose the move by consulting an evaluation function.

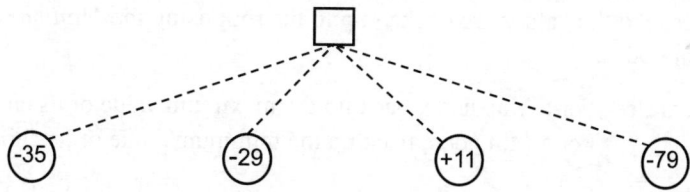

Figure 8.4 An evaluation function looks at a board position and returns a value that signifies how good the board position is for Max, with higher values being better. In this example, Max would choose the move with the value $+11$.

The evaluation function is a static function like the heuristic function that inspects a board position and returns a value. Let us consider the example of chess. How does a human expert look at a board position and evaluate it? The consensus is that there are two components to an evaluation function.

One computes material advantage, and the other evaluates positional advantage. In chess, for example, one can assign a relative value to each piece. John von Neumann is said to have given the following values. A king is valued at 200. This could be interpreted as Large = 200 because the absence of the king would mean the game is lost, irrespective of what other pieces are present. The queen is valued at 9. This would mean that if Black had the queen and White

---

[6] https://www.space.com/24054-how-old-is-the-universe.html, accessed November 2022.

did not, then one could subtract 9 from EVAL(J). Thus, every piece advantage would add or subtract something from the evaluation function. The rook has a value 5, the bishop and the knight 3, and each pawn has a value 1. Since in the initial position both players have all the pieces, the contribution of material to the evaluation function would be 0.

The other contribution to the evaluation is positional advantage. Chess experts have several features that they consider to be good, for example, connected pawns, doubled rooks, a protected king, control of the centre, the mobility of the pieces, and so on. The presence of each feature contributes to the evaluation function. Traditionally programmers have drawn upon human experts to devise evaluation functions. The evaluation function can be as complex as the detailed knowledge that an expert can articulate. The function used in the program *Deep Blue* has about 8,000 components (Campbell et al., 2002).

Is there a first move advantage? Should the evaluation function have a small positive value in favour of White? The jury is still out on this one.

If the evaluation function were perfect, then it would select the best move simply by looking at the options. But such functions are hard to find. Consider that one of the moves in chess captures an opponent bishop. Then it might look attractive because of the material gained. But if the opponent could capture your queen on the next move, then your move ends up looking worse than it appeared to be. A perfect evaluation function would take into account the effect of such future piece exchanges but is difficult to devise. To counter the imperfectness of the evaluation function most chess programmers implement a limited *lookahead*. The idea is that the inaccuracy in the evaluation function will be compensated by the look ahead. If the program is looking ahead k moves, we say that it is doing k *ply* search, and the tree it explores is k *plies* deep. The evaluation function is applied to the nodes on the *horizon* at k plies, and the values are backed up to the root using the *Minimax rule* bottom up from the horizon nodes.

Minimax rule: For a Max node, back up the maximum value of its children.
For a Min node, back up the minimum value of its children.

Figure 8.5 shows a 4 ply game tree with the leaves on the horizon labelled with the values of the evaluation function at each node or board position.

The reader is encouraged to apply the Minimax rule to the above tree and verify that the minimax value of this tree is 32, and to achieve this outcome Max must choose the rightmost move at the root.

Since we are not inspecting the entire tree, we do not know the final outcome of the game. Instead, what we have is a move that *appears to be* the best based on the k ply lookahead. After we make the move, the opponent responds, and we need to do another k ply search to choose our next move. The algorithm for playing a game is shown below.

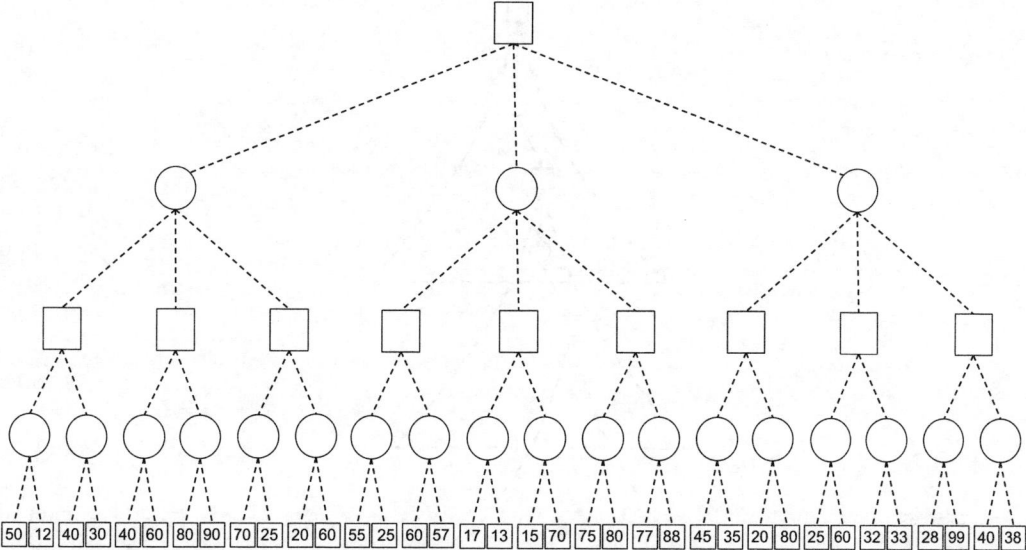

Figure 8.5 Most game playing programs look ahead to a certain *ply* depth and apply the evaluation function on the *horizon* nodes. In this figure the tree is 4 ply deep, and the values in the leaves on the horizon are the values returned by the evaluation function EVAL(J). Effectively the program is looking 4 moves ahead to decide the move for Max at the root.

Algorithm 8.1. The algorithm GAMEPLAY repeatedly calls k ply search for every move that it has to make. With every new call, it looks 2 plies further in the game tree.

GAMEPLAY(MAX)
1 **while game not over**
2   **call** k-ply **search**
3   **make move**
4   **get** MIN's **move**

In this way one can imagine the game playing algorithm pushing into the game tree with a limited lookahead at every step, as shown in Figure 8.6.

Remember our reason for not relying totally on the evaluation function with a 1 ply lookahead, that there may be unforeseen dangers beyond (Figure 8.4). This continues to be a case where the lookahead search may choose a poor move because it is oblivious of the danger lurking beyond the horizon. This is known as the *horizon effect*. We illustrate this with an example shown in Figure 8.7.

Let Max choose a move as shown with a minimax value backed up from the node Z on the horizon. A little earlier in the path let an intermediate Max node choose a value from its two successors marked X and Y. The node Y leads to a good position for Min in the node shown in black, which is a bad position for Max. The backed-up value from this node reduces the value of Y, which then is ignored. In the path via node X, Max has made an inconsequential move

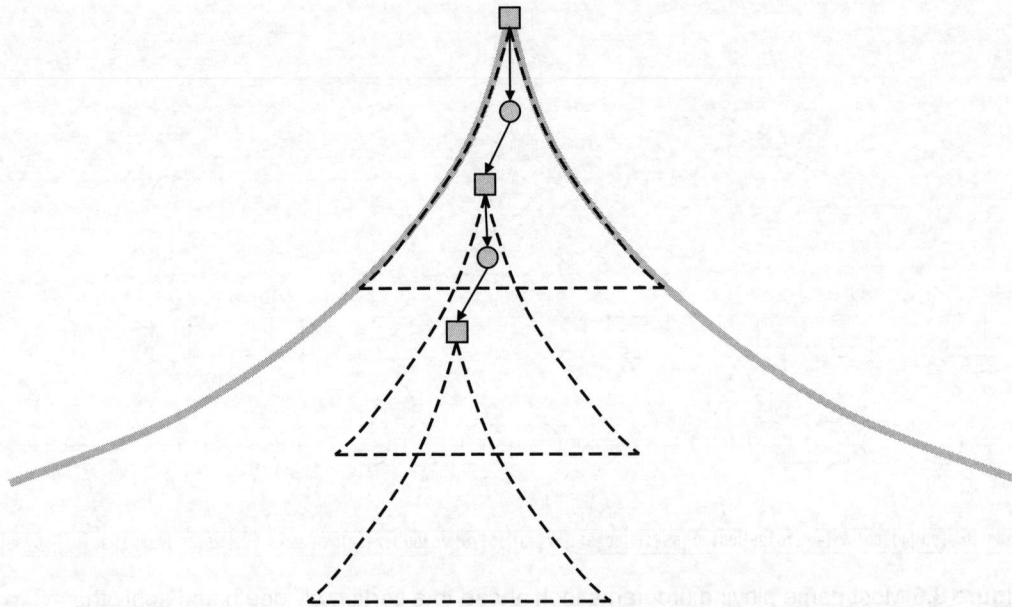

Figure 8.6 For every move it makes, a game playing program calls a *k* ply search and waits for the opponent's move. The figure shows three such calls. Figure adapted from Khemani et al. (2013).

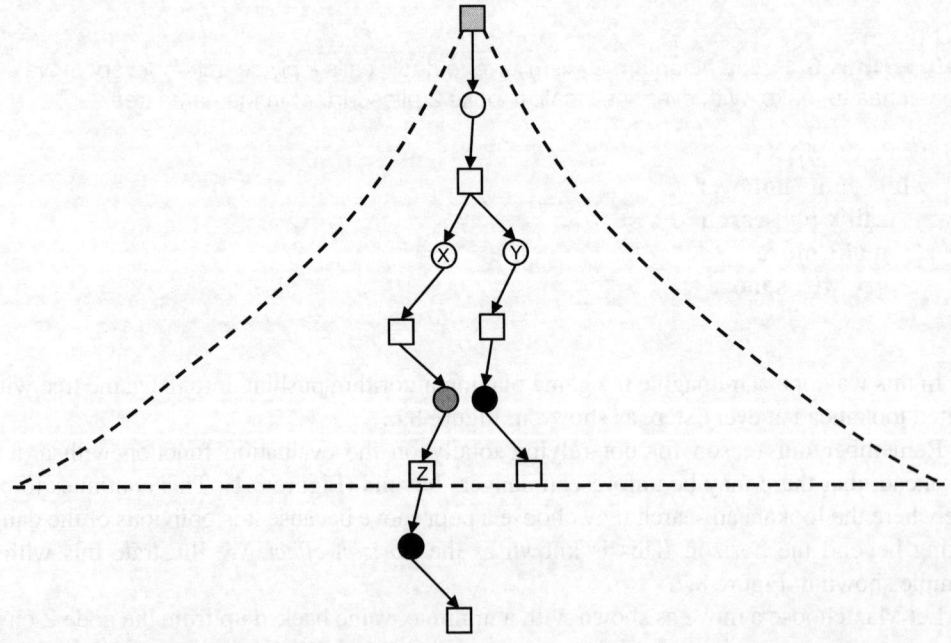

Figure 8.7 The horizon effect. Let the minimax value be from the node *Z* on the horizon. This happens because the Max node chooses the higher value that comes from its child *X*. The value from *Y* is poor because of the Min node shown in black. In the path from *X* there is an inconsequential pair of moves where Max moves to the grey coloured node. This inconsequential move pushed the black node beyond the horizon, making node *X* look better than it actually is.

shown in grey. One can always make such moves in a game, for example, pushing a nondescript pawn, or making a knight move and reversing it later. The effect of such moves is that they may push the black node beyond the horizon, as in the case of the chosen path, thus making X look better than it is.

The above example shows how the horizon effect may make a path look better than it is. One way to deal with this is to do a secondary search from the node Z before committing to the move. This could uncover any hidden danger. The other way is to not worry about it now, because the k ply lookahead for the next or a later move would anyway reveal the danger.

8.2.3 Algorithm Minimax

The simplest algorithm to compute the minimax value of a k ply game tree is to do a depth first search (DFS), sweeping the tree from left to right. We look at a recursive version of DFS in which the value of each child is evaluated recursively (Pearl, 1984). A Max node accepts the highest of the values that all its Min children supply, and a Min node accepts the lowest value. The recursive call to evaluate a Max initializes the value to −Large. It can only go higher from here, as Max receives values from its children one by one. Likewise, a call to evaluate a Min node begins with an initial value +Large. It can only get lower. The recursive calls end when the algorithm is evaluating a node on the horizon. We have assumed here that there is a function that can tell us if a node is terminal node or not. Such a function can be implemented by keeping track of depth in each call. Adding a depth parameter to the function MINIMAX and using that to determine whether the node being evaluated is a terminal node or not is left as an exercise.

Algorithm 8.2. The algorithm MINIMAX does a depth first search of the game tree. This version recursively calls itself to compute the value of each child. A node is a terminal node if it is on the horizon.

MINIMAX(N)
1 **if** N is a terminal node
2   value ← EVAL(N)
3 **else if** N is a MAX **node**
4   value ← -LARGE
5   **for each child** C of N
6   value ← max(value, MINIMAX(C))
7 **else** value ← + LARGE
8   **for each child** C of N
9   value ← min(value, MINIMAX(C))
10 **return** value

A critical question is how deep should the algorithm look ahead. The simple answer is as deep as the computational resources and the allotted time allow. How efficiently can one exploit the multicore processors that are available nowadays? Before the multicore era the *Hitech* chess machine by Hans Berliner, a computer scientist and a chess player of some repute, used a specialized sixty-four custom VLSI chip architecture (Berliner, 1987). One must keep in mind though that the tree grows exponentially, which is difficult to match even with increased computing power.

One approach seeks to fine-tune the ply depth to tournament conditions. In an international tournament one typically has to play 40 moves in 2 hours, and the next 20 moves in an extra hour. After this if the game is still in progress, one needs to play faster. Within each block of time, each player can decide how long each individual move takes, since the two players have individual clocks, only one of which runs at a time. This allows us to follow a flexible time schedule. The key is to decide the time one can devote to a move by the calling program, which keeps track of time. This can be done by employing the strategy of *depth first iterative deepening* (DFID) (Chapter 3), in which the calling program can call for the move anytime, and the DFID MINIMAX returns the best move found. The opening game is often handled by deploying well studied standard openings in which the moves can be made rapidly, leaving more time for other moves. Some implementations rely on quasi-stability in the fluctuation of the evaluation function with depth, usually during sequences of material exchanges. That particular call can then be allowed more time, but at the expense of time available for future moves.

The MINIMAX algorithm, however, does some unnecessary work traversing the entire k ply game tree. There are situations when a significant part of the tree can be pruned without affecting the correctness of the minimax value calculation. We look at two such algorithms below.

8.2.4 AlphaBeta pruning

The MINIMAX algorithm sweeps the k ply game tree from left to right doing depth first search, as it (recursively) goes down the plies. At any point, each node J being refined seeks the best value from its children, which it evaluates one by one. When J is refined, it in turn offers its value to its parent, who will then query J's right sibling, if there is one. In this way one can visualize the entire traversal in terms of a supply chain process in which Max nodes at each level seek the maximum of the values offered by their children, and Min nodes accept the minimum of what their children have to offer. During processing, each node has received some value from the partially traversed tree on its left and will only accept a better offer from below.

We adopt a nomenclature which gives a formal basis for the analysis that follows (Pearl, 1984). Let the Max nodes be called *Alpha* nodes and the values they store be α values. Likewise, Min nodes are called *Beta* nodes and they store β values. These values are the partially computed (minimax) values of the nodes received from the explored subtrees on the left. And these values can *only* improve. Alpha nodes will only accept higher values from its unexplored children on the right, and Beta nodes will only accept lower values. For each Alpha node, α is a lower bound on its value, and for each Beta node, β is an upper bound. Only if the values coming from below respect these bounds should one continue to explore them. Otherwise, they can be pruned. This is the essence of the ALPHABETA algorithm that prunes the game tree even as it is being explored.

Figure 8.8 is a snapshot showing the progress of depth first search sweeping from left to right. The root has the value α_0 and is looking for a higher value. Its child being explored has a value β_1 and is looking for a lower value. Lower down, node β_3 (we will conflate the name and the value of nodes) is being explored. It has generated and evaluated its first child α_3. Should it generate and evaluate its second child J?

The node β_3 has a value $\beta_3 = \alpha_3$ received from its first child. This becomes an upper bound for the node and it can only get lower if a smaller value is received from any of its other children. Moreover, this value must be higher than that of its parent α_2 if it is to be selected, and lower than β_2, higher than α_1, lower than β_1, and higher than α_0. Only then does it have a chance to be the minimax value of the root node. If all these conditions do not hold, then β_3 need not be refined any further and can be pruned.

Let $\alpha = \max\{\alpha_0, \alpha_1, \alpha_2, \alpha_3,\}$ and $\beta = \min\{\beta_1, \beta_2, \beta_3\}$. Then the node J should be explored only as long as $\beta > V(J) > \alpha$, where β is the lowest β ancestor, and α is the highest α ancestor. Else J can be pruned. A *cutoff* is said to have occurred at its parent node.

If J is an Alpha node and its parent has a β value that is lower than the highest Alpha ancestor value α, then J is not generated and its Beta parent not evaluated any further. We say that an α-cutoff has happened. An α-cutoff is an α-induced cutoff because some Alpha ancestor has a value α, and the Beta node which has a lower value cannot contribute to changing it.

Likewise, a β-cutoff happens when an Alpha node is ceased to be evaluated because it already has an α higher than some β ancestor, which will block that value.

The algorithm ALPHABETA is in fact a technique of pruning added to the MINIMAX algorithm and is said to have been introduced by many people at different times. The algorithm is described in Algorithm 8.3.

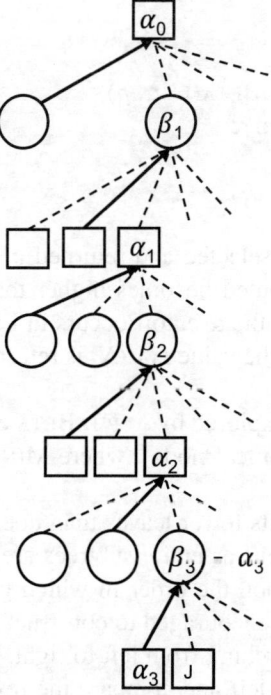

Figure 8.8 Should ALPHABETA generate and evaluate node J? Only if β_3 is greater than all the α ancestors and smaller than all the β_3 ancestors.

As can be seen, the algorithm operates in a window with an upper bound β and a lower bound α. Initially $\beta = +$Large and $\alpha = -$Large. Gradually, as the tree is traversed, better values are found and the α-bound increases and the β-bound decreases. One can see this as a shrinking window. If at any point an α-value becomes higher than the β-bound, the window shuts and a β-cutoff occurs. Likewise, if a β-value becomes lower than the α-bound, then an α-cutoff occurs.

> **Algorithm 8.3.** The AlphaBeta algorithm augments the Minimax algorithm with cutoffs, pruning parts of the game tree that cannot contribute to the minimax value of the root. The value α of a Max node cannot go higher than a β-bound received from an ancestor. Likewise, the β of a Min node cannot go lower than a α-bound imposed by an ancestor. Initially $\alpha = -$Large and $\beta = +$Large.
>
> AlphaBeta(N, α, β)
> 1   **if** N **is a terminal node**
> 2   **return** Eval(N)
> 3   **If** N **is a MAX node**
> 4   **for each child** C **of** N
> 5 $\alpha \leftarrow \max(\alpha, \text{AlphaBeta}(C, \alpha, \beta))$
> 6   **if** $\alpha \geq \beta$. **then return** β
> 7   **return** α
> 8   **else** ▷ N is a MIN node
> 9   **for each child** C **of** N
> 10 $\beta \leftarrow \min((\beta, \text{AlphaBeta}(C, \alpha, \beta))$
> 11   **if** $\alpha \geq \beta$ **then return** α
> 12   **return** β

In Line 5 a higher α-value is selected and returned in Line 7 unless a β-cutoff occurs in Line 6 if the α-value being computed becomes higher than the β-bound. Likewise, in Line 10 a lower β-value is selected and the α-cutoff occurs in Line 11 if it becomes lower than the α-bound. As before, in Minimax, the value $\textit{Eval}(N)$ is returned for a node on the horizon in the base case in Lines 1 and 2.

Figure 8.9 shows the subtree explored by AlphaBeta on the 4 ply game tree of Figure 8.5. AlphaBeta explores 26 of the 36 leaf nodes, where Minimax would have inspected all the 36 nodes.

In general AlphaBeta inspects fewer leaves than does Minimax, and the deeper the ply depth, the more the pruning happens as entire subtrees are discarded. The amount of pruning that AlphaBeta does depends upon the order in which good nodes are encountered during DFS. As an exercise, the reader is encouraged to construct a tree by filling in leaf node values such that *no cutoff* takes place searching from left to right. Then on the same tree simulate the algorithm searching from right to left and compare the result. One can then observe that the earlier the best moves are found in search, the greater will be the pruning. The intuition here is that the later moves are worse and hence cut off.

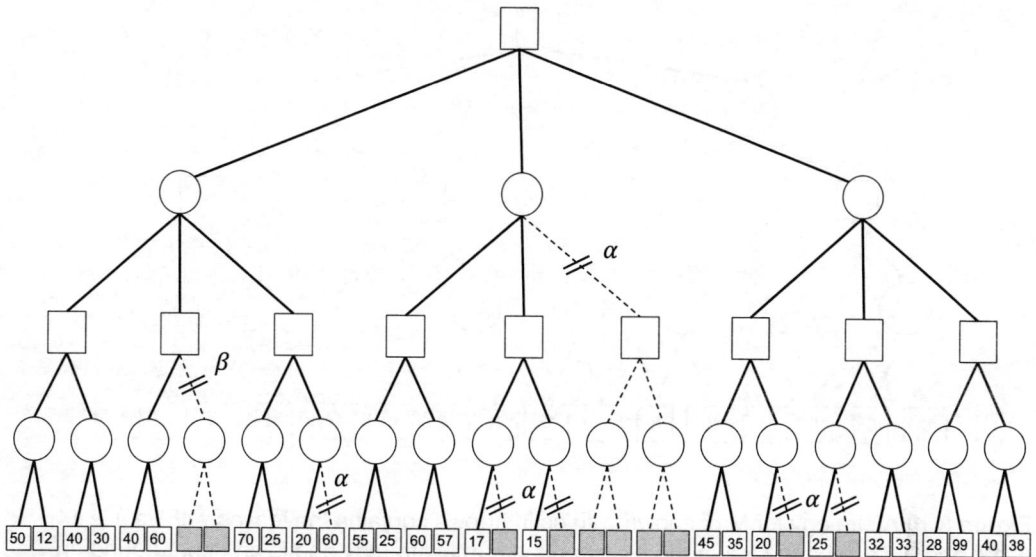

Figure 8.9 The ALPHABETA algorithm inspects 26 of the 36 nodes on the 4 ply game tree of Figure 8.5. The cutoffs are indicated by double lines. As can be seen, there is one β-cutoff and six α-cutoffs.

The algorithm can be further optimized as follows. While traversing the tree one also computes the (minimax) value for each of the internal nodes. These values can be utilized to order the moves in the *next* round, generating better children first, so that they appear earlier in the tree. This is expected to result in greater pruning. The same can also be done during the different passes if one is implementing a DFID version for a tournament. And finally, after having made a move, one can also "think in opponent's time" by doing another search to order the moves.

But instead of ordering the move generation, can one explore the tree with a sense of direction? It turns out the answer is yes, provided one does some preliminary work to get an estimate of the value resulting from each choice. We look at a best first approach next.

8.2.5 Algorithm SSS*: best first search

The algorithm SSS* is a *best first* version of the ALPHABETA algorithm (Stockman, 1979). It reasons only with choices for Max and embodies the following problem solving strategy from earlier chapters:

Refine the best candidate or partial solution, till the best candidate is fully refined.

Unlike our earlier heuristic search algorithms, the notion of best in algorithm SSS* is defined in a domain independent fashion, and not by a user defined heuristic function. To understand that, we need to return to the definition of a strategy. Recall that a strategy is a subtree that freezes the choices of one player. Figure 8.10 shows two strategies for Max in a tiny 4 ply game tree. The first strategy is shown with solid arrows in which Max chooses the

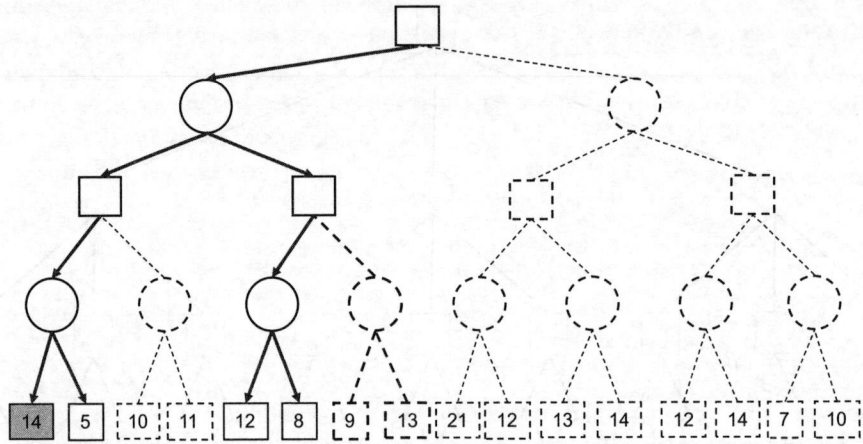

Figure 8.10 A strategy for Max shown with solid arrows contains the leaves with values 14, 5, 12, and 8. The shaded leaf node represents a partial strategy and is an upper bound on the value of the strategy. It is also part of another strategy where Max chooses leaves 9 and 13 shown in think dashed lines, instead of 12 and 8 in the first strategy. The shaded leaf node represents a cluster of two strategies with an upper bound of 11 on both.

leftmost branch at each choice point. The four leaves in this strategy have values 14, 5, 12, and 8. The *value* of this strategy is 5, the minimum of the values of the leaves. If Max were to adopt this strategy, Min would drive the game towards the node with value 5, since now only Min gets to choose.

The shaded leaf node in the figure is a *partial strategy*, with only one leaf identified. It is also part of the second strategy in which Max chooses the right branch at its level 3. This is marked with thick dashed edges and contains nodes 9 and 13 instead of 12 and 8. This strategy also has a value 5. The shaded node is part of both strategies and is itself an upper bound on both the strategies. The node thus represents a *cluster* of two strategies. The following observation is pertinent:

> Any leaf node in a game tree represents a cluster of strategies, and its value is an upper bound on the values of all the strategies in the cluster.

When we look at strategies *for* Max, we view the problem of playing the game essentially as a one person or single agent problem, where the task is to find the *optimal* strategy, the one with the *maximum* value. To guarantee finding the optimal strategy, we must consider *all* the strategies available for Max. But the number of strategies is huge. The above tiny tree has eight distinct strategies. Earlier in Figure 8.3 we had looked at a procedure to count strategies in an arbitrary game tree. For a game tree with all leaves at ply depth k and a constant branching fact b, the count can be expressed as a formula. If k is even, then let $n = k/2$, else $n = (k + 1)/2$. Here n represents the number of layers where Max has to choose. At level 1 there are b choices for

Max, at level 3 there are b^b choices, for $n = 5$ we have b^{b*b} choices, and so on. Adopting the sum-product procedure from Figure 8.3, we get the following expression:

$$\text{The number of strategies at depth } 2n = b^{(1 + b + b^2 + b^3 + \cdots b^{(n-1)})}$$
$$= b^{[(b^n-1)/(b-1)]}$$

Clearly, a brute force approach inspecting all strategies is not desirable. Instead, we follow a strategy analogous to the B&B for TSP from Section 6.1.2. We identify an exhaustive set of partial solutions, or partial strategies, each representing a *cluster* or set of strategies. We keep track of the lowest upper bound on each cluster and refine the best (highest value) cluster till the best cluster is a fully refined one. The following method identifies clusters of strategies that exhaustively cover the space of all possible strategies. Each cluster is represented by a single node. Starting with the root, we identify the clusters as follows:

If the node is a Max node, select all branches.
If the node is a Min node, select one branch.

How many clusters does Max need to consider in a game tree? If the branching factor is a constant b, then Max has b choices or clusters at the root which is at level or ply 1. Since we select only one branch at level 1, we still have b clusters for 2 ply search. For each of these b choices, there are b further choices at level 3, giving us b^2 choices. At level 5, each of these b^2 choices can be refined into further b choices, giving us b^3 strategies. In general, at level k, where k is odd, one has $b^{(k+1)/2}$ clusters. If k is even, the formula becomes $b^{k/2}$.

The game tree in Figure 8.10 has eight strategies in four clusters as shown in Figure 8.11. At the Max level we select all branches, and at the Min level we select one. Without loss of generality, we choose the leftmost branch at the Min level. The four clusters are named A, B, C, and D in the figure.

The value of each leaf is an upper bound on the strategies it represents and serves as the heuristic function that guides search. Each node is a partial strategy and a candidate for refinement. In the above tree, node 21 of cluster C happens to be the most promising cluster and is selected. On refinement, its sibling with value 12 is included in the cluster. But now the value of cluster C drops to 12, and SSS* shifts its attention to cluster A with the upper bound 14. This jump in the search space is characteristic of the best first search behaviour studied earlier in the book.

When cluster A is refined, its value drops to 5 and attention shifts to cluster D. The sibling of 13 is 14, and the value of D does not change. Figure 8.12 depicts the game tree at this stage. After refining clusters C and D, cluster C gets pruned because of the Max parent above them. This is like an alpha cutoff. The shaded nodes depict the remaining three contenders, with cluster D with value 13 still leading the pack. To refine cluster D, we have to shift to its Max sibling, and solve it as long as it does not become greater than or equal to 13. If it were to exceed 13 or become equal to it, then the Min parent would induce a beta cutoff.

Refining cluster D now entails recursively solving its sibling with an upper bound of 13. As shown in the figure, this results in two clusters D_1 and D_2 with value 12 and 7 respectively. The reader should verify that solving these will result in a minimax value of 12 for the sibling of D, which now takes on the mantle of cluster D. What is more, there are no more Max siblings and the Min parent is completely solved with a value 12.

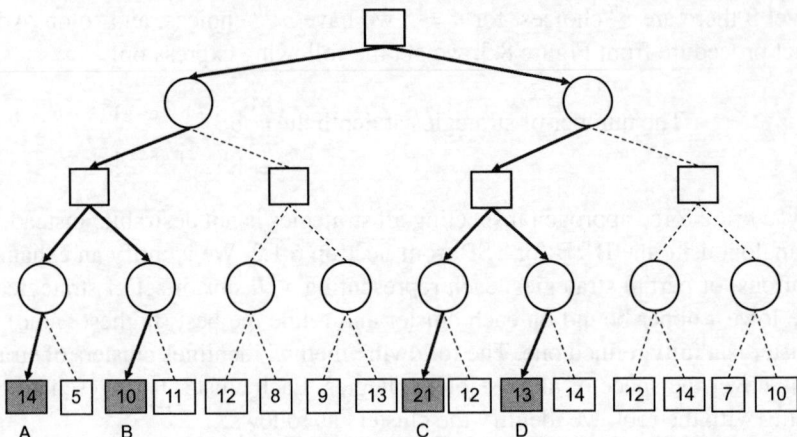

Figure 8.11 The 4 ply game tree with branching factor 2 has eight strategies in the four clusters named *A, B, C,* and *D*. The thick arrows show how these are identified. All choices at the Max level, and one choice at the Min level. Each shaded node represents a cluster of two strategies.

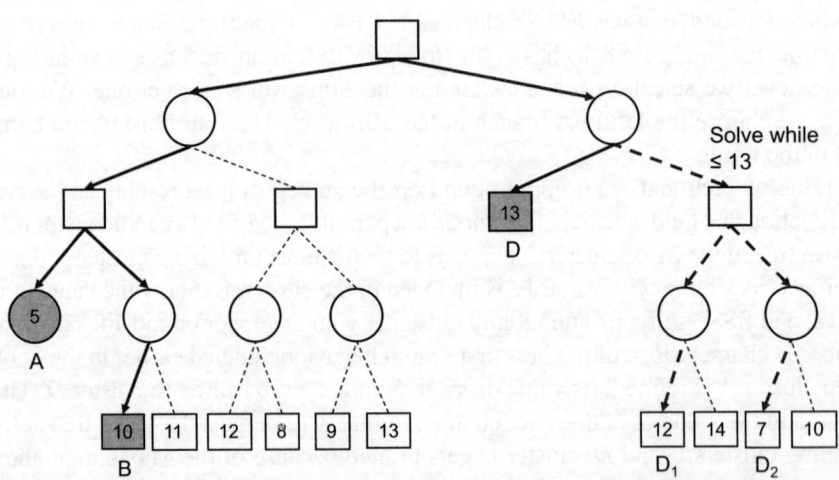

Figure 8.12 The game tree as seen by SSS* after inspecting three more leaf nodes. Cluster *C* has been pruned having a value 12, which is dominated by the value 13 from cluster *D*. Cluster *D* is the next cluster to be refined. This involves solving its sibling with an upper bound of 13. This involves recursively solving the Max sibling. Cluster *D* now splits into two, D_1 and D_2.

The tree as seen by SSS* is depicted in Figure 8.13. At this point the best cluster D_1 is fully refined, and the other clusters have lower upper bounds. The algorithm SSS* terminates with the minimax value 12. The reader should verify with Figure 8.10 that this is indeed the minimax value.

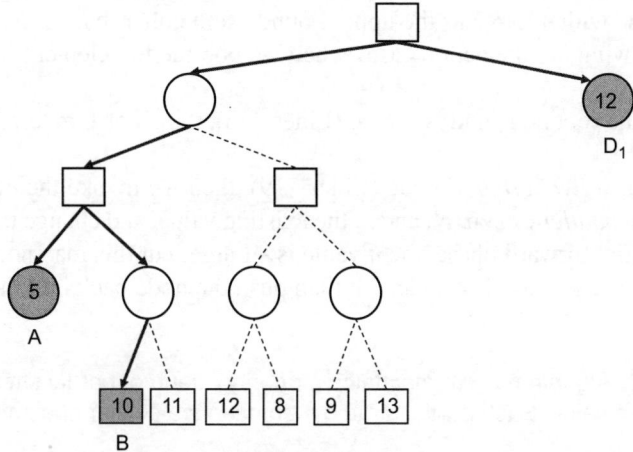

Figure 8.13 Algorithm SSS* terminates when cluster D_1 is fully refined with value 12, higher than other partial clusters, which have upper bounds lower than 12.

As is evident from this example, the algorithm has two phases. In the forward phase, we traverse the game tree up to the horizon and identify the clusters. Then as we refine the clusters, the values are propagated upwards. Having solved an internal Max node, we again embark on the forward phase to solve a sibling, whose value is again arrived at by propagation from the terminal nodes.

In the following section we describe an iterative version of the SSS* algorithm. The reader is encouraged to compare the algorithm with the *AO** algorithm in Section 7.32 and observe the similarities. The game tree as seen by SSS* is similar to an And–Or tree, with Max nodes being *Or* nodes, and Min being *And* nodes, from the perspective of Max. Interleaving the forward and backward phases happens because Max has to consider all responses from Min, which is like an *And* node. We even use the same terminology of classifying each node as *live* or *solved*, and also terminate when the root node is labelled *solved*.

8.2.6 Algorithm SSS*: an iterative version

The algorithm described below maintains a priority queue like the BESTFIRSTSEARCH algorithm. It picks the node from the head of the queue and terminates when that node is the solution. Each node represents a cluster of one or more strategies. A node is *solved* if its minimax value is known, else it is a *live* node whose value is an upper bound on its minimax value. Observe that we require the estimated value of a node to be an upper bound, as opposed to it being a lower bound in B&B and A* algorithms. This is because here we want the strategy with the maximum value for Max, and we can terminate if we know that partial strategies cannot be higher than their estimated costs.

The node structure is a triple <*name, status, h*> where *name* identifies the node, the *status* is *live* or *solved*, and *h* is the estimated upper bound value. In addition, we assume that we have a function that can identify when a node is *terminal*, or on the horizon of search. The algorithm is described in Algorithm 8.4.

We begin in Lines 1 and 2 by inserting the root node <root, live, +Large> in the priority queue as a live node with +Large as the upper bound. Remember that +Large stands for a win for Max. The following are the action cases when we pop the top element *N* from the priority queue:

If the node is the root node and is *solved* (Lines 4–6), the algorithm terminates and returns the minimax value.

If the node *N* is a *live terminal* node (Lines 7–9), then we invoke the evaluation function EVAL(*N*), choose the *smaller* of EVAL(*N*) and *h*, the existing value, and change the status to *solved*. Note that in the initial forward phase, the *h* value is +Large, but this may not be the case when we solve a sibling of a solved Max node. We then push the node back into the priority queue.

Algorithm 8.4. Algorithm SSS* maintains a priority queue of partial strategies sorted on their estimated value. Each partial strategy represents a cluster of complete strategies.

SSS*(root)
1 OPEN ← **empty priority queue**
2 **add** (root, LIVE, + LARGE) **to** OPEN
3 **loop**
4   (N, status, h) ← **pop top element from** OPEN
5   **if** N = root **and** status **is** SOLVED
6   **return** h
7   **if** status **is** LIVE
8   **if** N **is a terminal node**
9   **add** (N, SOLVED, min(h, EVAL(N))) **to** OPEN
10   **else if** N **is a** MAX **node**
11   **for each child** C **of** N
12   **add** (C, LIVE, h) **to** OPEN
13   **else if** N **is a** MIN **node**
14   **add** (**first child of** N, LIVE, h) **to** OPEN
15   **if** status **is** SOLVED
16   P ← **parent**(N)
17   **if** N **is a** MAX **node and** N **is the last child**
18   **add** (P, SOLVED, h) **to** OPEN
19   **else if** N **is a** MAX **node**
20   **add** (**next child of** P, LIVE, h) **to** OPEN
21   **else if** N **is a** MIN **node**
22   **add** (P, SOLVED, h) **to** OPEN
23   **remove all successors of P from** OPEN

Chess and Other Games | 245

If the node *N* is a *live* internal Max node, then we add *all* its children as *live* nodes with the *same h value*. For a Min node we only add one child. This is in effect the process of identifying the clusters in the forward phase (Lines 10–14).

If the popped node *N* is a *solved* node (Line 15), then three cases arise. If *N* is a Max node and also the last child of its parent, then its parent *P* is solved as well, and is added to the priority queue as a *solved* node (Lines 17–18), else the next sibling of *N* is added to the queue as a *live* node with the *h* value copied from *N* (Lines 19–20). This *h* value serves as an upper bound for solving the sibling recursively. Finally, if *N* is a *solved* Min node (Line 21), then, since it was at the head of the priority queue, it is better than all its Min siblings. We prune the siblings (alpha cutoff) and add the Max parent of *N* as a *solved* node with the same *h* value (Lines 22–23).

The algorithm SSS* thus identifies nodes on the horizon to form clusters of strategies and uses the value of each node to serve as a heuristic value to guide search. Since the value of a node is an upper bound on the strategies that it is a part of, the algorithm is guaranteed to have found an optimal strategy when it terminates with a complete strategy being picked from the priority queue.

Figure 8.14 shows the subtree explored by SSS* for the game tree of Figure 8.9, whose visited nodes are copied below for comparison. As can be seen, SSS* explores fewer nodes.

The reader is encouraged to simulate SSS* on the tree and verify that both ALPHABETA and SSS* find the same minimax value. The ALPHABETA algorithm visits the leaf nodes from left to right. What is the order in which SSS* visits the leaf nodes on this tree?

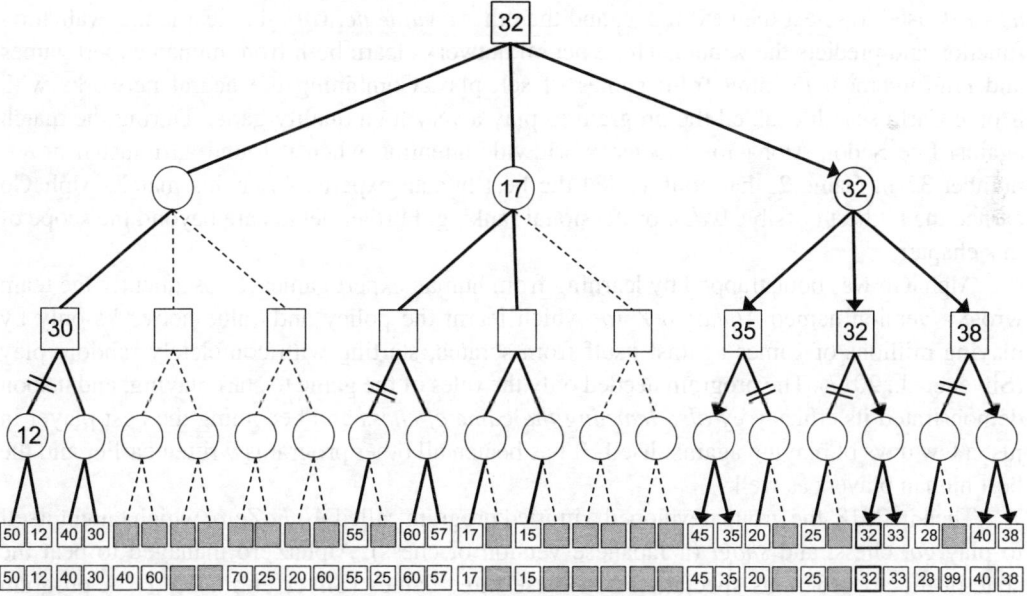

Figure 8.14 The game tree of Figure 8.9 as explored by Algorithm SSS*. The shaded leaves are not inspected. The cutoffs shown are the alpha cutoffs by SSS*. The optimal strategy on the right is marked by arrows. The leaves explored by ALPHABETA are also shown below the tree. SSS* does more pruning than ALPHABETA.

8.2.7 AlphaGo, AlphaGo Zero, and AlphaZero

Among the various recent advancements in playing board games, the ones made by the company DeepMind have been the most striking. It all started with the game of *go*, considered to be much harder than chess, primarily because of the much larger game tree. *Go* is said to have been invented in China about 3,000 years ago and is played on a 19 × 19 board on which two players alternately place *stones*, one black and one white, with black placing the first *stone*. Broadly the objective of the game is to capture territory, and in the process one may encircle and capture opponent pieces too. A comprehensive introduction to the rules and intricacies of the game can be found on the British Go Society website.[7]

Conceptually *go* is a simple game. Not very different from noughts and crosses or Othello, where the two players place 'stones' alternately. The board size makes *go* complex. As mentioned earlier, there are 10^{170} possible board positions. It was assumed for a long time that intuition and perception played a greater role in *go* than lookahead. So much so that thinking in *go* has been associated with Zen Buddhist practices (Cobb, 1997). The history section in the British Go Society website also states that 'the future of Tibet was once decided over a Go board when the Buddhist ruler refused to go into battle; instead he challenged the aggressor to a game of Go'.

It was not until the program AlphaGo from DeepMind beat the world's best player Lee Sedol in 2016 that a combination of search and deep neural network learning approach was used to master the game (Silver et al., 2016). AlphaGo employed two networks: one, a *policy network* used to select the next move, and the other a *value network* that learns the evaluation function and predicts the winner. These neural networks learn both from human expert games and reinforcement learning from games of self-play. Combining the neural networks with Monte Carlo search enabled the program to play a very high quality game. During the match against Lee Sedol, AlphaGo attracted worldwide attention when it found an unusual move, number 37 in Game 2, that confounded the best human experts. After this match, AlphaGo earned the highest possible 9 *dan* professional ranking. Further details are beyond the scope of this chapter.

AlphaGo was bootstrapped by learning from human expert games. Subsequently the team wrote a version named *AlphaGo Zero*, which learnt the policy and value networks only by playing millions of games against itself from scratch, starting with completely random play (Silver et al., 2017). The program needed only the rules of the game to start playing, and it soon demonstrated its efficacy by *also learning the game of chess* and becoming the best player in just a few days of playing against itself. It has beaten all other programs written earlier and the best human players as well.

Then in 2018, the group developed a unified program called *AlphaZero*, which taught itself to play *go*, chess, and *shogi* (a Japanese version of chess). AlphaZero managed to beat the state-of-the-art programs specializing in these three games (Silver et al., 2017). A program called *Leela Chess Zero*[8] made its appearance also in 2018. It was developed by Gary Linscott and used deep reinforcement learning using an open-source implementation of AlphaZero.

---

[7] https://www.britgo.org/intro, accessed November 2022.

[8] https://lczero.org/, accessed May 2023.

DeepMind in turn introduced its program *MuZero* in 2019, a model-based reinforcement learning algorithm that incorporates a learned model into the training procedure (Schrittweiser et al., 2020).

Clearly, learning evaluation functions is a job well suited for deep reinforcement learning.

8.3 Backgammon and Scrabble: Elements of Chance

The games studied so far have all been two-person zero-sum complete-information alternate-move deterministic games. We now look at games that involve an element of chance.

We first look at two games in which computer programs attained world champion status quite early. These are backgammon and Scrabble. We will not discuss these games in detail. The interested reader is directed to (Khemani, 2013) and other publications for more detail.

Backgammon is a board game in which two players are engaged in a race to the finishing line. A little bit like snakes and ladders and Ludo, two games popular in many households. The rules have some commonalities. The players have to move their coins to a destination location. In snakes and ladders the competing coins run on the same track. In Ludo they run on overlapping tracks starting and ending at their own locations. Imagine four runners starting and ending 100 metres apart on a 400 metre track. In backgammon they move in opposite directions on the same track. All these games have a capture rule, a bit like in chess, except that the piece goes back to its starting square. In backgammon and Ludo doubled pieces can block opponent pieces. Both in Ludo and backgammon a player has more than one piece on board, and hence there is an element of tactics in deciding *which* piece to move on one's turn. Snakes and ladders, on the other hand, is purely a game of chance not requiring any decision making.

Most importantly, the moves in all three games are dictated by the throw of a dice, two in the case of backgammon. As a consequence, the branches available to the player on move are dynamically determined by the throw of the dice. What is more important is that since this is true of future moves as well, including those of the opponent, a game tree cannot be constructed, and the algorithms discussed earlier are ineffective.

The champion program *TD-Gammon* written by Gerald Tesauro (1994, 1995, 2002) was one of the first programs to learn the evaluation function by playing against itself. It relied on the technique called reinforcement learning devised by Sutton and Barto (1998). Since the feedback from the outcome of the game had to be propagated back through the various moves made, a particular version of reinforcement learning called *temporal difference* (TD) learning was used, giving the program its name (Sutton, 1988). Like the case of AlphaGo Zero, Tesauro's first program called *Neurogammon* learnt only from the games of human experts (Tesauro and Scjnowski, 1989; Tesauro, 1989). Later versions of TD-Gammon combined both forms of learning and also did a limited amount of lookahead by doing many *rollouts* of the dice to decide on the current move (Woolsey, 2000).

Another game where computers have done well is the word game Scrabble. It is played by 2–4 solo players, each trying to win. The game is played with a hundred letter tiles in a bag, from which each player blindly draws seven and places them on a *rack* hidden from other players. Each letter has an associated score in the range from 1 to 10 with higher values for infrequent letters like *Q* and *Z*. Two of the tiles are blank with a zero score. They can stand for any letter. The game is played on a 15 × 15 board by placing letters to make legal words on the board. The first player can make any word which occupies the centre square. Subsequent words must be connected with existing words, like in a crossword puzzle. All sequences of letters on the board must be valid words. In addition to the different letters having their own score, certain squares on the board also amplify the score of the player, with double or triple letter squares which multiply the letter value correspondingly, and double or triple word scores that multiply the word score arrived at from the letter scores. Playing all seven tiles a player holds earns an additional bonus. Figure 8.15 shows a typical combination of words on a Scrabble board. Observe that there are no sequences of letters that are not words.

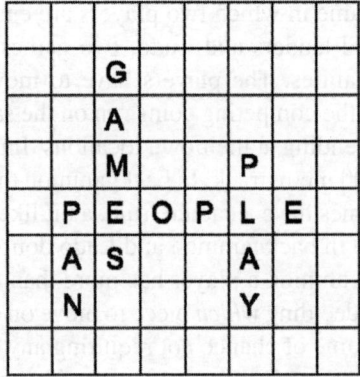

Figure 8.15 Word combinations on a Scrabble board. The next player must use the existing letters to form a new word and at the same time not leave meaningless letter combinations. For example, one can play HAD vertically on the bottom three squares on the left, but cannot make FAD because FAS would not be a word horizontally.

Scrabble is of course a game in which one player wins, and is zero-sum in that sense. But in another sense it is not adversarial because each player is trying to maximize her own cumulative score. In that sense it is more like a competition. One could even imagine playing it solo to try and maximize the score. But there is certainly a tactical element to the game, bringing in adversarial decision making. An advanced player may decide to minimize the number of openings available to the opponents, since new words have to be connected to existing ones. Or she may use up or block a triple word opening, employing a dog in the manger tactic.

The main objective though is to maximize one's own score. A large vocabulary of words here is advantageous, and clearly computers possess that. Selecting a subset of the seven tiles you hold and placing them at an appropriate place on the board is the main task. According to Brian Sheppard who implemented the program *Maven*, a typical rack and board position may have 700 possible moves (Sheppard, 2002). So the branching fact can be very high. It is the objective of maximizing one's score that makes it harder. It is often easy to find a word to make, but a little harder to exploit the letter scores and special squares on the board. A little bit like the fact that finding some tour in a TSP problem is much easier than finding good or optimal tours. Most humans adopt a greedy approach trying to maximize the points earned by making the current word. But carefully leaving useful letters, called the *rack leave*, for future moves requires more imagination. As the game proceeds and letters get used up, one can also make educated guesses about what letters are still in the bag, and perhaps what opponents hold. This is because we know the initial set of hundred tiles. These kinds of inferences are more common in contract bridge where we know that the pack has fifty-two cards, but more on that later. In their program called *Inference Player*, Richards and Amir (2002) do make such inferences about the opponent's rack. It infers that any letters that could have been used for high scoring words are not on the rack, because otherwise they could have been used. Like Sherlock Holmes, the fictional detective created by Arthur Conan Doyle, asking[9] 'why did the dog not bark?'

A *trie* data structure with common prefixes of different words combined together can be used to look for words. Such a network is known as a *directed acyclic word graph* (DAWG) and is a compact searchable structure containing the words in the dictionary (Aho et al., 1974). Figure 8.16 shows a part of a DAWG for a small collection of words beginning with the letters *G* and *H*. Shaded squares are those where a legal word ends.

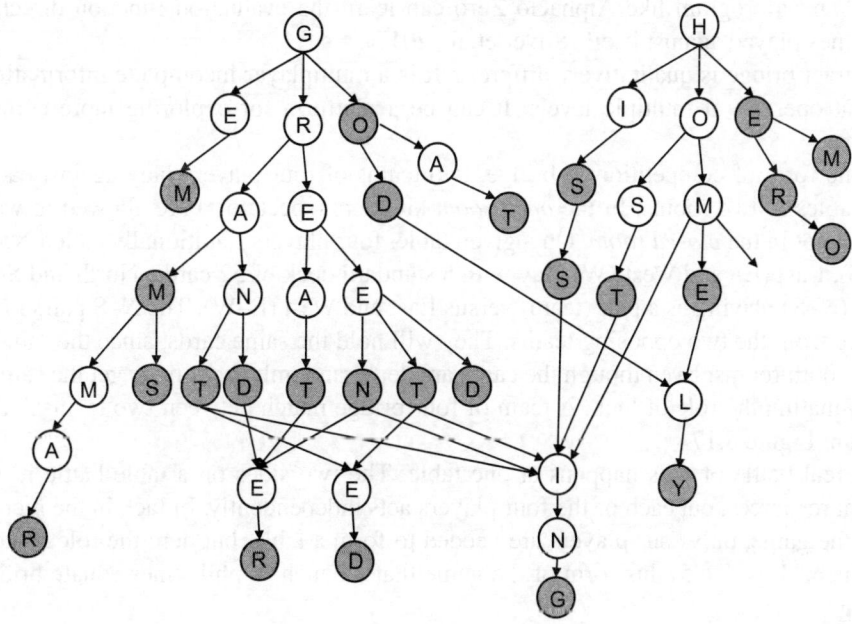

Figure 8.16 A compact searchable dictionary can be represented as DAWG. Here we have a few words beginning with the letters *G* and *H*. Shaded nodes represent the culmination of a legal word.

---

[9] In "The Adventure of Silver Blaze."

The problems to solve for a Scrabble player are the following. Given the existing words on the board and the letters on one's rack, identify the words that can be made such that the score is as large as possible, but leaves useful letters on the rack for the future, and prevents an opponent from making a high scoring word. The DAWG in Figure 8.16 is useful to look for words starting with a given letter. It would be interesting to devise a representation that can also look for words with a given letter somewhere in it, or at a specific location. How does one retrieve the words – like had, him, gin, or go – that can be formed in the location in the bottom left identified in Figure 8.15? We leave that as a point to ponder for the reader.

Next, we look at a game that is still a challenge for computer programs. Contract bridge is an incomplete-information team-game which is adversarial being ultimately a zero-sum game.

8.4 Contract Bridge: A Challenge for AI

The complexity of games like chess and *go* arises due to the huge search space they confront a search algorithm with. Conceptually, though, they are simple, being one dimensional in the sense that one can, in principle, look ahead in the game tree to determine the best move. And looking ahead is possible because the board position is visible to both players, as are the available moves. The strategy then is to look ahead deeper, perhaps more selectively, and to evolve better evaluation functions. The latter has been demonstrated by (deep) reinforcement learning, and a program like AlphaGo Zero can learn the evaluation function directly from many games played against itself (Silver et al., 2017).

Contract bridge is qualitatively different. It is a multiplayer incomplete information team game that operates at multiple levels. It can be a platform for exploring more complex AI algorithms.

In one form of competition in bridge, two teams of four players play against each other on two tables in two rooms. In the *open room* kibitzers (spectators) are allowed to watch the game, but not in the *closed room*. On a given table, four players traditionally called North (N), South (S), East (E), and West (W) play with a standard deck of 52 cards. North and South are partners (N–S) playing as a pair (team) versus East and West (E–W). The N–S pairs in the two rooms are from the two opposing teams. They will hold the same cards, since the same deal is played in both rooms. Even though the cards are dealt randomly, both pairs get the same cards, thus eliminating the role of luck. A team of four bridge match between two teams A and B is depicted in Figure 8.17.

The real battle of wits happens at one table. The two sides on a table battle it out with their joint resources, but each of the four players acts independently. In fact, in the recreational form of the game, only four players are needed to form a table, but here the role of lady luck is prominent. It is due to this form of the game that a bunch of philistines equate bridge with gambling.

The pack of cards is made of four *suits* – spades (S), hearts (H), diamonds (D), and clubs (C), each having 13 cards {ace, king, queen, jack, 10, 9, … 2}. The cards are ranked with the ace being the highest and the deuce, or the 2, being the lowest. On a given table each of the four players is dealt 13 cards randomly and the essence of the game is to play out 13 rounds, or tricks. In a trick each player in turn plays a card, and the rule is that the card must be from the same suit as the first card in that trick, unless the player does not have one, in which case she

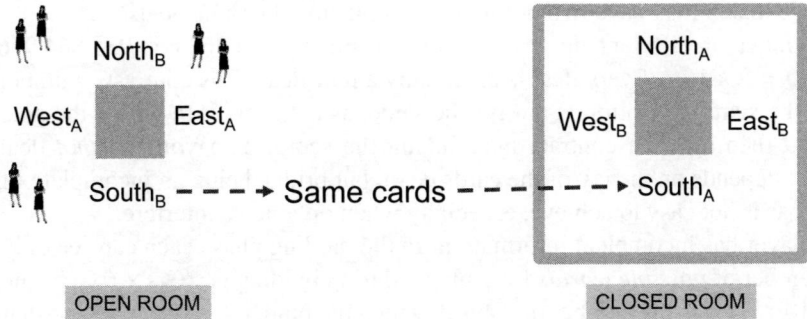

Figure 8.17 Two teams A and B play a bridge match. In the open room where kibitzers are allowed, East–West are from team A and North–South from team B. The same deal is played in the closed room with the opponents holding the same cards as their corresponding players in the other room.

can play any card. A trick is won by the side whose player has played the highest card in the suit led in that trick. That player gets to play the first card in the next trick.

In general, the more the tricks a side makes the better, but the actual score in a deal is governed by a *contract*, giving the game its name *contract bridge*. The contract is arrived at in a *bidding* phase preceding play, with the side bidding for the highest number of tricks winning the contract. The rules are a little more elaborate with the suits also being named as *trumps* during bidding, but are not of immediate interest to us here. In a team of four bridge match, the first obligation of the contracting side is to fulfil, or *make*, the contract. The other side aims to *defeat* the contract and if they succeed they would get a positive score. Contracts can also be challenged during bidding and, in turn, counter-challenged, to multiply the stakes. In addition, there are levels of contracts called a *game*, a *small slam*, and a *grand slam* which have incremental hefty bonuses motivating the sides to bid higher, but only when they have the combined strength in cards to make the contract. Elaborate *bidding systems* have been devised to exchange information in pursuit of the par contract. More on that later.

In the bidding phase, each player can see only her 13 cards. When bidding ends, one side wins the contract and has to make at least the contracted number of tricks. The proceedings for the contracting side are conducted by only one player, called the *declarer*, who decides the cards to be played from both hands, and her partner, called the *dummy*, has no further say in the matter. The other pair tries to defeat the contract by winning enough of the 13 tricks so that the declarer does not succeed. These two players are called *defenders* and are said to *defend* the contract (but that really means they are trying to wreck it). Both the defenders have to make independent decisions. The dummy's cards are kept face up on the table and are visible to all. Each player can thus see two hands, her own and the dummy's.

The following features make contract bridge a much more complex game than board games like chess:

- On a table there are four players, forming teams of two each. This requires cooperation between partners, which entails exchange of information via legal codes and signals.

- A bridge match may have two or more tables playing the same deals.
- The number of different deals or starting positions is 5,364,659,935,864,916,575,237, 440,000 = 5×10^{27}. Every deal is practically a new deal. This contrasts with board games where the starting position is always the same, as is the goal. In bridge the goal is to first bid, and then make, a contract maximizing the score, or payoff, on *that* deal. The par contract depends on the lay of the cards dealt, but bridge being an incomplete information game, par is not easy to achieve, especially when opponents interfere.
- Each player has incomplete information. In the bidding phase each can see only 13 cards. The number of *possible worlds* for a player during bidding is $^{39}C_{13} \times {}^{26}C_{13}$ which is about 8.4×10^{16}. This is the number of ways in which the hidden cards can be distributed among the other three players. Given that the contract is decided in this phase, each partnership needs to exchange information about their hands. This is done by means of an encoding known as a bidding system.
- When play begins after bidding, each player can see two hands. Their own and the dummy's. The remaining 26 cards can be divided in $^{26}C_{13} = 10,400,600$ ways. In practice, one defender plays a card *before* the dummy is put down, called the *opening lead*, so for the other two players, there are only 25 hidden cards when their turn comes, which can be divided in 5,200,300 ways.
- The declarer can see the combined assets of her side, but the defenders have incomplete information of what cards their side holds. Thus the two sides face different *kinds* of problems.

The *key* to the game is information exchange over an open channel, by means of public announcements. This happens both in the bidding phase and in the play phase, and players attempt to *reconstruct and imagine* the card holdings of other players. Since information is exchanged over an open channel, the opponents are listening in too, and this in turn leads to the possibility of deception. If one knew all four hands, one could analyse the par contract along with the par line of play, determining the Nash Equilibrium. In practice, bridge players often attempt to capitalize on the lack of information the opponents face, and try and beat the par. The following is the kind of information exchange during the two phases of the game.

Bidding

On the face of it, a *bid* is a contract *proposed*. If no one bids higher, then that bid becomes the contract. In bridge the contracts range from 7 tricks to 13 tricks, along with the specification of a trump suit. The idea of trumps is that even the smallest trump card is higher than all cards of other suits. Both sides naturally strive to name a suit they have more cards of as trump. A contract of 7 tricks is the lowest you can bid for, because that leaves 6 for the other side, and the opponents will have to make 7 tricks too to defeat your contract. For higher contracts (of n tricks) they need fewer ($14 - n$) tricks to defeat the contract. A bid of 1S, read as 'one spade', says that one proposes to make $1 + 6 = 7$ tricks with spades nominated as trumps. The four suits themselves are ranked. Clubs is the lowest suit, followed by diamonds, hearts, and spades. One can even propose no trump (N), saying that there are no trumps. There are thus 35 possible bids for contracts, starting with 1C, 1D, 1H, 1S, 1N, 2C going up to 7D, 7H, 7S, and 7N.

Like in any auction, one can only make a higher bid than the last one, and as the bidding proceeds, the contracts being bid for get higher and higher. A player can also *pass* on her turn. A scoring system determines the *payoff* for each contract. If the contract is fulfilled the side gets the payoff, else there is a *penalty* which is the gain for the other side. Thus the goal is to bid as high as possible for high payoff, but *only* as high as being able to fulfil the contract. In addition to these 35 bids, there is a bid called *double* which can only be made if the last bid for a contract is *by either opponent* and it challenges the contract (though there have been cases reported when furious players have felt like doubling their partner's bid!). The *double* also raises the stakes for both the payoff and the penalty. The contracting side can counter the challenge with a *redouble*, which raises the stakes even further. Bidding ends when three players pass in succession, except in the opening sequence when after three passes the fourth player still has a bid.

The objective of bidding is to reach a *makeable* contract with the *highest* payoff. Since both sides may have suits in which they have more cards, they strive to buy the contract with their suit as trumps. The problem is that each player can see only her cards, whereas the makeable contract depends upon the combined strength of the side. Hence they need to exchange information about their holdings. But the only language they can use is the bids. In this way the bids are also used to encode information to be conveyed. The coding scheme is known as a *bidding system*, and there are numerous bidding systems that have been devised. These are not secret codes, but have to be revealed to the opponents.

A bid, thus, has two facets. One is the encoded information that it seeks to convey to the partner via a public announcement. The other is the literal meaning of the bid, which is the contract it specifies. The latter comes into effect only for the final contract.

Bidding systems are designed to encode information to be conveyed. This encoding is context sensitive. This means that the meaning of a bid depends upon the sequence of bids that precede the bid, including the bids made by the opponents. Thus bidding systems are quite elaborate. Even *double* and *redouble* have encoded meaning based on the context in which they appear.

Most commonly a bid encodes two kinds of information. One, the *length* of a particular suit held by a player. Consider the 1S bid. If it is the first bid made by a side, an *opening bid*, then in most bidding systems it promises five or more cards in spades. But in the sequence when a player is responding to the partner's opening bid, say 1D, then it shows four or more spades. If a player opens with, say, 1C and then bids 1S in the next turn, it usually shows exactly four cards in spades. The more cards a partnership has in a suit, the better the suit is as a choice for naming as trumps.

The other is information about high cards. High cards are important because they can win tricks. Almost all bidding systems use the following measure of *high card points* (HCPs). An ace is counted as 4 HCPs, a king 3, a queen 2, and a jack 1. These are known as *honour* cards. Each suit has 10 HCPs, and the full pack has 40. The more *a side* has in excess of the expected 20, the more they are likely to bid and make a high contract. The main goal of bidding is for each partner to create a picture of the holdings of the partner, and at the same time convey information about their own holding. This includes information about shape (how many cards of each suit) and HCPs. The catch is that as one makes more bids, the contract to be made gets higher, and there is a danger that the contract may not be makeable. One would like to convey as much as one can in the limited and shrinking bandwidth. A corollary of this constraint is that a side which perceives itself to be weaker, but with a long suit, can make a tactically high bid, known as a *pre-empt*, to consume the shared bandwidth and force the stronger side to guess with less information.

Bids do not just describe suit length and points as described above. The need for creating an accurate picture of the combined hands has necessitated the invention of bidding languages that are like dialogues, with questions being asked by *asking bids*, answers being given in *response bids*, inviting partner to bid higher with *invitational bids*, showing a red flag with a *sign-off bid*, or showing *control* in a suit by a *cue bid*, and other mechanisms too numerous to be described here.

As bidding proceeds, all players are privy to the encoded meaning of bids. It is the ability to imagine and reconstruct the hidden hands that is the hallmark of an expert. This involves making inferences from the bids made by other players about *their* cards. As bidding proceeds, the level gets higher and higher, each side being careful to not venture into a zone where the penalty they could concede is prohibitive, and bidding ends when three players pass. It is then the time to play the cards.

Play

The play of cards essentially entails planning, and counter-planning. The declarer makes plans to make the contract, and the defenders plan to defeat it. Both sides have to operate with incomplete information, and algorithms like ALPHABETA are not directly applicable.

In contrast, though, most bridge playing programs currently adopt a Monte Carlo approach in which they generate a set of sample hands, and then treat each instance as a complete information game (Ginsberg, 1999; Browne et al., 2012). And they do this for *every* card they have to play. By generating a sufficient number of samples, they are able to choose an action that often works. There have been cases, however, where, because each generated sample has complete information, the programs can always choose the winning card for that particular sample. In doing so, the program may choose an inferior line of play which may not work on the actual hand, where the human expert would choose a different line that would work *whatever* the lay of the opponent's cards is. A typical example is when an opponent queen needs to be *finessed*, and the sampling approach gets it right every time knowing all the cards, but a human expert would adopt an *end-play* that obviates guessing where the queen is. The Monte Carlo approach has achieved some success, but it is not clear whether it can perform at the top human level. Also, being based on sampling and not on planning and reasoning, such programs cannot explain their decisions.

A human declarer would adopt a planning approach. First she would count her top tricks, and determine how many more need to be *generated* for the contract to be fulfilled. Then she would delve into her repertoire of *thematic techniques*, learnt over a period of time by reading bridge columns and bridge books, and also learning from experts, and even from her own mistakes. These techniques are *conditional plan parts* retrieved from memory, adapted for the current problem, and strung together with other interleaved *thematic acts* (Khemani, 1989, 1994). A simple example is the *finesse*, say of an opponent held queen. This involves making the opponent commit first, to play or not play the queen, because her turn is earlier. If she does play it you can play a higher card, say the king, and if she does not, then you can play the jack. The finesse will work if the targeted opponent does indeed hold the queen. Without any other information, the *a priori* probability of this being true is 50 per cent. But with extra information, the *a posteriori* probability could change, and it is the expert bridge player that can *draw inferences* from bidding and play to choose the best line of play from the different options that present themselves.

There are numerous books written on the subject describing play techniques, for example, Kelsey (1995) and Love (2010), that describe techniques like *safety play, trump control, cross ruff, dummy reversal, elopement, double squeeze,* and many others, and even more exotic techniques like the *backwash squeeze* and *entry shifting squeeze* described in the bible of card play by Geza Ottlik and Hugh Kelsey (1983).

Defence

The play phase in bridge is not symmetric. One side is trying to fulfil the contract and has the advantage of fully knowing what is in their armoury, since the declarer can see her hand as well as the dummy's. The defenders too can see the dummy's cards, but cannot see their partner's cards. The two of them also have to take independent decisions, and neither knows the combined strength of their hands, whereas the declarer alone decides what cards the dummy will play.

To facilitate meaningful cooperation they have to resort to *signalling*. But the only medium available to them are the cards that they can play, even while obeying the rules of the game, and pursuing their goal of defeating the contract. Typically the signals involve the choice of otherwise equivalent cards. For example, having the two small cards, the 2 and 3 of spades, the order in which one plays them when the suit is led can be used to encode a message. A high-low, for example, might signal an even number of cards. The card that one starts a trick may be informative as well. Very often a low card signals possession of an honour card, while a 7 or an 8 may deny one, and an honour card lead indicates a touching honour. Such signals are not secret and have to be disclosed, and the declarer too can draw inferences from them. Defenders also follow certain heuristics like second-man-low (for information concealment) and third-man-high (to avoid giving away a trick cheaply). In addition, they also need to plan how to defeat the contract, albeit individually, coordinated only by the signals. As with declarer play, such planning draws upon the inferred knowledge that the players have, with their own set of techniques like the *uppercut* and *trump promotion*. Defence is widely considered to be harder than declarer play, and often requires greater imagination, as illustrated in the deal below.

The fact that both declarer and defence make plans means that their adversaries can try and recognize their intent and adopt countermeasures. As mentioned briefly above, the ability to make inferences contributes to choosing plans with higher probability of success. Likewise, the ability to reason about what others know, known as the 'Theory of Mind', gives more ammunition to an intelligent player to win the battle of wits (Leslie, 2001). Reasoning about what other agents know and believe is studied in the field of epistemic logic (Fagin et al., 2004) and is beyond the scope of this book. But we present below an example of such reasoning and planning from the real world as a motivation for researchers to take up programming bridge (Khemani and Singh, 2018).

Deception

Deception during bidding constitutes mainly of announcing some features that one does not have in the hope of deflecting the opponents from their best contract or dissuading defenders from the winning defence later in the play. Deception in play, likewise, is aimed to veer the opponent away from a winning line of play. In either case, the goal is to do better than par.

The following deal is a real life example described in Truscott and Truscott (2004). Maurice Gray (1899–1968) was a dispatch rider with the British Army during World War I and a keen bridge player. He was sitting West on the following hand which we analyse from the perspective of the declarer sitting South. The contract was to make 9 tricks with no trumps, as shown in Figure 8.18. On the left we see the cards as seen by the declarer, while on the right is the view that Gray sitting West had. Our discussion below focuses on the manner in which Gray *anticipated* the declarer's plan and spun a web of illusion. We hope that the discussion will be accessible to even readers not familiar with the game.

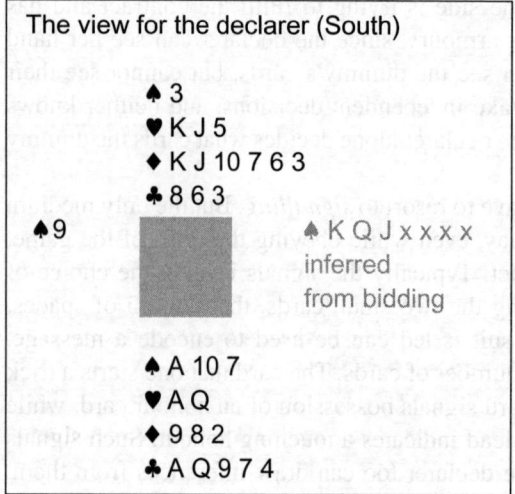 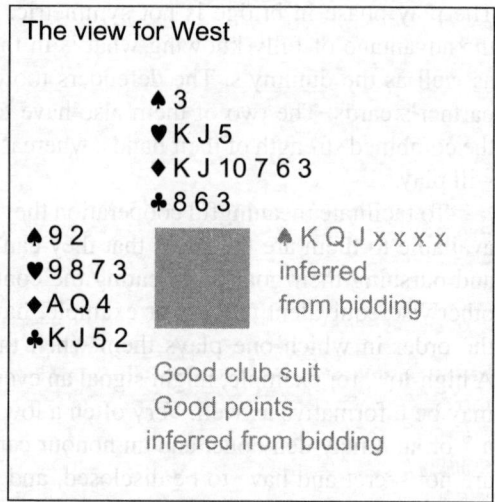

Figure 8.18 Maurice Gray, sitting West, led the ♠9 sitting West against 3N. The figure on the left shows the cards as seen by the declarer after the lead. On the right we have the cards as seen by Gray. Both have inferred the long spade suit with East.

There was no trump suit and spades were led. East had bid spades and was known to have a stack of them that could win many tricks if he got the chance after the ace was out of the way. The first goal of the declarer was to keep East away from getting the lead again. The declarer adopted the standard play of ducking (letting East win) the first two rounds of spades so that West would not have any left to play later. South identified the main source of tricks as the diamond suit, and he had to hope that the ace would be with West. His plan was the following – after winning the third round of spades with the ace, he would play a small diamond from the South hand. If West played the ace, he would play small from dummy. And if West played small, he would play the king from dummy, a kind of finesse, and then play another diamond which West would be forced to take (either with the ace or the queen if he had that too) but would have no spades left to play. South could then *set up* his diamond suit and make the contract.

This line of play is the best for the given diamond suit and the known information. Four cards of diamonds are out, and the declarer does not know how they are distributed. He has to hope and pray that West held the ace. There are eight possible ways the four outstanding diamonds can be divided when West has the ace, as shown in Figure 8.19. Of these the declarer's play would succeed in seven, shown in unshaded rectangles on the left. It would fail only in the *possible world* shown as a shaded rectangle at the bottom, because even though the ace is with the West as hoped, East still has the queen along with two small cards and would command a trick later. But the actual case (as in Figure 8.18) was the one shown in a solid line rectangle, and the declarer was destined to succeed.

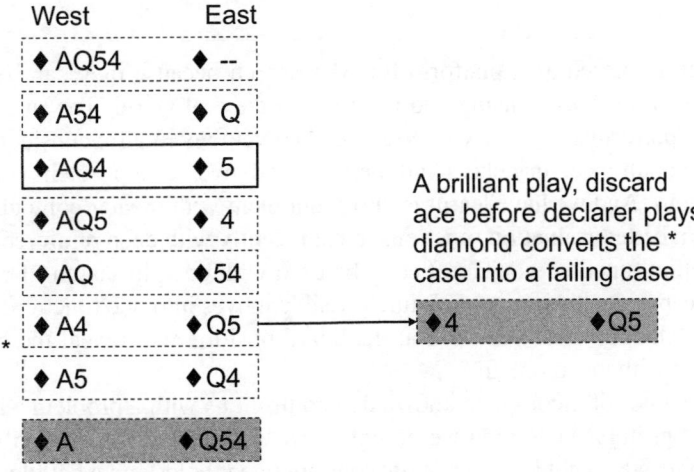

Figure 8.19 There are eight possible ways the diamonds could be divided assuming West had the ace, as shown on the left. Of these, on the last one in the shaded rectangle was the only case when the declarer's plan would fail. It would succeed in the other seven including the solid line rectangle. However, *if* the cards were in the possible world marked with an *, then West had a brilliant play available, of discarding the ace of diamonds earlier, and convert that into a failing case.

Maurice Gray, however, had other plans. He imagined a possible world in which the queen was actually with his partner. The imagined case is marked with an * in the above figure. Now, *if that were the real world,* a brilliant player could jettison the ace of diamonds on the third round of spades. Then he would be only left holding diamond 4 as shown on the right, and the declarer could no longer set up his diamond suit without yielding a trick to the queen with East, who would then run his spade winners. But *only if* Gray's diamonds were the ace and the 4, as in the marked case. Nevertheless, Gray *pretended* that *that* was indeed the case and discarded his diamond ace on trick three when spades were continued.

This would have had no impact, except the loss of a diamond trick, had the declarer embarked upon his plan in the diamond suit. But the declarer made the inference that Gray wanted him to make – that the real world was as in the starred case, and that Gray had made the brilliant discard. Otherwise, why would he discard the ace? The declarer then abandoned his diamond suit plan and went after the club suit, which, to the discerning reader, should be clear was destined to fail. The declarer fell into a trap because he was a thinking player, who could make inferences. A lesser player would have failed to draw the intended inferences.

Contract bridge then is a complex multidimensional game that combines various kinds of reasoning – communication, multi-agent reasoning with incomplete information, planning, epistemic reasoning with possible worlds and probabilities, plan recognition, and counter-planning. The most important is the ability to imagine possible worlds, the ability to imagine what the opponent knows (the Theory of Mind), the ability to draw inferences and augment one's imagined construction of the hidden hands, and monitor plans dynamically as play proceeds. This is complex reasoning. There are no simple or 'neat' ways of solving complex problems.

Summary

Games have been proposed as a platform for AI research because they can be implemented easily with minimal interface sensing and acting with the real world. The rules in games are well defined, and performance is easy to measure. And yet they require considerable skill.

For more than half a century chess and then *go* have given us a problem where the search space is humungous. And we have learnt to surmount them with greater computing power and the ability to learn better evaluation functions, most recently with deep reinforcement learning. The ideal evaluation function would suffice to be used with one ply search. If we can achieve that, it would encapsulate the impact of future possible moves into one function. Will we attain that stage of knowledge where one look at the board position will reveal the best move to a program? We wait with bated breath.

Meanwhile, games of incomplete knowledge confront us with a problem where the future is covered with fog. In backgammon we do not know what moves the dice will offer the two players. In Scrabble we do not know what tiles our opponent holds and what tiles are still in the bag. And yet these games have been conquered with a combination of reinforcement learning and efficient dictionary search.

It is perhaps time to move on to the next challenge. The multiplayer knowledge rich game of contract bridge, which requires players to communicate with their teammates, make inferences from the actions of others including what plans they have, and employ probability and deception in pursuit of a win. The state-of-the-art programs have relied on Monte Carlo techniques in the lay of cards eschewing the human approach of bringing in tactical knowledge and reasoning. Will these approaches be sufficient for playing better than us? They have been in board games. Edward Feigenbaum has been credited with a quote saying that just as aerodynamics enables us to build aeroplanes without mimicking how birds fly, as Daedalus is reputed to have attempted, artificial intelligence will find solutions that do not mimic human thought. Well, contract bridge does pose a challenge for AI.

Exercises

1. Does the following game styled like the prisoner's dilemna have a stable equilibrium? If yes, show how? In no, why not? What would happen in the payoffs in the bottom right square were to be $(-30, -30)$? Can you design the payoffs in which there is a stable equilibrium in which A always betrays and B always cooperates?

(A, B) payoff	B cooperates	B betrays
A cooperates	(1, 2)	(−9, 0)
A betrays	(0, −8)	(−3, −3)

2. What is the size of the game tree for noughts and crosses?
3. Draw the game tree for noughts and crosses and compute its minimax value. You can simplify your task by eliminating rotated and mirror positions. For example, instead of the nine first moves, you can consider only three distinct moves.
4. Solve the game tree of Figure 8.1.
5. [Baskaran] The figure shows a 4 ply game tree with evaluation function values at the horizon. The nodes in the horizon are assigned labels $A, B, C, ...,P$. Use these labels when asked to enter a horizon node or a list of horizon nodes.

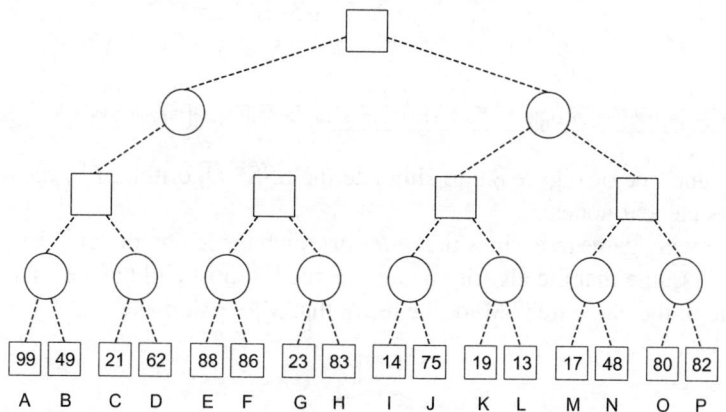

List the horizon nodes in the best strategy for Max. Enter the nodes in the ascending order of node labels.

6. For the above game tree, show the subtree explored by the ALPHABETA algorithm.
7. Modify the MINIMAX algorithm in Algorithm 8.2 to explicitly keep track of depth. Hint: The original call must have the depth k as a parameter, and recursive calls must reduce the value of k. A terminal node will have $k = 0$.
8. Flip the game tree of Figure 8.5 and simulate the ALPHABETA algorithm. Alternatively, simulate it from right to left on the same tree. How many leaf nodes does ALPHABETA inspect now?
9. Take the game tree of Figure 8.5 and enter new values for the leaves such that ALPHABETA does no pruning searching from left to right. For the same tree try simulating ALPHABETA from right to left.
10. Flip the game tree of Figure 8.9 and simulate the ALPHABETA algorithm. Alternatively, simulate it from right to left on the same tree. How many leaf nodes does ALPHABETA inspect now?
11. Simulate the ALPHABETA algorithm *from right to left* on the game tree from Figure 8.5 with 36 leaf nodes below. The values in the boxes are the values returned by the evaluation function. How many nodes does this traversal visit?

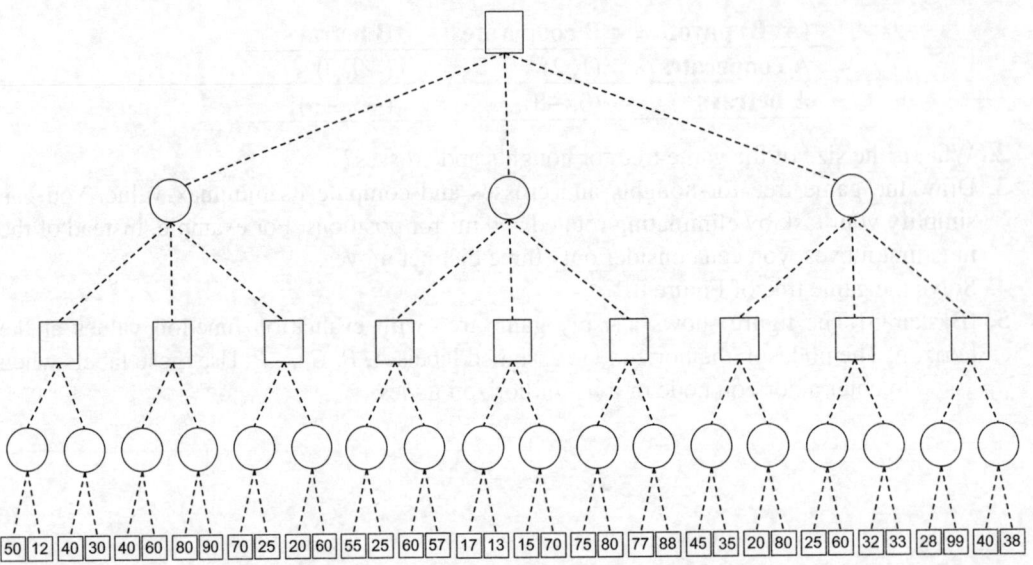

12. Take the game tree of Figure 8.9 to simulate the SSS* algorithm. *List the order* in which SSS* visits the leaf nodes.

13. For the following game tree, show the *order* in which the leaf nodes are visited by the SSS* algorithm. Assume that the algorithm chooses the leftmost child at a decision point. What is the value of the game tree? Mark the move that Max will make.

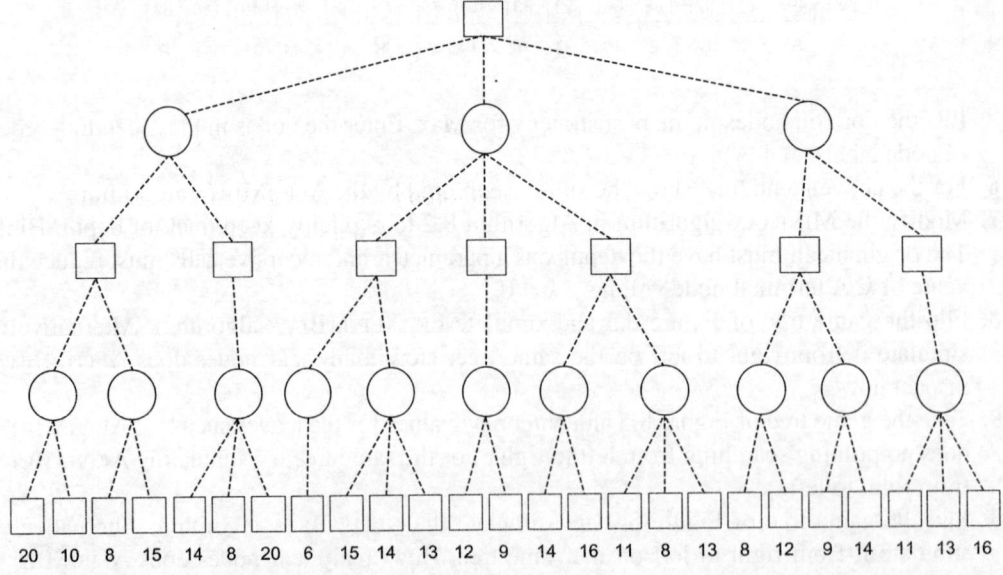

14. Draw the above game tree after algorithm ALPHABETA explores it searching from left to right. Clearly mark the alpha and beta cutoffs.

15. Fill in the leaf values in the following game tree such that there is maximum pruning done by algorithm ALPHABETA searching from left to right. Choose your date of birth in DD format as the value of the leftmost leaf. How many leaves does ALPHABETA inspect?

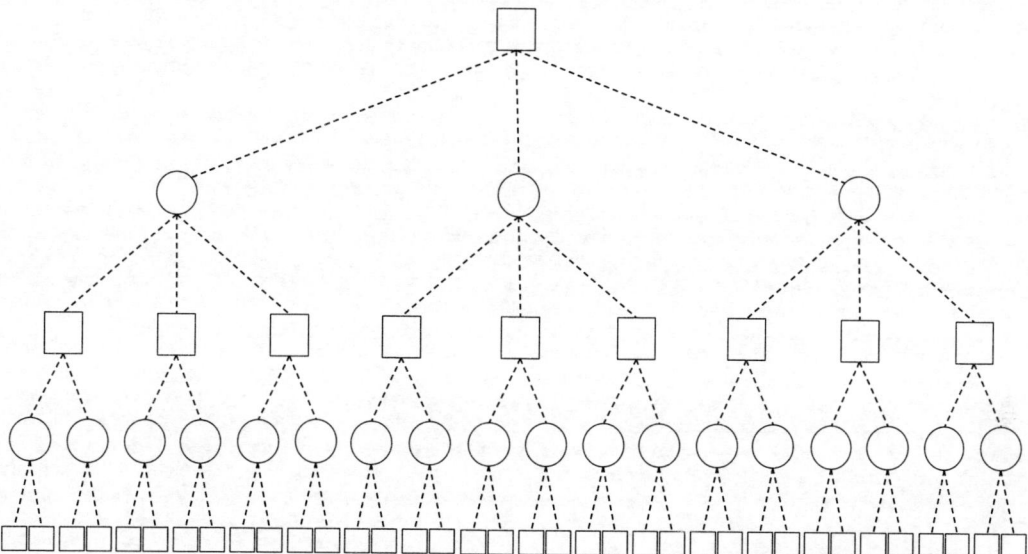

16. For the above tree fill in the leaves such that algorithm SSS* does maximum pruning.
17. For the above tree fill in the values such that algorithm ALPHABETA searching from left to right does no pruning.
18. Is it possible to fill in values in the leaves such that algorithm SSS* does no pruning? If yes, fill in the values such that that happens. If your answer is no, justify your answer.
19. Implement a program to play noughts and crosses with a user on the computer screen.

CHAPTER 9

Automated Planning

So far in this book we have not thought of plans as explicit representations. True, we have referred to the path from the start node in the state space to the goal node as a plan, but that has been represented as a sequence of states. When we looked at goal trees we could also think of the solution subtree as a plan. Likewise, the strategy found by the SSS* algorithm is also a plan. But even here the intent of the problem solving agent is captured in terms of what state or board position will the player move to.

In this chapter we see problem solving from the perspective of actions. We represent plans explicitly, and the agent goes about the task of synthesizing a plan. At the simplest level, a plan is a sequence of named actions designed to achieve a goal. We begin with planning in the state space and move on to searching in the plan space. We also look at a two stage approach to planning with the algorithms GRAPHPLAN and SATPLAN.

We end with a brief look at some directions in planning in richer domains.

An intelligent agent acts in the world to achieve its goals. Given the state of the world it is in, and given the goals it has, it has to choose an appropriate set of actions. The process of selecting those actions is called planning. Planning is the reasoning side of acting (Ghallab, Nau, and Traverso, 2004). Planning and acting do not happen in isolation. A third process is an integral part of intelligent agency – perceiving. An agent senses the world it is operating in, deliberates upon its goals to produce a plan, and executes the actions in the plan. This is often referred to as the *sense–deliberate–act* cycle. The entire process may need to be monitored by the agent. Since the world may be changing owing to other agencies, it may even have to modify its plans on the fly.

There has been considerable work on autonomous agents that plan their activity. This became necessary in space applications where communication with Earth takes too long, necessitating autonomy. This was the case with the Mars rovers experiments by NASA, and even ten years after landing on Mars the rover Curiosity is still active.[1] Likewise, autonomous submersible vehicles were designed to explore the deep oceans on Earth with a system called

---

[1] https://mars.nasa.gov/news/9240/10-years-since-landing-nasas-curiosity-mars-rover-still-has-drive/, accessed 4 October 2022.

Teleo-Reactive Executive (T-REX) (McGann et al., 2008; Rajan et al., 2009). Controlling an underwater autonomous vehicle (UAV) requires the system to follow the *sense–deliberate–act* cycle of an autonomous agent. Autonomous robots are being sent into volcanoes to study eruptions, in preparation for exploring Jupiter (Caltabiano and Muscato, 2005; Andrews, 2019). Teams of tiny robots are being employed for search and rescue missions (Murphy et al., 2008). More recently in 2022, driverless taxi services are being experimented with in San Francisco.

We focus on the deliberation part, planning, of this autonomous activity.

A *planning problem* is a triple $P = <S_0, G, O>$ where
- S_0 is a set of propositional sentences $\{s_1, s_2, ..., s_p\}$ describing the start state. The start state is completely described.
- G is a set of sentences $\{g_1, g_2, ..., g_k\}$ that form the goal description. The description may be partial and then it would be satisfied by more than one state.
- O is a set of operators that prescribe the actions that can be executed in a state. An operator has variables for elements. An action is an instance of an operator with specific elements substituted for variables.

All the three are expressed in a language suitably chosen from a family of languages for domain representation as described below. The languages in the family vary in expressiveness.

The goal of planning is to produce a plan which when applied in the start state will result in a goal state. In the simplest case, a plan is a sequence of actions.

9.1 Representation

The goal in *domain independent* planning is to devise general algorithms that work in diverse domains. The first step in this exercise is for the user to describe the domain, that is, the states and the actions. The planning community has come up with a family of languages to do so. These are called *planning domain definition languages* (PDDL), which can describe domains with varying degrees of richness (McDermott, 1998; Fox and Long, 2003; Edelkamp and Hoffmann, 2004; Gerevini and Long, 2005). The common theme in the family is to adopt the use of predicates from logic to describe the world by a set of propositional sentences, each describing some aspect of the state. This is like the working memory in rule based systems from Chapter 7. Each sentence in the language is like a working memory element, except that we adopt the syntax from first order logic. For example, the sentence *Filled(cup12, tea)* may be used to express the fact that cup12 contains tea, and *Holding(Mahsa, cup12)* may express the fact that Mahsa is holding cup12. These facts can change with activity, and we call them *fluents*, sentences whose truth value can change with time.

9.1.1 Time and change

Classical logics are not concerned with time. Rather they address timeless statements like theorems in mathematics. Planning has to deal with fluents since actions change the world. If Mahsa drinks her tea, the cup would no longer be full. How do we capture this changing world?

One attempt was to extend logic predicates with time as another attribute. *Holding(Mahsa, cup12)* would become *Holding(Mahsa, cup12, t1)* where *t1* is a time stamp. With this approach one would have to describe the world at each time step. A variation is to extend first order logic itself to handle time and change. In *event calculus* one does this with higher order predicates that take a fluent as one of the arguments, the other being time (Shanahan, 1999). Then the fact that Mahsa was holding the cup at time $t1$ would be expressed as *HoldsAt(Holding(Mahsa, cup12), t1)* where *HoldsAt(f,t)* is an event calculus predicate that asserts that fluent f is true at time t. Another *event calculus* predicate *Initiates(e,f,t)* asserts that event e happening at time t results in fluent f becoming true after that. This establishes a causal relation between an action and a resulting fluent. For example, one might say that if Mahsa is holding the cup and puts it down at time t, then it will be on the table, and she would not be holding it thereafter. This is done by two statements of the form *consequent ← antecedants*.

Initiates(Putdown(Mahsa, cup12), Ontable(cup12), t)
    ←HoldsAt(Holding(Mahsa, cup12), t)

Terminates(Putdown(Mahsa, cup12), Holding(Mahsa, cup12), t)
    ←HoldsAt(Holding(Mahsa, cup12), t)

The above two statements capture the relationship between actions and their effects in the domain. Another event calculus predicate *Happens(e, t)*, which states that event e has happened at time t, enables one to infer the statements that actually become true when actions happen.

HoldsAt(Ontable(cup12),t2) ←
    Happens(Putdown(Mahsa, cup12),t1)
    $\wedge$ *Initiates(Putdown(Mahsa, cup12), Ontable(cup12),t1)*
    $\wedge$ $t1 < t2$ $\wedge$ $\neg$*Clipped(t1, Ontable(cup12), t2))*

This inference rule states that *if* it *happens* that Mahsa put the cup down at time $t1$, and nothing happens to *clip* (undo) the resulting fluent, then at time $t2$ greater than $t1$ it would be on the table.

An early planner developed by Green (1969) was based on a logic based theorem prover. The logical approach to planning allows an agent to know when fluents (statements) are true, and what caused them to become true. However, it poses certain bookkeeping problems even assuming that time is discrete. First, one may need to assert, or infer, the truth value of fluents at all times resulting in the working memory becoming bloated. Second, and more important, how does one conclude that a fluent which was true at time t remains true at the next time instant and thereafter? This is known as the *frame* problem in literature (McCarthy and Hayes, 1969; Hayes, 1987; McDermott, 1987; Shanahan, 1997). In event calculus this is hidden in the predicate *Clipped(t1, f, t2)*. Beneath the hood of *Clipped* is the statement that there was some event that happened in the intervening period that made the fluent false. As a consequence, one has to add *frame axioms* to the knowledge base that say that fluents not affected by any action continue to retain their truth value. We will look at frame axioms later in this chapter when we encode a planning problem as *SAT*.

The planning community has circumvented these problems by doing away with time itself, and describing only the current state in the working memory. The planning operators are like the rules from Chapter 7, which *add* the fluents made true by an action, and also *delete* fluents made false by the action. And like rules, they have preconditions for the actions to be applicable. Actions are instances of operators and are also called ground operators. An operator has the following format:

>ActionName(arguments)
>>Preconditions of the action.
>>Effects of the action.

If the preconditions of the action are true in a given state, then the action is *applicable* in the state. If applied, then its effects describe the changes in the state. Different versions of PDDL allow one to describe both the preconditions and effects at different levels of expressivity. The algorithms described here will work with the simplest domain PDDL 1.0, where both the preconditions and effects will be a conjunction of simple relation free atomic formulas. Towards the end we will comment upon richer domains.

9.2 Simple Planning Domains

The simplest planning domains have the following properties.

The world is completely known. This means that the planner knows the world completely and does not have to contend with incomplete information.

The state is described in terms of a conjunction of propositions, each of which is an atomic propositional sentence that is true.[2] Each proposition is an instance of a predicate schema in which variables are substituted with constants that stand for elements in the domain. For example, the proposition *Holding(Mahsa, cup12)* is an instance of *Holding(Agent, Object)*.

Actions are instances of operators that are action schemas. For example, *Putdown(Mahsa, cup12)* is an instance of *Putdown(Agent, Object)*.

Actions are instantaneous. There is no notion of time. This means that either two actions can happen simultaneously or they occur in sequence. If actions had a duration, then one could talk of overlapping actions, or one action happening during the duration of another action.

Actions are deterministic. This means that the plan can be executed flawlessly, and the planner does not need to monitor its execution. Unlike trying to execute a hole-in-one in golf.

The agent is the only one acting in the domain and is the only one responsible for any change that happens in the world.

9.2.1 STRIPS

It all began with the Stanford Research Institute Problem Solver (STRIPS), one of the earliest planning programs developed at Stanford (Fikes and Nilsson, 1971). The authors say that their work was motivated by the frame problem that Green's program ran into. Their goal was to separate the logical reasoning part from the search component which they associated with means–ends analysis (see Chapter 7). The program was written for one of the first autonomous robots, Shakey, that roamed the corridors of Stanford from 1968 and currently is on display in the Computer History Museum at Mountain View, California. Shakey could perform tasks that required planning, route-finding, and the rearranging of simple objects, and has been called the

---

[2] A sentence is something which can be true or false. An atomic sentence has no logical connectives. A proposition is an atomic sentence without variables.

'great-grandfather of self-driving cars and military drones' by John Markoff, historian at the Computer History Museum.[3]

The paper by Fikes and Nilsson describes the domain in which a robot has access to some operators that enable it to move around, climb on boxes, push boxes around, and toggle light switches. The following operators are adapted from the paper. The meaning of the predicates used is sself-evident.

Robot goes from location m to location n in room r
  GoTo(m,n)
    Precondition: Room(r) $\wedge$ Loc(m,r) $\wedge$ Loc(n,r) $\wedge$ AtRobot(m)
    Delete list: AtRobot(m)
    Add list: AtRobot(n)

The above operator says that if the robot is at location m in room r, it can go to location n in the same room using the operator *GoTo(m,n)*. The operator is applicable if the precondition *AtRobot(m)* is true, and when the source and destination are in the same room. If the operator is applied, then the change is effected by deleting *AtRobot(m)* and adding *AtRobot(n)*. The following are some of the other operators:

Robot pushes box b from location m to location n in room r
  PushTo(b,m,n)
    Precondition: Room(r) $\wedge$ Loc(m,r) $\wedge$ Loc(n,r) $\wedge$ Box(b)
                    $\wedge$ AtRobot(m) $\wedge$ At(b,m)
    Delete list: AtRobot(m), At(b,m)
    Add list: AtRobot(n), At(b,n)

Robot climbs on box b
  ClimbOn(b)
    Precondition: Box(b) $\wedge$ AtRobot(m) $\wedge$ At(b,m) $\wedge$ OnRobot(floor)
    Delete list: AtRobot(m) $\wedge$ OnRobot(floor)
    Add list: OnRobot(b)

Robot climbs down from box b
  ClimbOff(b)
    Precondition: Box(b) $\wedge$ At(b,m) $\wedge$ OnRobot(b)
    Delete list: OnRobot(b)
    Add list: OnRobot(floor) $\wedge$ AtRobot(m)

Robot goes from room m to room n through door d
  GoThroughDoor(d,m,n)
    Precondition: AtRobot(d) $\wedge$ InRoom(m) $\wedge$ Connects(d,m,n))
    Delete list: InRoom(m)
    Add list: InRoom(n)

---

[3] https://www.sri.com/hoi/shakey-the-robot/, accessed 6 October 2022.

Robot switches on light *l* after climbing on box *b*
  SwitchOn(*l*)
    Precondition: Light(*l*) ∧ Box(*b*) ∧ At(*b*,*l*) ∧ OnRobot(*b*)
                     ∧ Status(*l*,off)
    Delete list: Status(*l*,off)
    Add list: Status(*l*,on)

Given these operators, it is possible to devise a plan that the robot that is in room 3 can go to room 1 in which a light needs to be switched on, push a box in room 1 to the location of the light, climb on to the box, and switch the light on. Observe that we have treated the light as also being a location. The following might be the start state, described as a set of propositions:

{Room(1), Room(2), Room(3), Connects(*D*1,1,2), Connects(*D*2,2,3), Loc(*M*,1), Box(*B*), At(*B*,*M*), AtRobot(*N*), Loc(*N*,3), Box(*C*), At(*C*,*O*), Loc(*O*,2), Light(*L*1), Loc(*L*1, 1), Light(*L*2), Loc(*L*2,3)}

This says that there are lights in rooms 1 and 3, there are boxes in rooms 1 and 2, and there is a robot in room 3.

Clearly the operators allow a plan in which the robot goes to room 1, pushes the box to the location of the light, and switches on the light by climbing on the box. But how can it switch on the light in room 3 where there is no box? The reader is encouraged to add an operator or operators that will enable the robot to fetch a box from another room.

9.2.2 The blocks world domain

The blocks world domain that we discussed in Chapter 4 has been used extensively to illustrate planning problems. We take it up again in this chapter and will use it throughout to illustrate the planning algorithms. The domain has a table large enough to hold any number of blocks. The blocks are named, one block can be placed on another block, and a tower of any height can be constructed. There is a one armed robot that can move blocks around one at a time. Figure 9.1 depicts a typical problem in the blocks world domain.

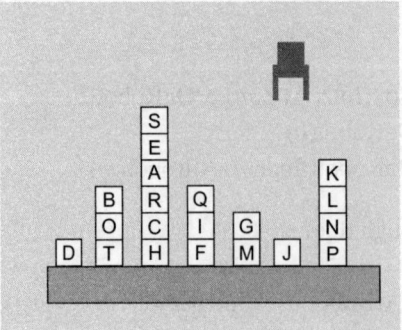 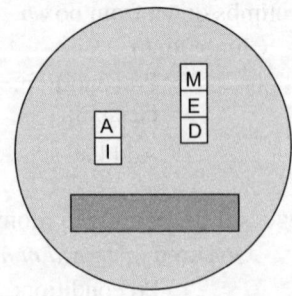

Figure 9.1 The start state on the left has twenty blocks in some configuration on a boundless table. There is one robot arm that is empty. The goal description on the right only says that block *A* should be on block *I*, block *M* should be on *E*, and *E* should be on *D*. Nothing else.

In the figure the start state is shown on the left and is completely specified. The goal description as illustrated on the right is a partial description of a state. This means that there may be more than one state that satisfies the goal description. The start state and the goal state are described by two sets of atomic sentences. Each set stands for a conjunct of propositions. If a proposition is present in the set, it means that it is true. If not present, then it is false. The following is the predicate schema for describing states and goals:

AE	The robot arm is empty.
holding(*X*)	The robot arm is holding block *X*.
onTable(*X*)	Block *X* is on the table.
clear(*X*)	There is nothing on block *X*.
on(*X,Y*)	Block *X* is placed on block *Y*.

When variables in the above predicates are substituted with constants (block names), then they become *propositions*. If a proposition is present in a given state description, then it is true. If a proposition is present in a goal description, then it has to be made true. A state S satisfies a goal G if the goal propositions are true in the state. That is, $G \subseteq S$.

The following is the start state description for the problem in Figure 9.1.

Start state = {*AE*, onTable(*D*), clear(*D*), onTable(*J*), clear(*J*),
  clear(*B*), on(*B,O*), on(*O,T*), onTable(*T*),
  clear(*S*), on(*S,E*), on(*E,A*), on(*A,R*), on(*R,C*), on(*C,H*), onTable(*H*),
  clear(*Q*), on(*Q,I*), on(*I,F*), onTable(*F*),
  clear(*G*), on(*G,M*), onTable(*M*),
  clear(*K*), on(*K,L*), on(*L,N*), on(*N,P*), onTable(*P*)}

The goal description specifies that the following propositions must be true in a goal state:

Goal description = {on(*A,I*), on(*M,E*), on(*E,D*)}

The following operators are defined in the domain:

Pickup(*X*)	Pickup block *X* from the table.
Putdown(*X*)	Putdown block *X* on the table.
Unstack(*X,Y*)	Unstack block *X* that is on block *Y*.
Stack(*X,Y*)	Place block *X* on block *Y*.

The operators are defined in STRIPS-like manner as follows. An operator where the variables are replaced by constants (block names) is an action. Each operator has a set of preconditions that need to be true in a given state for the action to be applicable in that state. Each operator has an add list that has propositions to be added, and a delete list that specifies the propositions to be deleted.

Pickup(X)
    Precondition: onTable(X) ∧ clear(X) ∧ AE
    Add list: holding(X)
    Delete list: onTable(X), AE

Putdown(X)
    Precondition: holding(X)
    Add list: onTable(X), AE
    Delete list: holding(X)

Unstack(X, Y)
    Precondition: on(X,Y) ∧ clear(X) ∧ AE
    Add list: holding(X), clear(Y)
    Delete list: on(X,Y), AE

Stack(X, Y)
    Precondition: holding(X) ∧ clear(Y)
    Add list: on(X,Y), clear(X), AE
    Delete list: holding(X), clear(Y)

The same operators are written again below in PDDL 1.2 which introduces types for objects. The other changes are that variables are identified by a '?' prefix, and the logical formulas are written in a list notation introduced by Charniak and McDermott (1985). For example, *on(X,Y)* is now written as (on ?x – block ?y – block). Further, the entire domain definition is assembled in a structured form. Finally, the add list and the delete list are replaced by one effect formula with deleted items prefixed with a negation (not).

```
(define (domain Blocks)
    (:requirements :typing)
    (:types block)

    (:predicates    (on ?x – block ?y – block)
                    (onTable ?x – block)
                    (holding ?x – block)
                    (clear ?x – block)
                    (AE))

    (:action pickup
        :parameters (?x – block)
        :precondition (and (onTable ?x) (AE) (clear ?x))
        :effect (and (not (AE)) (holding ?x) (not (onTable ?x))))
```

(:action putdown
    :parameters (?x – block)
    :precondition (and (holding ?x))
    :effect (and (not (holding ?x)) (AE) (onTable ?x)))

(:action stack
    :parameters (?x – block ?y – block)
    :precondition (and (holding ?x) (clear ?y))
    :effect (and (not (holding ?x)) (AE) (on ?x ?y) (not (clear ?y))))

(:action unstack
    :parameters (?x – block ?y – block)
    :precondition (and (on ?x ?y) (clear ?x) (AE))
    :effect (and (not (on ?x ?y)) (holding ?x) (clear ?y) (not (AE))))
)

A planning problem is defined by choosing a domain, specifying the set of objects in that problem, specifying the initial state, and specifying the goal conditions. Consider the following planning problem, specified in PDDL:

(define (problem tinyProblem1)
    (:domain Blocks)
    (:objects
        A – block
        D – block
        F – block
        I – block
        M – block
        O – block)
    (:init
        (onTable A)
        (onTable M)
        (onTable F)
        (on I M)
        (on D I)
        (on O F)
        (clear A)
        (clear D)
        (clear O)
        (AE))

    (:goal (and (on F A) (on A D)))
)

The goal description only states that block *F* must be on block *A*, which in turn must be on block *D*. Nothing is said about where the other blocks are and whether the arm is holding anything.

$S_0 = \{$onTable(A), onTable(M), onTable(F), on(I,M), on(D,I), on(O,F), clear(A), clear(D), clear(O), $AE\}$

$G = \{on(F,A), on(A,D)\}$

This tiny problem is depicted in Figure 9.2.

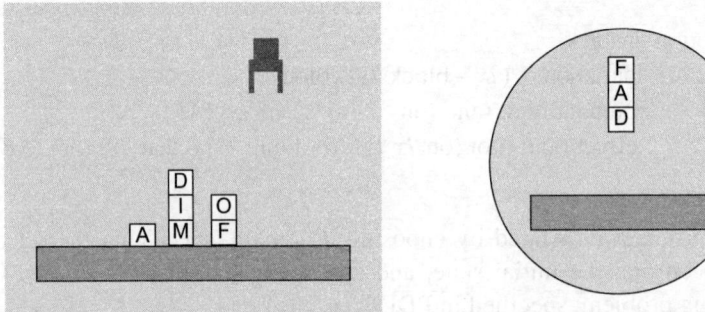

Figure 9.2 A tiny planning problem. The given state is on the left, and the goal description is on the right.

Some more domains are described in exercises, and the reader is encouraged to formulate them in PDDL or in a STRIPS-like language.

We will look at various algorithms for planning and we will use the blocks world always to illustrate the algorithms. We begin with state space planning.

9.3 State Space Planning

State space planning is just a different perspective on the state space search algorithms we have studied in earlier chapters. It standardizes the process with the formal definition of a domain, and the user can make choices of representation in a more informed manner. Having defined a domain, the MoveGen function is composed as a collection of actions that can be applied in a given state. And the GoalTest function applied to a state *S* tests if it satisfies the goal description, $G \subseteq S$.

There is one significant difference though. Planning actions change a given state into a new state. A MoveGen collects moves for a given state and can be used in forward search. Searching from the goal for a path from the start state requires a different MoveGen function. We will investigate that after we have dealt with searching from the start state to the goal state.

9.3.1 Forward state space planning

Let $P = <S_0, G, O>$ be a planning problem. Forward state space planning (FSSP) searches from the start state S_0 towards a goal state. A move in FSSP is implemented as follows:

Let a be an action that is an instance of an operator $o \in O$. Let pre(a) be the preconditions of a, let effect$^+$(a) be the positive effects of a corresponding to the add list, and let effects$^-$(a) be the negative effects of a corresponding to the delete list.

The simplest domains can be modelled as a *state transition system* which is defined as a triple = $(\mathbb{S}, \mathbb{A}, \gamma)$, where

- $\mathbb{S}$ is a finite set of *states*.
- $\mathbb{A}$ is a finite set of *actions* that the actor may perform.
- $\gamma: \mathbb{S} \times \mathbb{A} \rightarrow \mathbb{S}$ is a partial function called the *state transition function*.

An action $a \in \mathbb{A}$ is *applicable* in a state $S \in \mathbb{S}$ *iff* pre(a) $\subseteq S$.

If action a is applicable in S, then $\gamma(S,a)$ is the resulting state S'. If an applicable action a is applied to a state S, then the state *progresses* to a new state S' defined as

$$S' = \gamma(S, a)$$
$$= \{S \cup \text{effects}^+(a)\} \setminus \text{effects}^-(a)$$

One adds the elements in effects$^+$(a) to S and deletes the elements in effects$^-$(a) from S to get S'.

A plan π is a sequence of actions $<a_1, a_2, ..., a_n>$. A plan π is *applicable* in a state S_0 if there are states $S_1, ..., S_n$ such that $\gamma(S_{i-1}, a_i) = S_i$ for $i = 1,..., n$. The final state is $S_n = \gamma(S_0, \pi)$.

FSSP begins with the empty plan and incrementally constructs the plan by adding new actions at the end. In the algorithm for FSSP an action is added to a plan by the assignment

$$\pi \leftarrow \pi \circ a$$

It progresses over a to the state S' as described above and continues searching from there, till it finds a valid plan.

Let G be a goal description. Then a plan π is a valid plan in a state S_0 if

$$G \subseteq \gamma(S_0, \pi)$$

Algorithm FSSP searches from the start state, finding applicable actions, applying one and progressing to the resulting state. The main drawback of forward search is the high branching factor, since the number of applicable actions increases in general with the number of objects in the state. Figure 9.3 shows the set of actions applicable in the start state in Figure 9.1 and also in a state resulting from progressing over the action *Pickup(J)*.

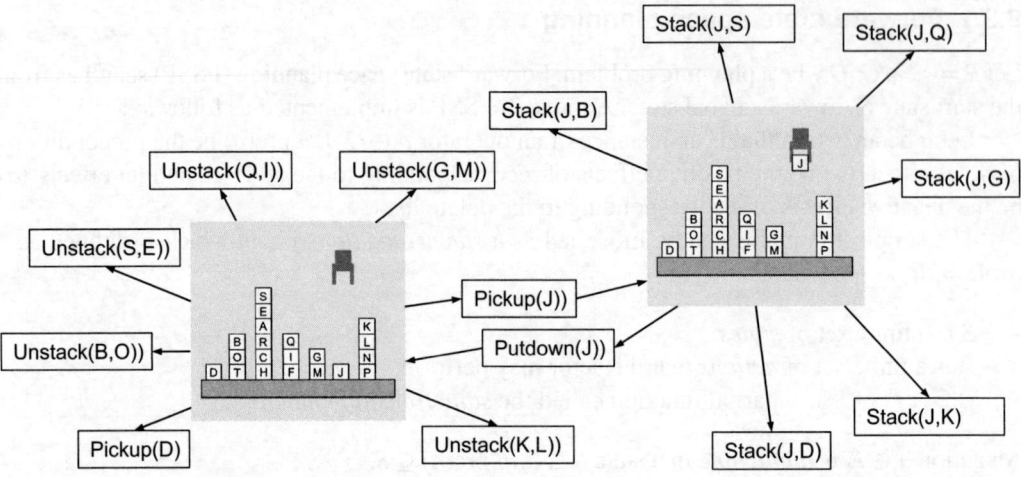

Figure 9.3 The applicable moves in the start state of Figure 9.1, and in the state after progressing over the action Pickup(J).

Even in the simplest domains, it has been shown that the complexity of planning is in PSPACE (Bylander, 1994). That means the algorithm uses space bounded by a polynomial function of the size of the input instance. It can, however, take time bounded by an exponential function of the size of the input instance. This has spurred research in many approaches to alleviate the complexity. One of the approaches has been to devise heuristic functions in a domain independent manner.

9.3.2 Domain independent heuristics

One way to address the issue of algorithms taking exponential time due to high branching factor is to employ a heuristic function. The algorithms we are studying are domain independent. Hence, one needs to look for domain independent heuristics as well. The key idea here is to *relax* the planning problem by removing negative effects from operators. The *relaxed problem* is often defined by ignoring the delete lists in planning operators.

A relaxed planning problem $P' = <S_0, G, O'>$ is defined by replacing the set of operators with operators without the delete lists. That is, for each $o' \in O'$, the set effects⁻(o') is empty. Finding an optimal plan for the relaxed problem is also hard. Researchers have therefore devised heuristics to estimate the cost of this optimal solution that is sought by the planner. The cost can be in terms of the number of actions in the plan. Individual actions may have associated costs as well. While such heuristics are not of constant time complexity like the domain specific heuristic functions, they are often effective in reducing the complexity of planning.

One of the first planners to employ domain independent heuristics was the *heuristic search planner* (HSP), which estimated the cost of each goal proposition in G and derived a heuristic $h'(S)$ from these costs (Bonet and Geffner, 2001a, 2001b). Let p be a goal proposition. The estimated cost of achieving p from a given state S is denoted by $g_s(p)$ and is computed as

$$g_s(p) = 0 \quad \text{if } p \in S$$
$$= \min_{a \in O(p)} [1 + g_s(precond(a))] \quad \text{otherwise}$$

where $O(p)$ stands for the set of actions that add p. The authors use a simple forward chaining procedure in which the measures $g_s(p)$ are initialized to 0 if $p \in S$ and to ∞ otherwise. Then, every time an operator op is applicable in S, each proposition $p \in$ effect$^+(op)$ is added to S and $g_s(p)$ is updated to

$$g_s(p) = \min[g_s(p), 1 + g_s(pre(op))]$$

These updates continue until $g_s(p)$ does not change. The procedure is polynomial in the number of propositions and actions, and corresponds to a version of Dijkstra's algorithm (Chapter 6).

The estimated cost of reaching a set of propositions, for example, $g_s(precond(a))$, is aggregated from the individual costs for each proposition. The following aggregation functions defined for the set of goal propositions G are popular. The estimated cost is adopted as the heuristic value of a state in FSSP.

$$h(S) \stackrel{\text{def}}{=} g_s(G)$$

The *additive heuristic* $g^+{}_s(G)$ computes the value as a sum of the costs of each proposition in the goal.

$$h^+(S) = g^+{}_s(G) = \Sigma_{p \in G}\, g_s(p)$$

This heuristic is known to be well informed but is not admissible. The additive heuristic may overestimate the cost because it assumes that the individual propositions are independent and achieved individually. That means that if it were used for the A^* algorithm, then the plan found may not be optimal.

The *max heuristic* instead picks the maximum of the estimates for individual propositions and is clearly admissible, but not well informed. That would imply searching more of the space.

$$h^{max}(S) = g^{max}{}_s(G) = \text{Max}_{p \in G}\, g_s(p)$$

The original version of HSP used the *additive heuristic* with HILLCLIMBING (see Chapter 4). A later version, HSP2, uses the wA^* algorithm (see Chapter 6) and can be made an admissible algorithm with an appropriate choice of heuristic function and the weights. HSP2 also used a variation on the heuristic function called $h^2(S)$. Instead of taking only the costliest proposition as is done in $h^{max}(S)$, the $h^2(S)$ function looks at the costliest *pair of propositions* that are achieved at the same time. Let us say that the two propositions are called p and q. Then the heuristic function $h^2(S)$ is defined as (Ghallab, Nau and Traverso, 2004)

$$g^2{}_s(p) = 0 \quad \text{if } p \in S$$
$$= \min_{a \in O(p)} [1 + g^2{}_s(precond(a))] \quad \text{Otherwise}$$

$$g^2{}_s(\{p,q\}) = 0 \quad \text{if } p,q \in S$$

$$\begin{aligned}
&= \min \{\min_a [1 + g^2{}_s(precond(a))] \mid \{p,q\} \subseteq \mathit{effect}^+(a)], \\
&\quad \min_a [1 + g^2{}_s(\{q\} \cup precond(a))] \mid p \in \mathit{effect}^+(a)], \\
&\quad \min_a [1 + g^2{}_s(\{p\} \cup precond(a))] \mid q \in \mathit{effect}^+(a)]\}
\end{aligned}$$

$$g^2{}_s(G) = \max_{p,q} \{g^2{}_s(\{p,q\}) \mid \{p,q\} \subseteq G\}$$

and

$$h^2(S) = g^2{}_s(G)$$

Another planner that deploys domain independent heuristics in FSSP is the algorithm FASTFORWARD, abbreviated as FF, which also introduces another variation in forward search (Hoffmann and Nebel, 2001). For the heuristic estimate, it relies on applying the algorithm GRAPHPLAN (discussed later in Section 9.6) on the relaxed problem. Unlike the additive heuristic of HSP, this does not assume that actions are independent and is more likely to be admissible.

What is pertinent is that GRAPHPLAN returns a plan which is sequence of sets of actions <Set$_1$, Set$_2$, ..., Set$_k$> where each $S_i = \{a_{i1}, a_{i2}, ..., a_{ip}\}$ contains actions that could execute in parallel in step i. Then if the goal propositions first appear after Set$_m$, the heuristic function is computed as

$$h(S) = \Sigma_{i=1,m} |\mathrm{Set}_i|$$

The following two heuristics can be used to prune the search space in any forward search planning algorithm. The first is identifying the set of *helpful actions* for any state S. These are the actions that are in *Set$_1$* of a *relaxed plan* from the state S. Recall that the relaxed plan is the plan (found by GRAPHPLAN) for the relaxed planning problem $P' = <S, G, O'>$. For the planning problem in Figure 9.2, the relaxed plan is <{Pickup(A), Unstack(O,F)}, {Stack(A,D), Putdown(O)}, {Pickup(F)}, {Stack(F,A)}>. Observe that <{Pickup(A), Unstack(O,F)}, {Stack(A,D), Stack(O,A)}, {Pickup(F)}, {Stack(F,A)}> is a *relaxed plan* too because the *relaxed action Pickup(A)* does not delete *onTable(A)* and *clear(A)*. In both cases *Set$_1$* = {Pickup(A), Unstack(O,F)} and these two actions are called *helpful actions*, and one of them can be selected during search. Observe that this plan could be executed in four time steps by a two armed robot in the blocks world. We say that the *makespan* of the plan is 4. How long would a three armed robot take to solve this problem?

The second heuristic involves deleting actions that achieve a certain goal that is destined to be undone later in the plan. In the same example from Figure 9.2, there are two goals to be achieved, {on(F,A), on(A,D)}. Clearly achieving *on(F,A)* first will require it to be undone when the robot needs *clear(A)* as a precondition for *Pickup(A)* needed for achieving *on(A,D)*. The *added goal deletion* heuristic works as follows. Let us say that a forward planning algorithm has reached a state S in which *on(F,A)* is true. Then GRAPHPLAN produces the relaxed plan <{Unstack(F,A)}, {Pickup(A)}, {Stack(A,D)}> which undoes the earlier goal on(F,A) in the *original unrelaxed* problem. If that happens, one can prune the state S from the search space, and instead find a plan in which *on(A,D)* is achieved before *on(F,A)*. The authors, however, point out that such pruning may result in incompleteness in some domains and a plan may not be found.

More interestingly, FF introduces a new search algorithm called *EnforcedHillClimbing*, which reverts to *breadth first search* when on a (local) optimum. Once a better neighbour is found, the algorithm resumes HillClimbing. This is based on Hoffman and Nebel's observation that in many domains the heuristic functions they are using tend to have small plateaus and local optima, and the algorithm finds a gradient to traverse in a few steps. The interested reader is encouraged to look at their paper for a detailed description and performance analysis.

9.3.3 Backward state space planning

A high branching factor directly contributes to exponential search spaces. The branching factor in FSSP is high because the start state and the states that the algorithm progresses to are completely described. This means that many actions are applicable. The goal description, on the other hand, is only partial. Could that result in a low branching factor? A similar choice presents itself in the domain of logical reasoning, as described in the next chapter. *Backward state space planning* (BSSP) pursues this line of investigation.

Searching backwards from the goal requires one to identify actions that will add one or more goal propositions in the final step. This can be done by identifying *relevant actions*.

An action a is *relevant* to a goal $G = \{g_1, g_2, ..., g_k\}$ *iff* at least one positive effect of a is a proposition in the goal, and none of its negative effects is. The planning community also refers to each such proposition in $\{g_1, g_2, ..., g_k\}$ as a goal, so when we say 'goal', we will mean a proposition too. That is,

$$\{\text{effect}^+(a) \cap G\} \neq \phi \wedge \{\text{effects}^-(a) \cap G\} = \phi$$

When a relevant action is applied to a goal G, then the algorithm regresses to a subgoal G' (also referred to as a goal G').

$$G' = \gamma^{-1}(G, a) = \{G \smallsetminus \text{effects}^+(a)\} \cup \text{pre}(a)$$

As before, a plan π is a sequence of actions $<a_1, a_2, ..., a_n>$. A plan π is *relevant* to a goal G_n if there are goals $G_0, ..., G_{n-1}$ such that $G_{i-1} = \gamma^{-1}(G_i, a_i)$ for $i = 1, ..., n$.

The final goal is $G_0 = \gamma^{-1}(G_n, \pi)$.

BSSP begins with the empty plan and incrementally constructs the plan by adding new actions at the front of the current plan. In the algorithm for BSSP, this is done by the assignment

$$\pi \leftarrow a \circ \pi$$

It regresses over a to the given goal G' as described above and continues searching from there, till it finds a plan. Let S_0 be the start state. BSSP ends when $G_0 \subseteq S_0$. The test for a valid plan still needs to be done by progression over the actions in the plan. The plan π is a valid plan if $G_n \subseteq \gamma(S_0, \pi)$.

The BSSP algorithm may reach the condition $G_0 \subseteq S_0$ but the plan found may not be a valid plan. This is because regression over operators is not sound as illustrated in Figure 9.4. This, in turn, is because of the way relevant actions are defined. They only define what goal propositions are required to be true in goal G'. And as illustrated in Figure 9.4, this can result in actions that are not applicable in the state which satisfies G'. In fact, for such spurious actions, the regressed goal is not a (valid) state at all!

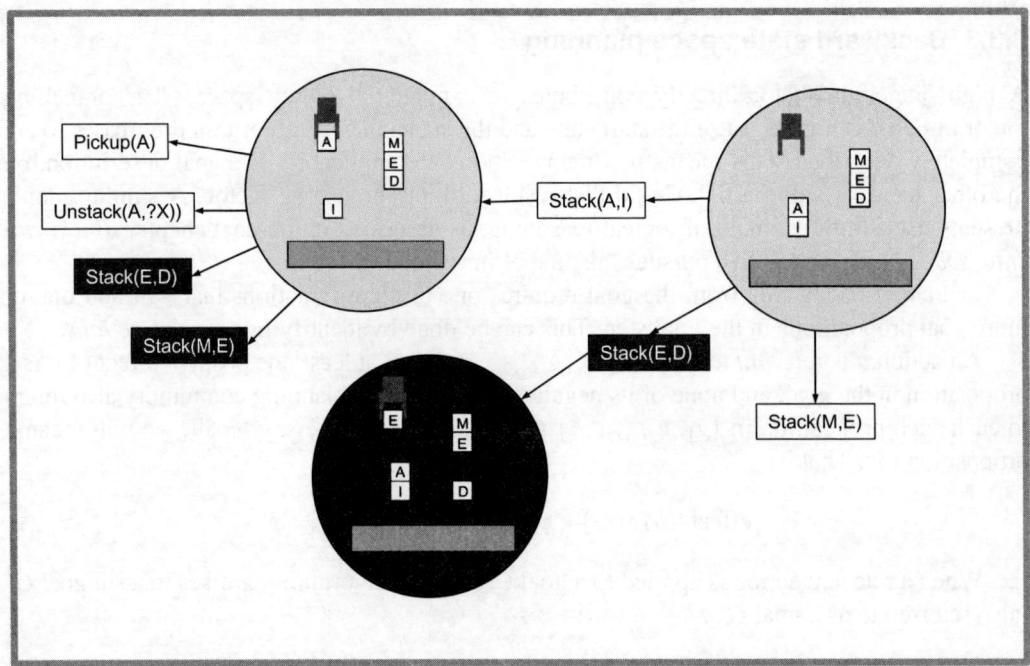

Figure 9.4 An illustration of regression as done by *backward state space planning*, on the goal description in Figure 9.1. The actions in the shaded rectangles would not be applicable, even though they are relevant, because the goals regressed to are not valid states. The regressed goal for *Stack(E,D)*, for example, requires *holding(E)* and *on(M,E)* to be true in the same state.

This is because regression is going against the arrow of time defined by the planning operators. An action is relevant if it achieves a certain goal (every fluent is a goal). It simply proposes what needs to be true in the regressed goal if *that* action is to be applicable. Adding those conditions may lead to a description that is not a valid state. This is because goal descriptions are incomplete and the regressed goals could be spurious. In progression, on the other hand, an action is applicable if its preconditions are true in a given state, and states are completely described. The resulting description is a valid state, and progression is a sound step.

Regression is thus not a sound step. BSSP is not aware of this danger, and could thus propose the final three actions as <Stack(*E,D*), Stack(*M,E*), Stack(*A,I*)> which clearly are not all feasible. It is for this reason that the termination criterion for BSSP cannot just be ($\gamma^{-1}(G_n,\pi) \subseteq S_0$) but will need a validity check for the plan π as well.

A planner that searches in the backward direction from the goal is the *heuristic regression planner* (HSPR) which is a variation on HSP (Bonet and Geffner, 2001a, 2001b). A major advantage of HSPR is that it does not have to recompute the heuristic value repeatedly. In HSP the value $g_s(p)$ for a goal proposition p has to be computed from every state S that the planner is looking at. In HSPR $g_{s0}(p)$ is computed once and for all from the start state S_0 for every proposition p. Then, when the planner regresses to a goal G', the heuristic distance of achieving G' is simply the sum of the distances for each goal $p \in G'$. For any goal G the heuristic estimate is

$$g^+_{s0}(G) = \Sigma_{p \in G}\, g_{s0}(p)$$

Figure 9.5 depicts the space for HSPR after it has found the relevant actions for goal G in the same problem. There are three actions, *Stack(A,I)*, *Stack(M,E)*, and *Stack(E,D)*, that the algorithm can regress over producing the three subgoals G_1, G_2, and G_3 it has to choose from. It will use the above heuristic function to make the choice.

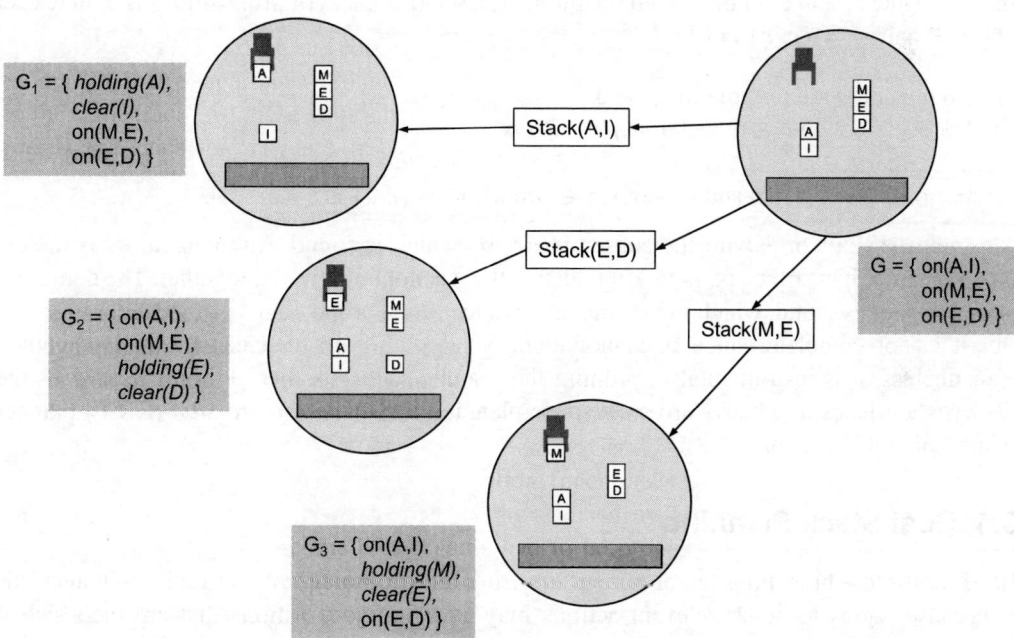

Figure 9.5 Given the goal G in the problem from Figure 9.1, HSPR has to choose between three actions it can regress over to reach one of G_1, G_2, and G_3. It will use the heuristic $g^+_{s0}(G_1)$, $g^+_{s0}(G_2)$, and $g^+_{s0}(G_3)$ that sums up the precomputed heuristic estimates for the constituent propositions in the three subgoals to make the choice.

As discussed above, not all goal sets that backward search regresses to are feasible. This has been illustrated in Figure 9.4 which shows that the subgoal G_2 is not feasible since it requires both *on(M,E)* and *holding(E)* to be true in the same state. Taking a cue from GRAPHPLAN Bonet and Geffner introduce the notion of *mutex pairs* of propositions that can never be achieved together starting from S_0. This would mean that if any such pair of propositions occurs in any goal, then that goal is not feasible, and that goal can be pruned. In the above example, goal G_2 can be pruned because it contains $on(M,E)$ and $holding(E)$. Algorithm HSPR constructs a set M of mutex pairs as follows. It begins my first constructing a set M_0 which is the set of potentially mutex pairs (Bonet and Geffner, 2001a, 2001b). Instead of starting with all pairs of proposition which would work too, HSPR starts with a smaller set defined as

$$M_0 = M_A \cup M_B \text{ where}$$

- M_A is the set of pairs $P = \{p, q\}$ where some action a adds p and deletes q. That is, $p \in \textit{effects}^+(a)$ and $q \in \textit{effects}^-(a)$.
- M_B is the set of pairs $P = \{r, q\}$ such that for some pair $P' = \{p, q\}$ in M_A, there is an action a, such that $r \in pre(a)$ and $p \in \textit{effects}^+(a)$.

From this set M_0 one can extract a subset M^* by removing 'bad pairs' in M_0 which may not in fact be mutex. The bad pairs are those pairs that *do not* satisfy the following conditions. Given an initial state S_0 and a set of ground operators A, a set M of pairs of propositions is a mutex set *iff* for all pairs $R = \{p, q\}$ in M,

1. Both p and q are not true in S_0, and
2. For every action $a \in A$ such that $p \in \textit{effects}^+(a)$,
   a. either $q \in \textit{effects}^-(a)$
   b. or $q \notin \textit{effect}^+(a)$ and for some $r \in pre(a)$, $R' = \{r, q\}$ is a pair in M.

Algorithm HSPR works with the subset M^* of M_0 which is sound. Any pair in M^* is mutex, though there might still be pairs lurking outside that cannot be achieved together. The definition of mutex pairs is sound, which means that the two identified propositions cannot occur together. But it is not complete, since it cannot identify all such pairs that cannot occur together. Nevertheless, it is instrumental in pruning the search space. Another algorithm that uses the FF *style* heuristics in a backward state space planning is the *rapid regression* (RR-F) planner (Kumashi and Khemani, 2002).

9.4 Goal Stack Planning

BSSP has a low branching factor but suffers from the occurrence of spurious subgoals. This is because regression over relevant actions may add goal propositions that are inconsistent with the existing goal propositions. This can lead to the planner constructing plans that are not valid plans, requiring backtracking after a validity check. Progression over actions in FSSP is sound, and the resulting set of propositions is always a valid state. However, FSSP suffers from high branching as there may be numerous applicable actions in any given state. An interesting algorithm that combines the best features of the two was developed early in the history of planning. It is called *goal stack planning* (GSP) (Rich, 1983).

The key features of GSP are that it searches for a plan in the backward direction as in BSSP but constructs the plan by progressing over actions from the start state S_0 as in FSSP.

$$\pi \leftarrow \pi \circ a$$

9.4.1 Linear planning

Unlike BSSP, the algorithm GSP does not always have the freedom to choose any goal in the search tree being constructed. Rather, it chooses one goal proposition g_1 from the initial goal G and solves it before moving on the next goal g_2, and so on. In doing so, it does what is called *linear planning*, solving the goals in a linear order. It searches backward from each goal proposition till it reaches an action that is applicable in the *current* state. Then it applies the action and progresses over it to define a new current state. The algorithm begins with S_0 as the current state, and hence the first applicable action is in the start state. The plan is synthesized in a forward manner starting with the first action in the plan.

To facilitate this mix of backward search and forward plan synthesis, the algorithm uses a stack data structure in which it pushes goals and relevant actions. After pushing each relevant action it also pushes its preconditions onto the stack. Thus preconditions of an action are popped before the action, and consequently when it is the turn of an action to be popped, its preconditions must be true, and the action would be applicable. If any popped precondition is not true in the current state, then a relevant action and its preconditions are pushed onto the stack before proceeding further.

We illustrate the stack handling with the tiny problem in Figure 9.2. There are two goals to be achieved – $on(F,A)$ and $on(A,D)$. Clearly, it makes sense to achieve $on(A,D)$ first and then $on(F,A)$, and the reader is encouraged to simulate the algorithm described below. But without any heuristics how will the algorithm know? We follow the progress of the algorithm when it chooses the wrong order of selecting $on(F,A)$ first. This is done by pushing $on(A,D)$ first onto the stack, so that $on(F,A)$ is popped first. The goal $on(F,A)$ is not true in $S = S_0$. There is one relevant action, $Stack(F,A)$, and that is pushed onto the stack, along with its preconditions. The stack now has $holding(F)$, a precondition of $Stack(F,A)$, on the top.

$holding(F)$
$clear(A)$
$Stack(F,A)$
$on(A,D)$

GSP pops $holding(F)$. This is not true in the state S_0, and the action $Pickup(F)$ is pushed onto the stack. We have made this right choice non-deterministically, eschewing $Unstack(F,?X)$, but an implementation would have to rely on backtracking along with a heuristic that looks at the current state S to choose between the two. In this illustration, we will, from now on, make the choice of actions and the order of goals to be added non-deterministically. Action $Pickup(F)$ along with its preconditions are now pushed onto the stack, which now looks like,

$onTable(F)$
$clear(F)$
AE
$Pickup(F)$
$clear(A)$
$Stack(F,A)$
$on(A,D)$

Next, *onTable(F)* is popped. It is *true* in the current state, because *onTable(F)* ∈ S_0, so nothing needs to be done. Next, *clear(F)* is popped, but it is not true. The action *Unstack(O,F)* is pushed along with its preconditions.

> on(*O,F*)
> clear(*O*)
> *AE*
> Unstack(*O,F*)
> *AE*
> Pickup(*F*)
> clear(*A*)
> Stack(*F,A*)
> on(*A,D*)

The next three goals, *on(O,F)*, *clear(O)*, and *AE*, are popped one by one and are true in S_0. Now comes the first action to be popped and added to the plan. The state progresses over the action *Unstack(O,F)* and we have

π   = <Unstack(*O,F*)>
S   = {on(Table *A*), onTable(*M*), onTable(*F*), on(*I,M*), on(*D,I*), clear(*A*), clear(*D*), clear(*O*), holding(*O*)}

The stack now is

> *AE*
> Pickup(*F*)
> clear(*A*)
> Stack(*F,A*)
> on(*A,D*)

The goal *AE* is popped next, and it is not true. GSP adds the action *Putdown(O)* to the stack, along with *holding(O)*. The latter is popped, turns out to be true, and the action *Putdown(O)* is popped next and added to the plan. After this, *Pickup(F)* is popped and added to the plan. Note that *holding(F)* is true in the resulting state.

π = <Unstack(*O,F*), Putdown(*O*), Pickup(*F*)>

The stack is

> clear(*A*)
> Stack(*F,A*)
> on(*A,D*)

Then goal *clear(A)* is popped and is true in the revised *S*. The action *Stack(F,A)* is added to the plan next, achieving the first goal *on(F,A)*.

π = <Unstack(*O,F*), Putdown(*O*), Pickup(*F*), Stack(*F,A*)>
S = {on(Table *A*), onTable(*M*), on(*F,A*), on(*I,M*), on(*D,I*), clear(*F*), clear(*D*), clear(*O*), clear(*F*), onTable(*O*)}

This state is shown in the centre in Figure 9.6. At this point, GSP is only concerned with the remaining goal *on(A,D)* which is the only element left in the stack. The reader should work out the details and verify that it could find the plan <Unstack(*F,A*), Putdown(*F*), Pickup(*A*), Stack(*A,D*)>. The state *S* at this stage is drawn on the right in Figure 9.6. Here we selected the action *Putdown(F)* as a relevant action for the goal *AE* which would be a precondition for *Pickup(A)*, but it could have stacked *F* onto *O* as well. Worse, it could stack it back on *A* and gone into a loop, or even on *D*, thus destroying the precondition *clear(D)* of *Stack(A,D)*.

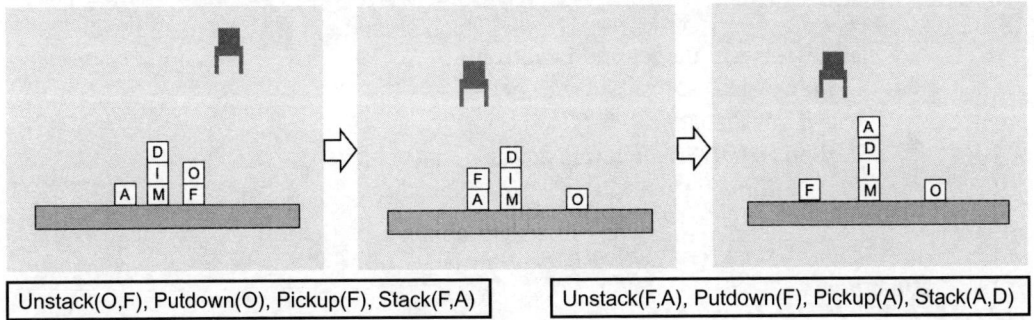

Figure 9.6 GSP started with two goals, *on(F,A)* and *on(A,D)*, on the tiny planning problem from Figure 9.2. After choosing *on(F,A)* and achieving it with the plan shown on the left, it reaches the state in the centre. From there it solves for the goal *on(A,D)* with the plan shown on the right ending in the state on the right. In the process it has undone the first goal *on(F,A)*.

After solving the second goal *on(A,D)* with the plan <Unstack(*F,A*), Putdown(*F*), Pickup(*A*), Stack(*A,D*)> we find that the first goal has been undone, as shown on the right in Figure 9.6. The plan found is not a valid plan. This a characteristic of planning problems in which the sub-goals are not serializable. They cannot be solved independently one by one. Other examples of such problems are the 8-puzzle and the Rubik's cube, in which one cannot set serial goals of solving one part and moving on to the next.

One way to address this is to add the compound goal that is to be solved as a conjunct before adding the constituent goal propositions. This is done by the function P∪sнSᴇт(*G, stack*) in the GSP algorithm described in Algorithm 9.1, and is also done when pushing the preconditions of an action. Then, after solving the goal propositions, the compound goal remains on the stack. When that is popped and if found not to be true in the current state *S*, it is simply added back to the stack. There is a danger though of the program going into an infinite loop with this feature if the compound goal does not have a solution. In Figure 9.6, when the algorithm again chooses the goal *on(F,A)*, it solves it by picking *F* and stacking it on to *A*. The second goal *on(A,D)* is already true in the resulting state, and a valid plan has been found, albeit not an optimal one. The reader should verify that if the algorithm had chosen the goals in a different order, solving for *on(A,D)* first and then solving for *on(F,A)*, this checkback step would not have been invoked.

Algorithm 9.1. Algorithm GSP starts by pushing the given compound goal G onto the stack. It then pushes each constituent proposition g in G onto the stack. If a popped goal g is not true, the algorithm pushes a relevant action onto the stack, along with its preconditions using PUSHSET. When an action is popped, it is added to the plan, and the state progresses over that action.

GSP(S_0, G, A)
1    $S \leftarrow S_0$; *plan* $\leftarrow$ < >; *stack* $\leftarrow$ [G]
2   **while** *stack* **is not empty**
3        $x \leftarrow$ **pop** *stack*
4   **if** $x \in A$
5   **then** *plan* $\leftarrow$ <*plan* $\circ$ x>
6                $S \leftarrow$ PROGRESS(S, x)
7   **else if** x **is a conjunct and is not true**
8   **then** *stack* $\leftarrow$ PUSHSET(x, *stack*)
9   **else if** x is a goal *proposition* and $x \notin S$
10   **then** CHOOSE a relevant action a that achieves g
11   **if** *none* **then return** FAILURE
12   *stack* $\leftarrow$ PUSH(x, *stack*)
13   *stack* $\leftarrow$ PUSHSET(*pre*(x), *stack*)
14   **return** *plan*

PUSHSET(G, *stack*)
1   PUSH(G, *stack*)
2   **for each** $g \in G$
3   PUSH(g, *stack*)
4   **return** *stack*

When GSP is applied to the problem in Figure 9.2, the plan found depends upon the order of tackling the two goals. Both orderings yield a plan, but the wrong order results in a longer plan. A plan is found nevertheless because the blocks world domain has only reversible moves. Contrast this with cooking, where often goals need to be tackled in a particular order. One cannot, for example, grind the rice and lentil into a batter for making *idli*s first, and then soak the rice and lentil mix. In the above example there was a right order for finding the shortest plan. But even in the blocks world domain there exist planning problems for which there is no right order of solving the goals.

9.4.2 Sussman anomaly

Gerald Sussman gave us a tiny blocks world problem with only three blocks and two goals in which neither order of choosing the goals results in a plan without doing extra work after solving the two independently (Sussman, 1975). The problem, known as the Sussman anomaly, is as follows:

S_0 = {onTable(A), onTable(B), on(C,A), clear(C), clear(B), AE}
G = {on(A,B), on(B,C)}

When GSP picks up *on(A,B)* first, it finds the plan <unstack(C,A), putdown(C), pickup(A), stack(A,B)> to reach the state S = {on(Table C), onTable(B), on(A,B), clear(C), clear(A), AE>}

Next it solves for *on(B,C)* with the plan <unstack(A,B), putdown(A), pickup(B), stack(B,C)> and the current state becomes

S = {on(Table C), onTable(A), on(B,C), clear(B), clear(A), AE>}

which as one can see on the left branch in Figure 9.7 is not the goal state.

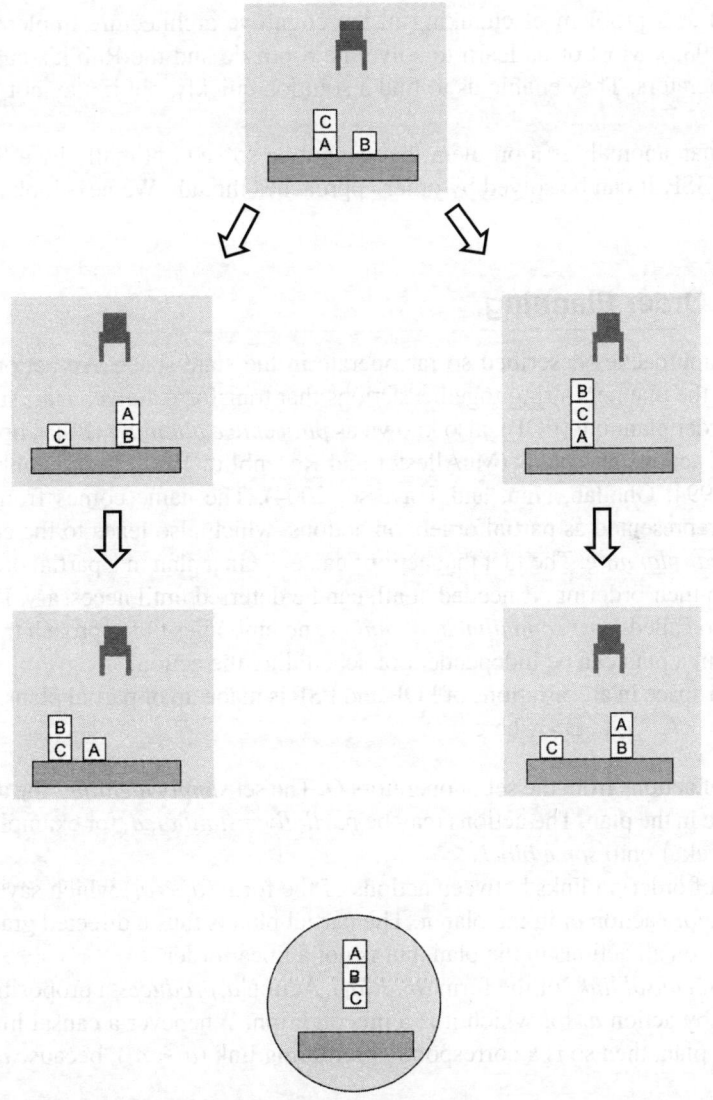

Figure 9.7 Neither choice of the goal to solve in the Sussman anomaly ends in a valid plan. On the left GSP chooses to solve *on(A, B)* first, and on the right *on(B, C)* first. In both cases the planner has more work to do.

In a similar manner, choosing the goal *on(B,C)* first leads to the non-goal state as shown in the right branch.

$$S = \{on(\text{Table } C), onTable(B), on(A,B), clear(C), clear(A), AE\}$$

We say that the goals are *non-serializable*. This is also the case in the context of the 8-puzzle (Korf, 1985b). Richard Korf attacked the problem by proposing a learning approach in which macro-operators could be learnt to move from one achieved sub-goal to the next one, where the first goal could be disrupted *en route* but would be restored subsequently. This was also addressed as a problem of chunking in the cognitive architecture implemented in *Soar* (Laird et al., 1985). Most of us learn to solve the 8-puzzle and the Rubik's cube by learning such macro-operators. They enable us to find a solution quickly, but it may not be the optimal solution.

The Sussman anomaly is a problem that cannot be solved optimally by a linear planning approach like GSP. It can be solved by other approaches though. We next look at searching in the plan space.

9.5 Partial Order Planning

The planning approaches described so far operate in the state space. An action defines state transition, and the planners string together actions that transform a given start state into a goal state. Partial order planning (POP), also known as *plan space planning* (PSP), operates entirely in the space of actions and plans (McAllester and Rosenblitt, 1991; Tate, Drabble, and Kirby, 1994; Weld, 1994; Ghallab, Nau, and Traverso, 2004). The name comes from the property that plans are represented as partial orders on actions, which also leads to the approach being called *non-linear planning*. The fact that actions can exist in a plan in a partial order means that commitment to their ordering, if needed at all, can be deferred until necessary. Due to this, the approach is also called *least commitment planning* and embodies the approach that selecting an action needed in a plan can be independent of scheduling the action.

The search space in all variations of POP and PSP is made up of partial plans. A partial plan is a 4-tuple $\pi = <\mathbb{A}, \mathbb{O}, \mathbb{L}, \mathbb{B}>$ where

- $\mathbb{A}$ is a set of actions from the set of operators *O*. The set *simply identifies* the actions that are somewhere in the plan. The actions may be *partially instantiated*, for example, *Stack(A,?X)* – stack block *A* onto *some block*.
- $\mathbb{O}$ is a set of ordering links between actions of the form $(a_i \prec a_k)$ which says that action a_i happens *before* action a_k in the plan π. The partial plan is thus a directed graph. It imposes some order on all actions in the plan, but is not a linear order.
- $\mathbb{L}$ is a set of *causal links* of the form (a_i, P, a_k). Action a_i *produces* a proposition *P* which is *consumed* by action a_k for which it is a precondition. Whenever a causal link (a_i, P, a_k) is added to a plan, then so is a corresponding ordering link $(a_i \prec a_k)$, because a_i must happen before a_k.
- $\mathbb{B}$ is a set of *binding constraints* that specify what values a variable can or cannot take. Thus a partially instantiated action may be added first, and a binding constraint can be added later. This conforms to the idea of least commitment.

Let $P = <S_0 = \{s_1, s_2, ..., s_n\}, G = \{g_1, g_2, ..., g_k\}, O>$ be a planning problem. The PSP algorithm always begins with an initial plan $\pi_0 = <\{A_0, A_\infty\}, \{(A_0 \prec A_\infty)\}, \{\}, \{\}>$ where A_0 is the *initial action* with no *preconditions* and *positive effects* $s_1, s_2, ..., s_n$, and A_∞ is the *final action* with *preconditions* $g_1, g_2, ..., g_k$ and no *effects*. One can think of π_0 as the set of all possible plans. One can *refine* a given partial plan by the following operators:

- add a new action or partially instantiated operator to $\mathbb{A}$
- add an ordering constraint to $\mathbb{O}$
- add a causal link to $\mathbb{L}$
- add a binding constraint to $\mathbb{B}$

We adopt a concise graphical notation as shown in Figure 9.8 to depict partial plans. The diagrams have shortened names for actions and predicates. Preconditions of actions are drawn above the action, and effects below. Negative effects are prefixed by the negation symbol ¬. In the style of Hasse diagrams for partial orders, actions which occur earlier are drawn above and later actions are drawn below them. Ordering links are drawn only when necessary. Causal links are drawn explicitly with dashed arrows from the producer to the consumer.

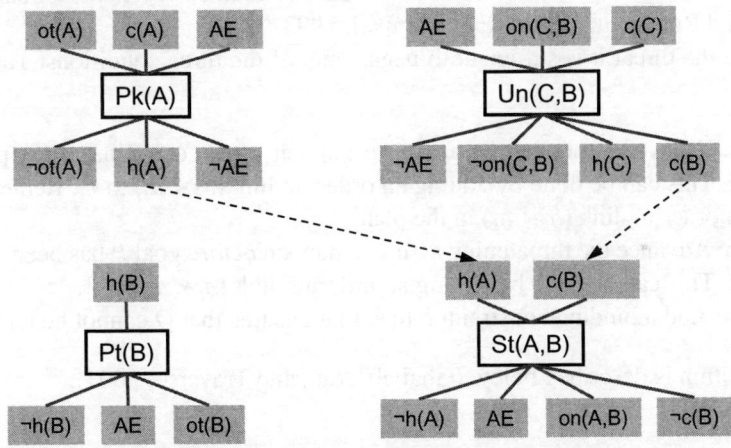

Figure 9.8 A concise notation for actions in partial plans. The action names have been shortened to Pk for *Pickup*, Pt for *Putdown*, Un for *Unstack*, and St for *Stack*. Predicate names have likewise been shortened: ot for on Table, h for holding, and c for clear. The preconditions are shown above the action and the effects below. Negative effects are marked by a negation sign ¬. For the sake of illustration two causal links are drawn as dashed arrows. As far as possible ordering links are not drawn, and actions occurring earlier in the partial plan are drawn above.

The refinement process continues till a *solution plan* is found. Unlike in state space planning which tests if a given state is a goal state, POP identifies a solution plan as a plan without any *flaws*.

A partial plan can have two kinds of flaws. The first is if a partial plan has an *open goal*. An open goal is a precondition of some action in $\mathbb{A}$ which is not supported by a causal link. Let action a_k have an open goal P. This can be *resolved* in two ways.

- If there exists an action $a_i \in \mathbb{A}$ which has a positive effect P, and it is consistent to add the ordering link $(a_i \prec a_k)$ to the plan, then add the causal link (a_i, P, a_k) to $\mathbb{L}$ and the ordering link $(a_i \prec a_k)$ to $\mathbb{O}$.
- Add a new action a_n that has a positive effect P to $\mathbb{A}$ along with the causal link (a_n, P, a_k) to $\mathbb{L}$ and the ordering link $(a_n \prec a_k)$ to $\mathbb{O}$.

The second kind of flaw is a *threat*. An action $a_t \in \mathbb{A}$ is a threat to a link (a_i, P, a_k) in $\mathbb{L}$ if the following three conditions hold.

- Action a_t has a negative effect Q such that Q can be unified with P.
- Action a_t can happen before action a_k.
- Action a_t can happen after action a_i.

If *all* three conditions are met, then the threat will *materialize*. That is, the threat actions happens after the producer of P and deletes P before the consumer can consume it. For example, A_t might be *Stack(B, ?X)* which deletes *clear(?X)*. If there is a causal link (Unstack(M,N), clear(N), Pickup(N)), then if *?X* were to be N the causal link would be destroyed. One may *even* treat an action a_t as a threat if it produces P, because it threatens to make action A_i redundant (McAllester and Rosenblitt, 1991; Kambhampati, 1993).

To resolve the threat it is sufficient to negate any of the three conditions. The three threat resolvers are

1. Demotion: Delay the threatening action to happen *after* goal P has been produced and consumed. This can be done by adding an ordering link $(a_k \prec a_t)$ to $\mathbb{O}$. Remember there is already an ordering link $(a_i \prec a_k)$ in the plan.
2. Promotion: Advance the threatening action to happen *before* goal P has been produced and consumed. This can be done by adding an ordering link $(a_t \prec a_i)$ to $\mathbb{O}$.
3. Separation: Add a binding constraint b to $\mathbb{B}$ that ensures that Q cannot be unified with P.

The PSP algorithm is described below (Ghallab, Nau, and Traverso, 2004).

Algorithm 9.2. The plan space planning (PSP) procedure attempts to resolve one flaw at a time. Function RESOLVE returns the set of resolvers for f in the plan π. CHOOSE is a non-deterministic operator that chooses the appropriate resolver r. The REFINE procedure applies the chosen resolver, and the algorithm PSP is called recursively to address the next flaw.

PSP(π)
1   *flaws* $\leftarrow$ *OpenGoals*(π) $\cup$ *Threats*(π)
2   **if empty** *flaws*
3   **return** π
4   **else**
5   select and remove some $f \in$ *flaws*
6   *resolvers* $\leftarrow$ *Resolve*(*f*, π)
7   **if empty** *resolvers*

```
  8            return FAIL
  9       else CHOOSE r ∈ resolvers
 10            π' ← Refine(r, π)
 11            return PSP(π')
```

Given a set of flaws to resolve, anyone can be chosen since all flaws have to be resolved. But having chosen a flaw to resolve, one has to choose an appropriate resolver. We use a nondeterministic *CHOOSE* operator in the description. In practice, one may have to resort to search. Procedure REFINE applies the chosen resolver to the plan. It might in turn introduce new flaws. This happens, for example, when one adds a new action a_n to $\mathbb{A}$, resulting in its preconditions being added as own open goals.

It may be possible that when a threat is resolved by promotion or demotion, another causal link may in fact be broken. This could be if the two actions in the new ordering had a common precondition which only one of them could consume. We illustrate the algorithm with the example below, when one is forced to impose an ordering on two actions applicable in the start state because they both consume the same proposition. Consider the following planning problem:

S_0 = {onTable(C), onTable(D), on(A,C), on(B,D), clear(A), clear(B), AE}
G = {on(A,B), on(B,C)}

This tiny problem is depicted in Figure 9.9.

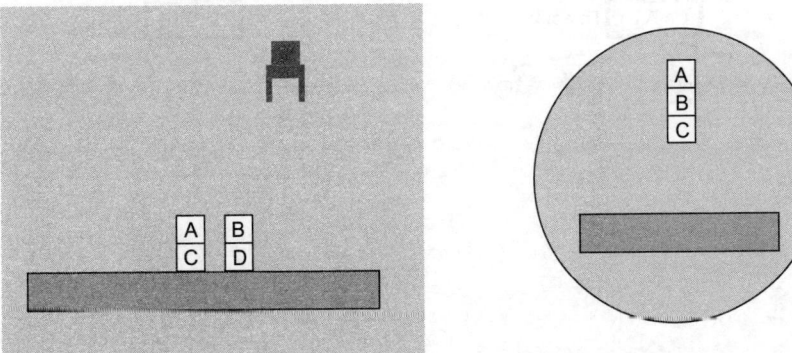

Figure 9.9 Another tiny planning problem. The given state is on the left, and the goal description is on the right.

The algorithm begins with $\pi_0 = \langle \{A_0, A_\infty\}, \{(A_0 \prec A_\infty)\}, \{\}, \{\} \rangle$ where

effects$^+(A_0)$ = {onTable(C), onTable(D), on(A,C), on(B,D), clear(A), clear(B), AE},
and pre(A_∞) = {on(A,B), on(B,C)}

Here is a description of how PSP might solve the problem. For the sake of illustration, we have assumed that the planner will somehow make the right choice of which flaw to resolve to minimize search. This is to illustrate the kind of reasoning that could happen. An augmented planner doing such reasoning could perhaps generate an explanation of the process. In practice, a wrong choice would lead to backtracking in search.

There are two open goals *on(B,C)* and *on(A,B)*. Let us say that the planner chooses the former, and adds the action *Stack(B,C)*, represented as *St(B,C)* as described in Figure 9.8, to resolve it. This action in turn introduces its own open goals, *holding(B)* and *clear(C)*. Let us say it selects the latter and adds the action *Unstack(A,C)* to resolve it. The open goals introduced by *Unstack(A,C)* can all be resolved because they are produced by the A_0 action. That is, they are true in the start state S_0. Assume that the planner next adds *Unstack(B,D)* to resolve the open goal *holding(B)*. That has three preconditions – *on(B,D)*, *clear(B)*, and *AE*. All three are true in the start state as well, but there is a two way threat now because the actions *Unstack(A,C)* and *Unstack(B,D)* both consume *AE* and both delete it. The partial plan with the two threats is shown in Figure 9.10.

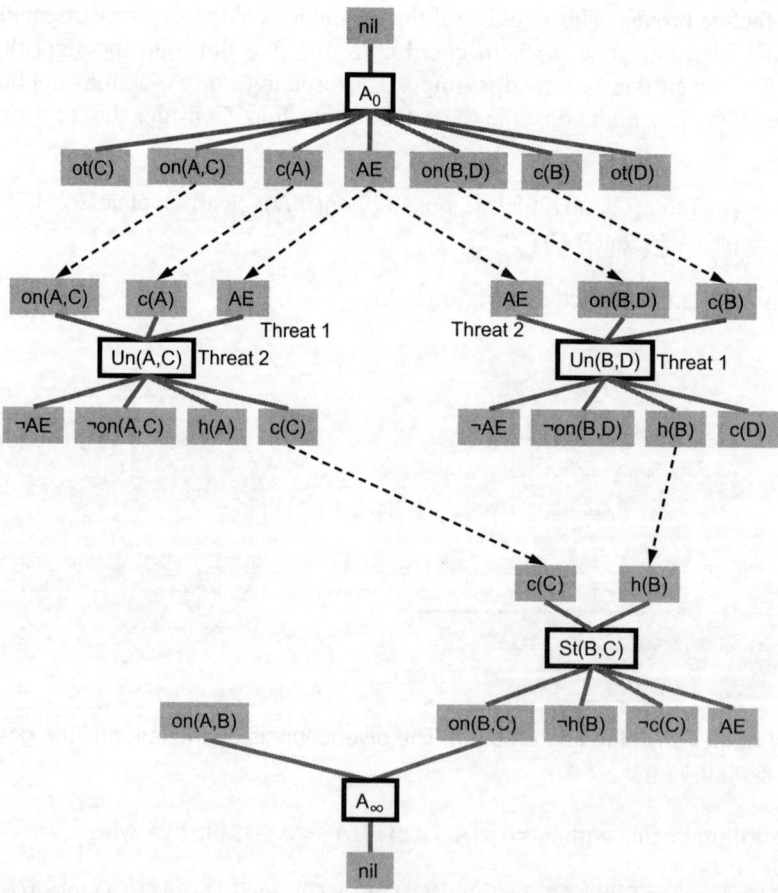

Figure 9.10 The partial plan after PSP has added three actions to solve one original goal *on(B, C)*. However, in the process, it has introduced two *threats*. Both *Un(A,E)* and *Un(B,D)* have *AE* as a precondition supported by a causal link from A_0, and both delete *AE*. One of them will have to be *demoted*.

An ordering has to be imposed on the two threatened actions to resolve the threat. Let us say that the planner fortuitously demotes *Unstack(B,D)* to happen after *Unstack(A,C)*. This results in the causal link (A_0, *AE*, Unstack(B,D)) being *clobbered* (Tate, Drabble, and Kirby, 1994) and it means that *AE* is again an open goal for *Unstack(B,D)*. This threat resolving action has undone the resolution of an earlier open goal. The next algorithm we look at, GRAPHPLAN, defers this kind of potential interaction as constraints to be resolved later. The ordering link (Unstack(A,C) ≺ Unstack(B,D)) due to demotion is shown as an explicit arrow in Figure 9.11. There are other ordering links too, but have not been drawn explicitly. Instead, one relies on the convention that actions drawn above happen before actions drawn below.

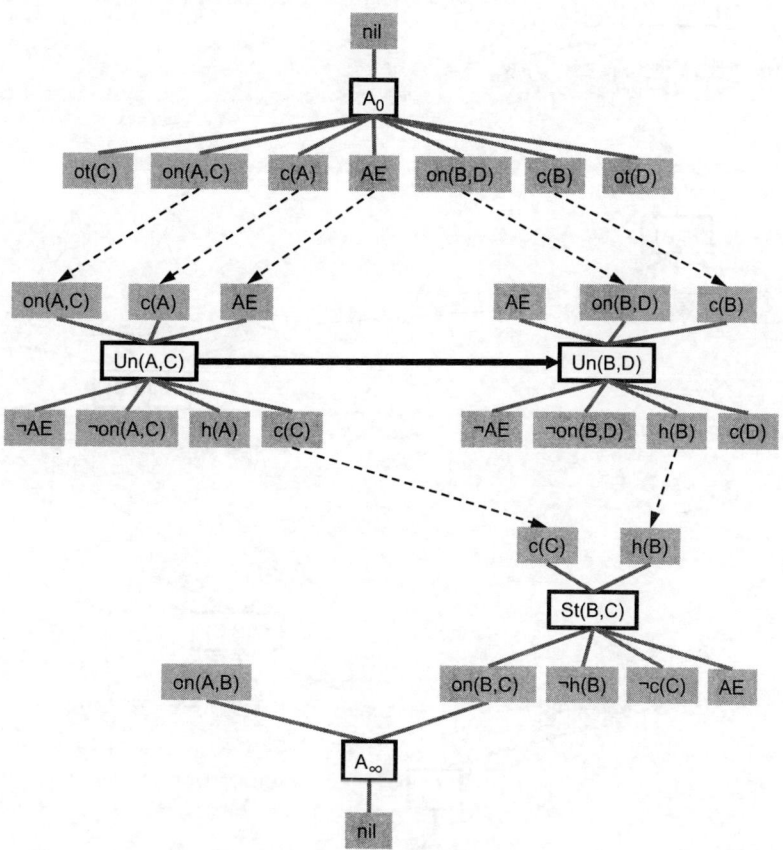

Figure 9.11 The two threats in Figure 9.10 are both resolved by demoting the action *Unstack(B,D)*, by adding the ordering link (Unstack(A,C) -≺ Unstack(B,D)) as shown with the solid arrow.

At this point there are two open goals in the plan, *on(A,B)* and *AE*, as can be seen in Figure 9.11. Both can in fact be achieved by one action *Stack(A,B)* added after *Unstack(A,C)* but that would disrupt *clear(B)* which is a precondition for *Unstack(B,D)*. Again let the planner non-deterministically choose the action *Putdown(A)* which produces *AE* to be consumed by *Unstack(B,D)*. The situation is shown in Figure 9.12.

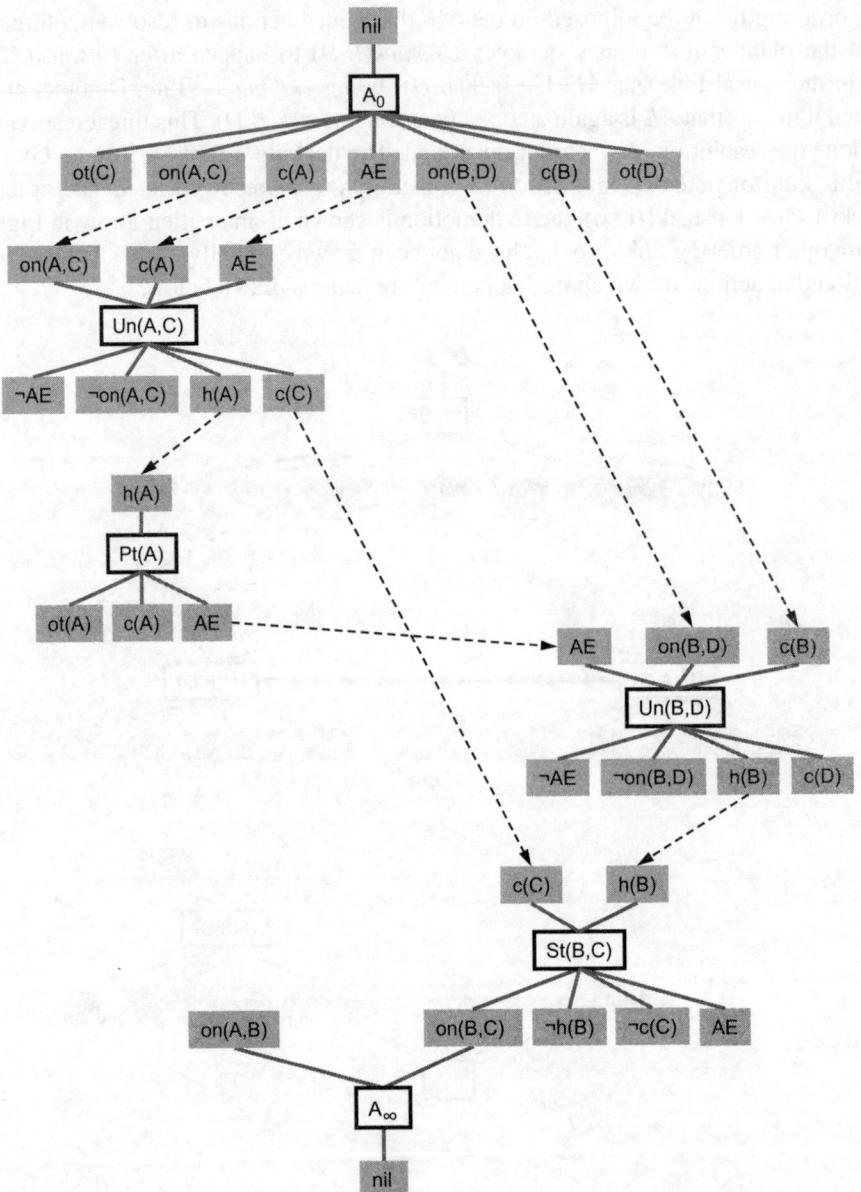

Figure 9.12 PSP next tackles the open goal *AE* in Figure 9.11. It non-deterministically chooses the action *Putdown(A)* to be added after *holding(A)* is made true by *Unstack(A, C)*. This pushes the action *Unstack(B, D)* further down because it has to consume *AE* produced by *Putdown(A)*.

The planner is now in a happy situation. It has only one open goal to solve which is *on(A,B)*. It adds the action *Stack(A,B)* as a resolver for that, and in turn it adds the open goal *holding(A)* which can be resolved by *Pickup(A)*. *Pickup(A)* can be achieved because *AE* is available as an effect of *Stack(B,C)*, and the other preconditions *clear(A)* and *onTable(A)* were added by *Putdown(A)* added earlier. Since a causal link (Stack(B,D), AE, Pickup(A)) will be added, *Pickup(A)* must happen after *Stack(B,D)* and the last action in the plan would be *Stack(A,B)*, at which point the partial plan would have no flaws.

The observant reader would have noticed that the solution plan is a linear plan. But what else can one expect from a one armed robot solving the problem?

9.5.1 A two armed robot

Clearly, a one armed robot can only handle one block at a time and, consequently, the plans are linear sequences of actions. But what if there is a multi-armed robot capable of multitasking? Are more hands better than one? Let us consider a two armed robot. For the sake of simplicity, we begin with independent actions and predicates for the two arms, *arm1* and *arm2*. The actions are *Pickup1*, *Pickup2*, *Putdown1*, *Putdown2*, *Unstack1*, *Unstack2*, *Stack1*, and *Stack2*. The predicates concerned with the robot are *AE1, AE2, holding1,* and *holding2*. An alternative, left as an exercise for the reader, is to include another parameter where needed, for example, *Pickup(N,X)* where *N* is the arm number.

Figure 9.13 adds another arm to the problem from Figure 9.9. One can immediately see that each of the arms can concurrently execute unstack actions. Will PSP find a plan which exploits this simultaneity of actions?

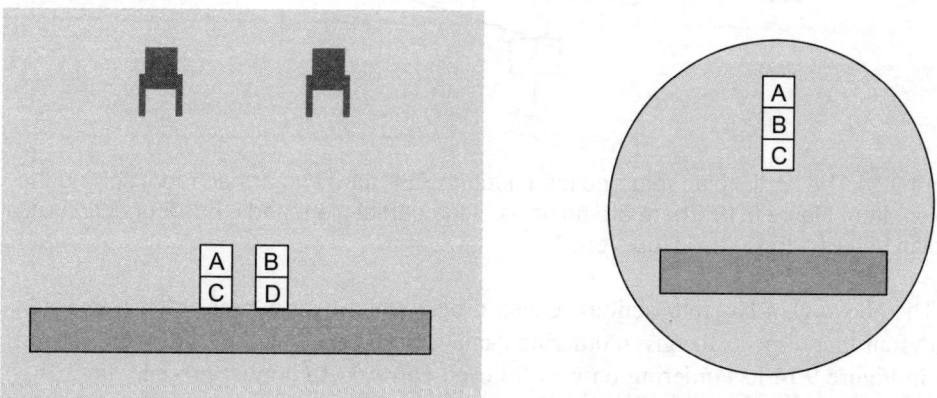

Figure 9.13 A two armed robot solving the problem from Figure 9.9. The two arms, *arm1* and *arm2*, have their own set of predicates, for example, *holding1* and *holding2*, and actions, for example, *Pickup1* and *Pickup2*.

We assume the simple modification of adding distinct predicates and actions for each arm, as described above. We continue to represent predicates and actions by their abbreviated notation. The partial plan that PSP terminates with is shown in Figure 9.14. The first thing to notice is that it has only four actions, which is fewer than the six actions it found for the one armed robot. The second is that it is not a linear plan. We have not drawn the ordering links as before and adopted the convention that actions higher up precede actions lower in the figure.

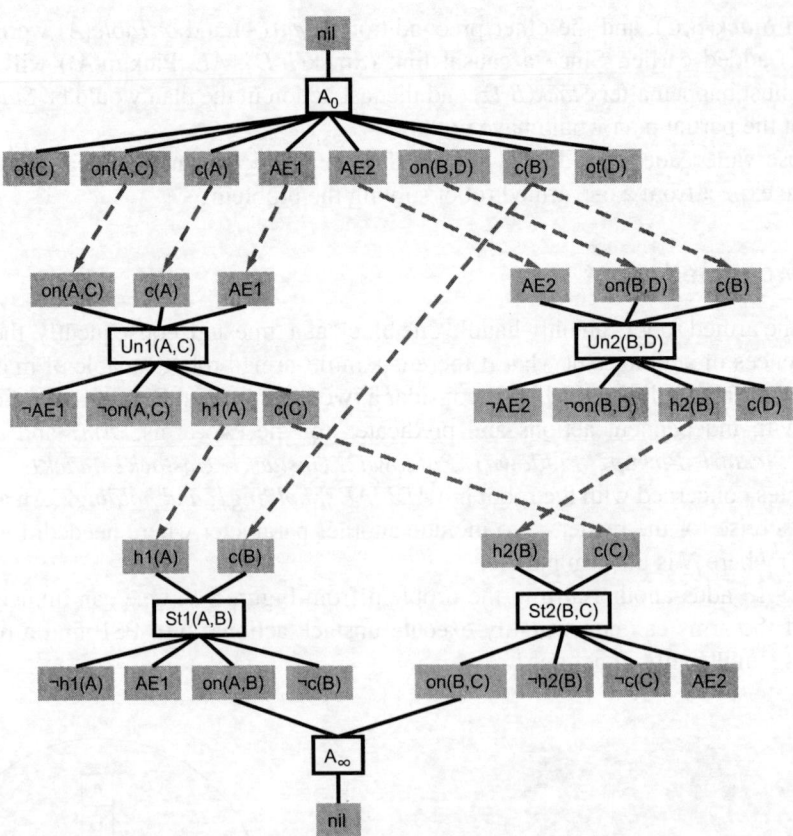

Figure 9.14 The partial plan returned by algorithm PSP for a two armed robot solving the problem from Figure 9.13. There are no flaws in the partial plan, and it has four actions. Can the plan be executed in two time steps?

The above plan has four actions. Given a plan which is a partial order, a linear plan of length four that respects the given ordering can always be generated by a topological sort. The plan in Figure 9.14 has ordering only as follows. *Unstack(A,C)* must precede *Stack(A,B)* and *Stack(B,C)*, and likewise *Unstack(B,D)* must precede *Stack(A,B)* and *Stack(B,C)*. The first two actions can clearly be done in parallel by the two arms, but what about the last two actions? The solution plan suggests that they can be done in parallel too. Imagine one arm stacking A on B, *even while* the other arm is stacking B on C. Needs a bid of dexterity perhaps, but there is no reason why it cannot be done. The plan can therefore be executed in two time steps. We say that the *makespan* of the plan is 2.

One would expect that while *arm2* is holding B, it should not be possible for *arm1* to stack A on B. But the preconditions for *Stack(A,B)* – *holding(A)* and *clear(B)* – are true after the *Unstack(A,B)* action. The reason for this is that while *clear(B)* is a precondition for *Unstack(B,D)*, it is not in the negative effects of the action. The rationale for not including *clear(X)* in the negative effect of either *Pickup(X)* or *Unstack(X,Y)* was that the *only* thing that a *one armed* robot could do next was to either put it on the table or stack it on another block. In either case *clear(X)* would be a positive effect, so why delete it in the first place? Consequently, even when *arm2* is holding *B*, *clear(B)* remains true, and so another block can be stacked on top of *B*. A little bit of thought should convince the reader that a multi-armed robot holding N blocks in N hands can create a tower of N blocks in one time step!

If the user does not want such jugglery, then a simple modification of the *Pickup(X)* or *Unstack(X,Y)* operators by adding *clear(X)* to the negative effects will do the trick. This is left as an exercise for the reader. What is the makespan of the plan now for the problem in Figure 9.13? Draw the solution partial plan.

We have assumed that there is a centralized planner. In real multiagent scenarios, for example, in search and rescue teams, a certain amount of autonomy would be necessary. Moreover, with more than one robot each acting independently but in a coordinated manner, can we address tasks like two robots holding a large table at each end and moving it to a new location? We discuss this briefly in the last section.

None of the algorithms described so far guarantee an optimal plan, one with the shortest makespan. We now describe a couple of algorithms that do. Both algorithms adopt a two stage approach, first converting the planning problem into an intermediate representation, and then solving for the plan on that representation. We begin with algorithm GRAPHPLAN.

9.6 Algorithm Graphplan

In the last decade of the twentieth century, researchers came up with a few algorithms that could tackle much larger planning problems than was possible earlier. One of them is the algorithm GRAPHPLAN by Blum and Furst (1997).

Like some of the others, GRAPHPLAN adopts a two phase approach. The first stage, which needs polynomial time, does reachability analysis of the goals by constructing a *planning graph*. It assembles a disjunction of resolvers for achieving intermediate goals. Only when it identifies a distinct possibility of the goal G being met does it attempt to extract a feasible plan from the planning graph. Bryce and Kambhampati (2007) attribute the sizable scaleup of planning algorithms since the mid-1990s to the *reachability* heuristics largely derived from planning graphs.

The planning graph is constructed by exploring all possible actions simultaneously instead of exploring each path from the start state separately. In doing so, the algorithm coalesces the different reachable states into a set of propositions that is a union of all possible future states at each time step.

9.6.1 The planning graph

The planning graph is a layered graph with alternating layers of propositions and actions. Each layer has relations or constraints between elements within the layer that prohibit simultaneity of the related elements, whether actions or propositions. These relations are *mutex* relations, defined later, with the term standing for mutual exclusion. In the original implementation, they are binary relations, eschewing the harder computation required for constraints between more elements.

The initial or the zeroth layer P_0 is a proposition layer containing all the propositions in the start state.

Then comes A_1, the first action layer, which is the union of all actions applicable in the preceding proposition layer P_0. An action in an action layer is applicable if its preconditions are non-mutex in the preceding proposition layer. The presence of many actions in a layer does not mean they can be executed in parallel. Only actions that are non-mutex can go together into a plan. The *solution* returned by GRAPHPLAN is a sequence of sets of actions <Set$_1$, Set$_2$, ..., Set$_k$> where each Set$_i$ = {a_{i1}, a_{i2}, ..., a_{ip}} contains actions that could execute in parallel in step i.

A proposition layer P_i follows every action layer A_i. Like an action layer, a proposition layer is the union of all propositions that are the effects of all the actions in the action layer. Any state S_i reachable in i steps would be a subset of the proposition layer P_i. If the subset is non-mutex, then the state can possibly be reached by actions in the preceding layers.

A goal G is reachable from the start state S_0 if there is a plan that achieves G. However, computing this by state space search is the planning problem itself. Algorithm GRAPHPLAN employs a weaker notion of reachability, which is a lower bound approximation. If the goal G occurs in a layer in the planning graph, it is said to be reachable. While this is a necessary condition for the goal to be reachable, it is not a sufficient condition. Consider, for example, achieving the goal propositions *on(C,A)* and *on(D,B)* from the start state in Figure 9.9. Both goals would appear in proposition layer P_4 since each tower can be inverted in four moves each, but both are achievable by a one armed robot only in layer P_8.

Between any two consecutive layers, the following sets of edges connect the nodes. These are shown in Figure 9.15 for the problem from Figure 9.9. Observe that the planning graph is constructed from S_0 without paying any heed to the goal state, except for the signal to stop. This process has also been called disjunctive refinement (Ghallab, Nau, and Traverso, 2004).

- Precondition edges link an action in a in A_i to its preconditions in layer P_{i-1}. As observed earlier, the preconditions must be non-mutex for a to find a place in A_i. These are shown by solid edges in the figure.
- Positive effect edges link an action a in A_i to its positive effects in the following layer P_i. These are shown by dashed arrows in the figure.
- Negative effect edges link an action a in A_i to its negative effects in the following layer P_i. These are shown by dashed arrows with rounded arrowheads in the figure.

Automated Planning | 297

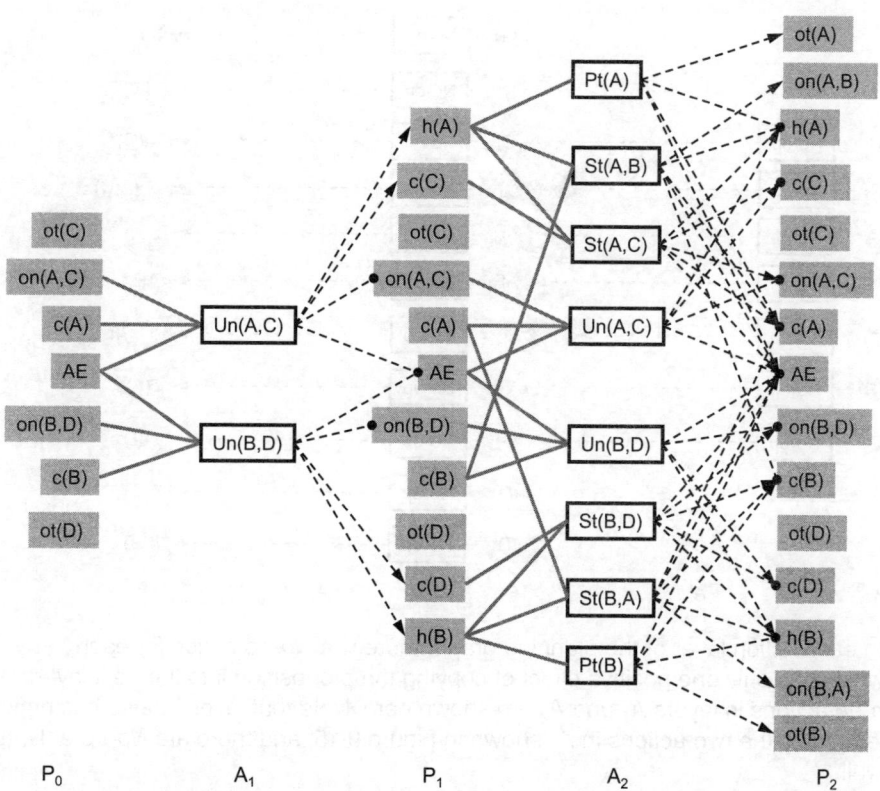

Figure 9.15 The proposition and actions layers in a planning graph, extended up to two levels. Each layer is a union of actions or a union of propositions. An action is connected by precondition links to its preconditions. It also has positive effects linking it to the next proposition layer with dashed arrows, and negative effects drawn as dashed arrows with rounded heads. The next action layer will be A_3.

Layer P_0 contains the start state. Layer A_1 contains all the actions that are individually applicable in the start state, and layer P_1 includes all the effects of these actions. The reader would have observed that there are propositions, like *onTable*(D) shown in abbreviated form as *ot*(D), that are present in layers P_1 and P_2. The reason for that is that we are not implementing the progression of states over actions, but only *incorporating* the effects of these actions. We do this because we do not know at this stage which actions will be included in the plan. Consider, for example, the proposition *on*(B,D). Given the one armed robot if *Unstack*(B,D) were to be the first action, then this would be deleted, but if *Unstack*(A,C) were to be the first action, this would be true in the state after the first action. Algorithm GRAPHPLAN includes a *No-op* action which says that nothing happens to the proposition *on*(B,D) and it should be included in the next layer as is. In fact, it does this for *all* propositions. Every proposition in a layer P_i is thus carried forward to layer P_{i+1}. The set of *No-op* actions is shown in Figure 9.16. This allows for the possibility that no robot action is executed in some layers of the plan. The STRIPS planning domain does not have a notion of time. But in richer domains there could be, and then a *No-op* action could be instrumental in meeting time constraints.

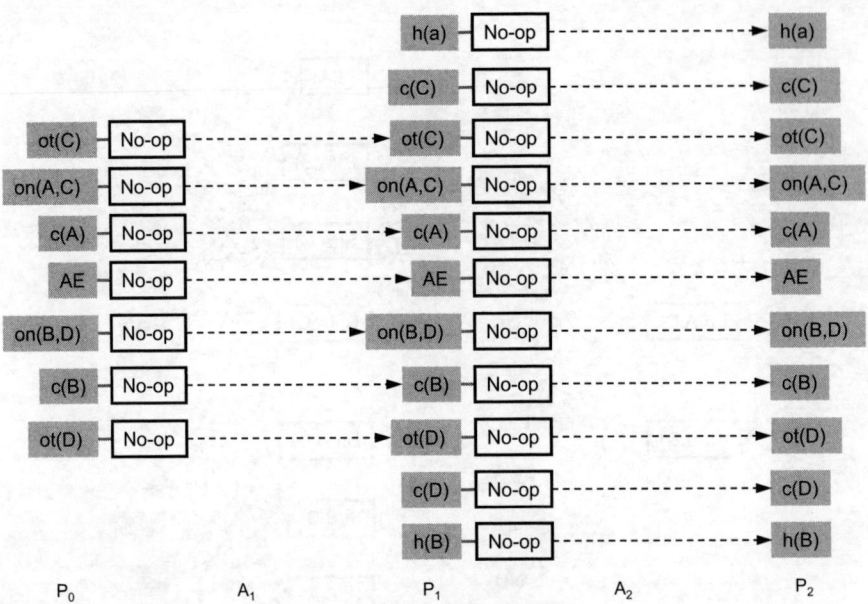

Figure 9.16 In each action layer of the planning graph, there is a *No-op* action for each proposition, which has only one positive effect of copying the proposition into the next layer. The set of *No-op* actions in layers A_1 and A_2 are shown here. Note that layer P_1 also has new propositions added by the two actions in A_1 shown in Figure 9.15, and there are *No-op* actions for these as well.

The *No-op* action takes one proposition p as an argument, which is the only precondition for the action and is the only positive effect, for example, *No-op*($on(A,C)$). The planning graph includes the *No-op* actions for every proposition in a layer along with the other planning actions. If a planning action has a negative effect, for example, $on(A,C)$ for *Unstack*(A,C) in A_1, then $on(A,C)$ has two effect arrows impinging upon it, one as a negative effect of *Unstack*(A,C) and the other as a positive effect of *No-op*($on(A,C)$). In the final plan, only one of them can exist, and they are mutually exclusive or *mutex*. Since *Unstack*(A,C) and *No-op*($on(A,C)$) are mutex, so are their positive effects, for example, $on(A,C)$ and $holding(A)$. We define the set of mutex relations below.

Two actions $a \in A_i$ and $b \in A_i$ are mutex if one of the following conditions holds. All the mutex relations between actions are stored in the set μA_i as pairs.

- *Strong interference:* There exists a proposition p such that $p \in pre(a)$, $p \in effects^-(a)$, $p \in pre(b)$, and $p \in effects^-(b)$. In the blocks world p could be AE.
- *Weak interference:* There exists a proposition p such that $p \in pre(a)$ and $p \in effects^-(b)$. Then only one linear order of the two would be possible.
- *Competing needs*: There exist propositions $p_a \in pre(a)$ and $p_b \in pre(b)$ such that p_a and p_b are mutex in P_{i-1}.
- *Inconsistent effects:* There exists a proposition p such that $p \in effects^+(a)$ and $p \in effects^-(b)$. Then the semantics of the two actions in parallel is not defined. If they are linearized, the semantics will depend upon the order.

Two propositions $p \in P_i$ and $q \in P_i$ are mutex if *all* pairs of actions $a \in A_i$ and $b \in A_i$ such that $p \in \text{effects}^+(a)$ and $q \in \text{effects}^+(b)$ are mutex. All the mutex relations between propositions are stored in the set μP_i.

Figure 9.17 shows some of the mutex pairs in the planning graph for the problem in Figure 9.9.

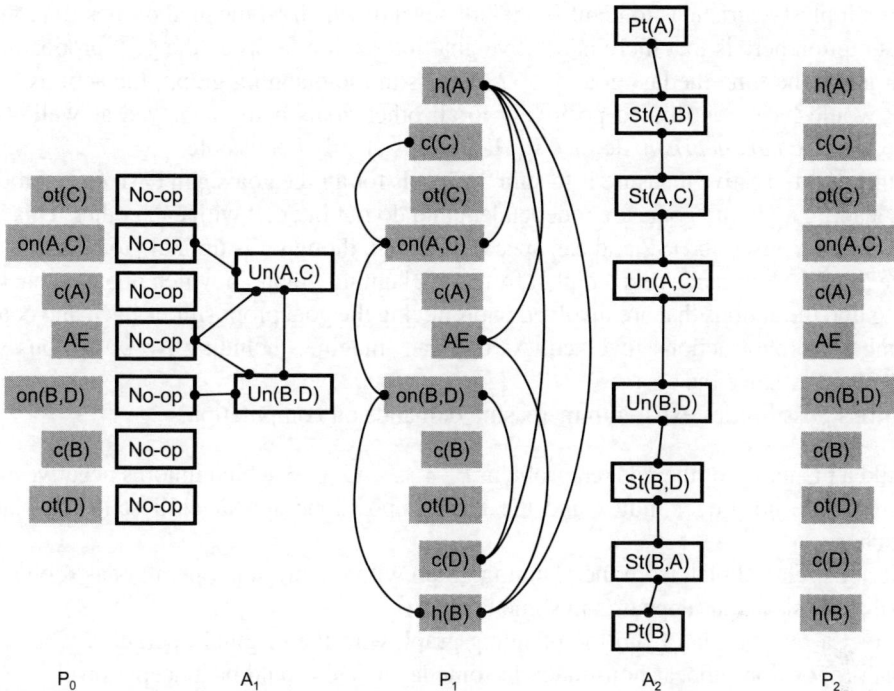

Figure 9.17 Some of the mutex relations in the planning graph being constructed. The *No-op* actions are shown in layer A_1 along with their mutexes with the two actions in the same layer A_1. All actions in layer A_2 are mutex with each other, though the links have not all been shown. Some mutexes for the proposition layer P_1 are shown as well.

Not all mutex relations have been shown in the figure to avoid cluttering. Also not shown in the figure are the precondition and effect links. Once they are added, the planning graph will be complete, represented by the following sets – proposition layers, action layers, precondition links and effect links across consecutive layers, and mutex links within each layer.

A one armed robot allows for only linear plans. Hence all the actions that the robot can do in each layer are mutex with each other, though not shown in the figure. This would not be the case for domains where parallel actions are possible. The reader is encouraged to draw two layers of the planning graph for the problem in Figure 9.13.

Two actions a and b are said to be independent if

- $\text{effects}^-(a) \cap \{\text{pre}(b) \cup \text{effects}^+(b)\} = \phi$ and
- $\text{effects}^-(b) \cap \{\text{pre}(a) \cup \text{effects}^+(a)\} = \phi$

9.6.2 Heuristics and solutions from the planning graph

The planning graph is a structure that compresses an exponentially growing state space into a polynomial graph. When we reach a layer in which all the goal propositions are present, the planning graph at that point can be a source of computing a heuristic value for the state S which is the state from where the planning graph is constructed (Bryce and Kambhampati, 2007).

The simplest heuristic is to identify the first level in which *all* the goal propositions appear. The assumption here is that there is positive goal interaction between the goal propositions in G. That is, by the time the last goal $g \in G$ appears in the planning graph, the actions leading up to it would have contributed positively to all other goals being achieved as well. This is analogous to the *max heuristic* defined in HSP and is clearly admissible.

A more conservative measure is to sum the levels for all the goals g in G. This assumes that actions leading up to the goals are independent and do not interact with each other. This is like the *sum heuristic* used in HSP and can speed up search, though it is unlikely to be admissible.

The third is to extract a relaxed plan from the planning graph, in which one uses backward search to find the actions that are involved in producing the goal propositions in G, and counting the number of robot actions involved. As one can imagine, including *No-op* actions where possible would reduce the cost.

Various possibilities exist with increasing demands on computation.

- Build a planning for the relaxed problem $P' = <S_0, G, O'>$ which ignores negative effects. Actions are no longer mutex and the goal propositions appear earliest in the planning graph.
- Extract a relaxed plan from the planning graph with the original operators as soon as each goal proposition has appeared in some layer.
- Extract a relaxed plan from the planning graph with the original operators, when all the goal propositions appear non-mutex in some layer. This could be a deeper level.

What is important is that since all actions are considered in parallel at each level, the heuristic estimate is a more accurate one, being somewhere between the optimistic *max* heuristic, which assumes positive goal interaction and may be grossly underestimating, and the conservative *sum* heuristic, which assumes that actions are independent and is in most cases overestimating. The backward search algorithm RELAXEDPLANEXTRACTION is described below (Bryce and Kambhampati, 2007). The algorithm returns a layered plan $<Set_1, Set_2, ..., Set_k>$ where $Set_i = \pi_i$ is the set of actions in the layer A_i.

Algorithm 9.3. Algorithm RELAXEDPLANEXTRACTION extracts a relaxed plan from a planning graph $PG(S)$ constructed starting from the state S. Starting with the last layer when the goal predicates, it traces back the actions that produced the goals and the subgoals.

RPE($PG(S)$, G, n)
1    $\pi \leftarrow []$
2    $G_n \leftarrow G$
3   **for** $i \leftarrow n$ **down to** 1

```
4           π_i ← []
5           G_{i-1} ← []
6           for all p ∈ G_i
7               Pick some a ∈ A_i such that p ∈ effects^+(a)
8               π_i ← a : π_i
9               G_{i-1} ← pre(a) ++ G_{i-1}
10          π ← π_i : π
11      return π
```

Observe that the above algorithm to extract the relaxed plan is silent on mutexes. Taking mutexes into account is what transforms it into the backward phase of GRAPHPLAN described below.

Thanks to the *No-op* actions, when a proposition p appears in a proposition layer, it will appear in all succeeding layers. As a corollary, any action a that appears in an action layer will also appear in all succeeding layers. So the number of propositions and actions grows in a strictly non-decreasing manner. Mutex relations, on the other hand, can appear and then disappear. In P_0 there are none, but they appear quickly, and they can disappear later as well. If two blocks A and B in Figure 9.9 are to be unstacked and placed on the table by a one armed robot, then $onTable(A)$ and $onTable(B)$ will be mutex in P_2 but will be non-mutex in P_4. Given a goal $G = \{g_1, g_2, ..., g_k\}$, the algorithm has to wait till the set of propositions appear non-mutex in some layer, before the search for a plan even begins.

The algorithm begins with the initial layer P_0 which contains the propositions in the start state S_0. Then from every proposition layer P_{i-1} the following sets are constructed.

- the set of actions in layer A_i along with the precondition links to P_{i-1} and effect links to P_i.
- the mutex relations between actions in layer A_i in μA_i.
- the set of propositions in layer P_i. First, all the propositions are copied from layer P_{i-1} due to the *No-op* actions, and then new propositions are added as the effects of the new actions in layer A_i.
- the mutex relations between proposition in layer P_i in μP_i.

In GRAPHPLAN the forward process of building the planning graph happens till one of the following two conditions is met:

1. All the goal propositions $g \in G$ appear mutex free in some layer P_n.
2. The planning graph has levelled off. A planning graph is said to level off when two consecutive proposition layers and mutex layers do not change. That is, $P_i = P_{i+1}$ and $\mu P_i = \mu P_{i+1}$. This means that no new actions can appear, and the layers cannot change any more.

The second condition says that no plan exists. In the blocks world this could be if the goal description is inconsistent, for example, $on(A,B)$ and $on(C,B)$. In other domains there could be goals that are not reachable from the start state, for example, if there is no path for a robot to go from one building to another.

As soon as all goal propositions appear non-mutex in a layer, the algorithm switches to phase two and searches backwards for a plan. Observe that this can be later than a layer where they occur together but where some are mutex, because mutexes can disappear. The algorithm to find a plan is backward search in the spirit of Algorithm 9.3, except that it must take mutex relations into account. Starting with the non-mutex goals, it must ensure that the actions chosen in Line 7 are non-mutex, and that the combined set of preconditions of these actions which form the goals at the preceding layer are non-mutex as well. In the absence of nondeterminism, GRAPHPLAN algorithm employs depth first search in the backward direction to search for mutex free actions leading back to the start state. Subsequently, other researchers have proposed other approaches to extract a plan, for example, treating the planning graph as a constraint satisfaction problem (CSP) (Do and Kambhampati, 2001). We will study CSPs in Chapter 12.

When the algorithm finds a plan it must be the shortest makespan plan, because this was the first occasion when the goal propositions appeared non-mutex and a plan could be extracted. If a plan is not found, GRAPHPLAN extends the planning graph by one more level. But before that it creates a memory of goal sets and the level at which they failed. This process is called *memoization*. The next time it embarks upon backward search, it will know not to proceed beyond the memoized goal sets. When it succeeds, the plan that algorithm GRAPHPLAN returns has the following structure:

$$\pi = <\{a_{11}, ..., a_{1p}\}, \{a_{21}, ..., a_{2q}\}, \{a_{31}, ..., a_{3r}\}, ..., \{a_{n1}, ..., a_{ns}\}>$$

That is, it contains n ordered sets of actions. The first set $\{a_{11}, ..., a_{1p}\}$ contains actions that can be executed in parallel in the start state, the second set $\{a_{21}, ..., a_{2q}\}$ in the state after that, and so on. A linear plan can always be extracted from it by topological sorting.

A key feature of GRAPHPLAN is that it is a two phase algorithm. In the first phase, which is computationally inexpensive, the planning problem is converted to another problem which is then solved for the solution. At around the same time, other planning algorithms were devised that too adopted a two phase approach, most notably *planning as CSP* (van Beek and Chen, 1999) and *planning as satisfiability*. We take a brief look at the latter next.

9.7 Planning as Satisfiability

The planner SATPLAN was one of the first programs to adopt a two stage process (Kautz and Selman, 1992; Kautz, McAllester, and Selman, 1996). With advancements in *SAT* solvers (Cook and Mitchell, 1997), there was renewed interest in logic based planners, and satisfiability based planners began performing very well in planning competitions. The idea behind *planning as satisfiability* is to pose the planning problem as a satisfiability problem, and then use efficient *SAT* solvers to find a solution.

Recall from Section 2.3.3 that a *SAT* problem is to find an assignment of *true* or *false* (or 1 or 0) to a set of propositional variables such that a given formula on those variables evaluates to *true*. For example, the following formula is on five variables $\{a, b, c, d, e\}$.

$$(b \vee \neg c) \wedge (c \vee \neg d) \wedge (\neg b) \wedge (\neg a \vee \neg e) \wedge (e \vee \neg c) \wedge (\neg c \vee \neg d)$$

An *interpretation* of the formula is an assignment to the variables, for example, $a = 1$, $b = 1, c = 0, d = 0, e = 0$, also represented succinctly as the binary string 11000. This is not a solution but 10000 is. An interpretation that makes a given formula true is called a *model* for the formula. We look at how a planning problem can be expressed as a *SAT* formula, which can be fed to a *SAT* solver, such that any model for the formula has a plan in it.

9.7.1 Direct encoding

Like GRAPHPLAN, expressing the problem as *SAT* too involves time to be associated with a proposition or an action being true. The variables in the simplest representation are the propositions describing the state at any given time point and the actions that are applicable. Time is added as a parameter to each. Like in GRAPHPLAN, time is discrete and progresses in units of 1 from the initial $t = 0$. In the original paper by Kautz and Selman, an action and its preconditions both happen at time t, and the effects at time $t + 1$. In a variation described in Weld (1999) actions occur at odd time points and propositions at even time points starting with 0. We will stick to the times as described in our GRAPHPLAN description in which the preconditions of actions hold at time $t-1$ and the action and its effect at time t.

The propositional variables are all instances of the predicates in the planning problem. For the problem in Figure 9.9 they are a collection of sets of the following type for each time point t.

{on(A,B,t), on(B,A,t), on(A,C,t), on(C,A,t), on(A,D,t), on(D,A,t), on(B,C,t), on(C,B,t), on(B,D,t), on(D,B,t), on(C,D,t), on(D,C,t), onTable(A,t), onTable(B,t), onTable(C,t), onTable(D,t), holding(A,t), holding(B,t), holding(C,t), holding(D,t), clear(A,t), clear(B,t), clear(C,t), clear(D,t), AE(t)}

The action variables likewise are all instantiations of the planning operators.

{Unstack(A,B,t), Unstack(B,A,t), Unstack(A,C,t), Unstack(C,A,t), Unstack(A,D,t), Unstack(D,A,t), Unstack(B,C,t), Unstack(C,B,t), Unstack(B,D,t), Unstack(D,B,t), Unstack(C,D,t), Unstack(D,C,t), Stack(A,B,t), Stack(B,A,t), Stack(A,C,t), Stack(C,A,t), Stack(A,D,t), Stack(D,A,t), Stack(B,C,t), Stack(C,B,t), Stack(B,D,t), Stack(D,B,t), Stack(C,D,t), Stack(D,C,t), Pickup(A,t), Pickup(B,t), Pickup(C,t), Pickup(D,t), Putdown(A,t), Putdown(B,t), Putdown(C,t), Putdown(D,t), No-op(t)}

Further, if a proposition P is *false*, it is represented as $\neg P$. An *interpretation* is an assignment of truth values to the propositions including negated ones. An interpretation is a *model* when the *SAT* formula evaluates to true. While encoding the planning problem as *SAT* one has to specify the number n of time points the plan extends to. This would be the makespan of the plan. Observe that we have included the *No-op* action as well that allows nothing to be done at a given time point. This would take care of the case when the specified n is more than the length of a solution plan. In practice, one iteratively poses a sequence of formulas with increasing n so that the first solution found is the shortest plan.

The task is then to create a *SAT* formula such that all models are valid plans. The *SAT* formula is expressed in conjunctive normal form (CNF) with the following types of clauses:

Clauses from the initial state: A set of clauses derived from the initial state S_0. This is a conjunct of all the propositions in S_0 and the negation of all propositions not in S_0. For the problem from Figure 9.9, the clauses are

{on(A,C,0), on(B,D,0), onTable(C,0), onTable(D,0), clear(A,0), clear(B,0), AE(0)}

and

{¬on(A,B,0), ¬on(B,A,0), ¬on(C,A,0), ¬on(A,D,0), ¬on(D,A,0), ¬on(B,C,0), ¬on(C,B,0), ¬on(D,B,0), ¬on(C,D,0), ¬on(D,C,0), ¬onTable(A,0), ¬onTable(B,0), ¬holding(A,0), ¬holding(B,0), ¬holding(C,0), ¬holding(D,0), ¬clear(C,0), ¬clear(D,0)}

The negated propositions are added to exclude models that do not represent valid states. The above clauses will form the first part of the CNF. If only the (positive) propositions in S_0 were to be included in the formula, then an interpretation with, for example, $on(A,B,0) = true$ would be a model too even though it is not true in S_0. The start state then contributes the following sub-formula:

[on(A,C,0) ∧ on(B,D,0) ∧ onTable(C,0) ∧ onTable(D,0) ∧ clear(A,0) ∧ clear(B,0) ∧ AE(0) ∧

¬on(A,B,0) ∧ ¬on(B,A,0) ∧ ¬on(C,A,0) ∧ ¬on(A,D,0) ∧ ¬on(D,A,0) ∧ ¬on(B,C,0) ∧ ¬on(C,B,0) ∧ ¬on(D,B,0) ∧ ¬on(C,D,0) ∧ ¬on(D,C,0) ∧ ¬onTable(A,0) ∧ ¬onTable(B,0) ∧ ¬holding(A,0) ∧ ¬holding(B,0) ∧ ¬holding(C,0) ∧ ¬holding(D,0) ∧ ¬clear(C,0) ∧ ¬clear(D,0)]

Clauses from the goal description: The goal description is incomplete, and only introduces the clauses explicit in G. For our example, the clauses are as follows where the formulation is of a plan of n steps. One does not care what else is true at time n.

[on(A,B,n) ∧ on(B,C,n)]

Clauses relating actions and propositions: For all values of t between 1 and n, each proposition p and each action a will be assigned a value in an interpretation. The interpretation will be a model if all the clauses are true in the interpretation. For the model to be a valid plan, these assignments must be consistent with the relation between actions and their preconditions and effects. This is achieved by adding the following clauses to the *SAT* formula, in which each action in the domain implies both its preconditions and its effects. If a is an action, then

$$(a \supset \text{pre}(a)) \land (a \supset \text{effects}^+(a)) \land (a \supset \text{effects}^-(a))$$

If $\text{pre}(a) = \{p_{a1}, p_{a2} ..., p_{ak}\}$ and $\text{effects}^+(a) = \{q_{a1}, q_{a2} ..., q_{al}\}$ and $\text{effects}^-(a) = \{r_{a1}, r_{a2} ..., r_{am}\}$, then

$(a \supset \text{pre}(a)) \equiv (a \supset (p_{a1} \land p_{a2} \ldots \land p_{ak})) \equiv (\neg a \lor p_{a1}) \land (\neg a \lor p_{a2}) \land \ldots \land (\neg a \lor p_{ak}).$

So the clauses added for the preconditions are $(\neg a \lor p_{a1}) \land (\neg a \lor p_{a2}) \land \ldots \land (\neg a \lor p_{ak})$. In a similar manner, the clauses for the effects are

positive effects: $(\neg a \lor q_{a1}) \land (\neg a \lor q_{a2}) \land \ldots \land (\neg a \lor q_{al})$
negative effects: $(\neg a \lor \neg q_{a1}) \land (\neg a \lor \neg q_{a2}) \land \ldots \land (a \lor \neg q_{am})$

Such clauses are added for *every* action instance for *every* time point t between 1 and n. For example, consider the actions *Stack(B,D,t)* at time point t. It contributes six clauses as shown below:

preconditions: $(\neg \text{Stack}(B,D,t) \lor \text{holding}(B, t-1)) \land (\neg \text{Stack}(B,D,t) \lor \text{clear}(D, t-1))$
positive effects: $(\neg \text{Stack}(B,D,t) \lor \text{on}(B,T, t)) \land (\neg \text{Stack}(B,D,t) \lor AE(t))$
negative effects: $(\neg \text{Stack}(B,D,t) \lor \neg \text{holding}(B, t)) \land (\neg \text{Stack}(B,D,t) \lor \neg \text{clear}(D, t))$

Then, if, say, *Stack(B,D,4)* is assigned a value 1 or *true*, the propositions *holding(B,3)*, *clear(D,3)*, *on(B,D,4)* and *AE(4)* *must* be all *true* as well, and *holding(B,4)* and *clear(D,4)* *must* be *false*, for the six clauses to be satisfied (evaluate to *true*).

The frame axioms: When an action happens, it has a set of effects that change the value of some propositions between two time steps. What about propositions not affected by actions in the plan? One needs to ensure that only actions in a plan change the values of propositions. This is done by adding *frame axioms* that assert that propositions that are not changed by an action retain their truth value at the next time point.

There are two approaches for doing this in the literature. The *classical frame axioms* assert that every action that does not affect a proposition leaves it unchanged for the next time step (McCarthy and Hayes, 1969). For *every* action a and for *every* proposition p that is *not* in the effects of a, one adds an axiom of the form $(p(t-1) \land a(t)) \supset p(t)$. For example, the following axiom, in CNF form, says that if block C is on the table when block B is stacked on to block D, then block C continues to be on the table.

$(\neg \text{Stack}(B,D,t) \lor \neg \text{onTable}(C,t-1) \lor \text{onTable}(C,t))$

An alternative set of axioms are the *explanatory frame axioms* as described in Haas (1987). These axioms enumerate the actions that could have led to a change in the value of a proposition p. There is one axiom for positive change, which lists all actions a that have $p \in \text{effects}^+(a)$ and likewise for negative change when $p \in \text{effects}^-(a)$. For example, if at some time step the arm is now holding block B, then the robot must have picked it up from the table or unstacked it from another block. Note that one such action must be present for every block it could have been unstacked from. In the Sussman anomaly example with three blocks A, B, and C, one has a CNF clause,

$(\neg \text{holding}(B,t-1) \land \text{holding}(B,t)) \supset (\text{Pickup}(B,t) \lor \text{Unstack}(B,A,t) \lor \text{Unstack}(B,C,t))$
$\equiv (\text{holding}(B,t-1) \lor \neg \text{holding}(B,t) \lor \text{Pickup}(B,t) \lor \text{Unstack}(B,A,t) \lor \text{Unstack}(B,C,t))$

Likewise, if the robot was earlier holding block B and is no longer holding it, then

$$(\text{holding}(B,t-1) \wedge \neg\text{holding}(B,t)) \supset (\text{Putdown}(B,t) \vee \text{Stack}(B,A,t) \vee \text{Stack}(B,C,t))$$
$$\equiv (\neg\text{holding}(B,t-1) \vee \text{holding}(B,t) \vee (\text{Putdown}(B,t) \vee \text{Stack}(B,A,t) \vee \text{Stack}(B,C,t))$$

Again, for every proposition in the planning problem, one will have to add an instance of such axioms for every time point between 1 and n.

Given the above set of clauses, every solution to the cumulative *SAT* will contain a valid plan of makespan n. The propositions in the initial and goal description are assigned a value *true*. Every action that is part of the plan is assigned a value *true* at the time step t when it happens, and *false* at other time points. The preconditions of the action at time $t-1$ are also assigned *true,* as are the positive effects at time t. The negative effects are assigned a value *false*. The remaining propositions are assigned truth values consistent with the frame axioms. If the solver finds that the formula is unsatisfiable, then a new encoding reflecting a one step longer plan length is generated.

9.7.2 The planning graph as SAT

The readers would have noticed the similarities between GRAPHPLAN and the *SAT* based approaches to planning. Both have a layered structure with increasing time. Both keep track of the time points when propositions are true or false, and when actions occur. The difference is that in the process of constructing the planning graph, GRAPHPLAN has already done some work in identifying the propositions that are reachable from the start state via actions, the actions that are instrumental in doing so, and the mutex relations in each layer. It has also identified the layer n in which the goal propositions first appear not mutex. The original GRAPHPLAN algorithm searches for a plan by doing depth first search in a backward direction. The goal of converting it into *SAT* is to exploit an efficient off-the-shelf *SAT* solver.

If we intend to convert the planning graph into a *SAT* problem, we also need to capture the relations between action layers and the preceding and succeeding proposition layers. This will enable choosing truth assignments for propositions that are consistent with the actions that produce them and the actions that they in turn make applicable. This is done as follows (Kautz, McAllester, and Selman, 1996). We begin with the layer P_0, and the layer P_n which contains the non-mutex goal propositions. Then we work backwards from the goals in P_n to encode the remaining clauses.

- For each proposition p in the domain, if $p \in P_0$, then the corresponding $p_0 = true$ else $p_0 = false$ where the subscript denotes the time point or layer number. The clauses here are the same as the clauses from the initial state in direct encoding.
- For the propositions from goal layer P_n, we similarly keep only the goal propositions with time stamp n. Like in the direct encoding, the two clauses for the Sussman anomaly are $on(A,B,n)$ and $on(B,C,n)$.
- Working backwards from the goal with $t = n$, every $p \in P_t$ such that $p \in G_n$ induces a disjunction of actions $a \in A_t$ in the planning graph that have p as a positive effect. The actions in the disjunction *must* be from the planning graph. For example, looking at the

planning graph in Figure 9.15, we have only two actions in the planning graph that could result in *holding(B,2)*,

$$holding(B,2) \supset Unstack(B,D,2) \vee No\text{-}op(holding(B),2))$$

This is different from the clause that direct encoding for the problem from Figure 9.9 would have generated.

$$holding(B,2) \supset (Pickup(B,2) \vee Unstack(B,A,2) \vee Unstack(B,D,2) \vee Unstack(B,C,2) \vee No\text{-}op(holding(B),2))$$

Observe that this is also similar to the explanatory frame axioms. This says that if *holding(B)* is to be true at time t, then one of the actions in the planning graph that produced it must have happened at time t as well. This translates to a smaller clause than in direct encoding,

$$\neg holding(B,2) \vee Unstack(B,D,2) \vee No\text{-}op(holding(B),2)$$

- If an action a happens at time t, then its preconditions must have been true at time $t-1$.

$$a_t \supset \text{pre}(a)_{t-1}$$

This is similar to encoding preconditions in the direct encoding, which for the same example is

$$(\neg Stack(B,D,t) \vee holding(B, t-1)) \wedge (\neg Stack(B,D,t) \vee clear(D, t-1))$$

- Actions that are mutex in the planning graph introduce corresponding clauses in the *SAT* encoding. For example, in Figure 9.15 we have

$$\neg Unstack(B,D,1) \vee \neg Unstack(A,C,1)$$

Observe that we do not need to encode mutex relations on propositions. This is because we begin backwards from the mutex free goal propositions and only regress to subgoal propositions that are preconditions of actions which are in the planning graph. And actions appear in the planning graph only when their preconditions are non-mutex in the planning graph.

In summary, by the time GRAPHPLAN is ready to search for a plan in the planning graph, it has already restricted the set of actions that are applicable and that lead to the goal eventually. For example, the action *Stack(C,D)* does not appear in the *SAT* encoding for the problem in Figure 9.15. Consequently, the *SAT* encoding is much smaller. This can be further reduced by doing a limited amount of fast inferences as reported in van Gelder and Tsuji (1996) and Kautz and Selman (1999).

9.8 Richer Planning Domains

The scope of planning in this chapter has been the simple STRIPS domain. This is characterized by a single agent in a static completely known environment, where the actions are instantaneous and deterministic that do not fail in the real world, and where the goals are propositions to be satisfied in the goal state. We have restricted ourselves to this simple domain to focus on the planning algorithms that have been developed. These algorithms have been extended to richer domains as well but are beyond the scope of this book. We end by presenting a brief description of the richer domains that are being addressed by the planning community.

9.8.1 Durative actions

Actions are rarely instantaneous in the real world. Boiling water for a cup of tea takes time, as does walking from home to the park. When an action is *durative* (has a duration), its relation to other actions can be complex. The language PDDL 2.1 (Fox and Long, 2003) introduces the notion of duration for actions. The duration could be static, or it could be determined dynamically during the course of planning. Durative actions invoke the notion of time, and planning with durative actions is called *temporal planning*.

When actions take time to execute, the preconditions and effects can apply at the start of the action, during the action, or at the end of the action. Consider the action of walking from home to the park. At the start the agent must be at home, at the end in the park, and is at neither location during the walk. If it is raining in addition, then an umbrella must be open during the walk, and in the park.

Consider a toddler playing, emptying one cup of water into another cup lying on the table. This could be represented in PDDL 2.1 as follows:

```
(:durative action transferWater
    :parameters    (?cup1, ?cup2 – cup
                    ?table – table
                    ?hand – hand)
    :duration      (= ?duration 5)
    :condition     (and  (at start (on ?cup1 ?table))
                         (at start (holding ?cup2 ?hand))
                         (at start (empty ?cup1))
                         (at start (full ?cup2))
                         (over all (on ?cup1 ?table))
                         (at end (on ?cup1 ?table)))
    :effect        (and  (at start (not (full ?cup2)))
                         (at start (non (empty ?cup1))
                         (at end (empty ?cup2))
                         (at end (full ?cup1))
)
```

This action optimistically assumes that the toddler has successfully transferred the entire water from the second cup to the first one. When planning with instantaneous actions, two actions can either be concurrent or one happens before the other. When actions have durations, then many more relations are possible. These are captured in Allen's interval algebra (Allen, 1983, 1991) and shown in Figure 9.18.

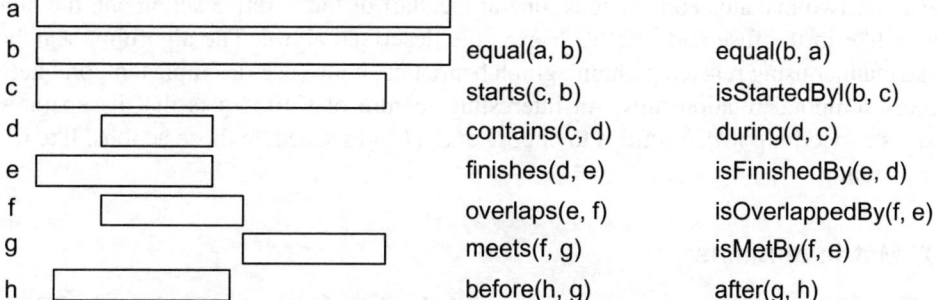

Figure 9.18 The thirteen relations in Allen's interval algebra, shown between eight intervals $\{a, b, c, d, e, f, g, h\}$. In the figure, the relations are shown from *b* onwards, and are between the interval and the one preceding it. The first relation *equal* is symmetric.

The STRIPS domain has no notion of time, only sequencing of actions one after the other. Parallel actions happen instantaneously at the same time point. Linear planning with durative actions is similarly a task of sequencing actions. It is when actions can be done in parallel that things become more interesting. A look at Allen's interval relations reveals why this is so. Having to deal with actions with different durations becomes tricky when there is interdependence between their preconditions and effects. When should two friends start walking from their homes to reach the park at the same time? Forward planning would have no basis, except lookahead, to choose the starting times. There may be situations when actions are required to be executed in parallel in one or more specific relations from the interval algebra.

The term *required concurrency* was introduced in Cushing, Subbarao Kambhampati, and Weld (2007) and Cushing (2012). These requirements are not stated explicitly but are implied by the preconditions and effects of durative actions. This is exemplified by the following example from Cushing's doctoral thesis. Consider the problem of repairing and inserting a broken fuse in a dark cellar, where the only potential source of light is one last matchstick in your matchbox. Assuming that you can repair the fuse in the dark, you will still need light while finally inserting the fuse to avoid getting electrocuted. Let $Fuse_{start}$ and $Fuse_{end}$ be the start and end times of the fuse repair action, and let $Match_{start}$ and $Match_{end}$ be the start and end times of the match lighting action. Then the plan to repair and insert the fuse would need to respect the following constraints:

$$Match_{start} < Fuse_{end} < Match_{end}$$

That is, the match must be lit before inserting the fuse, and its light must last till after the fuse has been inserted. The question is: when should the match be lit? A fielder on the boundary on a cricket field may similarly be required to time her jump accurately in order to catch a ball which would otherwise sail over the boundary.

Two of the earliest planners that handled durative actions are SAPA (Do and Kambhampati, 2003) and CRIKEY3 (Coles et al., 2008). The former is a metric temporal planner that maintains an event queue of durative actions and has two kinds of moves. The first kind selects and adds a new action to the plan, and the second advances the time to the next event in the event queue. The second is applicable when the event queue is not empty. CRIKEY3 splits the durative actions into two instantaneous actions, one at the start of the durative action and the other at the end, like in the fuse and matchbox example described above. The algorithm searches in FF-like manner using relaxed planning graph heuristics. It also uses a simple temporal network to record temporal relationships. An interesting feature of CRIKEY3 is that it separates the decisions concerning *which* actions to choose and *when* to schedule those actions, like in plan space planning.

9.8.2 Metric domains

Planning without having to use numbers can only be done in simple domains. The PDDL 2.1 domains incorporate reasoning with numbers. The use of numbers in a domain can provide a platform for measuring the consumption of critical resources and other parameters. An example of a metric domain problem is that fuel use must be minimized, or that overall execution time must be minimized, or both (Long and Fox, 2003).

The following example from Fox and Long (2002) is a variation of the toddler example described earlier. Here the jugs have capacities, and one can only empty one jug into another if there is enough space.

```
(define (domain jug-pouring)
        (:requirements :typing :fluents)
        (:types jug)
        (:functions
            (amount ?j – jug)
            (capacity ?j – jug))

        (:action pour
            :parameters (?jug1 ?jug2 – jug)
            :precondition (>= (- (capacity ?jug2) (amount ?jug2)) (amount ?jug1))
            :effect (and    (assign (amount ?jug1) 0)
                            (increase (amount ?jug2) (amount ?jug1)))
)
```

The above domain ignores the fact that pouring is a durative action. The reader is encouraged to add the temporal aspect to the action. Both *SAPA* and CRIKEY3 mentioned in the previous section are metric planners as well.

9.8.3 Conditional effects

Actions in *STRIPS*-like domains are comprehensive. The preconditions and effects are explicitly defined. But sometimes, for the sake of succinctness, one would like the effects of an action to be defined by the context in which the action occurs. Koehler et al. (1997) describe an extension to GRAPHPLAN that handles actions defined originally in the *action description language* (ADL) eventually absorbed into PDDL 2.1. The planner can handle conditional actions in which the effects can be universally quantified statements relating conditions to effects.

Consider the example of a bus driving from location A to location B. What should be the effects of this action, apart from the fact that the bus is at location B? Clearly, the driver and the passengers should be at location B as well. How does one include that in the new state? One way would be to add it as an effect of the alight action, in which when a passenger gets off the bus her location is the same as the location of the bus. But what if she does not alight from the bus? A similar example is of carrying books in a briefcase. Koehler defines the following conditional action, expressed in the ADL style:

name:	move-briefcase
par:	L_1:location, L_2:location
pre:	at-$b(L_1)$
eff:	ADD at-$b(L_2)$, DEL at-$b(L_1)$
	$\forall x$:object $[in(x) \supset$ ADD at(x,L_2), DEL at$(x,L_1)]$

The action says that when you move the briefcase from location 1 to location 2, then anything that is in the briefcase also gets transported to location 2. Clearly, this is a generic action that compresses all possible movement actions of individual objects into one. Further, if you have carried the briefcase to office and a colleague asks you whether you have a particular book in the office, you can reply in the affirmative without having to take the book out of the briefcase.

9.8.4 Contingent planning

Most real world planners have to face uncertainty. When planning for different possible situations, a planner is said to do *contingent planning*. Contingent planning is closely tied to monitoring of plans during execution and, having synthesized a plan, to cater to different scenarios. This is opposed to replanning when an ongoing plan fails, or even when an opportunity unexpectedly presents itself. That would be a lazy approach. Contingency planning is more conservative, aiming to cater in advance to the different possible scenarios. Humans do contingent planning all the time, generating plans where execution has checkpoints to choose between alternative plans. For example, a shopper may have a plan to visit a number of different stores in search of a product. Contingent planning is common in multi-agent scenarios where the moves by an opponent or a collaborator may influence your actions. Most of all, such planning may need to be done when the world is not known completely. A good bridge player would always have a backup plan up her sleeve were the lie of cards be different from the one she planned for. Often moves in a plan are specifically aimed at uncovering hidden information. A knowledge based planner that produces some contingent plans for declarer play in bridge was described in Khemani (1989, 1994).

David Smith and Daniel Weld (1998) describe the algorithm CONFORMANTGRAPHPLAN with an example of contingent planning posed by Drew McDermott (1987). The problem is concerned with bomb disposal. There are two packages one of which has a bomb in it. The bomb can be diffused by *dunking* the package in the only toilet in the building. Their algorithm constructs two parallel planning graphs for the two possible worlds with the bomb being in the two different packages in the two worlds, and returns a union of the two plans extracted from the two planning graphs. This involves dunking each package one after another since only one can be accommodated at a time. When dunking one package is the chosen action in one possible world (planning graph), it induces a mutex between the two dunking actions to delay dunking in the other possible world. If in addition the package can possibly clog the toilet, an unclog action is added after the first dunking, to clear the way for the second package.

9.8.5 Trajectory constraints and preferences

So far the validity of a plan has been met by the goal conditions being satisfied. However, there can be situations when one may also want the *intermediate states* to satisfy certain constraints. In the blocks world a fragile block may only be able to support two blocks above it. When driving cross country from one city to another one may require that one is never more than 20 kilometres from a hospital, or that one avoids steep gradients, or that the route must be as close to the sea as possible (the East Coast Road from Chennai to Mahabalipuram goes along the sea coast). While cooking one may have the condition that every time one opens the refrigerator door one must close it too. Trajectory constraints were introduced in PDDL 3.0 (Gerevini and Long, 2005). In addition to having a proposition true at the end of the plan, the following additional goals may be specified. In the following, <GD> refers to one or more propositions describing a goal.

- <GD> must be true *at the end* (like in the domains we have considered so far)
- <GD> must be true *at all times* during the plan
- <GD> must be true *at some time* during the plan
- <GD> must be true within N steps in the plan
- <GD> must be true *at most once* during the plan
- <GD1> must be true *some time after* <GD2> is true
- <GD1> must be true *some time before* <GD2> is true
- <GD1> must be true *within N steps after* <GD2> is true
- <GD> must be true *from step N to step M* in the plan
- <GD> must *hold after N steps* in the plan

The other feature introduced in PDDL 3.0 is *preferences*. Goals, whether on the end point or during the trajectory, that *must* be satisfied are hard constraints. Preferences, on the other hand, are soft constraints, which are desirable but not mandatory. For example, when going out for a dinner and a movie, one might want to buy some groceries on the way, but only if it is feasible. Soft constraints change the way we evaluate plans. There may be a penalty introduced if a preference is not met, or a reward added if it is. This changes the notion of what is a valid plan. Instead of looking for the stated goals to be necessarily satisfied, one thinks of it as an optimization problem in which as many soft goals as possible are satisfied.

9.8.6 Coordination in multi-agent systems

When multiple agents are collaborating to solve a problem, some agents may need to act in a coordinated fashion to do tasks that a single agent cannot do. For example, carrying a large table out onto a lawn may need two or more agents to synchronize their actions holding different ends of the table. While doing complex surgery in an operation theatre, a surgeon focussed on the task has only to reach out her hand and a colleague magically hands her the right instrument. Another place when coordination is visible is on the football field where, for example, a striker may rise to the occasion in a coordinated manner to intercept a cross from a teammate and head it into the opponent goal. Coordination between agents may also be required when multiple agents are working with shared resources. An example could be a children's hobby club working with a limited set of instruments that have be shared between them. Another example is when vehicles coming from different directions have to negotiate their movement across a roundabout. In some countries the drivers diligently observe the right of way conventions, while in others the drivers may have to rely on a keen eye and quick reaction, while in still others the civic authorities may install traffic signals, and in some even police personnel to monitor the traffic obeying the signals. A comparison of coordinated planning methods for rovers is given in Chien et al. (2000).

Coordination could be done by centralized planning in which the actions and schedules of the different actors are specified completely. In this situation one agent would be responsible for the planning and synchronization of actions.

Alternatively, the different agents may produce their own partial plans and send them to a central agent for reconciliation and synchronization.

Finally, the different agents may have common goals but act independently. Here too some amount of communication may be needed. This is clearly the case in team games like football and hockey.

Some domains involving coordination may require the agents to reason about what other agents know, and that would take us into the emerging field of epistemic planning.

9.8.7 Epistemic planning

The moment there is more than one agent in the domain, collaboration, or even competition, between agents requires reasoning about the goals, plans, and intentions of other agents. The planning problem then acquires another dimension, reasoning about knowledge (Fagin et al., 2004). Planning is no longer an isolated search algorithm, but has to interact with other kinds of reasoning processes. This takes us beyond the scope of this book, but we present a flavour of the kind of reasoning a planner will have to do in addition.

An *epistemic logic* extends a classical logic by adding a modality K for knowledge. If α is sentence, then $K_A \alpha$ stands for the fact that agent A *knows* α. For example, $K_{robot1} holding(robot2, D)$ says that $robot_1$ knows that $robot_2$ holds block D, and $K_{robot2} K_{robot1} holding(robot2, D)$ says that robot2 knows this. *Dynamic epistemic logic* (DEL) is a logic that adds actions of various kinds to epistemic reasoning (van Ditmarsch, van Der Hoek, and Kooi, 2007). We begin with an example from Bolander (2017) that introduces epistemic planning using DEL.

A father has ordered a gift for his daughter and needs to collect it from the post office before taking it home and wrapping it. The following simple plan can be found by a STRIPS planner:

π = go to post office; pick up present; go home; wrap present.

If there are two post offices, then the father needs to try both of them to pick up the gift packet. After visiting the first post office *he will know* whether he has the gift or not and can act according to the following plan where PO1 and PO2 are the two post offices:

Go(Father, H, PO1); TryPickUp(Father, Present, PO1);
if K_{Father}Has(Father, Present)
       then Go(Father, PO1, H); Wrap(Father, Present)
       else Go(Father, PO1, PO2); TryPickUp(Father, Present, PO2);
Go(Father, PO2, H); Wrap(Father, Present)

One can also introduce another agent, an employee in PO1 whom the father can ask, and who presumably knows if the post office has the present.

The following is a more complex problem involving epistemic coordinated actions from Engesser et al. (2017):

Bob would like to borrow his friend Anne's apartment while she is away. Anne is happy to lend it to him, and the plan is that she will leave the key below the door mat for Bob, who will use it to unlock her apartment.

The question is: what does she need to tell Bob (an epistemic action)? For this, Anne has to be able to view things from Bob's perspective. This is known as the 'Theory of Mind' which enables an agent to reason about what other agents know (Premack and Woodruff, 1978). She must realize that Bob needs to be told where the key is. She must also assume that Bob will himself synthesize the plan to retrieve the key from below the mat and use it to unlock and enter her apartment. The plan would then be:

Anne puts the key under the door mat; Anne calls Bob to let him know where the key is; when Bob arrives, Bob takes the key from under the door mat; Bob opens the door with the key.

This does qualify as an implicitly coordinated plan. Anne now *knows* that Bob will *know* that he can find the key under the door mat and hence will be able to reach the goal. Anne does *not* have to request or even coordinate the sub-plan for Bob (which is: take key under door mat; open door with key), as she knows he will himself be able to determine this sub-plan given the information she provides.

Finally, multi-agent card games are a fertile ground for epistemic planning. The complete pack is known to each player, but each can only see her own cards. Some inference can be made from the cards other players play. Contract bridge is probably the most sophisticated of all card games. This is due to two reasons. One, it is a partnership game requiring active communication between players. Second, in the play phase that follows the bidding phase, the cards of one player are exposed to all, which generates sufficient information to serve as fodder for complex reasoning.

Communication between partners is done publicly, and the opponents listen eagerly. And like in war and espionage, the communicators can target the opponent with false information (or fake news). This can give space to cloak and dagger operations. We have described one such example analysed in Khemani and Singh (2018) in Section 8.4 in which a player weaves a web of deception to score over his opponent. Logicians distinguish between *knowledge* and *belief*. Logically, knowledge can *only* be about what is true in the world, but beliefs have no such constraint, which leaves the door ajar for deception. Many computer programs employ Monte

Carlo methods to probabilistically choose a plan. For example, a program to play the simple game *Hannabi* is reported in Reifsteck et al. (2019).

Contract bridge, however, is still an open problem.

Summary

Planning is a critical activity for an intelligent agent. In this chapter we have studied different algorithms for domain independent planning. Starting with state space search, we moved on to searching in the plan space with POP, and algorithms GRAPHPLAN and SATPLAN. One reason that different kinds of algorithms have been explored is that planning is a hard problem, even for the simplest domains. Various approaches have been explored to mitigate complexity.

Backward state space planning was explored to exploit the fact that the goal description is sparse, leading to low branching. But it has a problem of generating spurious states. GSP was an attempt to combine the best features of forward and backward planning. The idea of domain independent heuristics was explored in FF, HSP, and HSPr, and it was observed that the planning graph is a source of such heuristics. PSP offers a separation of action selection and action scheduling, and could be the foundation of combining planning with reasoning about goals and actions. Both GRAPHPLAN and SATPLAN explore intermediate structures easy to compute, and where standard solvers could be deployed.

Mutex relations introduced logic and reasoning into the process of search. This also happens when planning in richer domains, which demand greater expressivity in describing domains and the associated reasoning. We will look at the interplay between search and reasoning again in Chapter 12. In the next chapter we look at logical reasoning and see how the process of reasoning also has an underlying search component.

Exercises

1. Extend the STRIPS operators described in Section 9.2.1 to allow the robot to move a box from one room to another room when the need arises.
2. The monkey and the banana is a popular variation of the STRIPS problem. In this variation there are some bananas hanging from a ceiling and a monkey needs to push a box to the location of the bananas, climb on the box, retrieve the bananas, climb down, and then eat the bananas. Devise STRIPS-like operators to solve the problem.
3. Combine the previous two domains so that the monkey has to push the box from another room to the room where the bananas are.
4. The Gripper domain is defined as follows. There are two rooms, four balls, and two robot arms. The predicates are – X is a room, X is a ball, X is inside Y, robot arm X is empty. The robot can move between rooms and pick up and drop one or two balls. In the initial state all four balls and the robot are in one room, and both robot arms are empty. The goal description is that all balls should be in the other room. Express the Gripper domain in PDDL.

5. The logistics domain is defined as follows (Bart Selman, Henry Kautz). There are several cities, each containing several locations, some of which are airports. There are also trucks, which can drive within a single city, and airplanes, which can fly between airports. The goal is to get some packages from various locations to various new locations. Express the domain in PDDL.
6. The Mystery domain was introduced in the international planning competition (IPC) 1998 by Drew McDermott. There is a planar graph of nodes. At each node are vehicles, cargo items, and some amount of fuel. Objects can be loaded onto vehicles (up to their capacity), and the vehicles can move between nodes, but a vehicle can leave a node only if there is a nonzero amount of fuel there, and the amount decreases by 1 unit. The goal is to get cargo items from various nodes to various new nodes. To disguise the domain, the nodes are called emotions, the cargo items are pains, the vehicles are pleasures, and fuel and capacity numbers are encoded as geographical entities. Express the planning domain in PDDL.
7. Given the planning problem described below, show how goal stack planning with STRIPS operators will achieve the goal. You may choose any order where a choice has to be made. What is the plan found?

    Start state = {onTable(A), onTable(C), on(B,C), on(D,E), onTable(F), on(E,F), AE, clear(A), clear(B), clear(D)}
    Goal description = {on(D,A), on(A,E)}

8. Given the planning problem described below, show how *goal stack planning* with STRIPS operators will achieve the goal. You may choose *any order* where a choice has to be made. What is the plan found?

    Start state = {onTable(A), on(B,A), on(C,B), on(D,C), onTable(E), on(F,E), AE, clear(D), clear(F)}
    Goal description = {on(D,B)}

9. Define blocks world operators for multi-armed robots. Modify the STRIPS operators to include another parameter for the arm number. For example, the *Stack(X,Y)* operator is modified to *Stack(N, X, Y)* where *N* is the arm number. Likewise for predicates, for example, *AE(N)*. Modify the planning algorithms studied to allow for plans with actions being done in parallel.
10. Modify the definition of the *Pickup(N, X)* and *Unstack(N, X, Y)* operators to add *clear(X)* to the negative effects of these actions. Show the solution found by PSP on the problem from Figure 9.13.
11. For the planning problem below but with a two armed robot, draw the partial plan found that is an optimal solution plan finishing the earliest possible. What is the makespan of the plan?

    Start state = {onTable(A), onTable(C), on(B,C), on(D,E), onTable(F), on(E,F), AE(1), AE(2), clear(A), clear(B), clear(D)}
    Goal description = {on(C,B), on(A,C), on(B,D)}

12. For the planning problem below but with a two armed robot, draw the partial plan found that is an optimal solution plan finishing the earliest possible. What is the makespan of the plan? State any assumptions you have made.

    Start state = {onTable(A), on(B,A), on(C,B), on(D,C), onTable(E), on(F,E), AE(1), AE(2), clear(D), clear(F)}
    Goal description = {on(C,D), onTable(D), on(F,C)}

13. Define the notion of a flaw in a partial plan represented in POP. How are the flaws addressed?
14. It was observed in Figure 9.17 that for the problem in Figure 9.9 all non-No-op actions are mutex with each other. What is the situation when GRAPHPLAN attempts the problem in Figure 9.13 which has two arms? Draw the first two action layers and identify the mutex actions.
15. Simulate the algorithm GRAPHPLAN on the planning problems from Figures 9.9 and 9.13. For each, take a large sheet of paper and draw the planning graph till the point when the backward phase succeeds.
16. Define the notion of 'mutex relations' used in algorithm GRAPHPLAN. When are two propositions mutex?
17. Define the mutex relations in GRAPHPLAN. Given the start state = {onTable(A), onTable(C), on(B,C), clear(A), clear(B) AE}, draw one level of the planning graph and show the mutex relations.
18. Consider the problem of inverting a stack of two blocks, A and B. Encode this as a planning graph with four layers.
19. Use the planning graph for encoding the two block inverting problem into *SAT*. Also generate the direct *SAT* encoding for the above problem, and compare the two.
20. Express the water jug problem from Section 2.4.3 as a metric planning problem. Extend it to a temporal problem by assuming that the time taken for a pouring action is proportional to the amount of water being poured.
21. Given that a durative action a can be written as two actions a_{start} and a_{end}, and likewise action b as actions b_{start} and b_{end}, express the thirteen Allen's relations in terms of constraints between these four instantaneous actions.

CHAPTER 10

Deduction as Search

An intelligent agent must be aware of the world it is operating in. This awareness comes mainly via perception. Human beings use the senses of sight, sound, and touch to update themselves. However, the entire world is not perceptible to any of us. Our senses have limitations. We cannot hear the dog whistle, or see the bacteria living on our skin or the mountain on the other side of the world. But through science and communication we know about the worlds beyond our sensory reach. Telescopes from Galileo to James Webb have delivered spectacular images of the universe, some taken in the infrared band in the spectrum. We augment whatever we know by making inferences. The conclusions we draw may be sound or they may be speculative yet useful. Evolution has preserved in us both kinds of inference making capability.

The world is dynamic and has other agencies making changes in the world too. If we observe something we may guess the cause or intention behind it. This kind of speculation is called *abduction*. The conclusion is possibly true, maybe even likely. If we see the local bully striding towards us, we may suspect ill intent on his part, and take evasive action. Better safe than sorry. If we develop a cough and fever, we may fear Covid and isolate ourselves from others. When we observe a few white swans, we may conclude that all swans are white. This is called *induction*. Neither abduction nor induction is always sound. Conclusions we draw may not always hold. But they are eminently useful.

In this chapter we study *deduction*, a form of inference that is sound. The conclusions that we draw using deduction are necessarily true. The machinery we use is the language of logic and the ability to derive proofs. We highlight the fact that behind deduction the fundamental activity is searching for a proof.

Logic and mathematics are often considered to be synonymous. Both are concerned with truth of statements. In this chapter we confine ourselves to the family of classical logics, also known as mathematical logics, in which every sentence has exactly two possible truth values – *true* and *false*. Nothing in between. No fuzzy concepts like tall and dark. Is a person whose height is 176 centimetres tall? What about 175 then? And 174? When does she become not tall? Or modalities like maybe. It is possible she loves him. Does that mean she loves him or does she not? Or values like don't know. Is it raining in Delhi? Classical logics would either say a sentence is true or it is false. Even a sentence which says that White always wins in chess is a sentence in classical logic, because *in principle* it must be either *true* or *false*. Even though we cannot find out given the size of the chess game tree.

The simplest classical logic is the logic of propositions, or *propositional logic* (PL). Every logic has an associated formal language. Every logic has a well defined vocabulary and syntax, which completely defines the language $\mathcal{L}$ as a set of sentences. And well defined semantics. There are two angles to the semantics of a language. One is truth functional semantics, which assigns a truth value to every sentence in the language $\mathcal{L}$. The other is denotational, which refers to what the sentences mean. Logicians are largely concerned only with the former, while the artificial intelligence (AI) community is also concerned with meaning. What does the sentence *represent*?

10.1 Logical Connectives

We begin with PL to introduce the key components in all logics, viz. logical connectives. It is logical connectives that allow us to construct new sentences from old.

The following is the vocabulary used to define the language $\mathcal{L}$ of PL:

- The set of atomic sentences A of $\mathcal{L}$ is a countable set of proposition symbols $\{P, Q, R, \ldots\}$. Propositional variables are also called Boolean variables after George Boole who invented Boolean algebra.
- The set of commonly used logical connectives of $\mathcal{L}$ are $\{\neg, \wedge, \supset, \vee, \oplus, \equiv, \downarrow, \uparrow\}$. Of these the first is a unary connective, and the remaining are binary connectives.
- The constant symbols '$\bot$' and '$\top$', called *Bottom* and *Top* respectively. These are atomic sentences whose truth value is known and constant, being respectively *false* and *true*.
- The set of punctuation symbols include the different kinds of brackets and parentheses.

The atomic sentence in PL is indivisible. We do not peer inside it but treat it as a unit. Logical reasoning itself is only concerned with *form* and not with meaning, though the meaning is defined by the user. Meaning lies in the mind of the beholder. We represent an atomic sentence in PL as a propositional symbol from a countable set of symbols $\{P, Q, R, \ldots, P_1, P_2, \ldots\}$. A propositional symbol can stand for or denote *any* sentence in a natural language. Even a complex sentence. For example,

P = All men are mortal
P = White always wins in chess.
P = Licypriya Kangujam is an Indian activist.
P = Disha Ravi is a hero.
P = The Earth is round.
P = The Earth is flat.
P = The tomato is a vegetable.
P = The tomato is a fruit.

The PL language $\mathcal{L}$ is defined as follows, where α and β are propositional variables that stand for any sentence of $\mathcal{L}$.

1. If α is an atomic sentence, then α is a sentence in $\mathcal{L}$.
2. If α is a sentence in $\mathcal{L}$, then $\neg \alpha$ is a sentence in $\mathcal{L}$.
3. If α and β are sentences in $\mathcal{L}$ and o is a binary connective, then $(\alpha \circ \beta)$ is a sentence in $\mathcal{L}$. Wherever there is no ambiguity, we can dispense with the parentheses and write it as $\alpha \circ \beta$.
4. Nothing else, except for the definitions 1–3, is a sentence in $\mathcal{L}$.

The set of sentences in $\mathcal{L}$ thus includes all the atomic sentences in the vocabulary and all sentences constructed using the logical connectives as defined above.

The truth values of atomic sentences are defined by a *valuation function V* which maps every atomic sentence to a two element set usually denoted by {true, false} or {T, F} or {1, 0}. For the rest of the formulas, the truth values are defined by structural induction. We as users associate these two values to the sentences being true or false. The following cases apply for the connectives described above:

- NEGATION $\neg$: If $V(\alpha) = $ true then $V(\neg \alpha) = $ false, else $V(\neg \alpha) = $ true.
- AND $\wedge$: If $V(\alpha) = $ *true* and $V(\beta) = $ *true* then $V(\alpha \wedge \beta) = $ *true*, else $V(\alpha \wedge \beta) = $ *false*.
- NAND $\uparrow$: If $V(\alpha) = $ *true* and $V(\beta) = $ *true* then $V(\alpha \uparrow \beta) = $ *false*, else $V(\alpha \uparrow \beta) = $ *true*.
- OR $\vee$: If $V(\alpha) = $ *false* and $V(\beta) = $ *false* then $V(\alpha \vee \beta) = $ *false*, else $V(\alpha \vee \beta) = $ *true*.
- NOR $\downarrow$: If $V(\alpha) = $ *false* and $V(\beta) = $ *false* then $V(\alpha \downarrow \beta) = $ *true*, else $V(\alpha \downarrow \beta) = $ *false*.
- IMPLIES $\supset$: If $V(\alpha) = $ *true* and $V(\beta) = $ *false* then $V(\alpha \supset \beta) = $ *false*, else $V(\alpha \supset \beta) = $ *true*. Here α is the antecedent and β is the consequent of the implication. In the literature, the symbols $\rightarrow$ and $\Rightarrow$ are also used in place of $\supset$.
- EQUIVALENCE $\equiv$: If $V(\alpha) = V(\beta)$ then $V(\alpha \equiv \beta) = $ *true* else $V(\alpha \equiv \beta) = $ *false*.
- XOR $\oplus$: If $V(\alpha) = V(\beta)$ then $V(\alpha \oplus \beta) = $ *false*, else $V(\alpha \oplus \beta) = $ *true*.

NAND is a short form for NOT-AND, and NOR is a short form for NOT-OR. XOR is short for EXCLUSIVE-OR and says that exactly one of its constituents is true, which is different from the inclusive OR where at least one is true. As can be seen, the following pairs of connectives are negations of each other: $\{\wedge, \uparrow\}, \{\vee, \downarrow\}, \{\equiv, \oplus\}$.

The truth value of combined statements hence depends upon the truth values of the constituents as well as the logical connectives used. The properties of the connectives described above are often illustrated by truth tables.

10.1.1 Truth tables

If the valuation function for atomic sentences is given, then the truth value for any arbitrary sentences can be determined by recursively computing the truth values of its constituent sentences as described above. This can also be done by constructing a *truth table*.

A truth table enumerates all permutations of the truth values propositional variables can take, one in each row. If there are N variables, then the truth table has 2^N rows. Moving inside out in the compound sentence, a column is added for every logical connective, and the corresponding truth value for each row is computed. The last column corresponds to the sentence whose truth value is desired.

Figure 10.1 shows two truth tables for the two formulas $((\alpha \land (\alpha \supset \beta)) \supset \beta)$ and $((\beta \land (\alpha \supset \beta)) \supset \alpha)$. Since there are two propositional variables in each, we have four rows, and since there are three connectives, three more columns are added.

α	β	$(\alpha \supset \beta)$	$(\alpha \land (\alpha \supset \beta))$	$((\alpha \land (\alpha \supset \beta)) \supset \beta)$
true	true	true	true	true
false	true	true	false	true
true	false	false	false	true
false	false	true	false	true

α	β	$(\alpha \supset \beta)$	$(\beta \land (\alpha \supset \beta))$	$((\beta \land (\alpha \supset \beta)) \supset \alpha)$
true	true	true	true	true
false	true	true	true	false
true	false	false	false	true
false	false	true	false	true

Figure 10.1 A truth table for a sentence with N variables and C connectives has 2^N rows and $N+C$ columns. The last column indicates when the sentence is *true*. The truth table on top shows that $((\alpha \land (\alpha \supset \beta)) \supset \beta)$ is always *true*, while the one below shows that $((\beta \land (\alpha \supset \beta)) \supset \alpha)$ is *false* when α is false and β is true.

As can be seen, the first sentence is always true whatever the truth values of α and β. Such sentences are called *tautologies*. Tautologies are of considerable importance since deduction is based on tautological sentences, like the first one in Figure 10.1.

If δ is a tautology, then $\neg \delta$ is a *contradiction* or is *unsatisfiable*. An unsatisfiable sentence is *false* whatever the truth values of the constituents.

The second sentence in the figure is an example of a *contingency*. A contingency is a sentence that is *true* for some valuations and *false* for others.

The set of all sentences of PL is partitioned into three sets – tautologies, contingencies, and contradictions. The reader is encouraged to ponder over the fact that all three sets are infinite, since a new sentence can always be constructed from old ones with any logical connective. Moreover, an injection can be constructed from any one to any other. This is because given a sentence from any set, a sentence in any other set can be produced. So, in this sense, the cardinalities of the three sets are the same (if you can compare the sizes of infinite sets).

The set of *satisfiable* sentences is the union of the sets of tautologies and contingencies. A sentence is said to be satisfiable if there is some valuation that makes the sentence true.

Constructing the truth table to determine the truth values of formulas is not practical. There are several reasons for this. First, the size of the table grows exponentially with the number of variables. Second, the method does not extend to richer logics, for example, the first order logic we look at later. And third, the most important of all, the valuation of the atomic formulas is rarely given to us. Instead, what we have is a set of formulas, not necessarily atomic, that are given as *true*. We will refer to this set as the knowledge base (KB). The KB can be the set of axioms that are always provably true or a set of premises that we accept as true.

The question then is: given a KB that is true, and given a query sentence α, is α true as well?

10.2 Entailment and Proof

Logic is the study of consistent arguments in the quest for truth. Absolute truth has been a goal professed by philosophers but that is elusive. Instead, logic focuses on conditional truth or relative truth. Given a body of knowledge held by the knower, what *else* can the knower conclude? Or equivalently, given a KB made up of true sentences in a given language, and a query sentence α in the same language, is α true? Is α a logical consequence of the KB?

Consider the following simple KB:

> Mahsa is a girl. Mahsa either likes to sing or she likes to fight. Mahsa does not like to fight.

Assuming the above sentences are true, is the following also true?

> Mahsa likes to sing.

In PL, this might be encoded as follows:

G = Mahsa is a girl.
S = Mahsa likes to sing.
F = Mahsa likes to fight.

Then, $KB = \{G, (S \oplus F), \neg F\}$
and Goal = S

We say that a KB is *true* if every sentence in the KB is true. The notion of entailment embodies the connection between the truth values of the KB and the goal α. Given a true KB, if a sentence α is *necessarily true*, we say that the sentence α is *entailed* by the KB. Informally we also say that α is *true*. This is expressed as

$$KB \models \alpha$$

Entailment is a *semantic* notion. It looks at truth values but does not provide us with a procedure to determine whether α is true. Instead, we turn to the *syntactic* notion of proof.

A proof procedure aims to *derive* the goal α from the KB via a sequence of derivation steps. This process is also known as *theorem proving*. A theorem is a true statement, and the term is used extensively in mathematics, where theorems are true once and for all. Each derivation step employs a *rule of inference* that allows a new sentence to be *added* to the KB. For example, the well known rule *modus ponens* (MP), which says that if the KB has sentences matching β and $(\beta \supset \alpha)$, then one can add α to the KB. This can be written as

$$MP: \beta, (\beta \supset \alpha) \vdash \alpha$$

It is also common to write the rule as a table of rows.

$$\beta$$
$$\underline{\beta \supset \alpha}$$
$$\therefore \alpha$$

The rule MP says that one can derive α from β and $(\beta \supset \alpha)$. Here β and α are *patterns* that can syntactically match *any* sentences in the language. If the implication statement and the antecedent of the implication are present, then one can add the consequent. Observe that we are no longer talking about truth values but only manipulating symbols structures.

If a sentence α can be derived from a KB by applying a sequence of rules, then we write

$$KB \vdash \alpha$$

We also say that α can be proved given the KB, or that we have deduced α from the KB, and the process is known as *deduction*. The set of starting sentences and the sequence of intermediate sentences leading up to the goal α constitute the *proof* of α. Finding the proof is the stuff mathematicians engage in, sometimes for long periods of time. Remember Fermat's Last Theorem? In our modern times we can deploy the computing power available to us to search for a proof. This is possible because deriving a proof is a purely syntactic activity. This chapter gives us a glimpse into this process.

We now have two different notions. The notion of entailment, which talks of truth, the subject of our interest, and the notion of proof, which is purely syntactic. Proofs, and programs to find proofs, would be useful only if the derived sentences are also entailed. There are two properties in logic that pertain to this issue.

10.2.1 Soundness

A logic is said to be *sound* if every sentence we can derive is a true sentence. Or every provable sentence is true (given that the KB is true).

Soundness: If $KB \vdash \alpha$ then $KB \vDash \alpha$

The soundness property of a logic system reflects on the rules of inference used in the logic. A rule of inference is *sound* or *valid* if it is based on a tautological implication. For example, *modus ponens* is sound because it is based on $((\alpha \land (\alpha \supset \beta)) \supset \beta)$, which is a tautology as shown in Figure 10.1. In our earlier definition we had had said that we derive α from β and $(\beta \supset \alpha)$. This should not cause any confusion because α and β are variables and can stand for any sentence. In fact, the following two are logically equivalent:

$$((\alpha \land (\alpha \supset \beta)) \supset \beta) \equiv ((\beta \land (\beta \supset \alpha)) \supset \alpha)$$

The reader is encouraged to construct a truth table for the above equivalence and verify that it is indeed a tautology. In fact, tautological equivalences are the basis for rules of substitution, where the left hand side can replace the right hand side and vice versa. This is not surprising because the equivalence is essentially a biconditional as the following sentence depicts:

$$((\alpha \equiv \beta) \equiv ((\alpha \supset \beta) \land (\beta \supset \alpha)))$$

The equivalence relation is really two implication statements and, if it is a tautology, the corresponding rule of substitution is essentially two rules of inference. The following are some common rules of inference. In addition, an arbitrary number of derived rules can be devised based on tautological implications (Manna, 1974; Stoll, 1979; Smullyan, 2009).

Modus ponens:	$\beta, (\beta \supset \alpha) \vdash \alpha$
Modus tollens:	$\neg \alpha, (\beta \supset \alpha) \vdash \neg \beta$
Conjunction:	$\alpha, \beta \vdash \alpha \wedge \beta$
Addition:	$\alpha \vdash \alpha \vee \beta$
Simplification:	$\alpha \wedge \beta \vdash \alpha$
Hypothetical syllogism:	$(\alpha \supset \beta), (\beta \supset \delta) \vdash \alpha \supset \delta$
Disjunctive syllogism:	$(\alpha \vee \beta), \neg \alpha \vdash \beta$
Constructive dilemma:	$((\alpha \supset \beta) \wedge (\gamma \supset \delta)), (\alpha \vee \gamma) \vdash \beta \vee \delta$
Destructive dilemma:	$(\alpha \supset \beta) \wedge (\gamma \supset \delta), (\sim\beta \vee \sim\delta) \vdash \sim\alpha \vee \sim\gamma$

The process of proving a given sentence is called theorem proving. Here theorem refers to a statement that is true. And in classical or mathematical logics, the truth values of sentences do not change with time. Remember the Pythagoras theorem?

In the most intuitive form of the algorithm the theorem prover picks a rule with matching antecedents in the KB and adds the consequent to the KB. This process is entirely syntactic in nature, based only on pattern matching. Let α = 'The Earth is round' and β = 'Roses are red'; the sentence $(\alpha \wedge \beta)$ stands for 'The Earth is round and roses are red'. If it is given that this (compound) sentence is true, then the rule *simplification* allows one to conclude α = 'The Earth is round'. But what about β = 'Roses are red'? That should follow logically too. But simple pattern matching allows only the first component α of $(\alpha \wedge \beta)$ to be inferred. Clearly one should be able to infer β too. Here is when the rules of substitution come into effect. One such rule says that $(\alpha \wedge \beta)$ is equivalent to $(\beta \wedge \alpha)$ and either one can replace the other. This allows us to add $(\beta \wedge \alpha)$ to the KB and then infer β using *simplification*. These kinds of situations abound in formal proofs. The following are some commonly used rules of substitution (Manna, 1974; Stoll, 1979; Smullyan, 2009):

$(\alpha \vee \text{true}) \equiv \text{true}$	domination
$(\alpha \vee \text{false}) \equiv \alpha$	identity
$(\alpha \wedge \text{true}) \equiv \alpha$	identity
$(\alpha \wedge \text{false}) \equiv \text{false}$	domination
$(\alpha \wedge \neg \alpha) \equiv \text{false}$	contradiction
$(\alpha \vee \neg \alpha) \equiv \text{true}$	tautology
$\alpha \equiv \neg(\neg \alpha)$	double negation
$\alpha \equiv (\alpha \vee \alpha)$	idempotence of $\vee$
$\alpha \equiv (\alpha \wedge \alpha)$	idempotence of $\wedge$
$(\alpha \vee \beta) \equiv (\beta \vee \alpha)$	commutativity of $\vee$
$(\alpha \wedge \beta) \equiv (\beta \wedge \alpha)$	commutativity of $\wedge$
$(\alpha \wedge (\beta \vee \alpha)) \equiv \alpha$	absorption

$(\alpha \lor (\beta \land \alpha)) \equiv \alpha$	absorption
$((\alpha \lor \beta) \lor \gamma) \equiv (\alpha \lor (\beta \lor \gamma))$	associativity of $\lor$
$((\alpha \land \beta) \land \gamma) \equiv (\alpha \land (\beta \land \gamma))$	associativity of $\land$
$\neg(\alpha \lor \beta) \equiv (\neg\alpha \land \neg\beta)$	DeMorgan's Law
$\neg(\alpha \land \beta) \equiv (\neg\alpha \lor \neg\beta)$	DeMorgan's Law
$(\alpha \land (\beta \lor \gamma)) \equiv ((\alpha \land \beta) \lor (\alpha \land \gamma))$	distributivity of $\land$ over $\lor$
$(\alpha \lor (\beta \land \gamma)) \equiv ((\alpha \lor \beta) \land (\alpha \lor \gamma))$	distributivity of $\lor$ over $\land$
$(\alpha \supset \beta) \equiv (\neg\beta \supset \neg\alpha)$	contrapositive
$(\alpha \supset \beta) \equiv (\neg\alpha \lor \beta)$	implication
$(\alpha \equiv \beta) \equiv ((\alpha \supset \beta) \land (\beta \supset \alpha))$	equivalence
$((\alpha \land \beta) \supset \gamma) \equiv (\alpha \supset (\beta \supset \gamma))$	exportation
$((\alpha \supset \beta) \land (\alpha \supset \neg\beta)) \equiv \neg\alpha$	absurdity

The reader is encouraged to verify that each of these rules is sound, by verifying that these are tautologies.

10.2.2 Completeness

A logic must be sound if the sentences it derives are to be believed. But to be useful it must also be able to produce all sentences that are entailed by the KB. This property is called completeness.

Completeness: If KB $\models \alpha$ then KB $\vdash \alpha$

Completeness proofs can be hard and are beyond the scope of this book. However, this is a laudable quest and we will illustrate it to some extent when we talk of proof methods in first order logic. Friedrich Ludwig Gottlob Frege (1848–1925) was a German mathematician, logician, and philosopher who worked at the University of Jena.[1] In 1879 he gave the first axiomatization of propositional calculus that was both sound and complete (Frege, 1879). He showed that all tautologies that can be expressed in the language of propositional logic can be derived from the following set of six axioms and one rule of inference, MP:

1. THEN-1 $\alpha \supset (\beta \supset \alpha)$
2. THEN-2 $(\alpha \supset (\beta \supset \gamma)) \supset ((\alpha \supset \beta) \supset (\alpha \supset \gamma))$
3. THEN-3 $(\alpha \supset (\beta \supset \gamma)) \supset (\beta \supset (\alpha \supset \gamma))$
4. FRG-1 $(\alpha \supset \beta) \supset (\neg\beta \supset \neg\alpha)$
5. FRG-2 $\neg\neg\alpha \supset \alpha$
6. FRG-3 $\alpha \supset \neg\neg\alpha$

Remember that α, β, and γ are propositional variables and can match any sentence. In an axiomatic system only the axioms are to be taken for granted. Any other true statement needs a proof. Even the seemingly obvious sentence $(P \supset P)$, where P is a propositional symbol.

---

[1] https://plato.stanford.edu/entries/frege/, accessed November 2022.

First, we introduce a derived rule of inference. A derived rule of inference is like a macro call that serves as a short cut. The rule we are interested in is *hypothetical syllogism* (HS) and is derived as follows:

1. $(\beta \supset \gamma)$ Premise
2. $(\beta \supset \gamma) \supset (\alpha \supset (\beta \supset \gamma))$ Then-1 (after appropriate substitution)
3. $(\alpha \supset (\beta \supset \gamma))$ MP, 1, 2
4. $(\alpha \supset (\beta \supset \gamma)) \supset ((\alpha \supset \beta) \supset (\alpha \supset \gamma))$ Then-2
5. $(\alpha \supset \beta) \supset (\alpha \supset \gamma)$ MP, 3, 4
6. $(\alpha \supset \beta)$ Premise
7. $(\alpha \supset \gamma)$

We have shown that given the two premises $(\alpha \supset \beta)$ and $(\beta \supset \gamma)$, we can derive $(\alpha \supset \gamma)$

$$\text{Hypothetical syllogism: } (\alpha \supset \beta), (\beta \supset \gamma) \vdash (\alpha \supset \gamma)$$

Now we can prove $(P \supset P)$.

1. $P \supset \neg\neg P$ FRG-3 (substituting P for α)
2. $\neg\neg P \supset P$ FRG-2
3. $P \supset P$ HS, 1, 2

The deduction theorem says that a given set of premises A, B, C, and D entail a conclusion E if and only if $((A \land B \land C \land D) \supset E)$ is a tautology.

$$\text{Deduction theorem: } A, B, C, D \vDash E \text{ iff } \vDash ((A \land B \land C \land D) \supset E)$$

The fact that Frege's axiomatic system is complete means that if $((A \land B \land C \land D) \supset E)$ is a tautology, then there is a proof for the formula. Thus, wherever the given premises A, B, C, and D entail a conclusion E, a proof can be found.

Frege's axiomatic system handles only two operators $\{\neg, \supset\}$ and only one rule of inference. Students of logic would recall that anything that can be expressed with any of the 16 binary operators and negation can be expressed with the set $\{\neg, \supset\}$. Only that the representations can tend to blow up in size. For example, we know that the connective $\land$ can be replaced as follows: $(\alpha \land \beta)$ is equivalent to $\neg\neg(\alpha \land \beta)$, which is equivalent to $\neg(\neg\alpha \lor \neg\beta)$, which is equivalent to $\neg(\alpha \supset \neg\beta)$. Also, given the definition of equivalence as two implications, we can rewrite $(\alpha \equiv \beta)$

$$(\alpha \equiv \beta) \equiv (\alpha \supset \beta) \land (\beta \supset \alpha)$$
$$\equiv \neg(\alpha \supset \beta) \supset \neg(\beta \supset \alpha)$$

In practice, we tend to use more logical connectives, which also require more rules of inference to make inferences, some of which have been mentioned in Section 10.2.1. The question of which combination of logical connectives yields a complete proof system has to be considered before choosing one. We will look at one such system later in the chapter. For now, we turn our attention to a more expressive language than PL, in which we can peer inside atomic sentences and view them as relations between elements in a domain.

10.3 First Order Logic

Propositional logic is pure logic. It is pure form. It is not concerned with content, which is the user's business. It is only concerned with how logical connectives determine the truth values of compound sentences. It also treats each atomic sentence as indivisible. For a conclusion to be derivable, it must be explicitly connected to sentences in the KB.

First order logic (FOL), also known as *predicate logic*, adds an element of knowledge representation to logic, even though the logicians may shy away from admitting it. It looks inside an atomic sentence and captures relations between *entities*. Entities are ontologically different from sentences. While a sentence is something that can be true or false, an entity or an element stands by itself. A symbol for an entity usually stands for a member of some domain. The relations between entities are expressed by encapsulating them under the banner of a predicate. It is the predicate along with the entities it talks about that now becomes an atomic sentence. But what makes FOL a game changer is that it can talk of many individuals in one breath. Take, for example, the iconic sentence from the logic of the Greek philosophers: *All men are mortal*.[2] As an atomic sentence the only thing it can connect to is another sentence like (*all men are mortal* ⊃ *no man lives forever*), but then the only conclusion one can draw is the consequence that no man lives forever. However, when seen as a universal relation between the categories *men* and *mortal*, one can argue that if any individual belongs to the first category then so does he belong to the second. This is the classic Socratic argument.

> All men are mortal
>
> Socrates is a man
>
> Therefore, Socrates is mortal

This is one of the many syllogisms in Aristotelian logic. A syllogism is a pattern of valid reasoning. It is like the rules of inference we have discussed earlier. The validity of the conclusion is based on the *form* of the argument, and not on the content of the constituent sentences. We say that logic is *formal logic*. The following argument has identical form, and the conclusion is equally valid:

> All children are full of life
>
> Mahsa is a child
>
> Therefore, Mahsa is full of life

The Greek syllogism was close to natural language. Modern logic adopts a more formal representation. As in the case of any formal logic, there are two parts to FOL. One, the syntax of the language itself and, two, the semantics. The semantics itself has two facets. One, denotational, which is concerned with the meaning of sentences. The other, truth functional, which deals with truth values. Instead of first describing the formal language and then its semantics, we will adopt a more informal approach weaving through both.

---

[2] Even women are.

Every FOL has a domain independent part or logical part whose vocabulary is the following (Fitting, 2013):

- Symbols that stand for connectives or operators: '$\wedge$', '$\vee$', '$\neg$', and '$\supset$'...
- Brackets: '(', ')', '{', '}'...
- The constant symbols: '$\bot$' and '$\top$'.
- A countable set of variable symbols: $\mathcal{V} = \{v_1, v_2, v_3, ...\}$ or $\{x, y, z, x_1, y_1, z_1, ...\}$
- Quantifiers: '$\forall$' read as 'for all', and '$\exists$' read as 'there exists'. The former is the *universal quantifier* and the latter the *existential quantifier*.
- The symbol '=' read as 'equals'. This is optional.

The domain specific part of the language $\mathcal{L}(\mathcal{R}, \mathcal{F}, \mathcal{C})$ is defined by three sets $\mathcal{R}$, $\mathcal{F}$, and $\mathcal{C}$. $\mathcal{R}$ is a set of relation or predicate symbols, $\mathcal{F}$ is a set of function symbols, and $\mathcal{C}$ is a set of constant symbols. An interpretation $\mathcal{I} = <D, \mathbb{I}>$ specifies a domain D for the language $\mathcal{L}(\mathcal{R}, \mathcal{F}, \mathcal{C})$ and a mapping $\mathbb{I}$ from each of $\mathcal{R}$, $\mathcal{F}$, and $\mathcal{C}$ to elements of D. In addition, an assignment $\mathbb{A}$ maps every variable in $\mathcal{V}$ to the domain D. Like in PL, the interpretation determines which sentences are true. In FOL it also determines what the expressions in the language stand for, at least the structural nature of the relation in set theoretic terms.

We first describe how the elements or entities in a domain are represented.

10.3.1 Terms and domains

The set of terms $\mathcal{T}$ in the language are defined as follows:

- If $t \in \mathcal{V}$, then $t \in \mathcal{T}$.
- If $t \in \mathcal{C}$, then $t \in \mathcal{T}$.
- If $t_1, t_2, ..., t_N \in \mathcal{T}$ and $f \in \mathcal{F}$ is an *n*-place function symbol, then $f(t_1, t_2, ..., t_N) \in \mathcal{T}$.

Terms are made up of variable symbols, constant symbols, or recursively defined using the function symbols. Each term of $\mathcal{L}(\mathcal{R}, \mathcal{F}, \mathcal{C})$ refers to an element in the domain. The mapping from the set of terms $\mathcal{T}$ to the elements of a chosen domain D is given by

- Every *n*-place function symbol f is mapped under $\mathbb{I}$ as follows: $\mathbb{I}(f) = f^{\mathbb{I}}$, where $f^{\mathbb{I}}$ is the image of f and is an *n*-ary function $f^{\mathbb{I}}: D^N \to D$.
- If $t \in \mathcal{V}$, then $t^{\mathbb{I}\mathbb{A}} = t^{\mathbb{A}}$. Every variable is mapped by the assignment $\mathbb{A}$.
- If $t \in \mathcal{C}$, then $t^{\mathbb{I}\mathbb{A}} = t^{\mathbb{I}}$. Every constant is mapped by the mapping $\mathbb{I}$.
- If $t_1, t_2, ..., t_N \in \mathcal{T}$ and $f \in \mathcal{F}$, then $f(t_1, t_2, ..., t_N)^{\mathbb{I}\mathbb{A}} = f^{\mathbb{I}}(t_1^{\mathbb{I}\mathbb{A}}, t_2^{\mathbb{I}\mathbb{A}}, ..., t_N^{\mathbb{I}\mathbb{A}})$.

Every *n*-place *function symbol* maps to an *n*-ary *function* in the domain. Variables are mapped to some element by the assignment $\mathbb{A}$, constant symbols are mapped to specific elements in the domain as specified by $\mathbb{I}$, and terms using function symbols use a combination of $\mathbb{A}$ and $\mathbb{I}$ as needed.

Given the domain of natural numbers, a variable x may map to some number, say 11, given an assignment $\mathbb{A}$. A constant *zero* or *sifar* or *0* may map to the number 0. We generally overload the use of a numeral to stand for both the *constant in the language* and *the number in the domain.* Thus, the constant symbol 7 stands for the number 7. Then *sum*(7,11) will map to the element 18, and *successor*(7) maps to the number 8.

Likewise in the domain of people, we use the name of a person both as a constant symbol in the language and as the person in the domain. For example, Mahsa, Zhina, Nika, Sarina, Hadis, and Neda *may* stand for persons having the said names. Then *father*(*Sarina*) would stand for the individual who is the father of Sarina. Remember that mathematically *father* is a 1-place function. Where there is ambiguity, we may suffix a name with a number. For example, the constants Nika16 and Nika21 refer to potentially two different individuals. The mapping could still map them to the same person, in which case the two symbols become aliases.

When writing programs we often adopt some convention to distinguish the sets $\mathcal{V}$ and $\mathcal{C}$. The convention we will follow is to use lower case words as constant and function symbols, and prefix a word with a '?' if it is a variable. Thus, variable names need not be restricted to the symbols *x, y*, and similar mnemonic, and a programmer may use variable names such as ?count, ?age, and ?girl also as variable names. The programming language Prolog adopts the convention of using capitalized words for variables and lower case for constants.

In an interpretation $\mathcal{I} = <D, \mathbb{I}>$, a term of $\mathcal{L}(\mathcal{R}, \mathcal{F}, \mathcal{C})$ points to an element of the domain D.

10.3.2 Atomic formulas

Atomic formulas are relations on the domain. They specify which entities are related to which. The set of atomic formulas $\mathcal{A}$ is defined as follows:

- If $P \in \mathcal{R}$ is an *n*-place predicate and $t_1, t_2, ..., t_N \in \mathcal{T}$ are N terms, then $P(t_1, t_2, ..., t_N) \in \mathcal{A}$.
- If $t_1, t_2 \in \mathcal{T}$ are terms, then $(t_1 = t_2) \in \mathcal{A}$.
- The constant symbols $\bot$ and $\top$ are also atomic formulas.

An *n*-place predicate $P \in \mathcal{R}$ is mapped by the interpretation $\mathbb{I}$ to an *n*-ary relation $P^{\mathbb{I}}$ in the domain $P^{\mathbb{I}} \subseteq D \times D ... \times D$ or $P^{\mathbb{I}} \subseteq D^N$.

Logicians tend to employ the symbols P, Q, R and P_1, P_2, P_3 as elements of $\mathcal{R}$, whereas AI researchers interested in representation use names such as Man, Mortal, Brother, Friend, and LessThan. Names which make sense to *us*. The *meaning* of an atomic formula is provided externally by the interpretation. For example, the formula Loves(X, Y)[3] where Loves is a predicate symbol in $\mathcal{R}$ could mean that *X loves Y* but it could also mean *X hates Y*. The same would be the case if we had said P(X, Y). What it does say is that there is a relation between two elements *X* and *Y* in the domain, which is captured by the mathematical relation Loves$^{\mathbb{I}}$ which is a subset of $D \times D$. It is *this* set that defines the relation between the two elements. The *name* of a predicate is only incidental. As Shakespeare said, a rose by any other name would smell as sweet. We can extend that to the meaning of sweet itself, which is defined in set theoretic terms as a subset of the domain *D* containing all things sweet. In this context, we tend to conflate our notions of *category* and *property* into unary predicates. *We* think of Man(X) and Rose(X) as categories and Mortal(X) and Sweet(X) as properties. But in FOL they are just the corresponding subsets of the domain.

---

[3] We could just as well have written P(X, Y).

An atomic formula is the smallest unit that can be assigned a truth value. The truth value is determined by the interpretation $\mathcal{I} = <D, \mathbb{I}>$. For propositional symbols $\{P, Q, R...\}$ the interpretation is simply a truth assignment by a *valuation function V*. In the case of FOL, an atomic formula is true if the corresponding relation holds in the domain D and interpretation $\mathbb{I}$. We can define a similar valuation function $Val: \mathcal{A} \rightarrow \{\text{true}, \text{false}\}$ as follows:

$Val(P(t_1,t_2, ..., t_N)^{\mathbb{I}\mathbb{A}}) = \text{true}$ iff $<t_1^{\mathbb{I}\mathbb{A}}, t_1^{\mathbb{I}\mathbb{A}}, ..., t_1^{\mathbb{I}\mathbb{A}}> \in P^{\mathbb{I}}$

$Val(t_1 = t_2) = \text{true}$ iff $t_1^{\mathbb{I}\mathbb{A}} = t_1^{\mathbb{I}\mathbb{A}}$

$Val(\top) = \text{true}$

$Val(\bot) = \text{false}$

The following examples use names of predicates and functions familiar to us:

- Val(Friend(Sakshi, Vinesh)) = true iff <Sakshi, Vinesh> ∈ Friends, where Friend$^{\mathbb{I}}$ = Friends.
- Val(Friend(father(Nika), Hadis)) is true if Nika's father is a friend of Hadis.
- Val(LessThan(7, 8)) = true if <7,8> ∈ LessThan$^{\mathbb{I}}$, the set of pairs in natural numbers where the first element is smaller than the second element.
- Val(LessThan(8, 7)) = false because <8,7> does not belong to LessThan$^{\mathbb{I}}$.
- Val(JacindaArden = pm(NZ)) = true because Jacinda Arden and the prime minister (pm) of New Zealand (NZ) are the same person.
- Val(sum(2, 2) = 5) is false because 2+2 stands for 4 and not 5.
- Val(sum(3, X) = Y) is not known unless the assignment 𝔸 is specified. But as we will see later, this formula can be a goal or a query where we ask if there are values of X and Y such that the sentence is true.

When we include atomic formulas of the kind $(t1 = t2)$, we say the language is FOL *with equality*. Then, as we will see later, we need to add some additional axioms for the logic to be complete.

When talking about truth values we often leave out the valuation function and make assertions like Friend(Hadis, Neda) = true. This is somewhat informal and also incorrect (because only terms can be equal), but since there is no ambiguity it is often used as a short form. We can write this correctly as Friend(Hadis, Neda) ≡ ⊤, where two atomic formulas are connected by a logical connective.

10.3.3 Quantifiers, formulas, and sentences

When we want to state that *all* men are mortal, or *all* roses are sweet, we need to take recourse to quantifiers. There are two commonly used quantifier symbols as described in the syntax of FOL earlier. They are the universal quantifier ∀, read as 'for all', and the existential quantifier ∃, read as 'there exists'. Both quantifiers quantify one variable, for example, $\forall x(\alpha)$ and $\exists x(\alpha)$, respectively, and the quantification applies to the formula α in the *scope* of the quantifier.

We now define the set of formulas $\mathcal{F}$ of a language $\mathcal{L}(\mathcal{R}, \mathcal{F}, C)$ for some common binary connectives (Fitting, 2013).

- If $\alpha \in \mathcal{A}$, then $\alpha \in \mathcal{F}$.
- If $\alpha \in \mathcal{F}$, then $\neg(\alpha) \in \mathcal{F}$. Where there is no ambiguity, we can write $\neg\alpha \in \mathcal{F}$.
- If $\alpha, \beta \in \mathcal{F}$, then $(\alpha \wedge \beta) \in \mathcal{F}$. Where there is no ambiguity, we can write $\alpha \wedge \beta \in \mathcal{F}$.
- If $\alpha, \beta \in \mathcal{F}$, then $(\alpha \vee \beta) \in \mathcal{F}$. Where there is no ambiguity, we can write $\alpha \vee \beta \in \mathcal{F}$.
- If $\alpha, \beta \in \mathcal{F}$, then $(\alpha \supset \beta) \in \mathcal{F}$. Where there is no ambiguity, we can write $\alpha \supset \beta \in \mathcal{F}$.
- If $\alpha, \beta \in \mathcal{F}$, then $(\alpha \equiv \beta) \in \mathcal{F}$. Where there is no ambiguity, we can write $\alpha \equiv \beta \in \mathcal{F}$.
- If $\alpha, \beta \in \mathcal{F}$, then $(\alpha \supset \beta) \in \mathcal{F}$. Where there is no ambiguity, we can write $\alpha \supset \beta \in \mathcal{F}$.
- If $\alpha \in \mathcal{F}$ and $x \in \mathcal{V}$, then $\forall x(\alpha) \in \mathcal{F}$. We often use $\forall x,y(\alpha)$ as a short form for $\forall x(\forall y(\alpha))$.
- If $\alpha \in \mathcal{F}$ and $x \in \mathcal{V}$, then $\exists x(\alpha) \in \mathcal{F}$. We often use $\exists x,y(\alpha)$ as a short form for $\exists x(\exists y(\alpha))$.

Let us look at some examples of quantified formulas, with meaningful (to us) predicate, function, and constant names.

A. $\forall x(\text{Man}(x) \supset \text{Mortal}(x))$ says that all men are mortal.
B. $\forall x(\text{Rose}(x) \supset \text{Red}(x))$ says that all roses are red. We have not mentioned colour here. One way to do so would be to write it as $\forall x(\text{Rose}(x) \supset \text{Colour}(x, \text{red}))$. But then what element does the term *red* refer to? One way around this is to treat 'red' as a *reified* element, which can then be related to the roses. Then the fact that roses are red can also be expressed as (colour(rose) = red) where *colour* is a function that maps its argument to its colour.
C. $\exists x(\text{Man}(x) \wedge \text{Mortal}(x))$ says that *some* men are mortal. Observe the different connective used here. We are saying that there is someone who is both a man and a mortal.
D. $\exists x(\text{Rose}(x) \wedge \text{Gave}(\text{Zhina}, \text{Neda}, x))$ says that Zhina gave Neda a rose. All we know about the gift is that it is a rose. If we wanted to say that Zhina gave Neda the rose, then we would express it as $(\text{Rose}(\text{rose21}) \wedge \text{Gave}(\text{Zhina}, \text{Neda}, \text{rose21}))$ where *rose21* is a named rose.
E. $\forall x(\text{Boy}(x) \supset \exists y(\text{Girl}(y) \wedge \text{Love}(x, y))$ say that every boy loves a girl. This could mean that for every boy there is a girl whom he loves. The English language can be ambiguous. If our intended meaning was that there is one particular girl, say Mahsa, whom every boy loves, we would express it as $\exists y(\text{Girl}(y) \wedge \forall x(\text{Boy}(x) \supset \text{Love}(x, y)))$.

The valuation for the formulas with the logical connectives is defined by the semantics of the connectives and is similar to the valuation in PL. The valuation for quantified formulas is defined below. An assignment $\mathbb{B}$ is said to be an *x*-variant of an assignment $\mathbb{A}$ if the two agree on all variables except x (Fitting, 2013).

- $\text{Val}(\exists x(\alpha)^{\mathbb{IA}})$ = true iff $\alpha^{\mathbb{IA}}$ is true for *some* assignment $\mathbb{B}$ that is an *x*-variant of $\mathbb{A}$. In other words, the formula α is true for some value of x.
- $\text{Val}((\forall x(\alpha))^{\mathbb{IA}})$ = true iff $\alpha^{\mathbb{A}}$ is true for *all* assignments $\mathbb{B}$ that are *x*-variants of $\mathbb{A}$. In other words, the formula α is true for all values of x.

Observe that on the right hand side of *iff* the formula is without the quantifier. Effectively, we are saying that $\exists x(\alpha)$ is true if one can find some assignment to the variable x such that the formula becomes true. For example, $\exists x(\text{Even}(x))$ is true because x can, for example, be 2. Likewise, $\forall x(\text{GreaterThan}(\text{successor}(x), x))$ is true because $x+1$ is greater than x for every x. Where there is no ambiguity, we can drop the outer brackets and write $\exists x\text{Even}(x)$ and $\forall x\text{GreaterThan}(\text{successor}(x), x)$.

Not every formula $\alpha \in \mathcal{F}$ can be assigned a truth value. In particular, formulas with *free* variables may not have a defined truth value. Consider, for example, the sentences LessThan (x, y) or $\exists x \text{LessThan}(x, y)$ or $\forall x \text{LessThan}(x, y)$. We cannot say anything about the truth values of any of these. That is because y is a free variable in all three, and x is a free variable in the first. A free variable is one that is not bound. A bound variable is one that occurs in the scope of a quantifier. In the second and third sentences, x is bound and y is free. A formula without any free variables is a *sentence* of $\mathcal{L}(\mathcal{R}, \mathcal{F}, \mathcal{C})$. A sentence can be assigned a truth value. The truth value of a sentence is not dependent on an assignment but does depend upon the interpretation $\mathcal{I} = <D, \mathbb{I}>$. For example, given the domain of natural numbers, the truth values of the following four sentences are:

- $\text{Val}(\exists x \, (\exists y \, \text{LessThan}(x, y))^{\mathbb{I}})$ = true because we can always find two number x and y such that $x < y$.
- $\text{Val}(\exists x \, (\forall y \, \text{LessThan}(x, y))^{\mathbb{I}})$ = false because there is no x such that for all y, x is less than y. Not even 0 because it is not less than itself. If the relation was *less than or equal to*, then the sentence would become true.
- $\text{Val}(\forall x \, (\forall y \, \text{LessThan}(x, y))^{\mathbb{I}})$ = false because we can find many counter-examples where $x > y$.
- $\text{Val}(\forall x \, (\exists y \, \text{LessThan}(x, y))^{\mathbb{I}})$ = false because we have a counter-example with $x = 0$.

A point to note. The name of a variable does not affect the truth value of a formula. Thus, the formula $\exists x \, (\forall y \, \text{LessThan}(x, y))$ is logically equivalent to $\exists y \, (\forall x \, \text{LessThan}(y, x))$, which is equivalent to $\exists z \, (\forall y \, \text{LessThan}(z, y))$. One can rename variables as long as the new name does not occur elsewhere in the sentence. As we will see, renaming variables *apart* in two formulas can sometimes be necessary.

10.4 Deduction in FOL

The bulk of inferences we make deal with quantified sentences. In addition to the rules from PL, we employ the following rules:

Universal instantiation (UI) allows us to infer specific sentences from general ones. This is also called $\forall$ *elimination* since the inferred formula does not have the quantifier.

$$\text{UI: } \forall x(\alpha[x]) \vdash \alpha[a]$$

We have introduced a new notation here. The formula $\alpha[x]$ is read as a formula α which has the variable x somewhere in it, and the rule UI is read as: Replace all instances of x in $\alpha[x]$ with a, where a is a term in the language. The following description of the rule is equivalent.

$$\text{UI: } \forall x\alpha \vdash \alpha\{x \mapsto a\}$$

Here are some examples of the use of UI.

- From $\forall x \text{Man}(x)$ infer Man(Socrates)
- From $\forall x(\forall y \text{Loves}(x, y))$ infer $\forall x \text{Loves}(x, \text{Nika})$
- From $\forall x(\forall y \text{Loves}(x, y))$ infer Loves(Sarina, Nika)

Generalization (G), also known as *existential generalization* or *existential introduction*, allows us to introduce the existential quantifier into a sentence, by replacing a constant with an existentially quantified variable. This is also called ∃ *introduction*.

$$G: \alpha[a] \vdash \exists x \alpha[x]$$

Some examples are

- From Even(6) infer $\exists x$Even(x)
- From Loves(Neda, Nika) infer $\exists x$Loves$(x, $Nika$)$ or infer $\exists x$Loves(Neda,$x)$
- $\forall x$Loves$(x, $Nika$)$ infer $\exists y \forall x$Loves(x, y)

It is useful to think of the quantified statements as closed form representations of longer formulas, possibly infinitely long formulas for infinite domains such as numbers. Consider a finite domain of four young women {Nika, Sarina, Hadis, Neda}. Then the following equivalences hold (assuming generalized connectives ∧ and ∨ that can connect an arbitrary number of formulas):

$\forall x$Brave$(x) \equiv ($Brave(Nika) ∧ Brave(Sarina) ∧ Brave(Hadis) ∧ Brave(Neda))

$\exists x$Brave$(x) \equiv ($Brave(Nika) ∨ Brave(Sarina) ∨ Brave(Hadis) ∨ Brave(Neda))

The first sentence $\forall x$Brave(x) says that all of them are brave, and the second one $\exists x$Brave(x) says that at least one of them is brave. Now one can relate rule *G* to *addition* ($\alpha \vdash \alpha \vee \beta$) from PL, and UI to *simplification* ($\alpha \wedge \beta \vdash \alpha$). This correspondence carries forward to DeMorgan's laws of substitution.

$$\neg \forall x\, \alpha \equiv \exists x\, \neg \alpha$$
$$\neg \exists x\, \alpha \equiv \forall x\, \neg \alpha$$

Logicians also introduce the following rules to facilitate reasoning, but that needs careful treatment.

Universal generalization or ∀ *introduction* produces a universally quantified formula from a formula with an arbitrary constant. Care must be taken that the constant name is not one of the existing constant names. When that is the case, and we start with a premise $P(k)$, where k is a new symbol, and can derive $Q(k)$, then we can introduce a tautological implication $\forall x(P(x) \supset Q(x))$. The intuition is that since k was arbitrarily chosen, the connection between $P(k)$ and $Q(k)$ must be universal.

$$UG: (\alpha[k] \vdash \beta[k]) \vdash \forall x(\alpha[x] \supset \beta[x]) \text{ where } k \notin C \text{ is an arbitrary name}$$

Existential instantiation essentially creates a new name for an unnamed entity that is known to exist.

$$EI: \exists x \alpha[x] \vdash \alpha[d] \text{ where } d \notin C$$

Now this entity can participate in a rule of inference, as shown in the example below.

The police thug: A police person murdered Mahsa. Anyone who murders someone is a murderer. All murderers must be prosecuted. Therefore, the police person must be prosecuted.

1. $\exists x(P(x) \wedge \text{Murdered}(x, \text{Mahsa}))$ Given
2. $P(\text{Bulli}) \wedge \text{Murdered}(\text{Bulli}, \text{Mahsa})$ EI, 1
3. $\text{Murdered}(\text{Bulli}, \text{Mahsa})$ Simplification, 2
4. $\forall x, y(\text{Murdered}(x, y) \supset \text{MustBeProsecuted}(x))$ Given
5. $(\text{Murdered}(\text{Bulli},\text{Mahsa}) \supset \text{MustBeProsecuted}(\text{Bulli}))$ UI, 4
6. $\text{MustBeProsecuted}(\text{Bulli})$ MP, 3, 5
7. $P(\text{Bulli})$ Simplification 2
8. $P(\text{Bulli}) \wedge \text{MustBeProsecuted}(\text{Bulli})$ Conjunction 7, 3
9. $\exists x(P(x) \wedge \text{MustBeProsecuted}(x))$ Generalization, 8

The argument essentially goes like this. Some police person murdered Mahsa. Let that person be called Bulli. Bulli is a murderer. All murderers must be prosecuted. Bulli must be prosecuted. Therefore, some police person must be prosecuted.

This kind of reasoning can be handled naturally with *Skolemization* described later.

10.4.1 Implicit quantifier form

A quantifier identifies the nature of a variable. If one could devise a convention to identify the nature of the arguments to predicates, then one could dispense with the quantifier symbols. This would make the representation more compact, and easier to write programs with. Given the two quantifiers in our FOL language, the following convention distinguishes between the three different entities (Charniak and McDermott, 1985).

- Any name without a marker is a constant symbol, except Skolem constants discussed below. For example, Neda, Mahua, Kailash, two, 2, Chennai, Narmada, Asia, and Jupiter. In general, any unadorned string stands for a constant. This includes the ones commonly used by logicians, such as a, b, c, but also x, y, z that logicians reserve for variables. In the programming language Prolog, constants begin with lower case letters.
- A universal variable, or a universally quantified variable, is prefixed with a '?'. For example, ?Neda, ?Mahua, ?Kailash, ?two, ?2, ?Chennai, ?a, ?b, ?x, and ?y. In the Prolog language, variables are capitalized.
- An existential variable, or an existentially quantified variable, begins with the prefix 'sk' or 'sk-'. This is to ensure that it does not come from the set $\mathcal{V}$ of variables or the set $\mathcal{C}$ of constants from $(\mathcal{R}, \mathcal{T}, \mathcal{C})$. For example, sk-Neda and sk 2. These are called Skolem constants after the logician Thoralf Skolem, and the process of $\exists$ elimination is called Skolemization. We saw an example of this in the previous section.
- An existential variable in the scope of a universal quantifier is represented as a Skolem function of the universally quantified variable. For example, sk12(?x). This could have been in a formula $\forall x \exists y \text{Loves}(x,y)$ which would be Skolemized as Loves (?x, sk12(?x)). The intuition here is that y cannot be any arbitrary element and is dependent on the value of x.

Identifying the real nature of a variable has to be done with care. If one is looking at a negated quantified formula, then pushing the negation sign inside can switch the quantifier and reveal its true colours, as per DeMorgan's laws described earlier. Then if, talking about numbers, one has to represent the sentence 'no element is both odd and even', one might begin by expressing it as

$$\neg \exists x (Odd(x) \wedge Even(x)) \equiv \forall x \neg(Odd(x) \wedge Even(x))$$

The equivalent formula on the right says that 'every element is not odd and even' where the variable x is a universally quantified variable and the implicit quantifier representation should be $\neg(Odd(?x) \wedge Even(?x))$.

One must also keep in mind that the antecedent in an implication statement contains a negation because $(P \supset Q) \equiv (\neg P \vee Q)$. Then a formula of the kind $\forall x(\exists y P(y, x)) \supset Q(x))$ is equivalent to $\forall x, y(P(y, x)) \supset Q(x))$. The sentence 'every person who has a friend is happy' can be represented as

$\forall x(Person(x) \wedge \exists y Friend(y, x)) \supset Happy(x))$
$\equiv \quad \forall x \neg (Person(x) \wedge \exists y Friend(y, x)) \vee Happy(x))$
$\equiv \quad \forall x (\neg Person(x) \vee \neg \exists y Friend(y, x) \vee Happy(x))$
$\equiv \quad \forall x (\neg Person(x) \vee \forall y \neg Friend(y, x) \vee Happy(x))$
$\equiv \quad \forall x \forall y (\neg Person(x) \vee \neg Friend(y, x) \vee Happy(x))$
$\equiv \quad \forall x \forall y (\neg (Person(x) \wedge Friend(y, x)) \vee Happy(x))$
$\equiv \quad \forall x \forall y ((Person(x) \wedge Friend(y, x)) \supset Happy(x))$

So now the variable y that was ostensibly in the scope of an existential quantifier turned out to be a universal variable which reads as 'For all x and y, if x is a person and y is a friend of x, then x is happy'.

Here is another example. Consider the definition of an aunt. On the one hand, one can say that 'for all A and X, if there exists a P such that A is a female and P is the parent of X and a sibling of A, then A is the aunt of X'.

$\forall a, x [(\exists p(Female(a) \wedge Parent(p, x) \wedge Sibling(a, p)) \supset Aunt(a, x)]$
$\equiv \quad \forall a, x [\neg (\exists p(Female(a) \wedge Parent(p, x) \wedge Sibling(a, p)) \vee Aunt(a, x)]$
$\equiv \quad \forall a, x [\forall p \neg (Female(a) \wedge Parent(p, x) \wedge Sibling(a, p)) \vee Aunt(a, x)]$
$\equiv \quad \forall a, x, p[\neg (Female(a) \wedge Parent(p, x) \wedge Sibling(a, p)) \vee Aunt(a, x)]$
$\equiv \quad \forall a, x, p[(Female(a) \wedge Parent(p, x) \wedge Sibling(a, p)) \supset Aunt(a, x)]$

Here, even though P is prefixed by $\exists$, it turns out to be a universal variable. This is because there is a $\neg$ hidden in the antecedent, which when pushed in turns the quantifier into $\forall$. This sentence is Skolemized as

$(Female(?a) \wedge Parent(?p, ?x) \wedge Sibling(?a, ?p)) \supset Aunt(?a, ?x).$

Interestingly, the converse treats the variable P differently. If A is an aunt of X then A must be female and there *must exist* a related individual P who is the parent of X and a sibling of A.

$\forall a, x [Aunt(a, x) \supset (\exists p (Female(a) \wedge Parent(p, x) \wedge Sibling(a, p)))]$

In this sentence, the variable P is a Skolem function of the universal variables A and X, because its value is dependent upon the values of A and X. This would be Skolemized as

Aunt(?a, ?x) ⊃ (Female(?a) ∧ Parent(sk31(?a, ?x)) ∧ Sibling(?a, sk31(?a, ?x)))

An interesting formula, named as the drinking formula (Smullyan, 2009), is ∃x(D(x) ⊃ ∀yD(y)), which brings out the nature of a quantifier in the antecedent. The reader is encouraged to show that this formula is a tautology, and is in fact logically different from (∃xD(x) ⊃ ∀yD(y)).

10.4.2 Unification

In PL a rule of inference is applicable when one can find matching patterns in the KB. For example, the rule *disjunctive dilemma* (DS) is applicable when we have {((P⊕Q) ∨ (R⊃S), ¬(P⊕Q)} in the KB. In the rule (α ∨ β), ¬α ⊢ β the propositional variable α matches the formulas (P⊕Q) and β matches (R⊃S), to produce (R⊃S) as the conclusion. Such direct pattern matching is not possible in FOL because variables have to match and be bound to constants, or even other variables, or functions, all being terms of different hues. Matching is accomplished by the UNIFICATION algorithm, which substitutes variables with terms to make the formulas identical, enabling the rule to be applied (Charniak and McDermott, 1985). Here are some definitions first.

A *substitution* θ is a set of <variable, value> pairs each denoting the value to be substituted for the variable. When we *apply* the substitution θ to a formula α, we replace every variable from θ in α with the corresponding value from θ. The new formula is denoted by αθ.

A *unifier* for two formulas α and β is a substitution that makes the two formulas identical. We say that α *unifies* with β.

A unifier θ unifies a set of formulas {α$_1$, α$_2$, ..., α$_N$} if

$$\alpha_1\theta = \alpha_2\theta = \ldots = \alpha_N\theta = \varphi$$

We call the common reduced form φ the *factor*.

A formula α is *more general than* a formula β if there exists a non-empty substitution λ such that αλ = β. A unifier is the *most general unifier* (MGU) of two formulas if no other unifier produces a more general factor.

Consider, for example, the two formulas Loves(?x, ?y), which says that everyone loves everyone, and Loves(?z, Nika), which says that everyone loves Nika. The two can be unified with a substitution λ = {<?x, Zhina>, <?z, Zhina>} to yield the factor Loves(Zhina, Nika). They can also be unified with the substitution θ = {<?x, ?z>} to yield the factor Loves(?z, Nika). Clearly this is more general than Loves(Zhina, Nika). In fact, it is the most general factor that can be found and θ is the MGU.

An MGU is preferred because the conclusions we can draw are the most general. If we have an implication in the KB that said {(Loves(?x, ?y) ⊃ LovedBy(?y, ?x)), Loves(?z, Nika)}, then we can conclude the *Nika is loved by everyone*, which is a more general inference than *Nika is loved by Zhina*. The latter can always be concluded from the former by UI. The Unification algorithm is described below.

The objective of unification is to construct the substitution θ. Two formulas can be unified if they have the same set of predicates connected by the same set of connectives in an isomorphic manner. The task is to ensure that the terms in the formulas are made identical. This involves identifying a common value for each corresponding pair of variables in each formula. Algorithm 10.1 works on formulas in implicit quantifier form, but can easily be extended to quantifiers as well. The main function UNIFY(arg1, arg2) accepts two formulas and returns a substitution that unifies the two formulas. It does so by incrementally traversing the two formulas looking for values for variables.

Algorithm 10.1. The UNIFICATION algorithm initializes the substitution to an empty list. It then parses the formula with calls to SUBUNIFY, ATOMICUNIFY, and TERMUNIFY. The synthesis of the substitution is done in VARUNIFY.

Unify(arg1, arg2)
1. **return** SubUnify(arg1, arg2, ())

SubUnify(arg1, arg2, theta)
1. **if** arg1 and arg2 are compound and have *same logical structure*
2.   **for each** atf$_i$ **in** arg1 **and** atf$_j$ **in** arg2 **call** AtomicUnify(atf$_i$, atf$_j$, theta)
3. **else return** AtomicUnify(arg1, arg2, theta)

AtomicUnify(functor1, functor2, theta)
1. **if** functor1 **is** P($t_1, t_2, ..., t_n$) **and** functor2 is P($s_1, s_2, ..., s_n$)
2.   **or** functor1 **is** f($t_1, t_2, ..., t_n$) **and** functor2 is f($s_1, s_2, ..., s_n$)
3.   **for each** t_i **and** s_i **call** TermUnify(t_i, s_j, theta)

TermUnify(term1, term2, theta)
1. **if** term1 **and** term2 are identical constants **return** theta
2. **if** term1 is a variable **return** VarUnify(term1, term2, theta)
3. **if** term2 is a variable **return** VarUnify(term2, term1, theta)
4. **return** AtomicUnify(term1, term1, theta)

VarUnify(var, term, theta)
1. **if** var **occurs in** term **return** NIX
2. **if** variable has a value <var, alpha> in theta **return** TermUnify(term, alpha, theta)
3. **return** <var, term>: theta

The function SUBUNIFY is initialized with the empty substitution, which is incrementally populated inside the VARUNIFY function that takes a variable and a term that could possibly be assigned as a value to the variable, and does so if it is consistent to do so. Observe that terms can be arbitrarily nested and calls between ATOMICUNIFY and TERMUNIFY can handle that case. The two formulas Friend(mother(mother(?x)), mother(mother(Zhina)) and Friend(mother(?y), ?z) will make such cross functions calls.

Care must be taken to separate the variables in the two formulas apart. This needs to be done since the variables in two quantified formulas should not have the same name. Consider the two formulas Loves(?x, Sarina) and Loves(Nika, ?x). The former says that everyone loves Sarina, and the latter says that Nika loves everyone. The two should be unifiable to yield the formula Loves(Nika, Sarin). But when they use the variable name ?x, Algorithm 10.1 will fail because the algorithm will first add <?x, Nika>. Then when trying to unify Sarina with ?x it will run into Line 2 of VARUNIFY which says that since ?x already has a value Nika, unify that value with Sarina. This cannot be done because they are different constants. Instead, if the second formula were to be Loves(Nika, ?y), which asserts the same fact, then <?y, Sarina> can be added to the substitution.

The occurs check in Line 1 of VARUNIFY prohibits the variable to be present in the term that we are assigning to it as a value. Clearly then the variable inside the term would have to be substituted as well, and that could get into an infinite loop.

10.4.3 Forward chaining

In forward reasoning we move from the given KB towards the desired conclusion, choosing and applying rules that are applicable. Every time we apply a rule, we add the newly inferred sentence to the KB. This process is also known as *natural deduction* since we naturally move from the given facts to the conclusions. Let us look at the Socratic argument.

In FOL the proof would be a two step process. For MP: $\beta, (\beta \supset \alpha) \vdash \alpha$ to be applicable, the formula β must be present as the antecedent in $(\beta \supset \alpha)$. In the Socratic argument, we need the two formulas Man(Socrates) and (Man(Socrates) $\supset$ Mortal(Socrates)) in the KB to infer Mortal(Socrates). But what we have is $\forall x(\text{Man}(x) \supset \text{Mortal}(x))$. The rule UI comes to the rescue and one can add (Man(Socrates) $\supset$ Mortal(Socrates)) as an instance of the universal statement and then Man(Socrates) is

1. $\forall x(\text{Man}(x) \supset \text{Mortal}(x))$ Premise
2. Man(Socrates)   Premise
3. Man(Socrates) $\supset$ Mortal(Socrates)   UI, 1
4. Mortal(Socrates) q.e.d   MP, 2, 3

The operational difficulty here is that moving from the KB towards conclusions one has to *guess* how to apply the rule UI. When working with the rule in the implicit quantifier form, the use of unification makes guesswork unnecessary because Man(Socrates) is the one being unified with the antecedent, using *modified modus ponens* (MMP).

1. Man(?x) $\supset$ Mortal(?x)   Premise
2. Man(Socrates)   Premise
3. Mortal(Socrates) q.e.d   MMP, 2, 3, <?x, Socrates>

The two versions of the proof are contrasted in Figure 10.2. Not only does the MMP rule obviate the need for guesswork, it also results in a shorter proof.

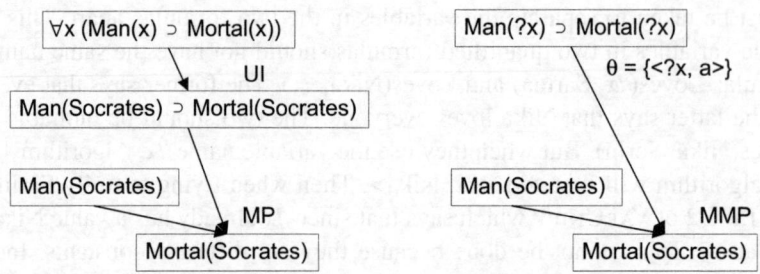

Figure 10.2 Forward chaining in FOL is a *two step* process with UI followed by MP as shown on the left. With the use of *implicit quantifier notation*, this collapses into a one step inference using the rule MPP as shown on the right. Moreover, one does not have to guess the value of *x* in the UI step.

The phrase 'forward *chaining*' alludes to the fact that a sequence of rules are instrumental in chaining the facts to the goal. This is most easily visualized with rules with one antecedent. For example, Man(Socrates) connects to Mortal(Socrates), which could in turn connect to FiniteLife(Socrates). In practice though, rules may have more than one antecedent, in which case the proof structure is a tree, as will be evident when we look at backward chaining.

The rule *existential instantiation* can naturally be adapted to the implicit quantifier form. The problem from the earlier section repeated below is solved with the existential variable being represented by a Skolem constant.

Some police person murdered Mahsa. Anyone who murders someone is a murderer. All murderers must be prosecuted. Therefore, the police person must be prosecuted.

1. $P(\text{sk-}x) \wedge \text{Murdered}(\text{sk-}x, \text{Mahsa})$ Given
2. $\text{Murdered}(\text{sk-}x, \text{Mahsa})$ Simplification, 1
3. $\text{Murdered}(?x, ?y) \supset \text{MustBeProsecuted}(?x)$ Given
4. $\text{MustBeProsecuted}(\text{sk-}x)$ MMP, 2,3
5. $P(\text{sk-}x)$ Simplification 1
6. $P(\text{sk-}x) \wedge \text{MustBeProsecuted}(\text{sk-}x)$ Conjunction 5, 4

We will use the adaptation of the MP rules to MMP in our further discussion. However, any rule of inference can be adapted to work with the implicit quantifier form. As long as the antecedents can be unified, the conclusion can be added by the forward chaining algorithm.

The forward chaining algorithm is akin to forward state space search as described in Chapter 7 and also Chapter 9. Theorem proving in FOL is a simpler process because sentences are only added, and never deleted. Otherwise, the algorithm essentially searches the space of theorems looking for the one it needs to derive. The *MoveGen* function in *forward chaining* is depicted in Figure 10.3, which is a reproduction of Figure 7.1. It lists the immediate inferences that can be made in a given state or KB.

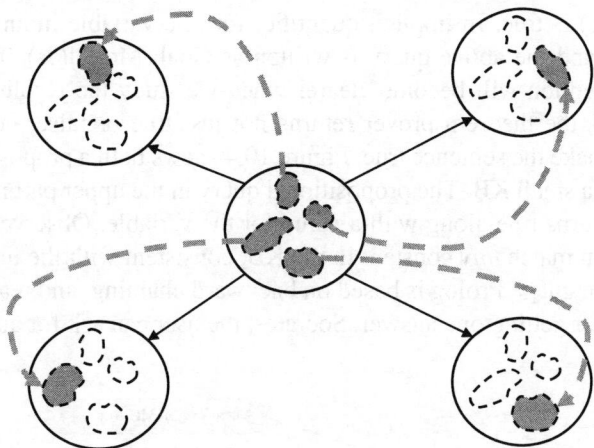

Figure 10.3 The MOVEGEN function for forward chaining is composed from the set of matching antecedent–consequent instances of rules of inference. The MOVEGEN is simply a collection of all instances of all applicable rules of inference.

Forward chaining is associated typically with forward reasoning, when we move from the given sentences towards the goals to be derived. It is also known as data driven, and has the trait of being *eager* computation. A forward chaining system makes inferences *when it can*. This contrasts with backward chaining or backward reasoning where inferences are made only when *they are needed*.

10.4.4 Backward chaining and deductive retrieval

Forward chaining, like forward state space planning, runs into a sea of possibilities. Backward chaining is like backward state space planning, focussed on achieving the goal. It is associated with goal directed reasoning, which has often been associated with human problem solving.

In forward chaining one works with the KB, which is a set of sentences that are (known to be) true. They may be the premises, or they may be the derived sentences. The target theorem only shows up in the goal test function. In backward chaining one needs to separate the goals from the facts. Because goals induce subgoals, and they must be distinguished from facts. We adopt a convention similar to the one proposed by Charniak and McDermott (1985) by prefixing a query with 'Goal:'. A goal Goal: γ is solved if $\gamma \in$ KB. If not, then we find a rule $\beta \supset \alpha$ such that γ unifies with α with the substitution θ, and we pose $\beta\theta$ as the subgoal Goal: $\beta\theta$. In this manner we chain backward from the consequent to the antecedent.

The ability to make inferences can facilitate *deductive retrieval* of implicit facts from a KB. This is an advancement over database management systems that can retrieve only explicit facts. One can, for example, ask if Socrates is mortal. Furthermore, like in a database, one can ask existential *queries*. The query does not have to be about an individual. One can even ask 'Is there someone in the KB who is mortal?' This would be tantamount to asking if the FOL sentence $\exists z Mortal(z)$ is true. In implicit quantifier form, a variable in an existential query is identified by a '?', and the above query is written as Goal: Mortal(?z). The intuition behind this change in convention will become clearer when we study the resolution method. Given an existential query, the theorem prover returns not just true or false, but when true it also returns values that make the sentence true. Figure 10.4 shows both a propositional query and an existential query on a small KB. The propositional query in the upper part returns true. The one in the lower part returns true along with a value for the variable. Observe that the variable ?z in Goal: Man(?z) can match *any* constant in the KB, consistent with the unification algorithm. The programming language Prolog is based on backward chaining, and readers familiar with it would know that after getting one answer, Socrates, the user can ask for another one.

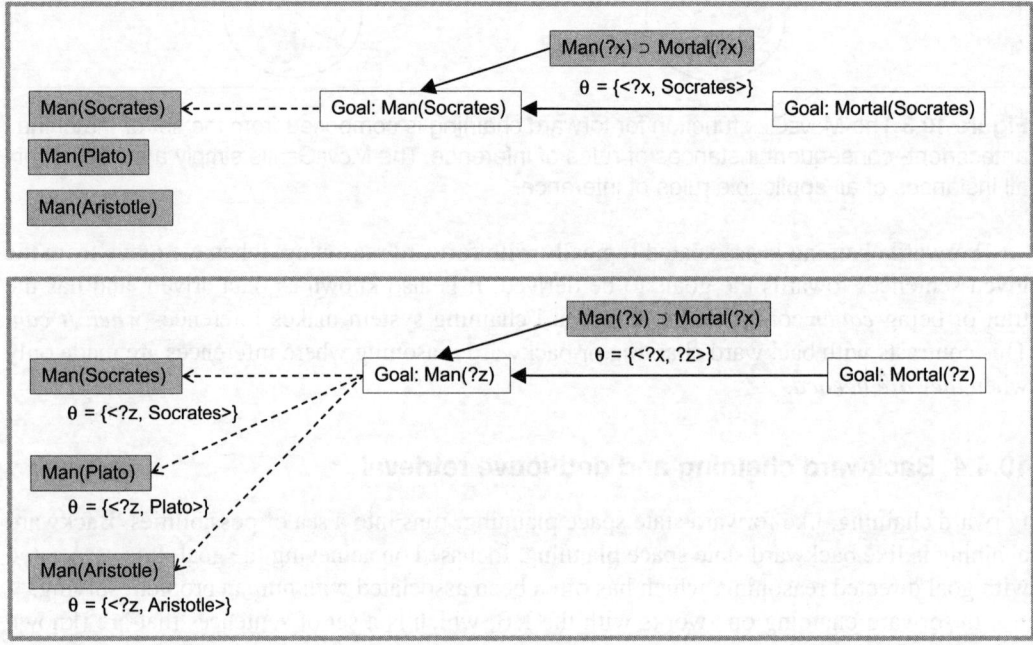

Figure 10.4 Backward chaining matches the consequent and moves to the antecedent, producing subgoals, till the goal is satisfied in the KB. In the figure the shaded boxes constitute the KB, and unshaded ones the goals. A propositional query Mortal(Socrates) on the top produces the subquery Man(Socrates) that has a matching fact in the KB. An existential query Mortal(?z) produces the subquery Man(?z) which can match any of the three facts.

A sentence that is true may have more than one justification or a proof. This could be because a goal maybe the consequent in more than one implication. This introduces search into the backward chaining process. Consider the following KB which defines *some* relations between humans:

KB1 = {(Sibling(?q, ?p) ∧ Parent(?q, ?y) ∧ Parent(?p, ?x)) ⊃ Cousin(?y, ?x),
          (Female(?a) ∧ Parent(?p, ?x) ∧ Sibling(?p, ?a)) ⊃ Aunt(?a, ?x),
          (Parent(?g, ?p) ∧ Parent(?p, ?x)) ⊃ GrandParent(?g, ?x),

Cousin(?x, ?y)) ⊃ Related(?x, ?y),
Aunt(?x, ?y)) ⊃ Related(?x, ?y),
GrandParent(?x, ?y)) ⊃ Related(?x, ?y),
Sibling(Sarina, Hadis), Parent(Sarina, Nika), Parent(Hadis, Zhina),
Female(Sarina), Female(Hadis), Parent(Neda, Hadis)}

We often refer to the implication statements as rules and the rest as facts. Both are part of the KB.

The backward chaining process explores an And–Or graph, similar to the ones described in Section 7.3. The search space for the query Goal: Related(?x, ?y), which asks if anyone is related to someone, is shown in Figure 10.5. Some edges are labelled with the rules that transform a query into a subquery. The formulas in the unshaded boxes are the goals, and the ones in the shaded boxes are the facts. As in Chapter 7, the solution is a subtree. The AND siblings are not independent, however. The shared variables must be bound to the same values. One solution for this query is marked by three solid edges connecting atomic queries to matching facts. In the solution, Hadis is related to Nika because she is her aunt, because Hadis is female and a sibling of Sarina who is Nika's parent.

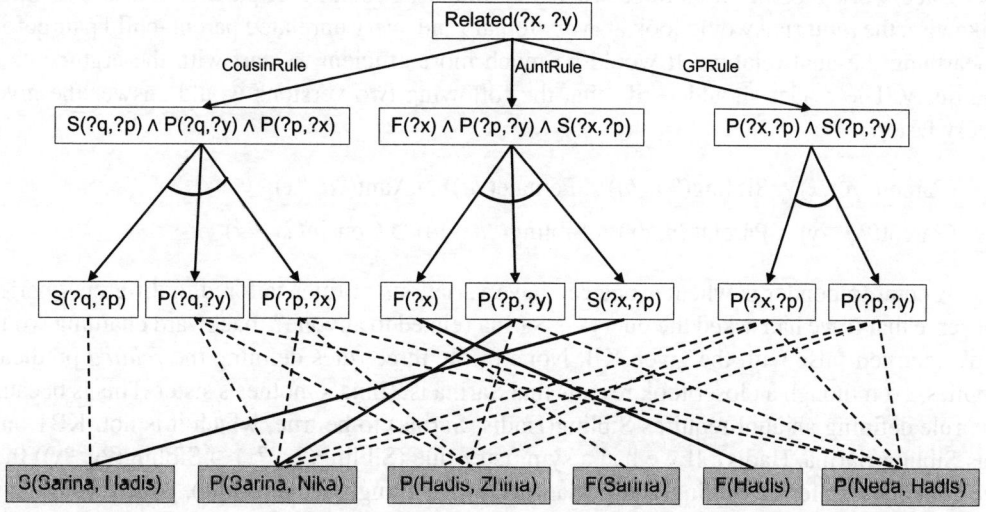

Figure 10.5 The goal tree for the query Goal: Related(?x, ?y). The shaded nodes depict one solution Related(Hadis, Nika) via the *aunt* rule. The predicates Sibling, Parent, and Female are represented by their first letters.

10.4.5 Prolog

The programming language Prolog does depth first search (DFS) on the goal tree or the *And–Or* graph. This amounts to inspecting the sentences in the *program* KB1 from top to down and left to right in the text notation. DFS would have returned the answer true with {<?x, Zhina>, <?y, Nika>} using the *cousin rule*. One can follow the computation via the sequence of goals that need to be solved. Starting with the query the sequence is as follows. On the left is the pending set of goals, and on the right the substitution employed. Doing DFS the leftmost goal is removed from the goal set, and replaced with the antecedents of a rule that the goal is a consequent in. If the goal matches a fact, then no new goal is added. The search may backtrack on the goal tree and terminates when the goal set is empty. In our example, the first path leads to a solution.

{Related(?x, ?y)}
{Sibling(?q, ?p), Parent(?q, ?y), Parent(?p, ?x)} { }
{Parent(Sarina, ?y), Parent(Hadis, ?x)}   {<?q, Sarina>, <?p, Hadis>}
{Parent(Hadis, ?x)}   {<?q, Sarina>, <?p, Hadis>, <y, Nika>}
{ }   {<?q, Sarina>, <?p, Hadis>, <?y, Nika>, <?x, Zhina>}

At this point there are no goals to solve and the original query Related(?x, ?y)} becomes true with ?x = Zhina and ?y = Nika. The goal is true and Zhina is related to Nika.

The perceptive reader would have noticed that order of writing the antecedents affects the performance of backward chaining with DFS. With the *cousin rule* Prolog would first look for a pair of siblings and *then* check if they have children who would be cousins. This could lead to wasted work looking at siblings when we should be looking at parents who are siblings. Likewise, the *aunt rule* would look at *every* female and *every* unrelated parent–child pair before unearthing the aunt relation. It would be much more efficient to start with the arguments in the query. The reader should verify that the following two versions would answer the given query faster:

(Parent(?p, ?x) ∧ Sibling(?p, ?a) ∧ Female(?a)) ⊃ Aunt(?a, ?x),

(Parent(?q, ?y) ∧ Parent(?p, ?x) ∧ Sibling(?q, ?p)) ⊃ Cousin(?y, ?x)

A point to ponder is whether one can have variations of rules tailored to different queries. Observe that if we had asked the query 'Is Sarina related to anyone?' backward chaining would have returned false with the given KB. None of the three rules defining the *related* predicate applies, even though a closer look reveals that Sarina is Zhina's mother's sister. This is because the rule defining an aunt requires Sibling(Hadis, Sarina) to be true, which it is not. KB1 only has Sibling(Sarina, Hadis). If we had a symmetry rule (Sibling(?x, ?y) ⊃ Sibling(?y, ?x)) then we could have inferred Sibling(Hadis, Sarina) from Sibling(Sarina, Hadis), which would have then matched the *aunt rule*.

The definition of grandparent given earlier is defined in terms of the Parent relation.[4] What about great grandparents? Or great-great grandparents? FOL allows one to recursively define the relation ancestor as follows, using two sentences:

1. ∀p,x (Parent(p,x) ⊃ Ancestor(p,x))
2. ∀a,x [(∃p (Parent(p,x) ∧ Ancestor(a,p)) ⊃ Ancestor(a,x)]
   ≡ ∀a,x,p [(Parent(p,x) ∧ Ancestor(a,p)) ⊃ Ancestor(a,x)]

---

[4] We use the terms 'relation' and 'predicate' synonymously. Strictly speaking, relations are in the domain while predicates are in the language of FOL.

The first clause is the base clause defining a parent as an ancestor. The second one says that the ancestor of a parent is an ancestor. The reader is encouraged to investigate variations of the second clause that may work for different queries when backward chaining does DFS.

Backward chaining goes beyond database retrieval. One can have recursive rules that can allow connecting elements that are arbitrarily far from each other in the number of inferences needed. This makes backward chaining a means of implementing a programming language. Robert Kowalski introduced the idea that logical reasoning is a way of doing any computation, an approach that is known as logic programming. In idealized logic programming the programmer or the user only needs to specify the relation between the input and the output in logic. The task of making the connection is left to the theorem prover. Backward chaining is the inference engine in Prolog (Sterling and Shapiro, 1994).

We look at a tiny example to define addition on the domain of natural numbers $NN = \{0, s(0), s(s(0)), s(s(s(0))), \ldots\}$, where 0 is a constant and s is a successor function. In maths we have adopted a naming convention for the numbers wherein $NN = \{0, 1, 2, 3, \ldots\}$. The following two statements define addition, and also serve as a program to add two numbers. The meaning of the sentences in the language of mathematics is shown alongside.

1. $\text{Sum}(0, ?n, ?n)$ $\qquad\qquad\qquad\qquad\qquad 0 + N = N$
2. $\text{Sum}(?x, ?y, ?z) \supset \text{Sum}(s(?x), ?y, s(?z))$ $\quad$ If $(X + Y = Z)$ then $((X + 1) + ?Y) = (Z + 1)$

As one can see, the two sentences give a recursive definition of addition. How can one check whether $3 + 2 = 5$ is true? We begin with the statement as a goal, and successively reduce it to subgoals till we reach a fact in the KB, in which case the set of goals to be solved becomes an empty set.

A. Goal: $\{\text{Sum}(s(s(s(0))), s(s(0)), s(s(s(s(s(0))))))\}$ $\qquad$ Query
B. Goal: $\{\text{Sum}(s(s(0)), s(s(0)), s(s(s(s(0)))))\}$ $\qquad$ A, 2
C. Goal: $\{\text{Sum}(s(0), s(s(0)), s(s(s(0))))\}$ $\qquad$ B, 2
D. Goal: $\{\text{Sum}(0, s(s(0)), s(s(0)))\}$ $\qquad$ C, 2
E. Goal: $\{\}$ $\qquad$ D, 1, $<?n, s(s(0))>$

One does not in practice use the above program for addition, because its complexity is proportional to the first argument, while we can get addition in constant time. We have mentioned it only to show that logic subsumes arithmetic.

An interesting feature is that with the same program we can both add and subtract. The query Goal: $\{\text{Sum}(s(s(s(0))), \text{sum}(s(s(0)), ?\text{sum})\}$ asks for the sum of $3 + 2$ while the query Goal: $\{\text{Sum}(s(s(s(0))), ?\text{diff}, \text{sum}(s(s(s(s(s(0)))))))\}$ asks for the difference between 5 and 3. One can even pose the query Goal: $\{\text{Sum}(?x, ?y, \text{sum}(s(s(s(s(s(0)))))))\}$ to find two numbers that add up to 5.

Figure 10.6 shows the proof for the statement $\text{Sum}(?n, s(s(0)), s(s(s(s(s(0))))))$, which is true when $?n = s(s(s(0)))$. Observe that the goal tree is linear because the implication has only one antecedent. The nodes in the unshaded rectangles are goals and subgoals, and the contents of the shaded rectangles are facts. The final subgoal $\{\text{Sum}(?x3, s(s(0)), s(s(0)))\}$ is satisfied by the fact $\text{Sum}(0, ?x, ?x)$.

346 | Search Methods in Artificial Intelligence

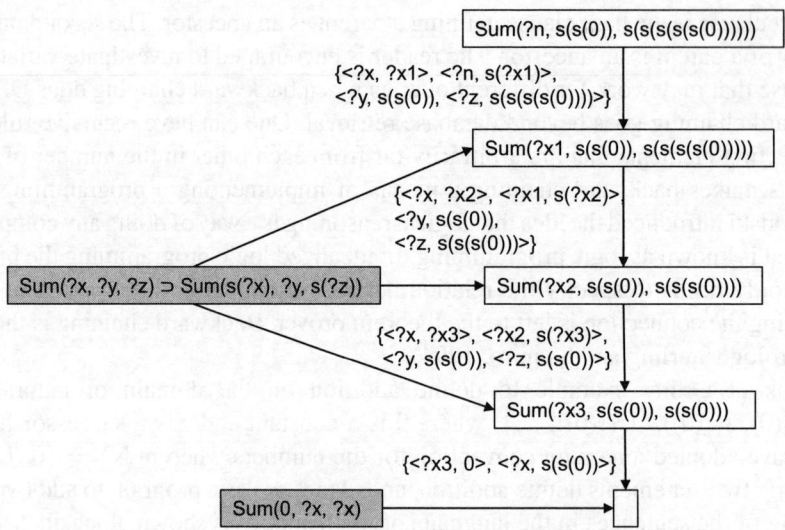

Figure 10.6 The Sum predicates for addition can be used for subtraction as well. The existential goal 'There is some number n such that $n + 2 = 5$' is shown to be true with ? $n = s(s(s(0)))$ via back substitution from ?$x3 = 0$. The unshaded nodes are the goals and the subgoals. The shaded nodes are the KB, which is also a program for addition and subtraction.

The sequence of goals and subgoals for the above problem is shown below. The clauses defining the *Sum* predicate are reproduced again. Only the relevant substitution is shown at each step. The others can be read off from the proof tree in Figure 10.6.

1. Sum(0, ?n, ?n)
2. Sum(?x, ?y, ?z) ⊃ Sum(s(?x), ?y, s(?z))

A. Goal: {Sum(?n, $s(s(0))$, $s(s(s(s(s(0))))))$}	Query
B. Goal: {Sum(?$x1$, $s(s(0))$, $s(s(s(s(0)))))$}	A, 2, <?n, s(?$x1$)>
C. Goal: {Sum(?$x2$, $s(s(0))$, $s(s(s(0))))$}	B, 2, <?$x1$, s(?$x2$)>
D. Goal: {Sum(?$x3$, $s(s(0))$, $s(s(0)))$}	C, 2, <?$x2$, s(?$x3$)>
E. Goal: { }	D, 1, <?$x3$, 0>

10.4.6 Incompleteness of forward and backward chaining

A deduction system is complete if all statements entailed by the KB can be derived. The following example shows that both forward chaining and backward chaining are incomplete. With this example we also show that the logic is formal. It is syntactic machinery that pays no heed to meaning. Only to form.

Consider the following KB with deliberately chosen opaque names for predicates and constants:

$$KB2 = \{P(c1, c2), P(c2, c3), Q(c1), \neg Q(c3)\}$$

The query is $\exists x, y\ (P(x, y) \land Q(x) \land \neg Q(y))$, which in the implicit quantifier form is written as

$$\text{Goal: } (P(?x, ?y) \land Q(?x) \land \neg Q(?y))$$

Neither forward chaining nor backward chaining has a derivation, primarily because the KB is just a set of propositions and has no logical connectives. Logic connectives are the basis of rules of inference, for example, MP relies on $\supset$. Yet the goal formula is entailed from the KB. Choosing a familiar domain makes that easier to accept.

Let the interpretation $\mathcal{I} = <D, \mathbb{I}>$ have D as the set of blocks from the blocks world, and let $\mathbb{I}$ be as follows:

$$P^{\mathbb{I}} = \text{On},\ Q^{\mathbb{I}} = \text{blue},\ c1^{\mathbb{I}} = A,\ c2^{\mathbb{I}} = B,\ c3^{\mathbb{I}} = C$$

In this interpretation the KB is

$$\text{KB2}^{\mathbb{I}} = \{\text{On}(A, B), \text{On}(B, C), \text{Blue}(A), \neg\text{Blue}(C)\}$$

and the existential query is 'There is a blue block on a non-blue block.' A little bit of thought reveals that this is indeed a true sentence. We do not know the colour of block B, but it is either blue or it is not. If it is blue, then then it is on C, and hence a blue block is on a non-blue block. If it is not blue, then the statement is again true because A is on B and A is blue. Thus, the existential statement is true, even though we cannot name the blocks that make it true. In fact, we do not know *which* two blocks make it true.

The interpretation is in the mind of the user. Consider another interpretation where D is the domain of women, and the mapping is as follows:

$$P^{\mathbb{I}} = \text{Like},\ Q^{\mathbb{I}} = \text{Married},\ c1^{\mathbb{I}} = \text{Sarina},\ c2^{\mathbb{I}} = \text{Hadis},\ c3^{\mathbb{I}} = \text{Mahsa}$$

In this interpretation the KB is

$$\text{KB2}^{\mathbb{I}} = \{\text{Like}(\text{Sarina}, \text{Hadis}), \text{Like}(\text{Hadis}, \text{Mahsa}), \text{Married}(\text{Sarina}), \neg \text{Married}(\text{Mahsa})\}$$

The query now is 'Does a married woman like an unmarried one?' and again the answer is yes, even though we do not know the two individuals that make it true. This lack of knowledge is characterized by disjunction, which has often been difficult to handle for computing, and there is a disjunctive sentence implicitly lurking in the KB – Hadis is married, or Hadis is not married.

Next, we look at a proof method that is complete for FOL. If a formula α is entailed by the KB, then the resolution refutation method will have a proof for α.

10.4.7 The resolution refutation method

The *resolution refutation method* was introduced by Alan Robinson (1965) and is a complete theorem proving algorithm that uses only one rule of inference, *the resolution rule*. It is an indirect method of generating a proof by contradiction. The goal or query is negated and added to the KB which, if the goal is entailed, will now become unsatisfiable. Robinson showed that given an unsatisfiable KB the empty clause, which stands for a contradiction and is always false, can always be derived.

The method works with the *clause form* which has the following structure:

$\forall x, y, z \; [C_1 \wedge C_2 \wedge \ldots \wedge C_k]$ where each C_i is a clause
$C_i = D_{i1} \vee D_{i2} \vee \ldots \vee D_{im}$ where each D_{ij} is an atomic formula (literal) or its negation

The Norwegian logician Thoralf Skolem had shown a century ago that any FOL formula can be converted into clause form (reproduced in Skolem (1970)). The procedure is as follows:

1. Standardize variables apart across quantifiers. Rename variables so that the same symbol does not occur in different quantifiers.
2. Eliminate all occurrences of operators other than $\wedge$, $\vee$, and $\neg$.
3. Move $\neg$ all the way in.
4. Push the quantifiers to the right. This ensures that their scope is as tight as possible.
5. Eliminate $\exists$. Skolemization.
6. Move all $\forall$ to the left. They can be ignored henceforth.
7. Distribute $\vee$ over $\wedge$.
8. Simplify.
9. Rename variables in each clause (disjunction).

We illustrate the process with the definition of an aunt.

1. $\forall a,x \; [\text{Aunt}(a,x) \equiv (\exists p \; (\text{Parent}(p,x) \wedge \text{Sibling}(a,p) \wedge \text{Female}(a)))]$
2. $\forall a,x \; [\text{Aunt}(a,x) \supset (\exists p \; (\text{Parent}(p,x) \wedge \text{Sibling}(a,p) \wedge \text{Female}(a))))]$ *Eliminate* $\equiv$ $\wedge$ $\forall a,x,p$
   $[((\text{Parent}(p,x) \wedge \text{Sibling}(a,p) \wedge \text{Female}(a)) \supset \text{Aunt}(a,x))]$
3. $\forall a,x \; [\text{Aunt}(a,x) \supset (\exists p \; (\text{Parent}(p,x) \wedge \text{Sibling}(a,p) \wedge \text{Female}(a)))]$ *Rename variables* $\wedge$
   $\forall a1,x1,p1 \; [((\text{Parent}(p1,x1) \wedge \text{Sibling}(a1,p1) \wedge \text{Female}(a1)) \supset \text{Aunt}(a1,x1))]$
4. $\forall a,x \; [\neg \text{Aunt}(a,x) \vee (\exists p \; (\text{Parent}(p,x) \wedge \text{Sibling}(a,p) \wedge \text{Female}(a)))]$ *Eliminate* $\supset$ $\wedge$
   $\forall a1,x1,p1 \; [(\neg \, (\text{Parent}(p1,x1) \wedge \text{Sibling}(a1,p1) \wedge \text{Female}(a1)) \vee \text{Aunt}(a1,x1))]$
5. $\forall a,x \; [\neg \text{Aunt}(a,x) \vee (\text{Parent}(\text{sk}(a,x),x) \wedge \text{Sibling}(a,(\text{sk}(a,x))) \wedge \text{Female}(a)))]$ *Skolemize* $\wedge$
   $\forall a1,x1,p1 \; [(\neg \text{Parent}(p1,x1) \vee \neg \text{Sibling}(a1,p1) \vee \neg \text{Female}(a1)) \vee \text{Aunt}(a1,x1))]$
6. $[\neg \text{Aunt}(a,x) \vee (\text{Parent}(\text{sk}(a,x),x) \wedge \text{Sibling}(a,(\text{sk}(a,x))) \wedge \text{Female}(a)))]$ *Discard* $\forall$ $\wedge$
   $[(\neg \text{Parent}(p1,x1) \vee \neg \text{Sibling}(a1,p1) \vee \neg \text{Female}(a1)) \vee \text{Aunt}(a1,x1))]$
7. $[\; (\neg \text{Aunt}(a,x) \vee \text{Parent}(\text{sk}(a,x),x)) \quad$ *Distribute* $\vee$ *over* $\wedge$
   $\wedge \; (\neg \text{Aunt}(a,x) \vee \text{Sibling}(a,(\text{sk}(a,x))))$
   $\wedge \; (\neg \text{Aunt}(a,x) \vee \text{Female}(a))$
   $\wedge \; (\neg \text{Parent}(p1,x1) \vee \neg \text{Sibling}(a1,p1) \vee \neg \text{Female}(a1)) \vee \text{Aunt}(a1,x1))]$
8. $[\; (\neg \text{Aunt}(a2,x2) \vee \text{Parent}(\text{sk}(a2,x2),x2)) \quad$ *Rename variables*
   $\wedge \; (\neg \text{Aunt}(a3,x3) \vee \text{Sibling}(a3,(\text{sk}(a3,x3))))$
   $\wedge \; (\neg \text{Aunt}(a4,x4) \vee \text{Female}(a4))$
   $\wedge \; (\neg \text{Parent}(p1,x1) \vee \neg \text{Sibling}(a1,p1) \vee \neg \text{Female}(a1)) \vee \text{Aunt}(a1,x1))]$

The clause form is a conjunction of a set of clauses. A literal is an atomic formula or the negation of an atomic formula. A clause is a disjunction of literals. The resolution rule takes two clauses as input. One of the clauses must have a positive literal and the other a matching negative literal. The inferred clause combines the remaining literals after removing the two opposite literals. Let C_i and C_k be two clauses with the structure

$$C_i = (L_1 \vee L_2 \vee \ldots \vee L_k \vee P_1 \vee P_2 \vee \ldots \vee P_n)$$
$$C_k = (\neg R_1 \vee \neg R_2 \vee \ldots \vee \neg R_s \vee Q_1 \vee Q_2 \vee \ldots \vee Q_t)$$

If θ is the MGU for $\{L_1, L_2, \ldots, L_k, R_1, R_2, \ldots, R_s\}$, then we can resolve C_i and C_k to derive the *resolvent*

$$(P_1\theta \vee P_2\theta \vee \ldots \vee P_n\theta \vee Q_1\theta \vee Q_2\theta \vee \ldots \vee Q_t\theta)$$

That is, we throw away all the positive literals L_j and negative literals $\neg R_i$, and combine the remainder after applying the substitution θ (Charniak and McDermott, 1985). We can illustrate the validity of the resolution rule by establishing the following equivalence for a simpler formula:

$$(R \vee Q) \wedge (P \vee \neg Q)) \equiv ((R \vee Q) \wedge (P \vee \neg Q) \wedge (R \vee P)$$

The reader is encouraged to show that the above formula is a propositional tautology. In the above formula each element in the disjunct – P, Q, R – is an atomic formula, and could be any formula including an FOL formula. A consequence of the soundness of the resolution rule is that adding the resolvent to the set of sentences does not change the truth value of the KB.

Robinson showed that the method is complete for an unsatisfiable set of clauses. We can make our KB unsatisfiable by adding the negation of the goal clause to the KB. Then the following holds. Given a set of premises $\{\alpha_1, \alpha_2, \ldots, \alpha_N\}$ and the desired goal β, we want to determine if the formula

$$((\alpha_1 \wedge \alpha_2 \wedge \ldots \wedge \alpha_n) \supset \beta)$$

is true. This follows from the well known Deduction theorem that asserts that

$$\{\alpha_1, \alpha_2 \ldots, \alpha_n\} \vdash \beta \quad \text{iff} \quad \vdash ((\alpha_1 \wedge \alpha_2 \wedge \ldots \wedge \alpha_n) \supset \beta)$$

Also $((\alpha_1 \wedge \alpha_2 \wedge \ldots \wedge \alpha_n) \supset \beta)$ is a tautology iff $\neg ((\alpha_1 \wedge \alpha_2 \wedge \ldots \wedge \alpha_n) \supset \beta)$ is unsatisfiable. To show that $\neg ((\alpha_1 \wedge \alpha_2 \wedge \ldots \wedge \alpha_n) \supset \beta)$ is unsatisfiable, one can add the negation of the goal to the set of premises, because the following are equivalent:

$$\neg ((\alpha_1 \wedge \alpha_2 \wedge \ldots \wedge \alpha_n) \supset \beta) \equiv \neg (\neg (\alpha_1 \wedge \alpha_2 \wedge \ldots \wedge \alpha_n) \vee \beta)$$
$$\equiv ((\alpha_1 \wedge \alpha_2 \wedge \ldots \wedge \alpha_n) \wedge \neg \beta)$$

The following equivalences establish that if we can add the *null* or *empty* clause to a KB, then the KB *must have been* unsatisfiable to start with. The null clause can be generated by resolving a literal P with its negation $\neg P$. Let the original KB be $\{P_1, P_2, \ldots, P_N\}$. Remember, this stands for a conjunction of the N clauses. To this we add a sequence of resolvents R_1, R_2, R_3, ... culminating with $\bot$. The databases at all stages are logically equivalent because the resolution rule is sound.

$$\begin{aligned} \{P_1, P_2, ..., P_N\} &\equiv \{P_1, P_2, ..., P_N, R_1\} \\ &\equiv \{P_1, P_2, ..., P_N, R_1, R_2\} \\ &\equiv \{P_1, P_2, ..., P_N, R_1, R_2, R_3\} \\ &\equiv \{P_1, P_2, ..., P_N, R_1, R_2, R_3, ..., \bot\} \end{aligned}$$

Now since the last set of clauses evaluates to *false* (because it contains the empty clause), the KB we started with, which is logically equivalent, also evaluates to *false*. Thus $\{P_1, P_2, ..., P_N\}$ is *unsatisfiable*. The procedure for finding a proof by the resolution refutation method is as follows:

1. Convert each premise into clause form.
2. Negate the goal and convert it into clause form.
3. Add the negated goal to the set of clauses.
4. Choose two clauses such that two opposite signed literals in them can be unified.
5. Resolve the two clauses using the MGU and add the resolvent to the set.
6. Repeat steps 4–5 till a null resolvent is produced.

The proof procedure uses only one rule of inference repeatedly till the null clause is derived. We look at our familiar example, the Socratic argument.

The three clauses for the Socratic argument are

$C_1 = \neg \text{Man}(?x) \vee \text{Mortal}(?x)$ premise
$C_2 = \text{Man (Socrates)}$ premise
$C_3 = \neg \text{Mortal(Socrates)}$ negated goal

The following resolvents are generated:

$R_1 = \text{Mortal (Socrates)}$ $C_1, C_2, \{<?x, \text{Socrates}>\}$
$R_2 = \bot$ C_3, R_1

Remember that applying a substitution $\{?x = \text{Socrates}\}$ is a kind of instantiation. And since $(\neg \text{Man}(?x) \vee \text{Mortal}(?x)) \equiv (\text{Man}(?x) \supset \text{Mortal}(?x))$, the first step in resolution derivation is emulating the MMP derivation. What is more, it can also emulate backward chaining. Consider the alternate derivation,

$R_1 = \neg \text{Man(Socrates)}$ $C_1, C_3, \{<?x, \text{Socrates}>\}$
$R_2 = \bot$ C_2, R_1

If we were to replace 'Goal': with '$\neg$', then Goal:Mortal(Socrates) becomes $\neg$Mortal(Socrates), which is C_3, the negated goal, and from that we derive $\neg$Man(Socrates), which is essentially the backward reasoning step deriving the subgoal Goal: Man(Socrates). If we are working with an existential goal $\exists x \text{Mortal}(x)$, then on negation the clause to be added to the resolution method would be $\neg \exists x \text{Mortal}(x) \equiv \forall x \neg \text{Mortal}(x)$, which would be Skolemized as ¬Mortal(?z), and the same derivation would apply. This gives us an intuition why in goals or queries the Skolemization convention is inverted and existential query variables are prefixed with a '?' and not replaced with Skolem constants or functions.

In this way, the resolution proof subsumes both forward and backward chaining, but can do more. The problem that both could not solve can be solved by the resolution method. The given KB2 = {$P(c1, c2)$, $P(c2, c3)$, $Q(c1)$, $\neg Q(c3)$} is already in clause form. The query is $\exists x, y(P(x, y) \wedge Q(x) \wedge \neg Q(y))$, which *after negation* is written as $(\neg P(?x, ?y) \vee \neg Q(?x) \vee Q(?y))$. The derivation of the null clause is shown in Figure 10.7 as a directed acyclic graph (DAG), as resolution refutation proofs are often depicted. To make it more intelligible to us we have used the first interpretation from the blocks world where $P(X,Y)$ denotes X is on Y, and $Q(X)$ denotes X is blue.

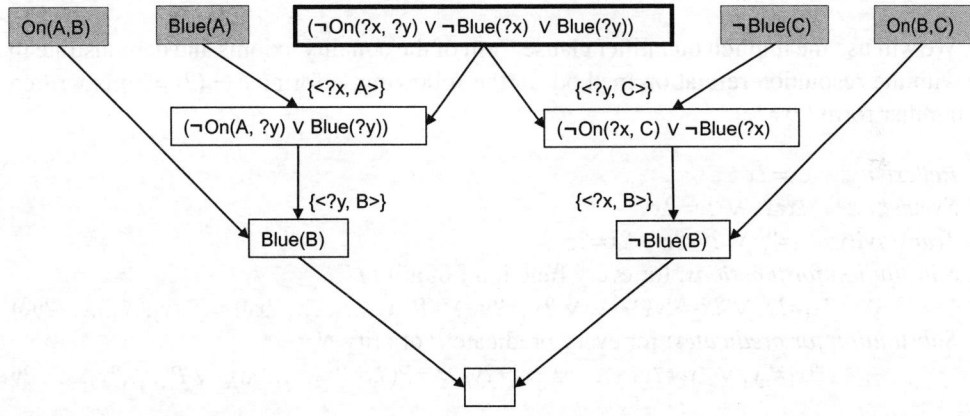

Figure 10.7 Given the KB {$On(A,B)$, $On(B,C)$, $Blue(A)$, $\neg Blue(C)$}, a resolution refutation proof of the statement 'A blue block is on a non-blue block' after adding the negated goal to the KB. The empty clause is derived by resolving $Blue(B)$ with $\neg Blue(B)$.

One can observe that the null clause or contradiction is derived from $Blue(B)$ and $\neg Blue(B)$, which cannot both be true at the same time. Block B is either blue, or it is not blue. We do not know whether it is blue or not. Reasoning by cases one can argue that if it is blue, then since it is on block C, a blue block is on a non-blue block. And if B is not blue, then the blue block A is on B, so the statement is again true. In either case the statement is true, even though we do not know which two blocks fit the bill.

10.4.8 FOL with equality

We have distinguished atomic formulas using equality from other formulas that use a predicate name to stand for a relation. If $t1, t2 \in \mathcal{T}$ are terms, then $(t1 = t2) \in \mathcal{A}$. Why did we not express this as $Eq(t1,t2)$? The reason is that the '=' relation comes with added knowledge about the nature of the relation. Without that added knowledge, reasoning in FOL will not be complete. Here is a simple example. Talking about numbers, if you know that $Eq(N, 5)$ and $Eq(M,N)$, you should be able to conclude $Eq(M, 5)$. But without adding transitivity as a property of the equality relation, one cannot arrive at the conclusion. Thus, when we say that our language is

FOL with equality, we implicitly state that the following *equality axioms* must be added to the KB (Brachman and Levesque, 2004).

- *Reflexivity*: $\forall x \, (x = x)$
- *Symmetry*: $\forall x,y \, (x = y \supset y = x)$
- *Transitivity*: $\forall x,y,z \, ((x = y \wedge y = z) \supset x = z)$
- *Substitution for functions*: for every function f of arity N
  $$\forall x_1,y_1, \, x_2,y_2, \, \ldots \, x_N,y_N \, ((x_1=y_1 \wedge x_2=y_3 \wedge \ldots \wedge x_N = y_N) \supset f(x_1,x_2,\ldots,x_N) = f(y_1,y_2,\ldots,y_N))$$
- *Substitution for predicates*: for every predicate P of arity N
  $$\forall x_1,y_1, \, x_2,y_2, \, \ldots \, x_N,y_N \, ((x_1=y_1 \wedge x_2=y_3 \wedge \ldots \wedge x_N = y_N) \supset P(x_1,x_2,\ldots,x_N) \equiv P(y_1,y_2,\ldots,y_N))$$

We will use the implicit quantifier clause form of the equality axioms and demonstrate their use with the resolution refutation method. In the following, a formula $\neg \, (?x=?y)$ is written in the familiar form $(?x \neq ?y)$.

- *Reflexivity*: $?x=?x$
- *Symmetry*: $?x \neq ?y \vee ?y=?x$
- *Transitivity*: $?x \neq ?y \vee ?y \neq ?z \vee ?x=?z$
- *Substitution for functions*: for every function f of arity N
  $$?x_1 \neq ?y_1 \vee ?x_2 \neq ?y_3 \vee \ldots \vee ?x_N \neq ?y_N \vee (f(?x_1,?x_2,\ldots,?x_N) = f(?y_1,?y_2,\ldots,?y_N))$$
- *Substitution for predicates*: for every predicate P of arity N
  $$?x_1 \neq ?y_1 \vee ?x_2 \neq ?y_3 \vee \ldots \vee ?x_N \neq ?y_N \vee \neg P(?x_1,?x_2,\ldots,?x_N) \vee P(?y_1,?y_2,\ldots,?y_N))$$

Given the above axioms and the KB = $\{?N = 5, ?M=?N\}$, the goal $?M = 5$ can be proved as follows.

1.	$?x=?x$	Reflexivity
2.	$?x1 \neq ?y1 \vee ?y1=?x1$	Symmetry
3.	$?x2 \neq ?y2 \vee ?y2 \neq ?z2 \vee ?x2=?z2$	Transitivity
4.	$?N = 5$	Premise
5.	$?M=?N$	Premise
6.	$?M \neq 5$	Negated goal
7.	$5=?N$	2, 4, $\{<?x1, ?N>, <?y1, 5>$
8.	$?N \neq ?z2 \vee 5=?z2$	3, 7, $\{<?x2, 5>, <?y2, ?N>$
9.	$?N=?M$	2, 5, $\{<?x1, ?M>, <?y1, N>$
10.	$5=?M$	2, 8, $\{<?z2, ?M>\}$
11.	$?M = 5$	2, 10, $\{<?x1, 5>, <?y1, ?M>$
12.	$\bot$	6, 11

The following shorter proof illustrates the fact that theorem proving involves search, and the choice of which two clauses to resolve can affect the length of the proof.

1. $?x = ?x$	Reflexivity
2. $?x1 \neq ?y1 \vee ?y1 = ?x1$	Symmetry
3. $?x2 \neq ?y2 \vee ?y2 \neq ?z2 \vee ?x2 = ?z2$	Transitivity
4. $?N = 5$	Premise
5. $?M = ?N$	Premise
6. $?M \neq 5$	Negated goal
7. $?x2 \neq ?N \vee ?x2 = 5$	3, 4, {<?y2, ?N>, <?z2, 5>}
8. $?M = 5$	5, 7, {<?x2, ?M>, <?N, 5>}
9. ⊥	6, 8

The shorter proof used only one of the three axioms listed as premises, but made a judicious choice as to which atomic formula in the clause to cancel. Clearly, search can run into combinatorial explosion here as well. The following two strategies have been used effectively to try and find shorter proofs:

- Unit clause strategy. Wherever possible use a unit clause (a clause with only one atomic formula) as one of the clauses to resolve. This ensures that the resolvent will be smaller than the other clause. In general if one is cancelling one literal each from two clauses of length N and M, the resolvent will be of length $(N - 1) + (M - 1)$. Then if $N = 1$, the length of the resolvent will be $M - 1$. Smaller resolvents are desirable because the empty clause is of length 0, and it can be derived from two clauses of length 1 each.
- Set of support strategy. Always choose a descendant of the negated goal as one of the clauses. This is because the null clause represents a contradiction. Remember it is derived from two unit clauses, say B and $\neg B$. And the augmented KB becomes unsatisfiable because we add the negated goal. If we keep resolving only among the clauses in the original KB, then a contradiction will never arise.
- SLD (selected literal, linear structure, definite clauses) resolution. This is described in Section 10.5.1.

We illustrate the utility of equality axioms with another example. The following puzzle has sometimes baffled some humans:

> A father and his son were crossing the road when they met with an accident. The father died on the spot, and the son was rushed to the hospital, and into the operation theatre. But the surgeon had one look at the boy and said 'I cannot operate on this boy. He is my son.'
>
> Who was the surgeon?

In the following we use upper case letters as variables in the style of Prolog. Let us start with the existential statement $\exists S Surgeon(S)$. The task is to find out more about who S is. Let $Son(X, Y)$ be read as 'The son of X is Y'.

Let $m(Y)$ and $f(Y)$ be two functions read as 'mother of Y' and 'father of Y'. Then the following statement is literally true: $\forall X,Y (m(Y) = X \equiv \neg f(Y) = X))$. Here $\neg f(Y) = X$ is another way of writing $f(Y) \neq X$. That is, X can be the mother of Y, or X can be the father of Y, but not both. This statement reduces to the following three clauses:

$$(\neg Son(X,Y) \vee m(Y) = X \vee f(Y) = X)$$
and $\quad (\neg m(Y) = X \vee Son(X.Y))$
and $\quad (\neg f(Y) = X \vee Son(X.Y))$

Let the statement 'He is my son' be represented as follows, with the additional piece of knowledge that a person who speaks must be alive. Observe that both the surgeon and the son are represented as Skolem constants.

$\qquad\qquad\qquad\qquad\qquad$ Alive(sk-S) $\wedge$ Son(sk-S, sk-Kian)
The father died $\qquad\quad$ Dead(f(sk-Kian))
Dead men are not alive $\quad$ $\neg$Dead(X) $\equiv$ Alive(X)

The last sentence is our knowledge of living beings and can be reduced to two clauses as follows:

$\neg$Dead(X) $\equiv$ Alive(X) $\quad \equiv (\neg$Dead(X) $\supset$ Alive(X)) $\wedge$ (Alive(X) $\supset \neg$Dead(X))
$\qquad\qquad\qquad\qquad\quad \equiv$ (Dead(X) $\vee$ Alive(X)) $\wedge$ ($\neg$Alive(X) $\vee \neg$Dead(X))

Then the equality predicates tailored to our KB are

Reflexivity: $X = X$
Symmetry: $X \neq Y \vee Y = X$
Transitivity: $X \neq Y \vee Y \neq Z \vee X = Z$
Substitution for functions:
$\qquad\qquad\qquad\qquad X \neq Y \vee (f(X) = f(Y))$
$\qquad\qquad\qquad\qquad X \neq Y \vee (m(X) = m(Y))$
Substitution for predicates:
$\qquad\qquad\qquad\qquad X1 \neq X2 \vee \neg$Alive($X1$) $\vee$ Alive($X2$)))

Finally, let us serendipitously choose the goal 'The surgeon is the boy's mother'. Figure 10.8 shows the resolution refutation when we add the negated goal to a set of relevant clauses from our KB. For the ease of reading we have maintained the variable names as upper case letters.

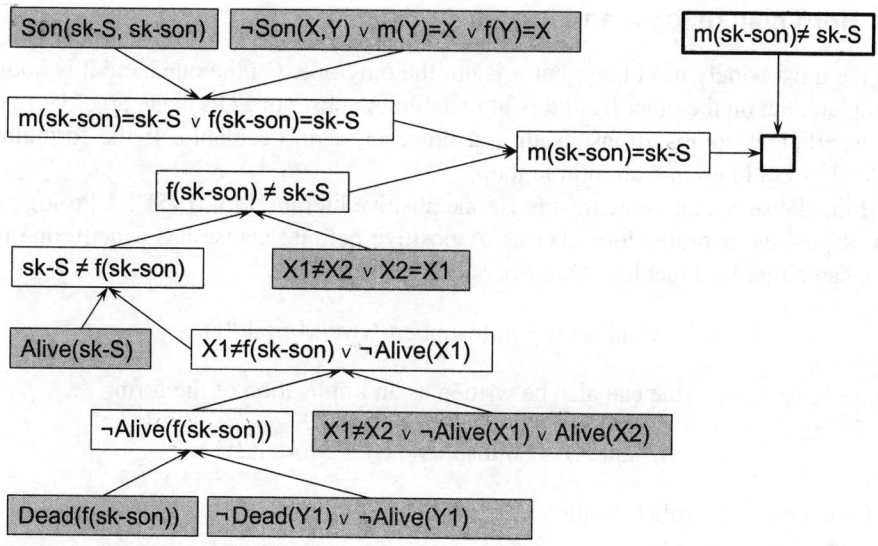

Figure 10.8 The answer to the question 'Who was the surgeon?' by resolution refutation. When the negated goal m(sk-son)# sk-S is added, the KB becomes unsatisfiable, and the null clause can be derived after adding the equality axioms.

Informally the argument is simple. If the surgeon said that the boy was the surgeon's son, then the surgeon must be the mother or the father of the boy. And it must be one of the two. Then since the father is dead, it must be the mother. If we take that as the goal and add its negation that surgeon is not the mother, then it will eventually lead to a contradiction.

10.5 The Family of Logics

In this section we take a brief look at other logics used for representation and reasoning. We begin with logics that are more restrictive than FOL, and then move on to more expressive logics. Each logic presents a trade-off between expressivity and computational complexity.

In FOL, we can build logic machines that are

- Sound: If you use valid rules of inference.
- Complete: With the resolution refutation method.
- Consistent: If the KB is consistent.
- Semi-decidable: Halts only if query is true. May loop otherwise.
- Intractable: Haken (1985) showed that *proofs in even in PL can be exponential* in length!

10.5.1 Horn clause logic and Prolog

FOL is the most widely used logic but it is not the only one. On the one hand it is both sound and complete, but on the other hand it is intractable. A subset of FOL made up of Horn clauses admits to efficient computations. Both are, however, semi-decidable. If the formula is not provable, they could go into an infinite loop.

A Horn clause is a clause with at most one positive literal (Horn, 1951). A Prolog program is a set of positive definite Horn clauses. A positive definite clause has exactly one positive literal. It can either be a fact like Man(Socrates) or a rule like

$$\neg Man(?x) \vee \neg Philosopher(?x) \vee Mortal(?x).$$

As we have seen, a rule can also be written as an implication of the form

$$(Man(?x) \wedge Philospher(?x)) \supset Mortal(?x).$$

In Prolog the same rule is written as

$$Mortal(X) \text{:- } Man(X), Philosopher(X)$$

All three variations represent the same sentence – If X is a man and X is a philosopher, then X is mortal.

Seen from the perspective of resolution, a goal is a negative clause in Prolog, obtained by negating the goal. Then, viewing the augmented KB as a set of (unsatisfiable) clauses, there exists a strategy, SLD resolution, that is equivalent to backward chaining that we saw earlier in this chapter. An SLD (selected literal, linear structure, definite clauses) derivation has the following structure:

- the first resolvent has one parent from the goal and the other from the program
- each resolvent in the derivation is negative (this is equivalent to doing backward chaining to generate a subgoal)
- the latest resolvent becomes one of the parents, and the other parent is a positive clause from the program

Every SLD resolution has the structure shown in Figure 10.9 (Brachman and Levesque, 2004). The shaded parent of every resolvent is a positive definite clause which generates the subgoal if it is a rule, or matches one of the subgoals if it is a fact. When the goal is atomic, then matching results in the null clause terminating the derivation. The reader is encouraged to verify that the proof in Figure 10.7 cannot be cast as an SLD derivation.

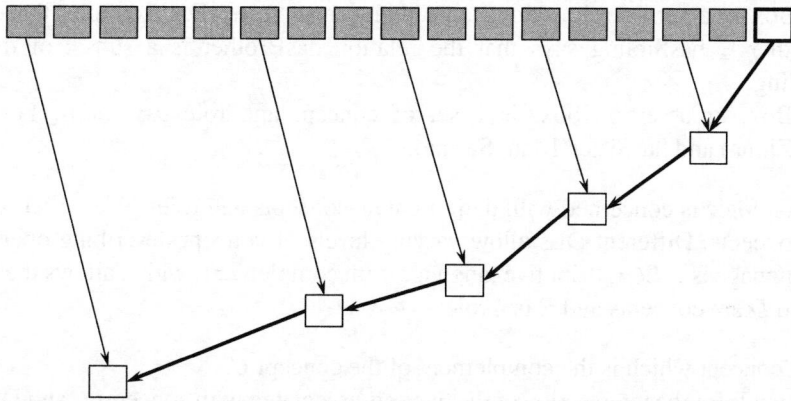

Figure 10.9 An SLD derivation with Horn clauses is essentially backward chaining. Starting with the (negated) goal on the top right shown as a thick lined rectangle, the resolvent at each step is a subgoal that backward chaining generates. In the final step a unit goal matches a fact in the program, resulting in the null clause.

The efficiency comes at a cost, though. Horn clauses cannot talk of disjunctions. One cannot say, for example, that every human is a man or a woman. This inability to handle disjunctions is also at the root of backward chaining being unable to solve the problem from Section 10.4.6. The goal there is existential, which is essentially a disjunctive statement.

10.5.2 Description logics

Another tractable subset of FOL is the family of description logics (DL) or the logics of noun phrases. A DL allows only unary and binary predicates of FOL. And it allows constants, but no variables.

Unary predicates stand for *concepts* or *categories*, which are essentially noun phrases. Except that instead of saying Man(X), DL uses names like Man, Teen, Person, Happy, Mortal, etc. Each such name stands for a concept or a category, and is a subset of the domain.

Binary predicates stand for relations. Thus, loves, hates, brotherOf, owns, hasSister are binary relations. In DL terminology they are called *roles*. Roles can be used to define new concepts.

A set of objects identifies named individuals in the domain. For example, Zhina, Nika, Sarina, Kian. Constants or object names can be arguments to concepts, for example, happy(Zhina), or roles, for example, hasSister(Kian, Sarina), when describing facts.

The KB constitutes a set of three *boxes*.

– The TBox, or terminological box, is a set of concept axioms that define new concepts from existing ones. For example, the sentence (Teen ⊓ ∃owns•Apple) ⊑ Happy. Here Teen, Apple, and Happy are concepts and 'owns' is a role. The expression (Teen ⊓ ∃owns•Apple) is a concept that is formed by the intersection (⊓) of two sets. One is Teen and the other is the set of individuals who own at least one Apple device. The sentence is equivalent to the FOL sentence $\forall x(\text{teen}(x) \land \exists y(\text{Apple}(y) \land \text{Owns}(x,y)) \supset \text{Happy}(x))$.

- The RBox is a set of role axioms that define new roles. For example, the expression hasBrother $\sqsubseteq$ hasSibling says that the relation hasBrother is a subset of the relation hasSibling.
- The ABox, or assertion box, is a set of concept and role *assertions*. For example, happy(Zhina) and hasSister(Kian, Sarina).

A description logic is concerned with defining new concepts and roles given a set of primitive roles and concepts. Different DLs allow varying levels of concept describing operators. One common language is $\mathcal{ALC}$ (attributive language with complement) which allows the following, where C and D are concepts and R is a role:

- $\neg C$ is a concept which is the complement of the concept C.
- $C \sqcap D$ stands for the intersection of the two sets associated with concepts C and D. Logically it is the relation AND. An object is in concept C and in concept D.
- $C \sqcup D$ stands for the union of the two sets associated with concepts C and D. Logically it is the relation OR. An object is in concept C or in D or both.
- $\forall R \cdot C$ is a concept that has objects whose *all* role fillers come from the concept C.
- $\exists R \cdot C$ is a concept that has objects which have at least one role filler from the concept C.

More expressive languages allow more precise descriptions of concepts and roles. For example, one can define a concept of people who have exactly one daughter who is a lawyer.

There are just three kinds of sentences that ALC is primarily concerned with.

- $C \sqsubseteq D$ says that the concept C is subsumed by concept D. For example, for the concepts CompetitionLawyer $\sqsubseteq$ Lawyer and TechCompany $\sqsubseteq$ Company.
- $C \equiv D$ says that the concepts C and D are equivalent. Such statements are often used to define new concepts from primitive ones. For example, ManAllDaughtersLawyers $\equiv$ Father $\sqcap$ $\forall$daughter$\cdot$Lawyer.
- $C(a)$ or $R(a, b)$, for example, YoungWoman(Mahsa) or daughterOf(Mahsa, Amjad).

The first two are sentences in the TBox while the third kind are assertions in the Abox. The following *kinds* of queries can be answered from a given KB:

- Is concept C subsumed by concept D?
- Is concept C equivalent to concept D?
- Is $C(a)$ true?
- Is $R(a,b)$ true?

The answers to the queries are determined structurally from the descriptions. In very simple languages a procedure called *structure matching* can answer the queries by scanning the concept descriptions once (Brachman and Levesque, 2004). In more expressive languages a proof procedure called the *tableau method* is used (Baader and Sattler, 2001). The tableau method has rules for each concept forming operator to break down complex formulas into simpler ones in search of an interpretation that is a model. The method is often used to decide that there is no model after adding the negative of the query to the KB, in the style of looking for a proof by contradiction. The tableau method, which applies to FOL as well, is beyond the scope of this text.

Once you have a procedure to compute subsumption, one can devise an algorithm to organize all concepts into a taxonomy. This can further speed up answering queries.

DLs provide a foundation for the Semantic Web (Antoniou and Harmelen, 2008). The Web Ontology Language (OWL) is based upon DL.

10.5.3 Default reasoning

Entailment and deduction pursue undeniable truth. When the KB is true, the theorem must be necessarily true. All men are mortal, and hence Socrates too must be. However, they say that the only certainties in life are death and taxes. Not much else.

Consider the statement 'All birds fly'. At the same time, Peppy is a bird but it cannot fly. Because she is a penguin. If that is the case, then the KB $\{\forall x(\text{Bird}(x) \supset \text{Flies}(x)), \text{Bird}(\text{Peppy}), \neg\text{Flies}(\text{Peppy})\}$ is inconsistent.

The culprit is the sentence 'All birds fly'. While it is true in general, it is not a universal truth. Penguins do not fly, emus do not fly, ostriches do not fly. There is a lot of *useful* knowledge about the world that does not translate into universal statements. Roses are red, but not all roses. Most leaves are green, but not all. Of course, there are many universally true statements. Parents are older than their children. Trapeziums are quadrilaterals, by definition.

Default reasoning is concerned with making inferences that are generally true, but are also *defeasible*. A conclusion may be true by default, but new information may defeat it. Default reasoning becomes necessary when information is incomplete (Brachman and Levesque, 2004). We look at a couple of approaches to make default inferences.

A simple approach to reasoning with incomplete information is to adopt the *closed world assumption* (CWA): that anything not known to be true or not derivable from the KB is false. This is the approach taken by Prolog and is known as *negation by failure* (to prove it). And in database systems. However, this has its own pitfalls when negative sentences are allowed with disjunction. For example, the tiny KB = $\{P \vee Q\}$ gets extended to $\text{KB}_{\text{CWA}} = \{(P \vee Q), \neg P, \neg Q\}$ because neither P nor Q can be shown to be true. But the extended KB is inconsistent.

The strategy behind CWA is to minimize the set of true statements. A more careful strategy is adopted in *circumscription* (McCarthy, 1980, 1986). The idea here is to identify a set of predicates as special and minimize the extent (or instances) of only those predicates. The customary example is $\forall x((\text{Bird}(x) \wedge \neg\text{Ab}(x)) \supset \text{Flies}(x))$. Here birds can fly provided they are *not abnormal* (in their ability to fly). However, given the fact Bird(Tweety), it is not straightforward to establish that it is not abnormal, and the reasoning process can be seen to be beyond FOL. Consider the KB

$\{\forall x\,(\text{Bird}(x) \wedge \neg\text{Ab}(x) \supset \text{Flies}(x)), \text{Bird}(\text{Tweety}), \text{Bird}(\text{Peppy}), \text{Peppy} \neq \text{Tweety}, \neg\text{Flies}(\text{Peppy})\}$

This KB has two birds Tweety and Peppy. Peppy is known to not fly and is therefore abnormal. We do not know whether Tweety is abnormal or not. In some models of the KB it is abnormal and in others it is not. Entailment minimizes the Ab predicate by choosing minimal models as shown in Figure 10.10.

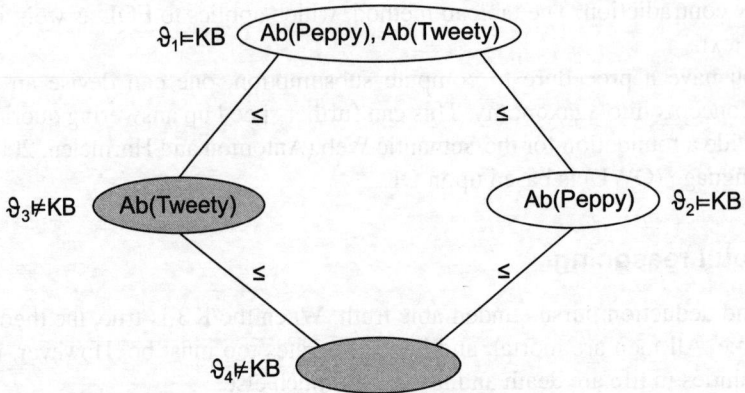

Figure 10.10 A lattice of interpretations. Of the four interpretations considered by circumscription, only two, ϑ_1 and ϑ_2, are models. The minimal model is ϑ_2 and in this only Peppy is abnormal. Tweety is not. Therefore, circumscription concludes that Tweety can fly.

There are two birds, and four different interpretations organized in a partial order. Of these, only ϑ_1 and ϑ_2 are models in which the KB is true. In the other two, Peppy is not abnormal and hence can fly. Which is not true. Of the two models, circumscription looks at only the smaller one, because its strategy is to accept conclusions only from minimal models.

Circumscription has its own difficulties. If we were to add a statement that penguins cannot fly, and are therefore abnormal, then reasoning with circumscription becomes entangled. Now in minimal models there are no penguins, which is not asserted in the KB.

10.5.4 Event calculus

Classical logic, also known as mathematical logic, does not have the concept of time and change. But the real world has. Change is the only constant, goes a popular saying. We have already described the approach taken in *event calculus* in Section 9.1.1. Change is the essence of planning and acting. Here we just look at a problem of reasoning about events and change, and highlight the issues arising out of incomplete information. The Yale shooting problem has been quoted as an example. The version below is from Shanahan (1999).

There are three types of action, Load, Sneeze, and Shoot. And three fluents, Loaded, Alive, and Dead. The following three Initiates and Terminates formulas describe the effects of actions:

$\quad$ Initiates(Load,Loaded,t)
$\quad$ Initiates(Shoot,Dead,t) $\leftarrow$ HoldsAt(Loaded,t)
$\quad$ Terminates(Shoot,Alive,t) $\leftarrow$ HoldsAt(Loaded,t)

The Yale shooting scenario comprises a Load action followed by a Sneeze action followed by a Shoot action.

Initially$_P$(Alive)	Initially the target is alive
Happens(Load,10)	At time 10 the Load action happens
Happens(Sneeze,30)	At time 30 the Sneeze actions happens
Happens(Shoot,60)	At time 60 the Shoot action happens

Show that

>HoldsAt(Dead,70)   At time 70 the target is dead.

The problem here is that we are working in the open world assumption. How do we know that everything that is relevant has been stated? What if the Sneeze action has the effect of making Loaded false? And how do we know that someone did not fire the gun at time 20 rendering Loaded false?

The answer is to resort to circumscription. We must restrict the effects of actions, and the happening of actions only to ones explicitly stated. Then we can say that *HoldsAt*(Dead, 70) is entailed under circumscription.

10.5.5 Epistemic logic

We have introduced epistemic reasoning in Section 9.7.7 on epistemic and coordinated actions in a multiagent scenario. Epistemic actions are the essence of communication. Here we present another example of epistemic reasoning.

Consider the following conversation between three perfectly rational agents:[5]

>*Neda:* Hello Kian and Nika! I have given you each a different natural number {0,1,2,...}. Who of you has the larger number?
>*Kian:* I don't know.
>*Nika:* I don't know either.
>*Kian:* Even though you say that, I still don't know.
>*Nika:* Aha! Now *I know which of us has the larger number*.
>*Kian:* In that case, I know both our numbers.
>*Nika:* And now I also know both numbers.

What are the two numbers?

To figure that out Nika and Kian have to reason about what the other person knows. Knowledge and belief are beyond the scope of FOL. One cannot have sentences like Knows(Kian, $N = 5$) or Believes(Nika, $N = 6$) because the arguments $N = 5$ and $N = 6$ are sentences in their own right, and arguments to predicates in FOL can only be terms. Moreover, how would one evaluate the truth values of such sentences?

The approach taken in epistemic logic is to use a modal operator $K_a\alpha$ where α is a formula in some logic. The sentence $K_a\alpha$ is read as 'Agent a knows α'. In a similar manner, doxastic logic employs a modal operator B where $B_a\alpha$ is read as 'Agent a believes α'. Logically the difference between knowing and believing is that one can only know what is true, while one can believe anything.

Without going into the details and formalization, the following epistemic reasoning takes place. When Neda makes the announcement, then the only case where Kian or Nika would know the answer is when their number is 0. But since Kian does not know, and announces it, his number must not be 0. Now Nika *knows*, because Kian said he does not know the answer,

---

[5] Adapted from http://jdh.hamkins.org/now-i-know/, accessed November 2022.

that Kian's number is not 0. If her number were to be 1, she could conclude that Kian has the higher number. But she does not know the answer. So now Kian knows that Nika's number is not 1. But he says he still does not know the answer. That means his number is higher than 2. *Now* Nika says that she knows who has the higher number! This can only be if *her* number were to be 2 or 3, because Kian can only have a higher number now. At this point we leave it to the reader to deduce what the two numbers are.

Summary

Since ancient times, logic has been concerned with uncovering the truth. The goal has been to identify valid forms of arguments. In modern times we look at the logic machine as operating on a formal language. We define the notion of entailment, associated with logical consequence, and proof as a means of arriving at conclusions. Our focus has been on illustrating the search that lies behind logical reasoning and deduction with algorithms that can connect facts to conclusions via rules of inference.

We have dwelt upon FOL because that is by far the most widely used language and includes our programming languages as well. We have described algorithms that are both sound and complete. But theorem proving in FOL is intractable. This leads to restricted languages like Horn clause logic and DL, which are computationally very efficient. Logic has come to be not just a machinery for deduction, but also a medium for knowledge representation. In the quest for agents being able to deal with real world problems, we mention more expressive languages. Default logic and its variations are designed to reason under the open world assumption when information available to the reasoner may be incomplete. We also mentioned event calculus, which models time and change, and epistemic logic, which allows an agent to reason about what other agents know. As we progress towards more expressive languages, the computational complexity increases. The search is on for efficient algorithms.

Doxastic logic is still in a nascent stage. But the need for being able to reason with beliefs is ever growing in this era of disinformation and fake news. Agents will need to check the consistency of their beliefs and perhaps weed out false beliefs. In strategic situations one may need to detect deception, or even indulge in it as well. That is a long way to go.

Deduction in logic is instrumental in digging out implicit facts from the KB. It does not generate any new knowledge. New knowledge comes from learning from induction. Machine learning (ML) has made great strides in extracting models from data. A model is experience distilled into knowledge. ML algorithms search in the space of models for ones that best fit the available data. We take a brief look at search in ML in the next chapter.

Exercises

1. Draw truth tables to show that the following sentences are tautologies.

    a. $((\alpha \wedge (\alpha \supset \beta)) \supset \beta) \equiv ((\beta \wedge (\beta \supset \alpha)) \supset \alpha)$
    b. $\alpha \supset (\beta \supset \alpha)$ (THEN-1)

  c. $(\alpha \supset (\beta \supset \gamma)) \supset ((\alpha \supset \beta) \supset (\alpha \supset \gamma))$ (THEN-2)
  d. $(\alpha \supset (\beta \supset \gamma)) \supset (\beta \supset (\alpha \supset \gamma))$ (THEN-3)
  e. $(\alpha \supset \beta) \supset (\neg \beta \supset \neg \alpha)$ (FRG-1)
  f. $\neg \neg \alpha \rightarrow \alpha$ (FRG-2)
  g. $\alpha \rightarrow \neg \neg \alpha$ (FRG-3)

2. Show that the following sentences are tautologies. These are the basis of the common rules of inference.

  a. $(\alpha \lor \text{true}) \equiv \text{true}$ domination
  b. $(\alpha \lor \text{false}) \equiv \alpha$ identity
  c. $(\alpha \land \text{true}) \equiv \alpha$ identity
  d. $(\alpha \land \text{false}) \equiv \text{false}$ domination
  e. $(\alpha \land \neg \alpha) \equiv \text{false}$ contradiction
  f. $(\alpha \lor \neg \alpha) \equiv \text{true}$ tautology
  g. $\alpha \equiv (\neg \alpha)$ double negation
  h. $\alpha \equiv (\alpha \lor \alpha)$ idempotence of $\lor$
  i. $\alpha \equiv (\alpha \land \alpha)$ idempotence of $\land$
  j. $(\alpha \lor \beta) \equiv (\beta \lor \alpha)$ commutativity of $\lor$
  k. $(\alpha \land \beta) \equiv (\beta \land \alpha)$ commutativity of $\land$
  l. $(\alpha \land (\beta \lor \alpha)) \equiv \alpha$ absorption
  m. $(\alpha \lor (\beta \land \alpha)) \equiv \alpha$ absorption
  n. $((\alpha \lor \beta) \lor \gamma) \equiv (\alpha \lor (\beta \lor \gamma))$ associativity of $\lor$
  o. $((\alpha \land \beta) \land \gamma) \equiv (\alpha \land (\beta \land \gamma))$ associativity of $\land$
  p. $\neg(\alpha \lor \beta) \equiv (\neg \alpha \land \neg \beta)$ DeMorgan's Law
  q. $\neg(\alpha \land \beta) \equiv (\neg \alpha \lor \neg \beta)$ DeMorgan's Law
  r. $(\alpha \land (\beta \lor \gamma)) \equiv ((\alpha \land \beta) \lor (\alpha \land \gamma))$ distributivity of $\land$ over $\lor$
  s. $(\alpha \lor (\beta \land \gamma)) \equiv ((\alpha \lor \beta) \land (\alpha \lor \gamma))$ distributivity of $\lor$ over $\land$
  t. $(\alpha \supset \beta) \equiv (\neg \beta \supset \alpha)$ contrapositive
  u. $(\alpha \supset \beta) \equiv (\neg \alpha \lor \beta)$ implication
  v. $(\alpha \equiv \beta) \equiv ((\alpha \supset \beta) \land (\beta \supset \alpha))$ equivalence
  w. $((\alpha \land \beta) \supset \gamma) \equiv (\alpha \supset (\beta \supset \gamma))$ exportation
  x. $((\alpha \supset \beta) \land (\alpha \supset \neg \beta)) \equiv \neg \alpha$ absurdity

3. Show that the following sentences are not true or not tautologies, by constructing a truth table or finding a falsifying assignment.
  a. $((\alpha \supset \beta) \lor \neg \delta)$
  b. $((\beta \land (\alpha \supset \beta)) \supset \alpha)$.

4. Show that there is an injection from the set of tautologies in PL to the set of contingencies in PL, and vice versa.

5. Show that there is an injection from the set of tautologies in PL to the set of unsatisfiable sentences in PL, and vice versa.

6. Show that there is an injection from the set of unsatisfiable sentences in PL to the set of contingencies in PL, and vice versa.
7. State the rule of UI and show using the resolution refutation method that it is a *sound* rule.
8. Negate the sentence $\forall x(\text{Boy}(x) \supset \exists y(\text{Girl}(y) \land \text{Love}(x, y)))$, use the substitution rules to simplify it, and express the negation in English.
9. Negate the sentence $\exists y(\text{Girl}(y) \land \forall x(\text{Boy}(x) \supset \text{Love}(x, y)))$, use the substitution rules to simplify it, and express the negation in English.
10. Find the MGU for the following sets of clauses, where 'a' is a constant and 'w, x, y, z' are variables. [Note. Each part requires an independent answer.]
    a. $P(a, x, f(g(y))) \lor P(z, h(z,w), f(w))$
    b. $P(a, x, f(g(x))) \lor P(z, h(z,w), f(w))$
11. Show that the following two formulas are logically different. What can you say about the truth values of the two formulas? Try and prove both of them.
    a. $\exists x(D(x) \supset \forall y D(y))$
    b. $(\exists x D(x) \supset \forall y D(y))$
12. Given the knowledge base KB1 in Section 10.4.3, the following facts cannot be deduced: Parent(Neda, Sarina), Sibling(Hadis, Sarina), Cousin(Zhina, Nika), GrandParent(Neda, Nika). Add rules to KB1 so that the above can be deduced. Do not add them as facts.
13. Given a KB {(Sibling(?x, ?y) $\supset$ Sibling(?y, ?x)) Sibling(Sarina, Hadis)}, how would backward chaining respond to the following queries:
    a. Goal: Sibling(Hadis, ?z)
    b. Goal: Sibling(Zhina, ?z)
14. An exercise in representation. Give the following *primitive* relations, define the other relations that you can name.
    a. Parent(X,Y): X is the parent of Y
        Schema: Parent(Parent, Child) helps in specifying the semantics for us when more than one element is involved.
    b. Male(X): X is male
    c. Female(X): X is a female
    d. Married(X,Y): X is married to Y

Hindi speakers will have a plethora of named relations like *dada, dadi, chacha, chachi, bua, tai, mama, nana,* and *nani* to choose from.

15. Consider the definition of ancestor given in the chapter, repeated below, and the alternatives to clause 2 given below.
    1. $\forall p,x\ (\text{Parent}(p,x) \supset \text{Ancestor}(p,x))$
    2. $\forall a,x\ [(\exists p\ (\text{Parent}(p,x) \land \text{Ancestor}(a,p))) \supset \text{Ancestor}(a,x)]$
       $\equiv \forall a,x,p\ [(\text{Parent}(p,x) \land \text{Ancestor}(a,p)) \supset \text{Ancestor}(a,x)]$

    $\forall a,x,p\ [(\text{Parent}(a,p) \land \text{Ancestor}(p,x)) \supset \text{Ancestor}(a,x)]$
    $\forall a,x,p\ [(\text{Ancestor}(p,x) \land \text{Ancestor}(a,p)) \supset \text{Ancestor}(a,x)]$
    $\forall a,x,p\ [(\text{Ancestor}(a,p) \land \text{Parent}(p,x)) \supset \text{Ancestor}(a,x)]$
    $\forall a,x,p\ [(\text{Ancestor}(p,x) \land \text{Parent}(a,p)) \supset \text{Ancestor}(a,x)]$
    $\forall a,x,p\ [(\text{Ancestor}(a,p) \land \text{Ancestor}(p,x)) \supset \text{Ancestor}(a,x)]$

Which of the above would be more efficient and for what kind of queries. What about the following queries?

Goal: $\exists x\, \text{Ancestor}(x, \text{Mahsa})$
Goal: $\exists x\, \text{Ancestor}(\text{Aishe}, x)$

16. Let the edges between two nodes Node1 and Node2 in a graph be stored in a KB with predicates Edge(Node1, Node2). Define the predicate Path(Start, Destination), which is true if there is a path in the graph from node Start to node Destination.

17. Construct a truth table to show that the following equivalence is a tautology:

$$(R \vee Q) \wedge (P \vee \neg Q) \equiv ((R \vee Q) \wedge (P \vee \neg Q) \wedge (R \vee P)$$

18. Show that the KB $\{\forall x(\text{Bird}(x) \supset \neg \text{Flies}(x)), \text{Bird}(\text{Peppy}), \neg \text{Flies}(\text{Peppy})\}$ is inconsistent.

19. Given the following set of sentences (where *a*, *b,* and *c* are constants),

    $\{\forall x \forall y\, (\text{proxy}(x,y) \supset (\neg \text{honest}(x) \wedge \neg \text{honest}(y))),$
    $\forall x\, (\text{sincere}(x) \supset \text{honest}(x))$
    proxy(a,b)
    sincere(c)}
    prove the following sentence:
    $\exists x \exists y\, (\text{honest}(x) \wedge \neg \text{honest}(y))$

20. *Prove* the following formula using the resolution refutation method:

    $[\forall x \exists y P(x, y) \wedge \forall y \exists z Q(y, z)] \supset \forall x \exists y \exists z (P(x,y) \wedge Q(y,z))$

21. Add the appropriate equality axioms and show using the resolution refutation method that the following set of statements is inconsistent. Here *nisha* and *t* are constants, while *P* and MPP are functions.

    nisha = $P(t)$
    $\forall x\, (\text{MPP}(x) = P(x))$
    nisha ≠ MPP(t)

22. Convert the Police Thug problem repeated here into clause form and derive a resolution refutation proof for the query.

    Some police person murdered Mahsa. Anyone who murders someone is a murderer. All murderers must be prosecuted. Therefore, the police person must be prosecuted.

23. Express the following facts in FOL. Ignore tense information. *Disha is a girl. Disha owns a dog. Curiosity is a policeman. Dem is a cat. Every dog owner is an animal lover. No animal lover kills an animal. Dogs and cats are animals. One of Disha and Curiosity killed Dem.*

    Use the resolution refutation method with the Answer Predicate to answer the query 'Does there exist a policeman who killed Dem?'

24. What are the two numbers in the puzzle in Section 10.5.5?

CHAPTER 11

Search in Machine Learning

Sutanu Chakraborti

The earliest programs were entirely hand coded. Both the algorithm and the knowledge that the algorithm embodied were created manually. Machines that learn were always on the wish list though. One of the earliest reported programs was the checkers playing program by Arthur Samuel that went on to beat its creator, evoking the spectre of Frankenstein's monster, a fear which still echoes today among some. Since then machine learning (ML) has steadily advanced due to three factors. First, the availability of vast amounts of data that the internet has made possible. Second, the tremendous increase in computing power available. And third, a continuous evolution of algorithms. But the core of ML is to process data using first principles and incrementally build models about the domain that the data comes from. In this chapter we look at this process.

The computer is ideally suited to learning. It can never forget. The key is to incorporate a ratchet mechanism *à la* natural selection – a mechanism to encapsulate the lessons learnt into a usable form, a model. Robustness demands that one must build in the ability to withstand occasional mistakes. Because the outlier must not become the norm.

Children, doctors, and machines – they all learn. A toddler touches a piece of burning firewood and is forced to withdraw her hand immediately. She learns to curb her curiosity and pay heed to adult supervision. As she grows up, she picks up motor skills like cycling and learns new languages. Doctors learn from their experience and become experts at their job – in fact, the words 'expert' and 'experience' are derived from the same root. The smartphone you hold in your hand learns to recognize your voice and handwriting and also tracks your preferences for recommending books, movies, and food outlets in ways that often leave you pleasantly surprised. This chapter is about how we can make machines learn. We also illustrate how such learning is intimately related to the broader class of search methods explored in the rest of this book.

Let us consider a simple example: the task of classifying an email as spam or non-spam. Given the ill-defined nature of the problem, it is hard for us to arrive at a comprehensive set of rules that can do this discrimination. Instead, we start out with a large number of emails, each of which has been marked as spam or non-spam by humans – this constitutes our set of observations. We hypothesize a set of three numeric features[1] that may be useful in discriminating spam emails from legitimate ones:

C: proportion of capitalized words in the email,

NL: proportion of words containing a mix of numbers and letters, and

SW: proportion of words in the email that appear in a precompiled list that are indicative of frauds, scams, or online advertisements.

Next, we represent each email in our dataset in terms of these three features and attempt to arrive at a hypothesis that successfully discriminates spam emails from non-spam ones. A template for such a hypothesis is as follows:

if $(C > p_1)$ AND $(NL > p_2)$ AND $(SW > p_3)$
   then MAIL is *SPAM*
   else MAIL is *NON_SPAM*

where p_1, p_2, and p_3 are constants in the closed interval $[0,1]$. Given this template, it is evident that each distinct combination of p_1, p_2, and p_3 leads to a distinct hypothesis, and there are infinite such hypotheses. Of these, we need to identify one that is expected to work the best not only over the given dataset, but also in terms of classifying emails outside those that are present in the dataset.

At the heart of *machine learning* (ML) is this ability to *search* over a space of candidate *hypotheses* and narrow down on one or more of the most promising ones that explain our *observations*. Approaches differ with respect to each other in terms of the *representation* of the hypotheses and the specific search mechanism. For instance, the if–then rule template above for spam classification could be generalized to allow both AND and OR connectives (disjunctions over conjunctions) in the antecedent. In Section 11.1 we look at decision trees that can discover such rules from data. Yet another possibility is to have a radically different template for representing our hypotheses, such as the one shown below.

if $w_1.C + w_2.NL + w_3.SW > threshold$
   then MAIL is *SPAM*
   else MAIL is *NON_SPAM*

We have a distinct hypothesis for each distinct choice of values of w_1, w_2, w_3 and *threshold*, which are referred to as *parameters*. Again, we have an infinite set of hypotheses. In Section 11.4, we turn our attention to *neural networks*, which use this template for representing hypotheses.

Irrespective of the diversity in the representation of hypotheses and the mechanisms used for search over the hypothesis space, a bottom-up (data driven) process that relies on *inducing* hypothesis from observations is central to ML, and this contrasts sharply with top-down (goal

---

[1] The example is solely for illustrative purposes. In practice, we would need a wider pool of more involved features for building an effective spam filter.

driven) approaches in classical (good old-fashioned) artificial intelligence. In order to situate inductive reasoning that drives ML in perspective, we recall from our discussion in Section 1.5.2 that humans and machines rely on three primary methods of reasoning: induction, deduction, and abduction. Let us say we have a cause C ('COVID-19'), an effect E ('body ache'), and a rule R ('COVID-19 $\rightarrow$ body ache'). Inferring E given C and R constitutes deduction. Induction arrives at R given C and E. Abduction postulates C given E and R. Unlike deduction, induction is not truth preserving (or sound) in the sense that the generalization that COVID-19 results in body ache is a 'leap of faith' based on data we have observed, but may fail to explain future instances we may encounter. Similarly, abduction is not sound either, since not all cases of body ache can be attributed to COVID-19. Having said that, induction and abduction have been shown to play a central role in aspects of human cognition like imagination, forming beliefs, and generating explanations.

Let us understand the nature of this inductive process using a simple hypothetical example. Let us say we have historical records on how the delay in treatment of cancer (D) affects the risk of death (R). These correspond to data points that are plotted in Figure 11.1(a). The task is to predict the risk of a new patient dying, given the delay in her treatment. In reality, there are several factors other than delay in treatment that determine the prognosis, such as the nature and severity of the disease when detected. Those details need not bother us now.

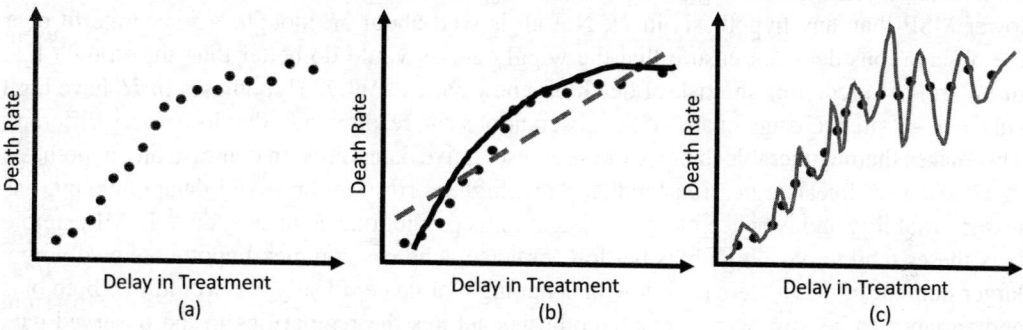

Figure 11.1 An example to illustrate regression

In order to solve a prediction task like this, we first need to build (induce) a theory of the world – in this case, a theory of how R varies as a function of D. We will define a space $\mathcal{H}$ of candidate hypotheses, say, the class of linear functions $R = a + bD$. Each hypothesis in this space $\mathcal{H}$ is characterized by a unique choice of *parameters*, a and b, and can be pictured as a straight line fit to the datapoints as shown by the dashed grey line in Figure 11.1(b). Since a and b are real valued, it is evident that we have an infinite hypothesis space. We now search for a hypothesis that *best fits* the given set of observations. This entails finding values of a and b that minimize a well defined criterion called the *objective function*, the formulation of which closely corresponds to our intuition of what constitutes a 'good' hypothesis. In the current context, the objective function should be low if the estimates of R produced by a hypothesis are close to the true values, given the values of D. For a datapoint $<D_i, R_i>$, a hypothesis with parameters a and b generates an estimate $\hat{R}_i = a + bD_i$. The following formulation called *mean squared error* (MSE), which aggregates the deviations of $\hat{R}_i$ with respect to R_i over all N datapoints while remaining agnostic to whether such deviations are positive or negative, is a common choice for objective function:

$$MSE = \frac{1}{N} \sum_{i=1}^{N} \left(\hat{R}_i - R_i \right)^2$$

We can find values of a and b that minimize MSE by equating the partial derivatives $\frac{\partial MSE}{\partial a}$ and $\frac{\partial MSE}{\partial b}$ to 0, and these values will characterize the 'best' hypothesis in $\mathcal{H}$. Later, we will encounter harder problems where such analytic closed form solutions are not possible. However, the essential idea of searching over a hypothesis space for finding parameters that optimize an underlying objective function forms the essence of most ML approaches.

The best hypothesis in $\mathcal{H}$ as obtained above may fail to correspond to the true unknown underlying functional relationship between D and R, which is shown pictorially as the bold curve in Figure 11.1(b). In this case, the reason is simple: no straight line can model such a curve. The inability of a simple model like linear regression to capture more complex underlying relationships is called *bias*. An obvious solution is to use a richer space of hypothesis $\mathcal{H}'$ that can accommodate higher order polynomials, involving a larger number of parameters that need to be estimated. Figure 11.1(c) shows the result of such a nonlinear regression on the same dataset. Clearly, the wiggly curve fits the given datapoints better and would result in a lower MSE than any hypothesis in $\mathcal{H}$. Not all is well about $\mathcal{H}'$ though. An accurate fit over the observations does not ensure that the wiggly curve would do better than the straight line fit in terms of predicting the risk of death of a new patient. Why? Hypotheses in $\mathcal{H}'$ have high *variance* – a small change in any of the observations can result in a fit that looks very different. This makes them vulnerable to noise or unrepresentative datapoints. In contrast, any hypothesis in $\mathcal{H}$ based on linear regression tends to have higher errors on observed datapoints, but has lesser variability and is hence more consistent in its predictions on unseen data. In ML, simple hypotheses tend to have high bias but low variance, whereas complex hypotheses involving a larger number of parameters have low bias but high variance. Ideally, we would like both bias and variance to be low, so that our hypothesis captures the regularities in the observed data and also leads to consistent predictions on unseen data. Finding such a sweet spot is hard in practice. This is referred to as the problem of *bias–variance tradeoff*. A wiggly curve from $\mathcal{H}'$ fits the observations accurately but fails to *generalize* well over unseen data – this is referred to as the problem of *overfitting*. In contrast, straight line fits drawn from hypotheses in $\mathcal{H}$ suffer from *underfitting*, because they lead to inaccurate estimates over observed data.

The datapoints used for inducing a hypothesis constitute the *training data*. In contrast, a separate set of datapoints called the *test data*, which was not exposed to the learner, is used for its ability to generalize and carry out prediction. In experiments, we are often just given a single dataset, a portion (say, 80%) of which is reserved for training purposes and the rest (say, 20%) used for testing. This is called a *train–test split*. In order to compare the hypotheses produced by ML approaches M_1 and M_2, we can create 10 such randomized train–test splits, and record the ten MSE values over training and test data for both hypotheses. A *statistical hypothesis test*, such as a paired *t*-test, is then carried over the 10 pairs of recorded values to arrive at a conclusive evidence of superiority of one of these hypotheses over the other, at a chosen level of confidence.

It is worth noting that no ML can happen unless we have some preconceived notion of the nature of hypotheses we intend to induce from observations. This preconceived notion is the learner's bias. When restricted to the hypothesis space $\mathcal{H}$, the learner is biased in favour of its supposition that R linearly depends on D. No learner is bias-free. Since a learner relies on induction, any prediction it makes outside the instances it has observed involves a 'leap of faith'. A learner that is empirically found to do well on classification or regression tasks can be deliberately shown to perform very poorly, simply by presenting test instances that do not abide by the regularities seen in the training data. This forms the key intuition behind the *no free lunch theorem*.

We have looked into two kinds of learning problems: classification and regression. Both fall under the broad category of *supervised learning*, where the training data is labelled. In classification, these labels are category names (such as *spam* or *non-spam* in our earlier example); in regression they are the values of dependent variables (such as risk of death). There are learning problems where no labels are available, and the goal is to infer hidden patterns or clustering tendencies in data. They fall under the category of *unsupervised learning*. There is a third kind of learning, called *reinforcement learning*, where an agent interacts with an environment, resulting in a change of state of the environment. In addition to having access to the *state* information, the agent also receives feedback in the form of a *reward* in proportion to the extent that its action positively contributed towards its goal – this determines its next action. Unlike supervised learning, there is no pre-labelled dataset – rather, the learner has to rely on intermediate reward values, and the action sequence it choses by way of exploring the environment determines the training examples. Reinforcement learning finds application in tasks like autonomous driving or playing board games like *go*.

In the rest of this chapter we take a closer look at four supervised learning approaches and one unsupervised clustering algorithm. The unifying thread is that they can all be viewed as realizing search over hypothesis spaces with the goal of inducing models from raw observations.

11.1 Decision Trees

A decision tree is a condensed representation of a set of rules that map attribute values or ranges of attribute values to class labels. Such a decision tree is induced from a given training dataset. An example is shown in Figure 11.2, where the goal is to classify a job applicant into one of three classes *Accepted, Rejected,* or *Waitlisted*, based on two attributes: her 10-point cumulative grade point average (CGPA) and her performance in the job interview. *CGPA* is a numeric attribute, whereas the performance in job interview is ordinal in that it can assume one of three symbolic values, viz. *Poor, Average,* and *Excellent*, with a natural ordering over these values. The input to the learning algorithm is the training data consisting of 600 instances, along with their attribute values and class label (the column *Decision*), as shown in Figure 11.2(a). The output is a decision tree as shown in Figure 11.2(b). The leaves of a decision tree correspond to class labels. All other nodes correspond to tests on attribute values. Decision trees are quite popular, since the hypothesis induced can be read out as a set of rules, which can be easily interpreted by humans.

Each path from the root to the leaf of a decision tree captures a rule, whose antecedent is a conjunction over specific tests on attribute values, and the consequent is a class label. For

Id	CGPA	Performance in Interview	Decision
1	9.5	Excellent	Accepted
2	7.2	Average	Rejected
3	8.8	Average	Waitlisted
4	8.3	Poor	Rejected
...	...	...	...
600	7.6	Excellent	Waitlisted

(a)

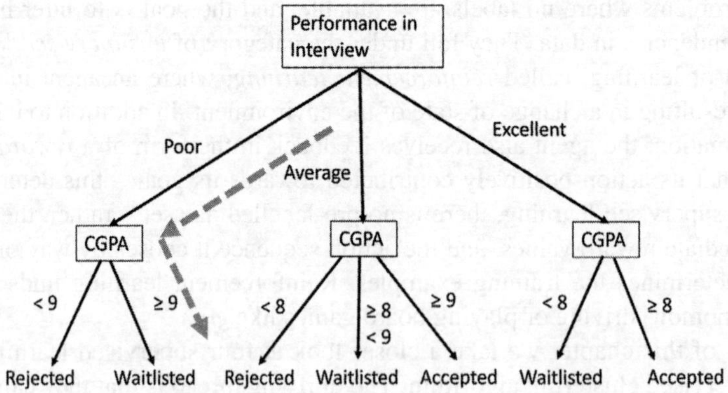

(b)

Figure 11.2 Illustration of a decision tree (b) induced from training examples in (a)

instance, the path corresponding to the grey shaded arrow in Figure 11.2(b) corresponds to the rule

if ((*Performance in Interview* is *Poor*) AND (*CGPA* $\geq$ 9))
    then (*Decision* is *Waitlisted*)

What is the hypothesis space of the decision tree classifier? It is the set of all candidate trees that fit the training data. These candidate trees differ from each other across one or more of the following criteria:

1. **Choice of attributes**: In a general classification setting, instances can have many attributes, and not all may be equally well suited for classification purposes. In our simple example, even if we had access to additional attributes like *Name* or *Gender*, they should ideally not play a role in deciding the outcome of a job interview. Candidate trees could differ with

respect to the attributes they use for classification. Figure 11.3 shows an unwieldy tree resulting from the test on the gender attribute at the root node. The same subtree rooted at the node 'Performance in Interview' is duplicated for both branches, since gender plays no role in deciding the class label (outcome).

2. **Ordering of attributes:** In the example in Figure 11.2(b), it is evident that as we read out a rule along a path from the root to the leaf, the test on the attribute *Performance in Interview* is done ahead of the test on *CGPA*. We could have obtained a different tree by doing this differently: testing on *CGPA* at the root node, and testing on *Performance in Interview* next. If the ordering is not done in a principled way, it can result in an unwieldy tree, where redundant comparisons are needed to classify a test instance.

3. **Attribute splits:** If an attribute is real valued or has a large number of categorical or ordinal values, the strategy used to split the attributes and the number of resulting splits will determine the nature of the induced tree. In our example, we have used three splits ($CGPA < 8$, $8 \leq CGPA < 9$, $CGPA \geq 9$) on the real valued attribute *CGPA*; choosing other values to split the *CGPA* attribute can result in trees that look very different.

The decision tree induction algorithm performs a search over the hypothesis space of all candidate trees to find a tree that is compact, based on objective measures that determine the attributes to be used, the order in which attributes have to be tested, and the splitting criterion.

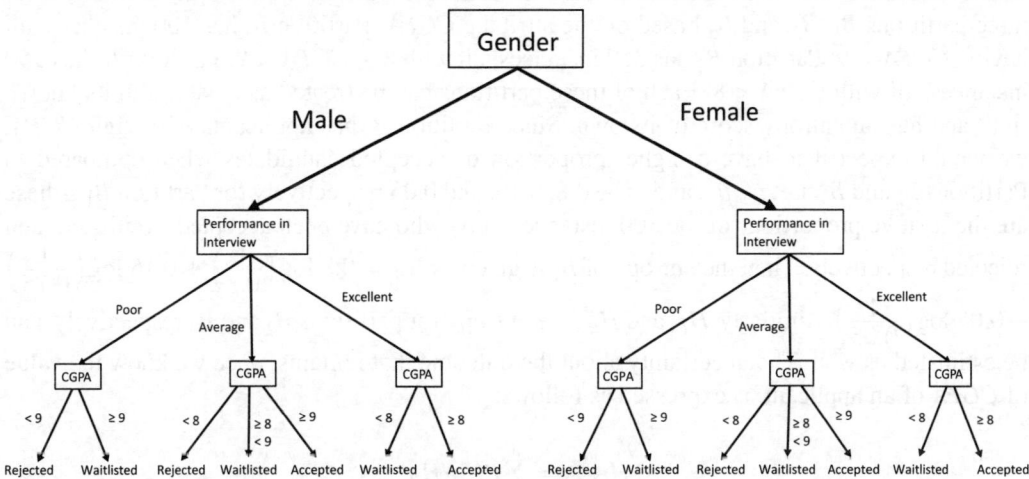

Figure 11.3 Illustrating the influence of choice of attribute on decision tree induction

In order to decide on the attributes we need to test at each node, we use a criterion called *information gain*. The intuition is as follows. Let us say, we have two datasets A and B, each having 600 instances of job applicants. In dataset A, the number of applicants accepted, waitlisted, and rejected are 20, 20, and 560 respectively. In contrast, the corresponding numbers in dataset B are 200, 200, and 200. It is evident that in the absence of any additional information about a candidate (her *CGPA* or *Performance in Interview*), we would be more uncertain about the outcome of a candidate in dataset B compared to that of a candidate in dataset A. *Entropy* is a measure of that uncertainty – the higher the uncertainty or randomness, the higher the entropy. Let p_a, p_w, and p_r be the probabilities of an applicant being accepted, waitlisted, and rejected respectively. For dataset A, $p_a = p_w = 0.03$ and $p_r = 0.93$; for dataset B, $p_a = p_w = p_r = 0.33$. Dataset A, which is skewed in favour of the class of rejected candidates, has lower entropy compared to dataset B where all outcomes are equally probable. Entropy is defined mathematically as

$$H = \sum_{i=1}^{C} p_i \log(1/p_i)$$

where C is the total number of classes and the p_i values correspond to probabilities of each class. In our example, $C = 3$, and p_1, p_2, and p_3 correspond to p_a, p_w, and p_r respectively. An interesting aspect of the entropy formulation above is that it attains its maximum value when all p_i values are the same. Therefore, it is in line with our intuition of entropy as a measure of randomness.

How does this help us in deciding the order in which attributes have to be tested in a decision tree? Note that the entropy H formulated above quantifies uncertainty in absence of any knowledge of attribute values. We would expect that knowledge of *CGPA* or of *Performance in Interview* should result in a *reduction* in the value of this uncertainty or a corresponding gain in information; this leads us naturally to the concept of *information gain* (IG). An attribute with a higher IG score should be preferred over another that has a lower IG score. Let us consider dataset B, which has 200 instances in each of the three classes. Let us split dataset B into three partitions B_1, B_2, and B_3 based on the attribute *CGPA*: Partition B_1 has 100 instances, all having $CGPA \geq 9$; Partition B_2 has 350 instances, all with $8 \leq CGPA < 9$; Partition B_3 has 150 instances, all with $CGPA < 8$. Each of these partitions can be treated as a dataset in its known right and has an entropy score of its own. Since Partition B_1 has applicants with high *CGPA*, we would expect it to have a higher proportion of accepted candidates when compared to Partitions B_2 and B_3. Let p_a, p_w, and p_r be 0.8, 0.15, and 0.05 respectively for Partition B_1 – these are the relative proportions of the 100 instances in B_1 who have been accepted, waitlisted, and rejected respectively. Then the entropy of B_1 is given as $H_{B_1} = 0.8 \log\left(\frac{1}{0.8}\right) + 0.15 \log\left(\frac{1}{0.15}\right) + 0.05 \log\left(\frac{1}{0.05}\right)$. Similarly, H_{B_2} and H_{B_3}, the entropy for partitions B_2 and B_3 respectively, can be estimated as well. The uncertainty about the outcome that remains, once we know the value of *CGPA* of an applicant, is expressed as follows:

$$H_{CGPA} = \sum_{i=1}^{3} P(B_i) H_{B_i}$$

H_{CGPA} is the weighted sum of H_{B_1}, H_{B_2}, and H_{B_3}, where the weights $P(B_1)$, $P(B_2)$, and $P(B_3)$ are the relative proportions of data in the three partitions. In our example, $P(B_1)$, $P(B_2)$, and $P(B_3)$ are (100/600), (350/600), and (150/600) respectively. H_{CGPA} evaluates to 1.39 bits. If the knowledge of *CGPA* alone is sufficient to classify all instances, the partitions B_1, B_2, and B_3 are *pure*, in that each of them contains instances of just one class. In that case, H_{CGPA} is minimum (0). On the other hand, knowledge of an attribute like *Gender* hardly helps in predicting the outcome, since it leads to two partitions corresponding to two values (*Male* and *Female*), each of which is expected to be impure. This is because each partition is expected to contain a mix of instances drawn from all three classes. Thus, H_{Gender}, which quantifies the uncertainty about the outcome that remains, once we know the gender of an applicant, is expected to be conspicuously higher than H_{CGPA}. Similarly, we can estimate H_F for any attribute F. The IG resulting due to F, denoted as IG_F, is defined as the difference between the original entropy H and the updated entropy H_F:

$$IG_F = H - H_F = H - \sum_{i=1}^{n} P(D_i) H_{D_i}$$

Note that the formula for H_F is a generalization over the 3-partition case, where the different values (or splits) of attribute F lead to n partitions D_1 through D_n, and H_{D_i} is the entropy of the ith partition.

The decision tree induction algorithm places nodes corresponding to attributes that have higher IG closer to the root, since they are better in terms of discriminating between classes. In the decision tree in Figure 11.2(b), the test on *Performance in Interview* is carried out at the root node, since the IG score of the attribute *Performance in Interview* is higher than the IG score of the attribute *CGPA*. Note that similar tests on IG are carried out recursively at each intermediate level of the tree as well. This is a straightforward extension of the idea described earlier. Each branch emanating out of a node represents a subset of data having certain values (or ranges of values) of the attribute corresponding to that node. The attribute that has the highest IG score with respect to that subset is the one that is tested against in the next node.

The process described above for the construction of decision trees is formalized in Algorithm 11.1. Given a data set D with attributes $\{A_1, A_2, ..., A_p\}$, the algorithm successively partitions the data set till some termination criterion.

Algorithm 11.1. Sketch of a simple algorithm to construct a decision tree with training data D with attributes $\{A_1, A_2, ..., A_p\}$. The algorithm chooses an attribute F∈ $\{A_1, A_2, ..., A_p\}$ which gives maximum IG, and adds a branch for each value V of F. For each child with high entropy, a recursive call is made to further partition the data to create subtrees.

DecisionTree(data: D, Attributes = $\{A_1, A_2, ..., A_p\}$)
1.   F ← attribute in Attributes with highest IG on partitioning D
2.   tree ← make F the root of the tree
3.   **if** termination criterion on entropy not met
4.   **then**
5.   **for each** value V of F
6. D_v ← select data elements from D with F = V
7.   subtree$_v$ ← DECISIONTREE(data: D_v, Attributes = $\{A_1, A_2, ..., A_p\}$)
8.   add subtree$_v$ below tree
9.   **else**
10.   **return** tree

As shown in the algorithm, a stopping criterion is needed to terminate the recursive process of growing the tree. The simplest criterion is to terminate when all instances in the subset of data represented by a branch belong to the same class, that is, the corresponding partition is pure. However, this criterion may lead to very deep trees that are likely to overfit the training data. To prevent this, the decision tree may be pruned after fully growing it, or, alternatively, the pruning can be done as the tree is being built. In the context of our earlier discussion on bias–variance trade-off, pruned decision trees are more compact, and have higher bias and lower variance, compared to fully grown trees. Yet another issue we have not discussed is the procedure for splitting continuous attributes like *CGPA*. In this context, while it seems intuitive to choose split points such that the gain in information is maximized, we also want to avoid too many splits since this can lead to overfitting.

Once the decision tree is constructed from training data, classifying any given test instance is straightforward: attribute tests are carried out at each node starting with the root all the way down till we reach a leaf node, from which we can read out the class label. Unlike some of the other classification approaches that we discuss next, decision trees are not black box in nature, in that they naturally produce easily interpretable rules, as long as they are short and compact.

11.2 *k*-Nearest Neighbour

The *k*-nearest neighbour (*k*NN) algorithm is an instance of a family of algorithms that goes by the name *instance based learning* (Fix and Hodges, 1989). The basic idea behind the *k*NN algorithm is fairly straightforward. Let us say we have training data, where each instance has a set of attribute values and a class label associated with it. We are interested in classifying a test instance whose attribute values are known but the class label is unknown. We use a distance function to find *k* training instances that are closest to the test instance. The test instance is assigned the label associated with a majority of these *k* training instances. The underlying premise of the *k*NN algorithm is that, given an appropriate distance function, instances that are close to each other tend to share the same class label.

Figure 11.4 shows examples of the *k*NN algorithm at work for $k = 1$ and $k = 3$, for a binary classification task with two class labels – square and circle. The test instance is shown with a cross. In Figure 11.4(a), $k = 1$, so only the closest training instance, which is *A*, is consulted. Since *A* has the class label circle, the test instance is declared to belong to the circle class as well. The same dataset is used for illustration in Figure 11.4(b) for $k = 3$, so the top three most similar cases to the test instance (the cross), viz. *A, B*, and *C*, are used for the classification. Since *B* and *C* belong to the square class, and *A* belongs to the circle class, the majority vote is in favour of the square class, which is assigned to the test instance. Interestingly, the examples in Figures 11.4(a) and (b) show that given the same training data, different choices of the value of *k* can lead to very different classifications of a test instance.

Yet another factor that plays a critical role in *k*NN classification is the choice of the distance function. Let us assume that all attributes assume numeric values. In such a case, the Euclidean distance between two instances can be used as a measure of distance. More complicated distance (or similarity) functions can be defined over attributes that are non-numeric, such as categorical variables or strings. These 'local' (attribute specific) similarities are aggregated to

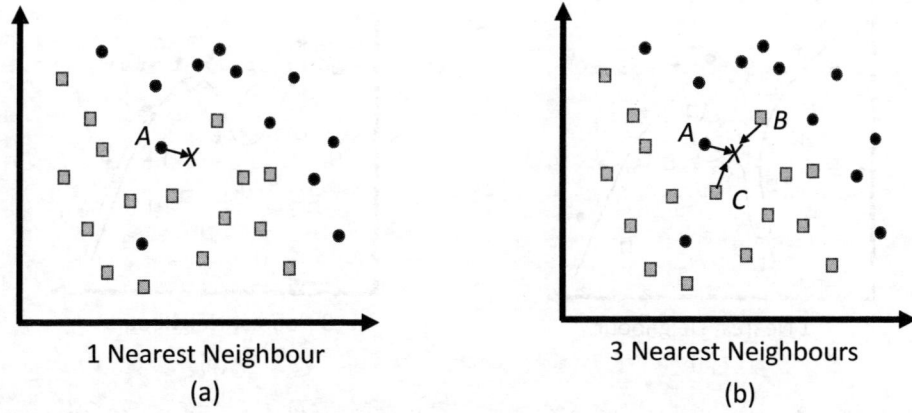

Figure 11.4 Illustrating the kNN algorithm for (a) $k = 1$ and (b) $k = 3$.

yield a 'global' measure of similarity between instances.[2] Also, weights can be associated with each attribute, such that attributes that play a more important role in the classification process can be weighed relatively higher than others.

Another simple improvement over the basic kNN approach self-suggests itself. Taking a majority vote over the near neighbours ignores their relative distances from the test instance. Ideally, we would like the closer neighbours to have a higher say in classification than the neighbours that are distant. This version is called the DISTANCEWEIGHTED-kNN algorithm (Dudani, 1976).

What is the nature of the hypothesis space induced by the kNN algorithm? We can answer this by investigating the nature of decision boundaries produced by the classifier for different values of k. When $k = 1$, the decision boundary closely fits the individual training instances – hence, a small change in the training instances has a pronounced influence on the decision boundary. Thus a low value of k leads to overfitting, with high variance and low bias. As we increase k, we have higher bias and lower variance, and the decision boundary becomes smoother. The influence of k on the nature of decision boundaries is illustrated in Figures 11.5(a) and (b).

Unlike many other classification techniques, the kNN algorithm does not explicitly construct a model out of training data ahead of time. Rather, it waits till a test instance comes in, and classifies the same based on the training data in its local neighbourhood alone. Hence, kNN is referred to as a *lazy* learner. This may be contrasted with *eager* approaches like decision trees, where the induced hypothesis is agnostic to the test instance being classified. Also, in comparison to eager learners, lazy learners are faster to train but slow in classifying fresh queries.

---

[2] Such generalized notions of similarity find application in the field of *case based reasoning* (CBR), a model of experiential reasoning where past problem solving episodes are recorded in the form of cases (refer to Section 1.5.1). Each case can be represented in terms of attribute–value pairs. Given a new problem, the closest cases are retrieved and reused to suggest a solution. CBR finds application in ill-defined domains in tasks such as diagnosis (helpdesks, for example), design, configuration, and planning.

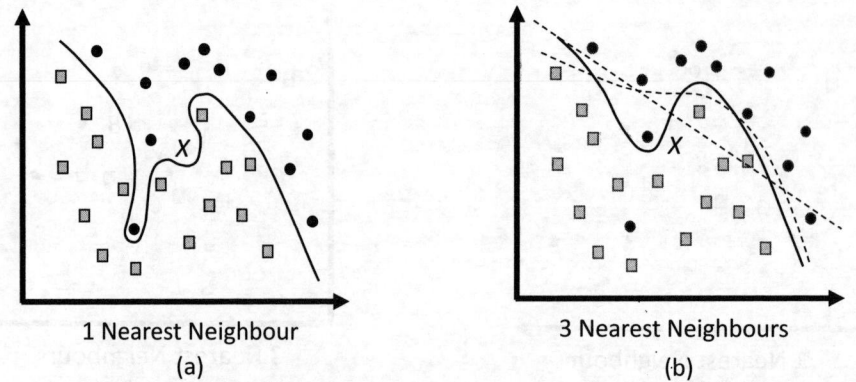

1 Nearest Neighbour
(a)

3 Nearest Neighbours
(b)

Figure 11.5 Implicit hypothesis spaces for $k = 1$ on the left and $k = 3$ on the right. Although no model has been explicitly constructed, one can imagine the decision boundaries. As the decision boundaries become smoother as shown by the dashed lines on the right, the classifier shows more error on the training data.

11.3 Bayesian Classification

The idea of a Bayesian classifier is based on the noisy channel model illustrated in Figure 11.6. Let us consider a patient visiting a doctor with a complaint of a sore throat. Given this symptom, the doctor has to identify which of several diseases, say, dengue, common cold, COVID-19, or throat cancer, could have caused it. The term 'channel' in Figure 11.6 refers to the human body, which expresses one or more symptoms given a disease. The term 'decoder' refers to the inferencing done by the doctor who uses the symptom(s) as the basis for her diagnosis. The decoder can alternatively be viewed as performing a classification. Given training data where symptoms are recorded as attribute values and diseases as class labels, the decoder uses Bayes' rule to perform classification:

$$P(D|S) = \frac{P(S|D) \cdot P(D)}{P(S)}$$

where S is the set of symptoms and D is the disease. The conditional probabilities $P(D|S)$ and $P(S|D)$ are referred to as *posterior probability* and *likelihood* respectively. $P(D)$ is the *prior* probability of the disease D, and $P(S)$ is referred to as the *evidence*. Given a test instance with symptoms (or attribute values) S, diseases are ranked on the basis of their posterior probabilities, and the disease with the highest posterior probability is the class to which the test instance is assigned. It is clear that the denominator $P(S)$ is independent of the disease and hence plays no role in classification. Therefore, in order to estimate rank classes based on their posterior probability, we need to estimate two terms: the likelihood and the prior. The likelihood term is central to the Bayesian formulation, as it answers the question: how likely is it for a disease to *generate* a given set of symptoms? The prior probability $P(D)$ is the probability of each disease and is agnostic to the symptoms observed. For example, throat cancer, being a relatively rare disease, has a much smaller prior probability compared to flu. Thus, even though both flu and throat cancer have comparable $P(S|D)$ terms and thus are almost equally likely to generate the symptom of sore throat, a patient with a sore throat is more likely to be suffering from flu than from throat cancer. This is because the much higher prior term associated with flu makes its posterior $P(D|S)$ higher than that of throat cancer.

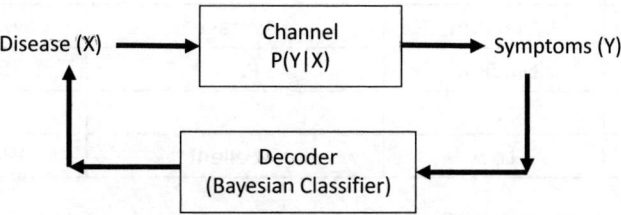

Figure 11.6 Illustrating the Noisy Channel Model

When generalized to a setting where several symptoms S_1 through S_n are used to diagnose the disease, the likelihood term takes the form $P(S_1, S_2, \ldots, S_n|D)$. The difficulty in estimating this term is that in order to get reliable estimates of this term, we need really large training data with the same instance repeated several times over. Hence, a simplifying assumption is often made, where the symptoms (attribute values) are assumed to be conditionally independent given the disease (class label). The likelihood term can be expressed as the product of individual attribute-specific conditional probabilities as shown below:

$$P(S_1, S_2 \ldots S_n|D) = \prod_{i=1}^{n} P(S_i|D)$$

Robust estimates can be obtained for each of the terms $P(S_i|D)$. A classifier based on the simplifying assumption above is called the Naïve Bayes classifier (Lewis, 1998).

Let us illustrate how the prior, likelihood, and posterior terms of a Naïve Bayes classifier are estimated from training data using an example. We use the same dataset as used in Figure 11.2(a) except that the *CGPA* scores are represented using ordinal values *High, Medium,* and *Low*. The form of the dataset is shown in Figure 11.7. The goal is to classify a test instance having attribute values *CGPA* and *InterviewPerf* as belonging to one of the three categories *Accepted, Rejected,* or *Waitlisted*.

$$P(Decision|CGPA, InterviewPerf) = \frac{P(CGPA, InterviewPerf|Decision) \cdot P(Decision)}{P(CGPA, InterviewPerf)}$$

The likelihood term is estimated in the Naïve Bayes classifier by assuming that the *CGPA* and *InterviewPerf* are conditionally independent of each other given the class label *Decision*. Thus,

$$P(CGPA, InterviewPerf|Decision) = P(CGPA|Decision) \cdot P(InterviewPerf|Decision)$$

Based on the training data, we need to estimate three quantities $P(CGPA|Decision)$, $P(InterviewPerf|Decision)$, and $P(Decision)$; these are *parameters* of the model. Let us say, the test data has attribute values *CGPA = High*, and *InterviewPerf = Average*. The goal is to assign one of the three class labels (*Decision*) to this instance. The posteriors

Id	CGPA	Performance in Interview	Decision
1	High	Excellent	Accepted
2	Low	Average	Rejected
3	Medium	Average	Waitlisted
4	Medium	Poor	Rejected
...	...	...	...
600	Low	Excellent	Waitlisted

Figure 11.7 Training Instances for the Naïve Bayes classifier

$P(Decision|CGPA = High, InterviewPerf = Average)$ are estimated for each class, and the test instance is assigned to the class having the highest posterior probability. For example,

$$P(Decision = Accepted|CGPA = High, InterviewPerf = Average) = \frac{P(CGPA = High|Decision = Accepted) \cdot P(InterviewPerf = Average|Decision = Accepted) \cdot P(Decision = Accepted)}{P(CGPA = High, InterviewPerf = Average)}$$

As observed earlier, the denominator term is ignored. So,

$P(Decision = Accepted|CGPA = High, InterviewPerf = Average) \propto$
$P(CGPA = High|Decision = Accepted) \cdot$
$P(InterviewPerf = Average|Decision = Accepted) \cdot P(Decision = Accepted)$

$P(CGPA = High|Decision = Accepted)$ is estimated as the relative proportion of training instances having $Decision = Accepted$ that have $CGPA = High$. Similarly, $P(InterviewPerf = Average|Decision = Accepted)$ is estimated as the relative proportion of training instances having $Decision = Accepted$ that have $InterviewPerf = Average$. The prior probability $P(Decision = Accepted)$ is simply the proportion of training instances that have the class label $Decision = Accepted$. Let us consider a concrete example. Consider a setting where, of the 600 training instances, 50 candidates are *Accepted*, 100 are *Waitlisted*, and 450 are *Rejected*. Of the 50 *Accepted* candidates, 40 have $CGPA = High$, 10 have $CGPA = Medium$, and none have $CGPA = Low$. Also, of the 50 *Accepted* candidates, 45 have $InterviewPerf = Excellent$, 5 have $InterviewPerf = Average$, and none have $InterviewPerf = Poor$. In such a case, the prior is

$$P(Decision = Accepted) = \frac{No. \text{ of applicants with class label Accepted}}{Total\ No. \text{ of Applicants}} = \frac{50}{600}$$

The two terms constituting the likelihood term are estimated as shown below:

$P(CGPA = High|Decision = Accepted)$

$$= \frac{No. \text{ of applicants with class label Accepted who have } CGPA = High}{No. \text{ of Applicants with class label Accepted}} = \frac{40}{50}$$

$P(InterviewPerf = Average \mid Decision = Accepted)$
$= \dfrac{\text{No. of applicants with class label Accepted who have InterviewPerf = Average}}{\text{No. of Applicants with class label Accepted}} = \dfrac{5}{50}$

Thus $P(Decision = Accepted | CGPA = High, InterviewPerf = Average) \propto \dfrac{40}{50} \cdot \dfrac{5}{50} \cdot \dfrac{50}{600}$
Similarly, we can estimate $P(Decision = Waitlisted | CGPA = High, InterviewPerf = Average)$ and $P(Decision = Rejected | CGPA = High, InterviewPerf = Average)$. The constant of proportionality is the same in all the three cases – the test instance is assigned to the class having the highest posterior value.

Bayesian reasoning can be used for two kinds of tasks: inference and prediction. As a simple example, if we toss a biased coin 10 times and obtain 4 heads, the inference problem is to estimate the parameter θ_H, the probability of heads. We can visualize a continuous hypothesis space, with an infinite set of hypotheses corresponding to θ_H values in the closed interval $[0,1]$. Denoting hypothesis and data (observations) as h and d respectively, the posterior distribution $P(h|d)$ is estimated using Bayes' rule as follows:

$$P(h|d) = \dfrac{P(d|h) \cdot P(h)}{P(d)}$$

Let us start with a prior which assumes that all hypotheses are equally likely, that is, the prior corresponds to a uniform distribution over probability of heads, as shown in Figure 11.8(a). After 10 flips with 4 heads, the posterior peaks around $p = 0.4$, as shown in Figure 11.8(b). Having observed 100 flips with 40 heads, the peak around $p = 0.4$ becomes even more pronounced (Figure 11.8(c)). The central idea here is: the posterior after each flip acts as the prior for the next flip. As we observe more and more data, we become more and more certain about the parameters that could have produced the data, given the underlying generative model that is encoded in the likelihood term.

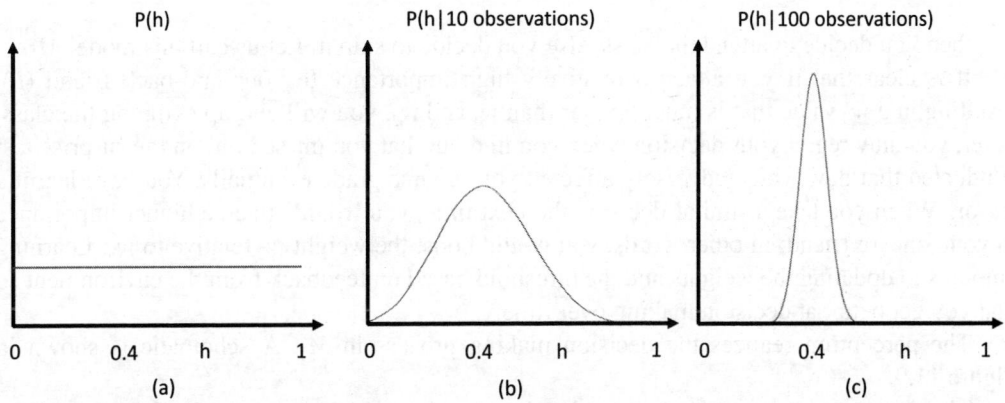

Figure 11.8 Illustrating the effect of number of observations on posteriors (figures are not drawn to scale)

In the prediction task, we make use of this posterior distribution to predict whether the next toss is likely to yield a head or tail: $P(\text{heads}|d) = \int P(\text{heads}|h)P(h|d)dh$. For discrete hypothesis spaces, the integral is replaced by a summation: $P(\text{outcome}|d) = \sum_h P(\text{outcome}|h)P(h|d)$. This can be viewed as a weighted aggregation over the predictions generated by each hypothesis h, and the weights of the hypotheses are their posterior probabilities $P(h|d)$. The hypotheses that are active (have non-zero posteriors) have a say in the final prediction. This is the Bayes Optimal setting.

11.4 Artificial Neural Networks

The human brain contains a large number of nerve cells called neurons which form part of a large interconnected network. Drawing inspiration from this, *artificial neural networks* (ANNs) also are made up of a network of nodes and directed links. In order to understand how ANNs work, we will start with a basic building block model called the *perceptron* (Rosenblatt, 1958).

Consider a simple setting that involves decision making: you are feeling lazy early one morning and thinking of skipping the early morning class at college. However, you suspect that the instructor may conduct a surprise test that day, which may count significantly towards the course grades. You consult two course-mates, one sincere and the other laid-back – the former suggests that you should attend; the latter advises you to skip it instead. How do you arrive at the final decision?

There are three factors that influence your decision: the importance of the class in terms of how it contributes to the grades (I), the advice of your sincere friend (S), and the advice of your laid-back friend (L). For simplicity, let us assume that each of I, S, and L can assume values $+1$ or -1, based on whether they contribute positively or negatively to the chances of attending the class respectively. In our case, $I = +1$, $S = +1$, $L = -1$. Let the importance you attach to the three factors be represented by weights w_1, w_2, and w_3 respectively; each weight is real valued in [0,1]. For a choice of threshold t, if

$$w_1.I + w_2.S + w_3.L > t,$$

then you decide to attend the class, else you decide to skip it. Let us call this model M_1.

It is clear that if you attach a relatively high importance to your laid-back friend (L), resulting in a w_3 value that is much higher than w_1 and w_2, you will end up skipping the class. Later, you may regret your decision when you find out that you missed out on the surprise test conducted that day, which adversely affected your course grade eventually. You have learnt a lesson. When you take a similar decision the next time, you would attach a higher importance to your sincere friend; in other words, you would boost the weight w_2 relative to w_3. Learning amounts to updating the weights and the threshold based on feedback from the environment so that you get better at decision making over time.

The perceptron realizes the decision making process in M_1. A schematic is shown in Figure 11.9.

The perceptron has two kinds of nodes: input nodes, which represent the factors used for arriving at the decision, and an output node, which produces the outcome. The figure shows three inputs x_1, x_2, and x_3 which correspond to I, S, and L respectively in the example above.

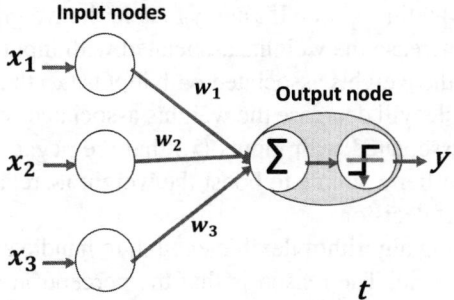

Figure 11.9 A perceptron

There is a directed edge from each input node to the output node. These edges carry weights, which are shown as w_1, w_2, and w_3 in the figure. The output node computes a weighted sum of the inputs and subtracts the threshold t. The sign of the result determines the value y as shown below:

$$y = \begin{cases} 1, & \text{if } w_1 x_1 + w_2 x_2 + w_3 x_3 - t > 0 \\ -1, & \text{if } w_1 x_1 + w_2 x_2 + w_3 x_3 - t \leq 0 \end{cases}$$

Note that the perceptron does exactly the same job as was formulated in model M_1. The idea can be extended to work over n input factors. Representing the threshold t as $-w_0$, and treating this as a weight attached to an additional (constant) input $x_0 = 1$, we obtain the more general formulation:

$$y = \begin{cases} 1, & \text{if } \sum_{i=0}^{n} w_i x_i > 0 \\ -1, & \text{if } \sum_{i=0}^{n} w_i x_i \leq 0 \end{cases}$$

Also, it is easy to see that the decision making problem we set out to solve can alternatively be viewed as a classification problem, where the two outcomes (*attend class* and *skip class*) correspond to class labels. Thus we have in a perceptron a general model of a classifier that can take in n attribute values of a test instance as an instance and classify it into one of several classes. The weight values w_0 through w_n are referred to as *model parameters*.

How does such a perceptron learn from training data in a classification setting? In model M_1, we used feedback from the environment to guide the process of updating weights and thresholds. The training data comes with this feedback about the desired outcomes corresponding to the attribute values of each training instance. The perceptron makes use of this to learn weights that achieve the desired classification.

The perceptron learning algorithm is simple. We start with random values of weights w_0 through w_n. The perceptron is tried on each training instance; each time it fails to correctly classify an instance, the weights are updated. The process is repeated over all the instances over several passes till all instances are correctly classified. The update rule for weight w_i associated with input x_i is as follows: $w_i \leftarrow w_i + \eta(l - y)x_i$, where l is the class label (either $+1$ or -1) associated with the current training instance, y is the perceptron output, and η is a positive constant. It is easy to see why this works. If l and y match, the weights stay the same. If l is $+1$ and y is -1, this rule will increase the weights associated with inputs (x_i) that are positive, and decrease the magnitude of the weights associated with inputs (x_i) that are negative. Conversely, if l is -1 and y is $+1$, this rule will decrease the weights associated with inputs that are positive, and increase the weights associated with inputs (x_i) that are negative. This is in line with the intuitions we used in our earlier example to boost the weight w_2 relative to w_3 when model M_1 failed to arrive at the correct decision.

Is the perceptron learning algorithm flexible enough to handle any classification problem? Unfortunately, the answer is no. The reason is that the perceptron algorithm can only model linear decision boundaries of the kind $\sum_{i=0}^{n} w_i x_i = 0$. Therefore, the perceptron can handle the class of *linearly separable* problems, where such a linear decision boundary can discriminate between classes. As shown in Figures 11.10(a) and (b), it can represent Boolean functions like AND and OR. Recall that OR(x_1, x_2) is 1 when at least one of x_1 and x_2 is 1. The dashed line in Figure 11.10(a) separates the three inputs that evaluate to 1 from the lone ($x_1 = 0$, $x_2 = 0$) that evaluates to 0. Likewise, a line separates the three 0s from the lone 1 ($x_1 = 1$, $x_2 = 1$) for AND (x_1, x_2) in Figure 11.10(b). The figure shows that the decision boundary induced by the perceptron splits the space into two parts (half-planes) which are labelled as $+$ and $-$. However, it fails to represent the XOR problem where no single straight-line separator can be found that separates the outcomes 0 and 1 (Figure 11.10(c)). Recall that XOR(x_1, x_2) is 1 when *exactly* one of x_1 and x_2 is 1, and the other is 0.

In linearly non-separable problems, we come up with a notion of error aggregated over all training instances and attempt to find weights that minimize this error. This leads to an alternate formulation called the *delta rule*, which uses the idea of *gradient descent* to find the best possible weights. We consider a perceptron model that is somewhat different from the one shown in Figure 11.9 in that the output node produces a weighted sum of inputs but does no thresholding. Let us call this perceptron a *linear unit*. We use the following formulation of error aggregated over training instances:

$$E = \frac{1}{2} \sum_{d \in D} (l_d - y_d)^2$$

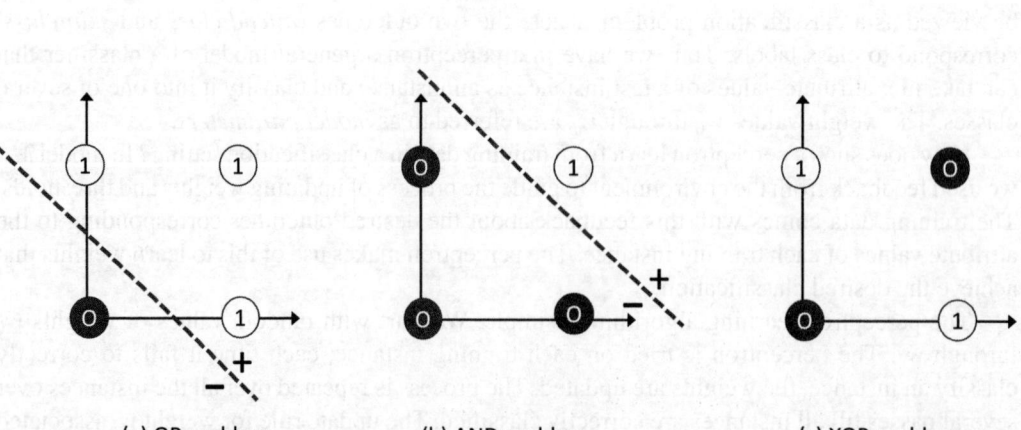

(a) OR problem   (b) AND problem   (c) XOR problem

Figure 11.10 OR, AND, and XOR, viewed as classification problems

where d refers to an individual training instance in the training data D, l_d refers to the class label of d as recorded in the dataset, and y_d is the corresponding output of the linear unit. The error is a function of the weights, and learning should involve adjustment of these weights. We start off with random initialization of weights, and iteratively update these weights with the goal of arriving at the combination of weights that minimizes error. We picture the error as a function of weight as shown in Figure 11.11. The gradient descent approach chooses a direction that leads to the *steepest descent* along the error surface.

For weight w_i the update formulation is as follows:

$$w_i \leftarrow w_i - \eta \frac{\partial E}{\partial w_i}$$

where η is a positive constant, and $\frac{\partial E}{\partial w_i}$ is the slope of the error surface with respect to w_i. Why does this work? Refer to Figure 11.11 which illustrates the simplest setting where the error is a function of just one weight w. Let us start with a random guess for w, say, w_A. Since the slope $\frac{\partial E}{\partial w}$ is positive at $w = w_A$, the update formulation above will lead to a decrease in the weight value (note that η is positive), and we move to a weight value to the left of w_A to w'_A as shown. On the other hand, consider an alternative starting weight w_B. Since the slope $\frac{\partial E}{\partial w}$ is negative at $w = w_B$, the update formulation above will lead to an increase in the weight value, which moves the weight value to the right of w_B to w'_B as shown. In either case, we move closer to the weight value w_A where the error is minimum.

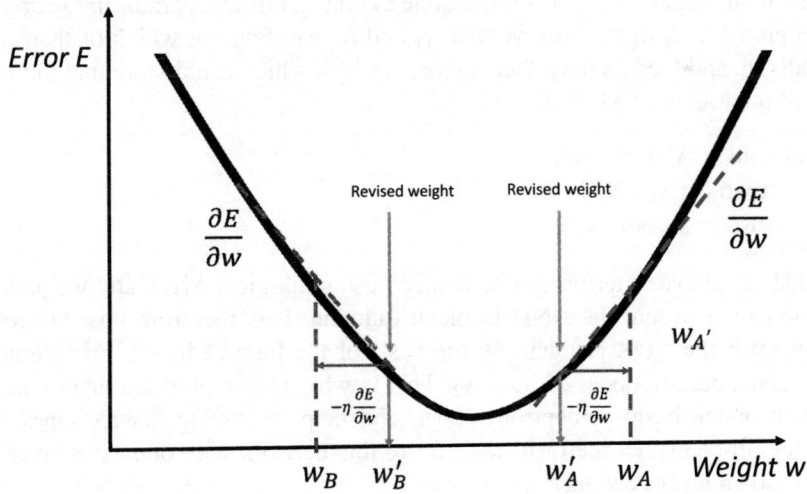

Figure 11.11 Illustration of the intuition behind the *gradient descent* algorithm

Is there a principled way to address the class of linearly non-separable classification problems? Yes. The solution is to use more than one layer of nodes. Each node is very similar to a perceptron; it produces a weighted aggregation of inputs and feeds the result to a nonlinear function called the *activation* function. While the perceptron uses a simple thresholding, accepting a value above a threshold completely, it is more usual to use nodes with a sigmoid activation function in multilayer networks. The sigmoid function is smooth and differentiable, and this helps in working out the mathematical derivation of the learning mechanism. The threshold and sigmoid activation functions are shown in Figure 11.12.

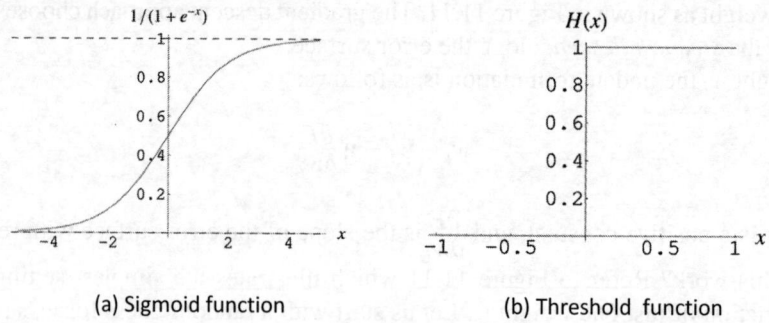

(a) Sigmoid function (b) Threshold function

Figure 11.12 The sigmoid and threshold activation functions

Source: https://mathworld.wolfram.com/SigmoidFunction.html and https://mathworld.wolfram.com/HeavisideStepFunction.html, both accessed November 2022

In order to see how using an additional layer can help, let us revisit the simplest instance of linearly non-separable problems that we have seen so far, the XOR problem. Consider the architecture in Figure 11.13(b), where nodes N_1 and N_2 feed their output to node N_3 that produces the final output. While it is impossible to draw a linear separator in the original space shown in Figure 11.13(a), we observe that N_1 and N_2 can find out which of their half-planes the input falls in, and feed their verdicts (+ or −) to N_3, which can do the final job by learning a criterion of the following kind:

if N_1 outputs + AND N_2 outputs +
    then class_label is 0
    else class_label is 1

Note that the above criterion can be realized using a logical AND, and we have seen that a perceptron can represent the AND Boolean function. It is therefore possible for node N_3 to correctly solve the XOR problem on the basis of the intermediate information regarding half-planes generated by nodes N_1 and N_2. This is why adding an extra layer can overcome the limitation of the basic perceptron model, and help in solving a wide class of linearly non-separable problems. Henceforth, we refer to this network with one extra layer as OALN (for one additional layer network).

Would we ever need more than one additional layer? Consider a multilayer network applied to the problem of handwritten digit classification. Each handwritten digit is pre-processed and represented in terms of n numeric feature values, constituting a feature vector. Each such vector is in an n-dimensional feature space. The training data contains several thousands of instances, along with the correct class labels. The multilayer network contains n input nodes and 11 output nodes: one for each of the digits and an additional dummy node for all images that could not be reliably assigned to any class. Feature vectors of images are expected to be close to each

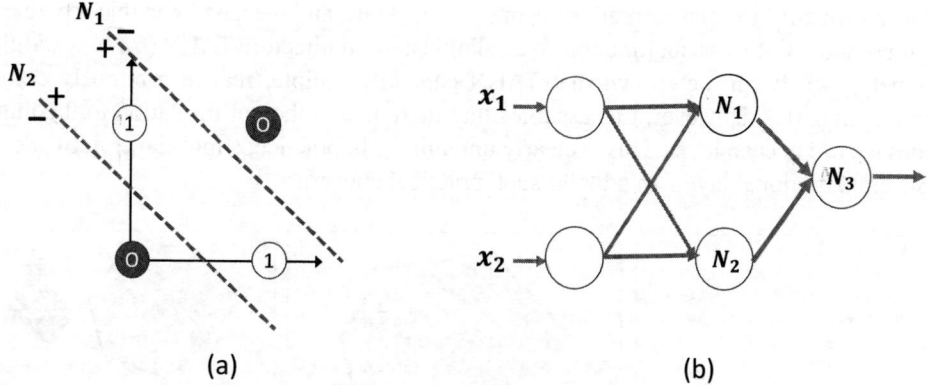

Figure 11.13 Using an additional layer to solve the XOR problem

other in the feature space if they belong to the same class (represent the same digit). Feature vectors of different classes are expected to be far from each other. There are certain practical challenges, though. The image in Figure 11.14(a) can be read as a 0 or a 6, and the one in Figure 11.14(b) can be a 1 or a 7. In either case, feature vectors of instances belonging to different classes are close to each other. There is yet another problem: the same digit 7 can be written in two different ways as shown in Figures 11.14(c) and (d). Here, feature vectors of instances that belong to the same class are far apart. The images shown in Figures 11.14(a) and (b) are hard to classify for humans as well. Hence, we will not be overly concerned if a machine goes wrong in such cases. However, we need to make sure that the neural network does well for the setting in Figures 11.14(c) and (d).

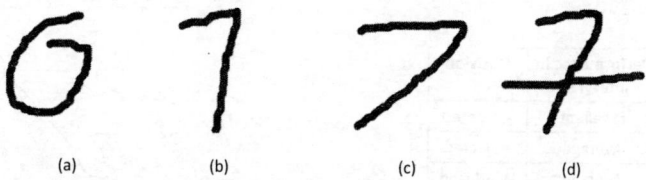

Figure 11.14 Illustrating challenges in handwritten digit recognition

As shown in Figure 11.15(b), OALN can extract regions in the feature space by realizing a logical AND over the half-plane information obtained from nodes in the previous layer. However, there are two distinct regions in the feature space corresponding to two different ways of writing the digit 7. These regions, shown as region A and region B, are well separated from each other, and yet represent the same class. In situations such as this where there is no neat correspondence between regions and classes, OALN is not sufficient. The output nodes in OALN need to feed the region information to yet another final layer which does the final job of classification. The final layer node learns a criterion of the following kind:

    if Region A OR Region B
        then Digit = 7

The above criterion can be realized using a logical OR, and we have seen that a perceptron can represent the OR Boolean function. We call this new architecture TALN (for two additional layers network). It can be shown that TALN can, in principle, realize arbitrarily complex decision boundaries. However, this assumes that there is no inherent limitation to the number of nodes we use in each layer. This is clearly unrealistic. In practice, a multilayer network often uses several additional layers to address such practical concerns.

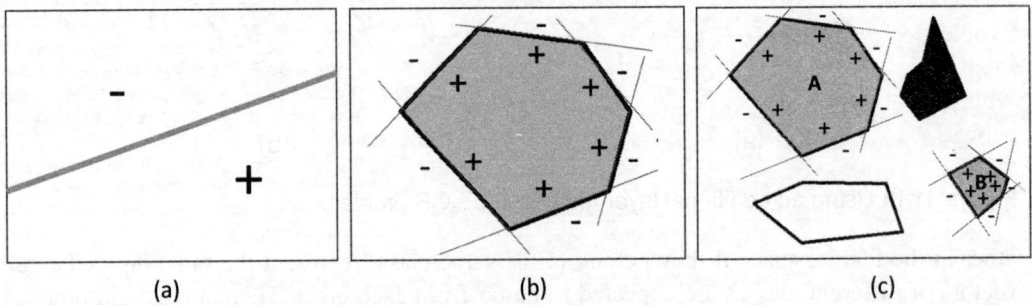

Figure 11.15 Illustrating why one additional layer may not always suffice

In Figure 11.16(b), we show a multilayer network used for solving the classification problem that we presented in Section 11.1. The training data is replicated in Figure 11.16(a) for ease of reference. The input nodes take in two features: *CGPA* and *Performance in Interview*, which are numeric and ordinal respectively. There are three nodes in the output layer based on the three classes *Accepted*, *Waitlisted*, and *Rejected*. The nodes $f_3, f_4,$ and f_5 in the intermediate layer realize linear separators and feed the half-plane information to $f_6, f_7,$ and f_8, which generate the final class labels.

Id	CGPA	Performance in Interview	Decision
1	9.5	Excellent	Accepted
2	7.2	Average	Rejected
3	8.8	Average	Waitlisted
4	8.3	Poor	Rejected
...	...	...	...
600	7.6	Excellent	Waitlisted

(a)   (b)

Figure 11.16 (a) An example dataset (b) A multilayer network trained on this data

How are the weights linking up nodes in successive layers learnt in such a multilayer network? The general idea of gradient descent can be applied in this setting as well. There is a key challenge, however. The measure of error is available only at the output nodes; for the intermediate nodes, we have no direct access to error information. A clever popular algorithm called BACKPROPAGATION that has revolutionized the field of ML solves precisely this problem

(Rumelhart, Hinton, and Williams, 1986). The idea is to propagate the errors back from the output nodes to the intermediate nodes, to facilitate weight updates across layers. Note that the signal travels forward; it is only the errors that propagate in the reverse direction.

11.5 K-MEANS Clustering

The objective of clustering is to discover natural groups in data, such that objects are more similar to other objects within its group than to objects in other groups. As shown in Figure 11.17, the input to one such clustering algorithm called the K-MEANS clustering is a set of instances along with the number of clusters desired (denoted by K), and the output is a clustering; the figure shows the objects in the three clusters coloured differently as grey, black, and white circles.

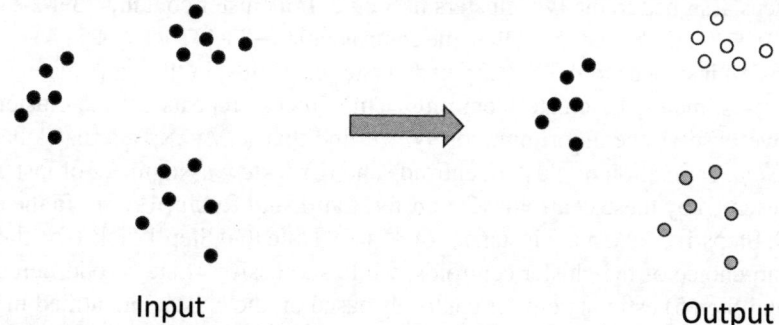

Input   Output

Figure 11.17 The clustering problem

The space of hypotheses is the set of possible clusterings given a dataset. For instance, given the data in Figure 11.18, three candidate hypotheses are shown in Figure 11.18. Visual inspection reveals that Clustering 3 is more representative of natural groups present in the data. We need an objective function to characterize this intuition so that we can use this function to guide the search over the space of hypotheses. Given a formulation of the objective function, the K-MEANS algorithm can be viewed as solving an optimization problem.

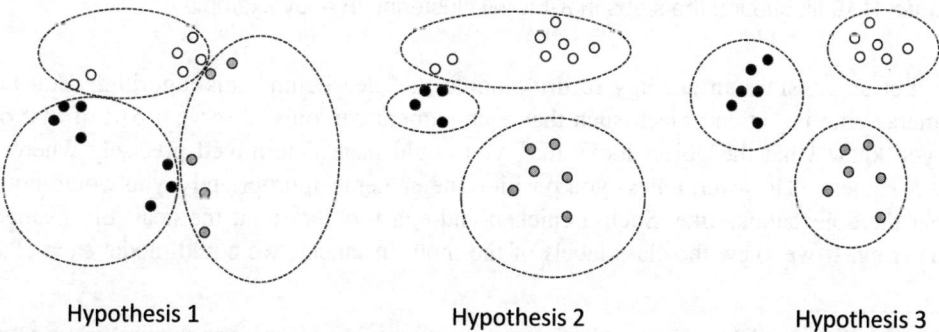

Hypothesis 1   Hypothesis 2   Hypothesis 3

Figure 11.18 Three candidate hypotheses, each representing a distinct clustering

A significant difference between clustering and classification techniques like decision trees, kNN, Naïve Bayes classifier, and neural networks is that clustering is unsupervised, while classification is supervised. This means that training data for classifiers come with category labels. In contrast, in unsupervised approaches such as K-MEANS clustering, the training instances have attribute values, but no class label.

K-MEANS is an iterative algorithm that needs a measure of distance between instances (MacQueen, 1967). Figure 11.19 illustrates five steps of K-MEANS on a toy dataset. Given a value of K, the algorithm starts by randomly identifying K instances as cluster centroids. In the example in Figure 11.19, $K = 2$, and Step 1 shows the two random initial centroids as boxed circles. In Step 2, each instance is assigned to the cluster centroid that it is closest to; this leads to a clustering where we have K clusters, two in our example. Step 2 in Figure 11.19 shows the two clusters using white and dark circles. In Step 3, the cluster centroids are estimated based on the instances assigned to the two clusters in Step 2. If a cluster contains four 2-dimensional instances (2,3), (3,5), (2,4), and (5,4), then the centroid is ((2+3+2+5)/4, (3+5+4+4)/4) = (3,4). Step 4 assigns all instances to the recomputed cluster centroids, in the same manner as Step 2. Step 5 then re-estimates the cluster centroids. This process repeats till the cluster centroids stabilize. If we observe the algorithm closely, we find that K-MEANS alternates between two steps: (a) *M step*: estimation of cluster centroids and (b) *E step*: assignment of instances to the clusters represented by these centroids (Dempster, Laird, and Rubin, 1977).[3] In the example in Figure 11.19, Steps 1, 3, and 5 are instances of *M* step. Note that Step 1 kickstarts the algorithm with a random choice of the cluster centroids, while successive *M* steps (odd-numbered steps like Step 3 and Step 5) estimate cluster centroids based on the clusters identified in the *E* step. Steps 2, 4, and all successive even-numbered steps are instances of *E* step.

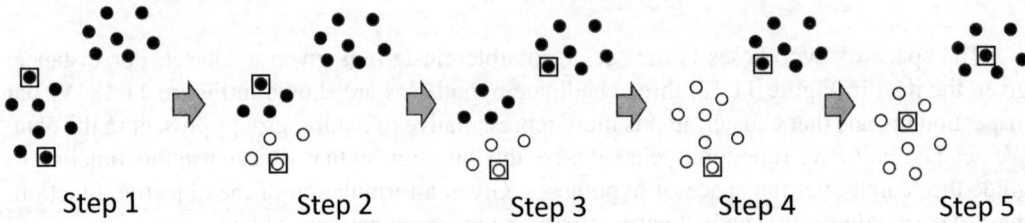

Figure 11.19 Illustrating the steps in K-MEANS clustering in a toy example

Let us consider an analogy to drive home the idea behind this algorithm. You have K cameras pointing at an object, such that each camera can only observe a part of that object. If you knew what the object looks like, you could have determined precisely where to fix the K cameras. However, unless you position the cameras appropriately, you would not know what the object looks like. Such a chicken-and-egg problem is at the heart of unsupervised clustering. If we knew the class labels of the input instances, we could model each class; on

---

[3] The terms 'E step' and 'M step' owe their origin to the words 'expectation' and 'maximization', which relate to the underlying optimization performed by K-MEANS. While the details need not bother us here, it may be worth noting that K-MEANS is a special case of a more general class of algorithms that goes by the name EM algorithm, which finds a lot of application in various contexts in ML literature.

the other hand, if we knew the models, we could have inferred the class labels. The *K*-MEANS objective function has two sets of parameters: the cluster centroids and affiliations of instances to the clusters. In the *M* step, the affiliations are held fixed and the centroids updated; on the other hand, the *E* step updates the affiliations given fixed centroids. It can be proved that we will never be worse than where we started off. The *K*-MEANS algorithm starts off with a random initial configuration of centroids. It may be noted that the clustering produced is sensitive to this random initialization, and it is typical in practice to try multiple initializations and pick the best result. This is somewhat like the *iterated hill climbing* algorithm described in Section 4.6.3.

Summary

ML aims to capitalize on the experience gleaned from problem solving from the past and create a representation that offers a quicker solution for a new problem. In this chapter we have seen a few examples of approaches in ML to learn from the past. The algorithms described are applied to solving a particular problem, which very often is classification. We have looked at four supervised learning algorithms in which the labels provided by the user in the past are used to generalize and create a hypothesis about how to describe and differentiate the different classes. The approaches all learn by optimizing some parameters that minimize the classification error in the training examples. If the training examples are sufficient in number and the model being learnt does not overfit, then the model is expected to perform well on the unseen test examples.

We also looked at unsupervised learning approaches to identify clusters in a dataset. The general idea behind clustering is that instances that are close to each other based on an appropriate distance measure are assigned to the same cluster. The *K*-MEANS clustering algorithm accepts *K* as a parameter and iterates through the data instances eventually forming *K* clusters.

In the next and last chapter we look at a unifying formalism of constraint satisfaction problems that allows a combination and search to be integrated into one problem solver.

Exercises

1. The algorithm for constructing decision trees selects the attribute whose values are used to partition the training data into subsets, on which a decision tree is constructed recursively. Suppose a tree similar to the one in Figure 11.2 is constructed from a large dataset for a company, and the attribute *Gender* shows up somewhere in the decision tree, what can one conclude about the algorithm and the dataset?
2. [Adwait] Section 11.2 illustrates how the classification of a test instance can potentially vary with the value of k. In practice, we typically restrict our attention to smaller values of k so that instances very far from (or equivalently, dissimilar to) the test instance do not influence its classification. Assuming that we have a finite set of candidate values of k, how would you choose a *suitable* value for k in a principled way?
3. [Adwait] When a test instance is presented to a *k*NN classifier, the classifier is required to compute the similarity of the test instance with all the labelled instances in the dataset. Apparently, this computation is time-consuming, and a boost in its efficiency is highly desirable. Consider the following dataset. Can you think of instances that can be *deleted*

without compromising the effectiveness of a 1-NN classifier? Also, argue that the time taken to classify a new test instance is lower when using this reduced dataset.

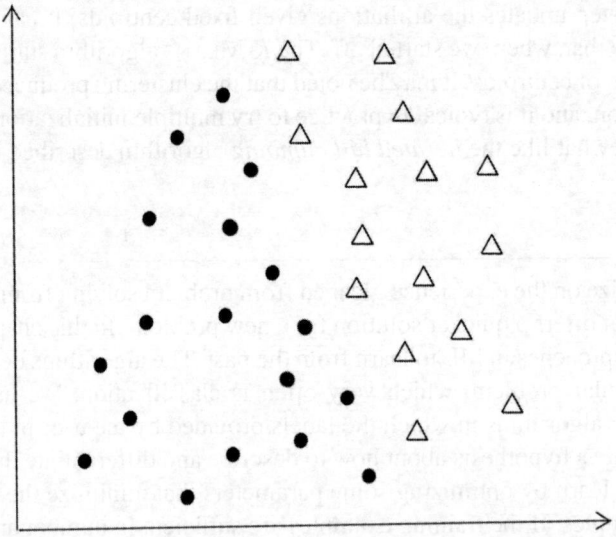

4. [Adwait] The Euclidean distance is a conventional distance measure used when dealing with numerical attributes. Consider a setting where a person is to be classified into one of two classes – *wealthy* or *not-wealthy* – based on her savings (S) and the number of houses (H) she owns. Notice that the ranges of these two attributes are drastically different, which might make using plain Euclidean distance formulation rather misleading. In particular, it is expected that S will dominate the value of the Euclidean distance and can very likely mask the contribution of H. Can you think of a way to overcome such an issue?

5. We have seen in Section 11.3 that the Naïve Bayes classifier assumes that the features are conditionally independent given the class label. Can you identify a real world situation where such an assumption makes sense and one where it does not?

6. [Adwait] The activation function in the intermediate nodes of an ANN plays a critical role in allowing the network to learn complex functions. Interestingly, if no activation function is used (equivalently, using a linear/identity activation function), the network is capable of only learning functions that are linear in nature. Provide supporting arguments for this claim.

7. [Adwait] We have seen that the clusters identified by the K-MEANS algorithm are often influenced by the choice of initial cluster centroids. Can you provide supporting evidence using a synthetic dataset such that different initial cluster centres lead to visibly different clusters?

CHAPTER 12
Constraint Satisfaction

What is common between solving a sudoku or a crossword puzzle and placing eight queens on a chessboard so that none attacks another? They are all problems where each number or word or queen placed on the board is not independent of the others. Each constrains some others. Like a piece in a jigsaw puzzle that must conform to its neighbours. Interestingly, all these puzzles can be posed in a uniform formalism, *constraints*. The constraints must be respected by the solution – the constraints must be *satisfied*. And a unified representation admits general purpose solvers. This has given rise to an entire community engaged in *constraint processing*. Constraint processing goes beyond constraint satisfaction, with variations concerned with optimization. And it is applicable on a vast plethora of problems, some of which have been tackled by specialized algorithms like linear programming and integer programming.

In this chapter we confine ourselves to finite domain constraint satisfaction problems (CSPs) and study different approaches to solving them. We highlight the fact that CSP solvers can combine search and logical inferences in a flexible manner.

A constraint network $\mathcal{R}$ or a CSP is a triple,

$$\mathcal{R} = <X, D, C>$$

where X is a set of variable names, D is a set of domains, one for each variable, and C is a set of constraints on some subsets of variables (Dechter, 2003). We will use the names $X = \{x_1, x_2, ..., x_n\}$ where convenient with the corresponding domains $D = \{D_1, D_2, ..., D_n\}$. The domains can be different for each variable and each domain has values that the variable can take, $D_i = \{a_{i1}, a_{i2}, ..., a_{ik}\}$. Let $C = \{C_1, C_2, ..., C_m\}$ be the constraints. Each constraint C_i has a scope $S_i \subseteq X$ and a relation R_i that is a subset of the cross product of the domains of the variables in S_i. Based on the size of S_i, we will refer to the constraints as unary, binary, ternary, and so on. A CSP is often depicted by a constraint graph and a matching diagram, as described in the examples to follow.

We will confine ourselves to finite domain CSPs, in which the domain of each variable is discrete and finite. We will also specify the relations in extensional form well suited for our algorithms. For example, given a common domain $\{1, 2, 3, 4\}$ for each variable, if we have a binary constraint between two variables x_i and x_k in which the value in x_i is smaller, then we represent it as

$$R_{ik} = \{<1, 2>, <1, 3>, <1, 4>, <2, 3>, <2, 4>, <3, 4>\}$$

The pairs in the relation are the allowable combination of values for the two variables respectively. For example, $x_i = 1$ and $x_k = 4$ are allowed. Note that we have adopted a naming convention for the relation as well, with the subscripts in R_{ik} referring to the subscripts of the two related variables. We shall focus largely on binary constraint networks (BCNs) in this chapter.

An *assignment* $\mathcal{A}$ is a set of variable–value pairs, for example, $\{x_2 = a_{21}, x_4 = a_{45}, x_7 = a_{72}\}$. We also say that the assignment is an *instantiation* of the set of variables. An assignment to a subset of the variables is a partial assignment. Wherever there is no confusion, we will represent the assignment as a tuple $\mathcal{A} = <a_1, a_2, \ldots, a_p>$ where it is understood that these are the variables $x_1, x_2, \ldots, x_p$ instantiated.

An assignment $\mathcal{A}$ *satisfies* a constraint C_i if $S_i \subseteq \{x_1, x_2, \ldots, x_p\}$ and $\mathcal{A}_{Si} \in R_i$ where $\mathcal{A}_{Si}$ is the projection of $\mathcal{A}$ onto S_i.

An assignment $\mathcal{A}$ is consistent if it satisfies all the constraints whose scope is covered by $\mathcal{A}$.

A solution to a CSP is a consistent assignment over all the variables in X. The CSP *expresses* the relation σ_X, also called sol($\mathcal{R}$), the solution relation, which is a relation on all the variables of X.

12.1 Constraints: Clearing the Fog

The solution sol($\mathcal{R}$) for a CSP is implicit in the network $\mathcal{R}$. Only that it is not explicitly specified. It is specified piecewise, like the description given by the blind men who are touching different parts of an elephant in an ancient parable in which each has only partial knowledge. The local constraints allow assignment of more values than the ones in the solution relation. There is a fog of possibilities that has to be cleared away for the solution to reveal itself. We begin with some examples to understand the problem.

12.1.1 The map colouring problem

The map colouring problem is a natural CSP. Political maps in school atlases demarcate the different regions using different colours. Such a colouring is the solution we seek. No two regions that share a boundary can have the same colour. This translates naturally into a set of binary constraints, and the map colouring problem is a BCN. Consider a small map of five regions A, B, C, D, and E, with the following pairs of regions sharing a boundary: $<A, B>$, $<B, C>$, $<B, D>$, $<C, D>$, and $<D, E>$. In our formulation, each region is a variable in the CSP. Each region has its own set of allowed colours as described in the domains below.

$\mathcal{R} = <X, D, C>$
$X = \{A, B, C, D, E\}$, $D = \{D_A, D_B, D_C, D_D, D_E\}$, $C = \{R_{AB}, R_{BC}, R_{BD}, R_{CD}, R_{DE}\}$
$D_A = \{b, g\}$, $D_B = \{r, b, g\}$, $D_C = \{b\}$, $D_D = \{r, b, g\}$, $D_E = \{r\}$
$R_{AB} = \{<b, r>, <b, g>, <g, r>, <g, b>\}$

$R_{BC} = \{<r, b>, <g, b>\}$
$R_{BD} = \{<r, b>, <r, g>, <b, r>, <b, g>, <g, r>, <g, b>\}$
$R_{CD} = \{<b, r>, <b, g>\}$
$R_{DE} = \{<b, r>, <g, r>\}$

Every CSP can be depicted as a *constraint graph*. The nodes in the graph are the variables in the CSP and an edge between two nodes says that the two variables participate in a constraint. This is true even when the constraint is ternary or higher. Constraint graphs are consulted by some algorithms in deciding the order of visiting variables.

Another diagram that is useful is the *matching diagram*. An edge in the matching diagram connects two *values* in two variables that together participate in some constraint. Figure 12.1 shows three views of the map colouring problem. On the left is the map showing the regions that share a boundary. In the centre is the constraint graph, where each region is represented by a node or a variable with an edge between two nodes that share a boundary. In the figure the nodes have the domains shown alongside, and the label on an edge represents the not-equal relation. The two related variables are only allowed different values. On the right is the matching diagram that makes the relation explicit, with every pair of *allowed* values being connected with an edge. Implicit in the matching diagram is the universal relation between nodes not connected in the constraint graph, for example, A and C. Any combination of values of such pairs of nodes is allowed, though not shown explicitly in the matching diagram.

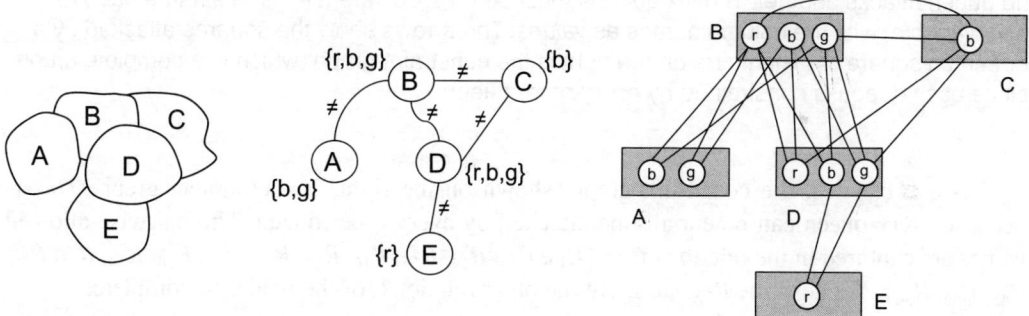

Figure 12.1 A map colouring problem on regions A, B, C, D, and E is on the left. The constraint graph is in the centre and the matching diagram on the right. An edge in the matching diagram stands for an allowable pair of colours. For regions that are not adjacent, the matching diagram has an implicit universal relation where any combination of values is allowed.

The matching diagram shows pairs of *values* that can *possibly* occur together in a solution. When the fog clears, only the pairs that are part of a solution are left. We illustrate this phenomenon with the 6-queens problem in the next section.

12.1.2 The *N*-queens puzzle

The general task is to place N queens on an $N \times N$ chessboard such that no queen attacks another. A queen attacks another in chess if the two are in the same row, same column, or the same diagonal. We can state this as a binary CSP by specifying constraints between any two queens.

Thinking of a physical chessboard, the first thought is to have N^2 variables for the squares with each possibly having a queen. But we can exploit the knowledge that only one queen can be in one row and one column. This suggests a compact representation that is commonly used. Each row (or each column) can be a variable which will have one queen identified by the column (or row) in which it is. Figure 12.2 shows the 6-queens problem in which a queen has to be placed in each row. The row number becomes the variable, and the column number the value. In this representation there are six variables $X = \{1, 2, 3, 4, 5, 6\}$ and each $D_i = \{a, b, c, d, e, f\}$.

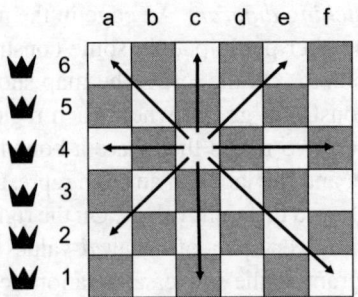 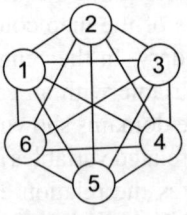

Figure 12.2 The 6-queens problem is to place the six queens on a 6×6 chessboard such that no queen attacks another. The six queens must be on six different rows. We name each row as a *variable*, with the column names as *values*. The arrows show the squares attacked by a queen on square c4. The figure on the right is the constraint graph, which is a complete graph since each queen is constrained by every other queen.

As one can see, the constraint graph, shown on the right, is a complete graph. This is because every queen can potentially be attacked by every other queen. The pairwise allowed values are captured in the relations $C = \{R_{12}, R_{13}, R_{14}, R_{15}, R_{16}, R_{23}, R_{24}, R_{25}, R_{26}, R_{34}, R_{35}, R_{36}, R_{45}, R_{46}, R_{56}\}$. We describe R_{12} and leave the other relations for the reader to complete.

$R_{12} = \{<a, c>, <a, d>, <a, e>, <a, f>,$
$\quad\quad <b, d>, <b, e>, <b, f>,$
$\quad\quad <c, a>, <c, e>, <c, f>,$
$\quad\quad <d, a>, <d, b>, <d, f>,$
$\quad\quad <e, a>, <e, b>, <e, c>,$
$\quad\quad <f, a>, <f, b>, <f, c>, <f, d>\}$

Figure 12.3 shows a part of the matching diagram. The relations covered in the diagram are R_{12}, R_{13}, R_{14}, R_{15}, R_{16}, R_{25}, and R_{36}. Even with this subset of relations, one can see that there is a large number of combinations to choose from. As one can see, there is verily a fog of connections for each variable.

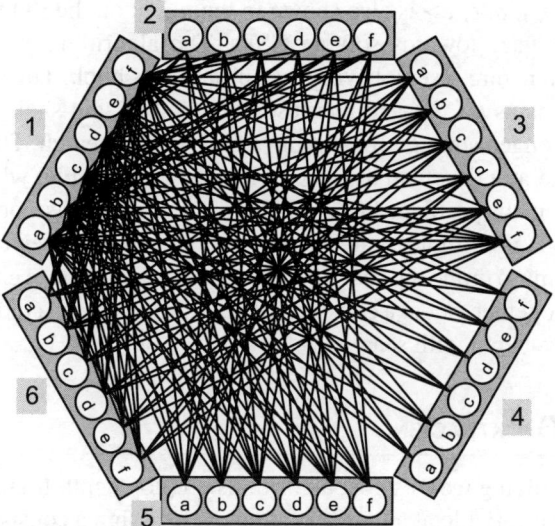

Figure 12.3 The matching diagram for the 6-queens problem. Only edges for the relations R_{12}, R_{13}, R_{14}, R_{15}, R_{16}, R_{25}, and R_{36} are drawn in the figure, giving rise to the higher density of edges on the left. A close scrutiny will reveal that there are three of four edges from a value for one variable to values in another variable.

In the solution, one value must be selected from the domain of each variable. Further, each value in each variable must have an edge connected to a value in every other variable that must be in the solution. The task of solving the CSP is to clear the fog and reveal the solution. Figure 12.4 shows one solution for the 6-queens problem.

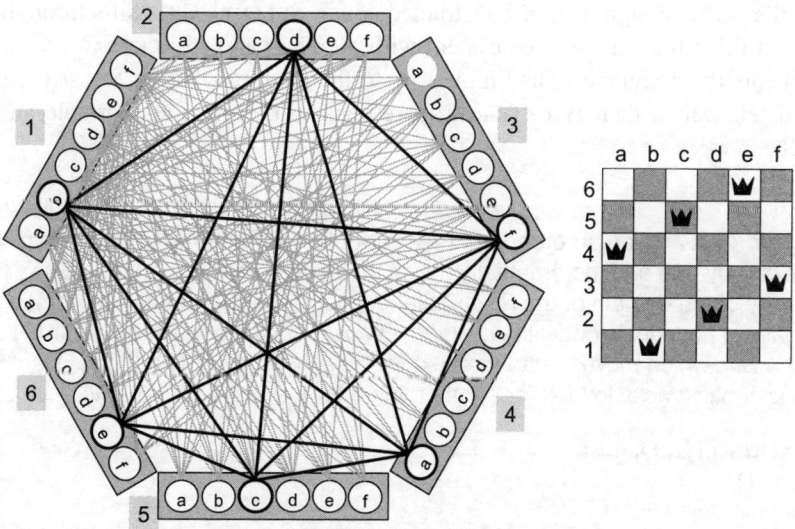

Figure 12.4 A solution <b, d, f, a, c, e> for the 6-queens problem highlighted on the matching diagram. The solution is also shown on the board on the right.

The solution <b, d, f, a, c, e> is also shown in the figure on the right.

Now we turn our gaze towards solving CSPs. The algorithms we are interested in are domain independent in nature, exemplifying the spirit of this book. The idea is again that users can pose their problems as a CSP, and then use a general off-the-shelf solver for solving the CSP. There is a two pronged strategy for solving a CSP. One is search. The idea here is that one picks the variables one at a time and assigns a value to the variable, which is consistent with earlier variables. The main problem faced by brute force search is combinatorial explosion, and we look at methods to mitigate that. The second is *consistency enforcement* or *constraint propagation*, which aims to prune the space being searched. Done to the extreme this can obviate the need for search altogether, but at a considerable cost. In practice, a judicious combination of the two works best. We begin with search.

12.2 Algorithm BACKTRACKING

Search is the soul of solving CSPs. Most algorithms employ depth first search (DFS) wherein the algorithm picks one variable at a time and attempts to assign a consistent value to it. Which variable to pick next and which value to try for the variable will be questions we will address as we go along. For the moment we assume that the order $(x_1, x_2, ..., x_N)$ of the variables is given in advance for a CSP with N variables.

The well known algorithm BACKTRACKING is described below. The inputs to the algorithm are the three constituents of the CSP, the set of variables X, their domains D, and the constraints C. It builds an assignment $\mathcal{A}$ incrementally starting from scratch. The value for the next variable must satisfy any constraints that are defined over that variable and its predecessors. That is, $\mathcal{A}$ must be consistent at all times. If it cannot be extended to the next variable, then the algorithm backtracks and tries a different value for the last variable it assigned a value to. When it considers a new variable, it makes a copy of its domain and passes it to function SELECTVALUE along with the partial assignment $\mathcal{A}$ constructed so far. SELECTVALUE sifts through the values in the domain till it finds a value consistent with $\mathcal{A}$. The parent BACKTRACKING accepts that value, augments the assignment, and moves on to the next variable. If SELECTVALUE cannot find a consistent value, then BACKTRACKING retreats to the previous variable and looks for another value.

Algorithm 12.1. Given an ordering of the variables, algorithm BACKTRACKING picks variables one by one and incrementally builds the assignment $\mathcal{A}$. Function SELECTVALUE takes a copy of the domain of the ith variable, the current assignment, and removes values that are not consistent with $\mathcal{A}$. When it finds a consistent value it returns the value to BACKTRACKING which moves on to the next variable. If it cannot find a consistent value, BACKTRACKING backtracks to look for another value for the previous variable.

BACKTRACKING {X, D, C}
1. $\mathcal{A} \leftarrow []$
2. $i \leftarrow 1$
3. $D'_i \leftarrow D_i$

```
4.    while 1 ≤ i ≤ N
5.        aᵢ ← SELECTVALUE (D'ᵢ, 𝒜, C)
6.        if aᵢ = null
7.            then  i ← i − 1
8.                  𝒜 ← tail 𝒜
9.            else  𝒜 ← aᵢ : 𝒜
10.                 i ← i+1
11.                 if i ≤ N
12.                     then D'ᵢ ← Dᵢ
13.   return REVERSE(𝒜)
SELECTVALUE(D'ᵢ, 𝒜, C)
1.    while D'ᵢ is not empty
2.        aᵢ ← head D'ᵢ
3.        D'ᵢ ← tail D'ᵢ
4.            if CONSISTENT(aᵢ : 𝒜)
5.                then return aᵢ
6.    return null
```

Figure 12.5 shows the progress on the tiny map colouring problem from Figure 12.1. The order of variables is alphabetic. The very first choices for the variables A, B, and C are accepted, but when it comes to variable D only the third choice g works.

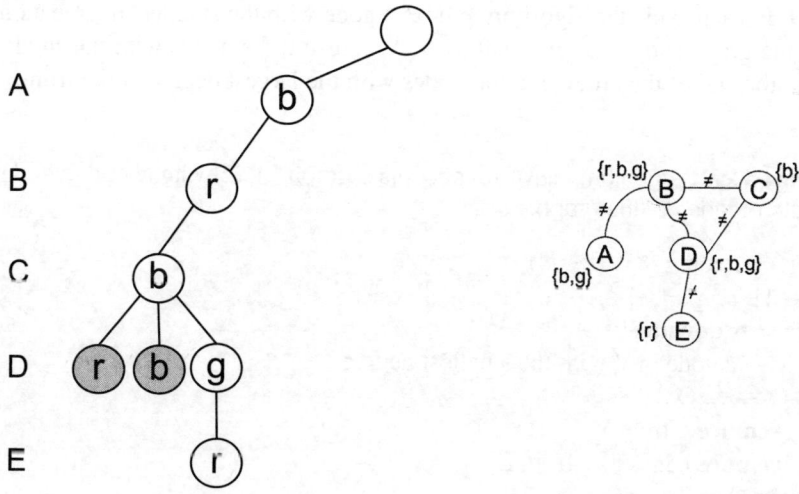

Figure 12.5 BACKTRACKING does depth first search on the problem from Figure 12.1 and finds the solution <b, r, b, g, r>. On the way SELECTVALUE has rejected the values $D = r$ and $D = b$. The constraint graph is shown on the right.

The order of processing variables will clearly impact the complexity of the search. There are essentially two approaches to deciding this order. One is a static approach that looks at the topology of the constraint graph to choose an order with fewer dead ends. We look at that next. The other is to dynamically choose the next variable to try, in tandem with constraint propagation. We will describe that after looking at the consistency enforcement algorithms.

12.2.1 Static variable ordering

The choice of a value for a variable is constrained by the other variables it is related to. If a variable X is connected to only one other variable Y, then the moment one chooses a value for Y, a value for X can be chosen, and that would be final. But imagine a variable U related to three other variables X, Y, and Z. Then choosing values for X, Y, and Z first may not leave a consistent value for Z. For example, if the domain of all four variables is $\{r, b, g\}$, then choosing $X = r$, $Y = b$, and $Z = g$ leaves no value for U. But choosing a value $U = r$ first allows for a choice of two values for each of X, Y, and Z. One can then hypothesize that variables of higher degree (connected to more other variables) should be assigned values earlier. This topological argument is the reason for choosing an ordering based on the degrees of the nodes. We begin with some definitions.

Given a CSP <X, D, C> and an ordering O of the variables $(x_1, x_2, ..., x_N)$, the *width* of a node in the ordering is the number of *parents* that it is connected to. A node x_i is the parent of a node x_k if the two have an edge between them in the constraint graph, and x_i precedes x_k in the ordering. The width of an ordering is the maximum of the width of all nodes in that ordering.

A min-width ordering of a graph is an ordering which has the lowest width (Dechter, 2003). A greedy algorithm described below produces the min-width ordering. It begins with an empty list O. In each cycle the algorithm plucks a node with the smallest degree along with its edges from the graph and concatenates it to O. As a result, the nodes with the smallest degree are placed in the end of the order, and the nodes with the largest degree at the front.

Algorithm 12.2. Algorithm MinWidth accepts a graph <V, E> with N nodes and returns a min-width ordering of the graph.

MinWidth (Graph = <V, E>, N)
1.   O ← []
2.   **for** i = N **downto** 1
3.   v ← a node in V with the smallest degree
4.   O ← v : O
5.   **remove** v from V
6.   **remove** edges to v from E
7.   **return** O

When a node has multiple parents, perhaps constraints can be imposed between them. Consider the example of variable U having variables X, Y, and Z as parents. Searching for values in the given order one can see that if X, Y, and Z were originally unrelated, then adding the constraints $X = Z$, $X = Y$, and $Y = Z$ would have made finding a value for U easier. One would not have to backtrack and try different values for X or Y or Z. As we shall see later, adding such constraints with the goal of minimizing backtracking is a strategy in consistency enforcement. It would be desirable to enforce enough consistency to make the search backtrack free. But the cost of achieving that could outweigh the savings.

In this context one can introduce the notion of an induced graph with an induced width, in which edges are added connecting parents of nodes in the order being imposed. Unfortunately, finding a min induced width ordering is NP-complete, but the following greedy algorithm often produces very good ones (Dechter, 2003). The greedy algorithm below is similar to Algorithm 12.2 except that before removing the selected node v from the graph, all its parents are connected pairwise.

Algorithm 12.3. Algorithm *MinInducedWidth* is similar to MinWidth except that before plucking the node *v* from the graph all its parents are connected pairwise.

MinInducedWidth (Graph = <V, E>, N)
1.   O ← []
2.   **for** i = N **downto** 1
3.   v ← a node in V with the smallest degree
4.   O ← v: O
5.   **connect** each pair of parents of v
6.   **remove** v from V
7.   **remove** edges to v from E
8.   **return** O

A variation that often performs better is to choose the node to be plucked using a different criterion. Instead of selecting the node with the lowest degree, one picks a node which has a minimum number of unconnected parents. Then only a few new edges will need to be added. This algorithm is called *MinFill*.

Figure 12.6 shows a few orderings for a small graph with seven nodes $X = \{A, B, C, D, E, F, G\}$ shown on the top. The first ordering in the figure is the alphabetic ordering (A, B, C, D, E, F, G). With this ordering BACKTRACKING would assign a value for variable A first and variable G last. The alphabetic ordering has a width 3, because node E has three parents A, B, and D. The second ordering is reverse alphabetic and has width 4 since A has degree 4. The third ordering is the one produced by the MinWidth algorithm and has a width 2. The last one is the one produced by the MinInducedWidth algorithm. It also has a width 2, but has an additional edge connecting D and G, the parents of F which occurs later in the ordering.

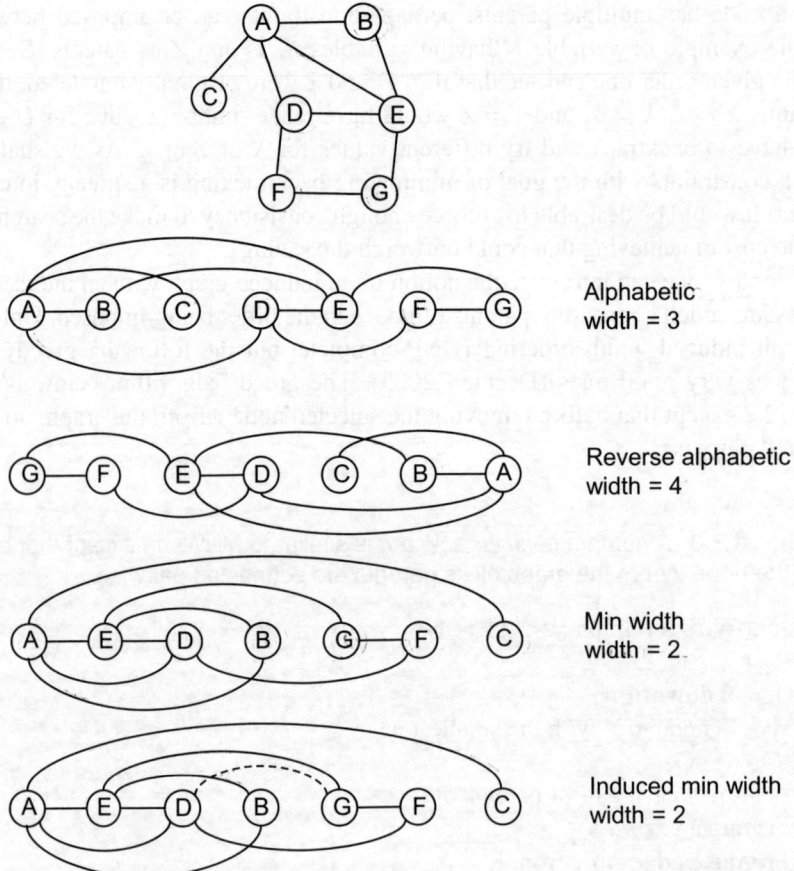

Figure 12.6 A graph and some orderings. Both the min-width and min-induced-width orderings have a width 2. The alphabetic ordering has a width 3, and the reverse alphabetic ordering has the maximum width possible 4.

The reader should verify that if the given graph were to be a tree, then both the algorithms will produce an ordering of width 1. When we have a CSP ordering of width 1, then it is possible to do backtrack free search. This is because each node is constrained by only one parent who already has a value.

If the graph has cycles, then the minimum width possible is 2. This is the case for the example above.

12.2.2 Dynamic variable ordering

The search algorithm backtracks when it cannot find a value for a variable consistent with the earlier variables. This is because *all* values available in the domain of the variable may be conflicting with values assigned to earlier variables. This could be because there are too many parents, like when X, Y, and Z are parents of variable U. But this could be also because there are too few values left in the domain of the current variable. For example, if the domain of U has only one variable, then it could easily conflict with earlier variables. One strategy would be to assign a value to this variable before considering the others. This is the approach taken in *dynamic variable ordering*, where the order in which variables are processed is decided on the fly.

This becomes even more relevant when the domains of future variables are pruned by the algorithm. We illustrate this with a cursory description of the algorithm FORWARDCHECKING discussed later in more detail. The crux of the algorithm is that when it considers a value for a variable, it deletes values from future variables that would become inconsistent with the current assignment. We illustrate this with the small map colouring example from Figure 12.1. The domains and the constraints are reproduced below.

$$D_A = \{b, g\}, D_B = \{r, b, g\}, D_C = \{b\}, D_D = \{r, b, g\}, D_E = \{r\}$$
$$R_{AB} = \{<b, r>, <b, g>, <g, r>, <g, b>\}$$
$$R_{BC} = \{<r, b>, <g, b>\}$$
$$R_{BD} = \{<r, b>, <r, g>, <b, r>, <b, g>, <g, r>, <g, b>\}$$
$$R_{CD} = \{<b, r>, <b, g>\}$$
$$R_{DE} = \{<b, r>, <g, r>\}$$

We begin with the variable C which is one of the two with the smallest domains.

1. $C = b$. Region C is adjacent to regions B and D. FORWARDCHECKING deletes b from their domains. Now $D_B = \{r, g\}$ and $D_D = \{g, r\}$ after pruning. Next, we consider E which has the smallest domain.
2. $E = r$. Region E is adjacent to D, and we prune the domain of D to get $D_D = \{g\}$. Now D becomes the smallest domain.
3. $D = g$. Only B is a future variable related to D. $D_B = \{r\}$ after pruning.
4. $B = r$. This value in B does not conflict with the values in A.
5. A can be either blue or green.

As seen here, dynamic variable ordering considers those variables first which have the fewest values to choose from. And deleting values from future variables removes potentially conflicting choices. In the process, if a future variable becomes empty, the search algorithm can backtrack from the current variable itself. We will illustrate this in Section 12.4.

12.3 Constraint Propagation

A typical CSP describes the constraints in parts. A search algorithm wades through the constraints looking for an assignment. *Backtracking* happens when a partial assignment that satisfies some constraints cannot be extended to another variable and another constraint. For example, given the CSP $<\{X, Y, Z\}, \{D_X = D_Y = D_Z = \{1, 2, 3\}\}, \{R_{XY} = X < Y, R_{YZ} = Y < Z\}$, then choosing $X = 2$ allows us to choose $Y = 3$ but we cannot choose a value of Z.

Constraint propagation or *consistency enforcement* is the endeavour to *tighten* the CSP so that these kinds of dead ends do not arise. This can be done by pruning domains of variables in the simplest case, or by adding constraints to limit the choices to values that can be part of a solution. Done to an extreme, consistency enforcement can make search backtrack free. But at a prohibitive computational cost. Very often the best approach is to adopt a combination of reasoning and search that is optimal. In this chapter we study a few algorithms for consistency enforcement.

We begin with the general notion of *i-consistency*. A network $\mathcal{R}$ is said to be *i-consistent* if every consistent assignment to any $i - 1$ variables can be extended to one more variable. A network is said to be *strongly i-consistent* if it is also *j-consistent* for all $j \leq i$.

A node is said to be 1-consistent or node consistent (NC) iff every variable x in the domain satisfies all constraints R_x on the variable. For example, if $R_x = \text{Even}(x)$, then there must be no odd value in any variable. Node consistency can be achieved by inspecting the domains of all variables and removing any values that do not satisfy some constraint.

12.3.1 Arc consistency

A variable X is said to be *arc consistent* (AC) with respect to a variable Y if there is an edge (X, Y) in the constraint graph and for every value $a \in D_X$, there exists a value $b \in D_Y$ such that $<a, b> \in R_{XY}$. We say that a supports b, and b supports a. A simple algorithm REVISE$((X), Y)$ makes X arc consistent to Y.

> **Algorithm 12.4.** Algorithm REVISE prunes the domain of variable X, removing any value that is not paired to a matching value in the domain of variable Y.
>
> REVISE$((X), Y)$
> 1.   **for** every $a \in D_x$
> 2.   **if** there is no $b \in D_Y$ s.t. $<a,b> \in R_{XY}$
> 3.   **then** delete a from D_x

The worst case complexity of REVISE is $\mathcal{O}(k^2)$ where k is the size of each domain. The worst case happens when no value in X has a matching value in Y. An edge (X, Y) in a constraint graph is said to be arc consistent iff both X and Y are arc consistent with respect to each other. A constraint network $\mathcal{R}$ is said to be arc consistent if all edges in the constraint graph are arc consistent. A node is said to be 2-consistent if an assignment to any variable can be extended to a consistent assignment to any other variable. Clearly, if a network is 2-consistent, it must be arc consistent as well. A simple brute force algorithm *AC-1* cycles through all edges in the constraint graph until no domain changes (Mackworth, 1977; Mackworth and Freuder, 1985).

> **Algorithm 12.5.** Algorithm *AC-1* cycles through all edges repeatedly even if one value is removed from one variable.
>
> AC-1 (X, D, C)
> 1.   **repeat**
> 2.   **for** each edge (x, y) in the constraint graph
> 3.   REVISE$((x), y)$
> 4.   REVISE$((y), x)$
> 5.   **until** no domain changes in the cycle

Let there be n variables, each with domain of size k. Let there be e edges in the constraint graph. Every cycle then has complexity $\mathcal{O}(ek^2)$. In the worst case, the network is not AC, and in every cycle exactly one element in one domain is removed. Then there will be nk cycles. The worst case complexity of *AC-1* is therefore $\mathcal{O}(nek^3)$.

Before improving upon the arc consistency algorithm, we look at how deduction with *modus ponens* can be seen as constraint propagation. Let the knowledge base be $\{P, P \supset Q, Q \supset R, R \supset S\}$. Working with Boolean formulas each propositional variable has two values in its domain, 1 (*true*) and 0 (*false*). The truth table of the binary relation $X \supset Y$ can be represented by the constraint $\supset_{XY} = \{<0, 0>, <0, 1>, <1, 1>\}$. The CSP can then be viewed as

<X, D, C> where
  $X = \{P, Q, R, S\}$,
  $D_P = D_Q = D_R = D_S = \{0, 1\}$
  $R_P = \{<1>\}$
  $R_{PQ} = R_{QR} = R_{RS} = \{<0, 0>, <0, 1>, <1, 1>\}$

First achieving node consistency prunes the domain of P to $\{1\}$. Then achieving arc consistency prunes the rest of the variables to also contain only 1. The process is illustrated in Figure 12.7 with matching diagrams.

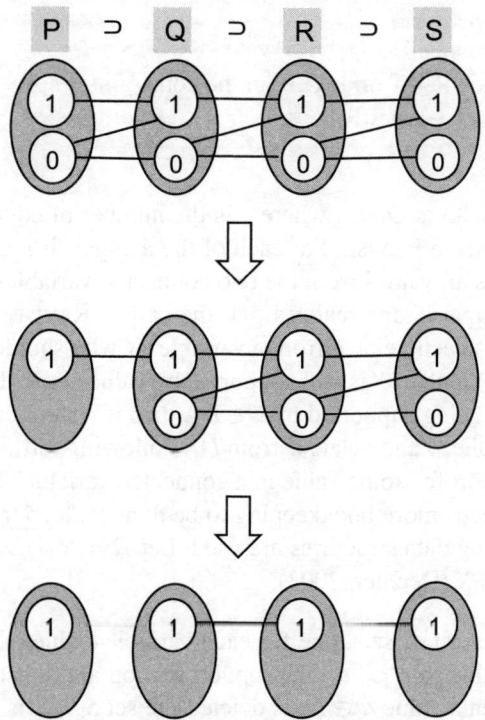

Figure 12.7 Logical deduction can be seen as consistency enforcement. Given the variables P, Q, R, and S and the constraints defined by the KB = $\{P, P \supset Q, Q \supset R, R \supset S\}$. Node consistency followed by arc consistency results in the domains of all variables having only 1. This amounts to deducing that Q, R, and S are true.

The algorithm AC-1 is an overkill. It makes unnecessary calls to REVISE. A better strategy is as follows. If REVISE((X), Y) removes some value v from the variable X, one need only check that all edges connected to X are still arc consistent. It is possible that the value v was the only support for some value w in a variable W. Then a call to REVISE((W), X) is needed. This is done by algorithm *AC-3* that pushes all such connected pairs of variables into a queue. A change in a variable is *propagated* to the connected variables. *Only those* are considered again for a call to REVISE.

> **Algorithm 12.6.** Algorithm AC-3 begins by invoking REVISE for all edges in the constraint graph. After that, if the domain of a variable *P* has changed, then consistency w.r.t. *P* is enforced for all neighbours of *P*.
>
> AC-3(X, D, C)
> 1.   Q ← []
> 2.   **for** each edge (N,M) in the constraint graph
> 3.   Q ← Q ++ (N,M) : [(M,N)]
> 4.   **while** Q **is not empty**
> 5.   (P,T) ← **head** Q
> 6.   Q ← **tail** Q
> 7.   REVISE((P), T)
> 8.   **if** D_p has changed
> 9.   **for** each R ≠ T **and** (R,P) in the constraint graph
> 10.   Q ← Q ++ [(R,P)]

The complexity of AC-3 is $\mathcal{O}(ek^3)$ where e is the number of edges and each domain is of size k. Of this, k^2 comes from REVISE. For each of the e edges, it makes $2k$ calls to REVISE in the worst case if it deletes all values from the two connected variables.

One can be more frugal if one realizes that the call to REVISE can itself be an overkill. Just because a value v has been deleted from a variable[1] X why should one make a call REVISE ((Y), X) to check if every value in Y is still supported by values in X? If one could keep track of the values in Y that were being supported by $v \in D_X$, then if v were the *only* support of a value $w \in D_Y$, then one can go ahead and delete w from D_Y. Following this, we will have to check if w in turn was the only support for some value in a connected variable. This is done by algorithm *AC-4* which, however, needs more bookkeeping to be done to keep track of individual support from values. The following data structures are used. Let $\mathcal{R} = <X, D, C>$ be the network, and let x and y be variables in X (Dechter, 2003).

–   The support set S is a set of sets, one for each variable–value pair $<x, a>$, named $S_{<x, a>}$. For each variable–value pair $<x, a>$ the support set contains a list of supporting pairs from other variables. When a value $a \in D_x$ is deleted the set $S_{<x, a>}$ is instrumental in checking which values in other variables might have lost a support.

---

[1] We often say 'from a variable X' as a short form for 'from the domain of a variable X'.

$S_{<x,a>} = \{ <y, b> \mid y \in X, b \in D_y \text{ and } <a, b> \in R_{xy} \}$

Given e constraints and domain sizes k, computing S is $\mathcal{O}(ek^2)$.

- Counter array *counter*. For each value $a \in D_x$, denoted by $<x, a>$, the counter array maintains the *number* of supports from a variable y. If the counter value becomes zero, then the value a has to be removed from D_x.

    Counter(x, a, y) = the number of values in D_y that support the pair $<x, a>$

    The counter array can be constructed along with S, adding a constant amount of computation for each label.

- A queue Q of labels without support that need to be processed. The unsupported variable–value pairs are added to this as and when they are identified.

The algorithm AC-4 begins by inspecting the network $\mathcal{R}$ setting up the records of links from the matching diagram in the set S, and a count of *how many* values from a variable y support a value $a \in D_x$. This requires visiting both ends of all the e edges in the network and all k^2 combination of values for the two connected variables. This is done in $\mathcal{O}(ek^2)$ time, and also requires $\mathcal{O}(ek^2)$ space. All this work is done upfront.

Algorithm 12.7. Algorithm AC-4 begins by inspecting all edges in the matching diagram and identifying the list of all supports for all variable–value pairs, and the count of number of supports for each value from another variable. It deletes a value with count 0 and then decrements the count of all connected values.

AC-4(X, D, C)
1. $Q \leftarrow []$
2.   **initialize** $S_{<x,a>}$ and counter(x, a, y) for each R_{xy} in C
3.   **for** each counter
4.   **if** counter(x, a, y) = 0
5. $Q \leftarrow Q \mathbin{++} <x, a>$
6.   **while** Q is not empty
7. $<x, a> \leftarrow$ **head** Q
8.   **delete** a from D_x
9.   **for** each $<y, b>$ in $S_{<x,a>}$
10.   counter(y, b, x) $\leftarrow$ counter(y, b, x) − 1
11.   **if** counter(y, b, x) = 0
12. $Q \leftarrow Q \mathbin{++} <y, b>$

After the initialization, if there is a missing support for a value a for variable x (from some variable y), then a is deleted from D_x and then the set S is inspected to decrement all counters for all related variable–value pairs $<y, b>$. If any counter becomes 0, then that variable–value pair is added to the queue Q of values destined for deletion. In this manner, the propagation is extremely fine grained. Whenever a value is deleted, the algorithm pursues the links in the matching diagram, effectively deleting each such link. Then if a value in some domain is left without a link, that is added to the queue for deletion as well.

The initialization step that creates the counters and the support pointers requires, at most, $\mathcal{O}(ek^2)$ steps. The number of elements in $S_{<x,a>}$ is of the order of ek^2 and each is accessed at most once. Therefore the time and space complexity of AC-4 is $\mathcal{O}(ek^2)$.

What can one say about a CSP on which arc consistency has been achieved and no domain is empty? If the constraint graph is a tree, then the CSP has a solution. This is because each variable is constrained by exactly one variable. Moreover, if one chooses the min-width ordering of the variables, the search will be backtrack free. If the constraint graph is not a tree, then it may be possible that the CSP has no solution. This is illustrated in Figure 12.8 where the network on the left has no solution even though it is arc consistent. But after removing one edge (BC) it becomes a tree, and this has two solutions.

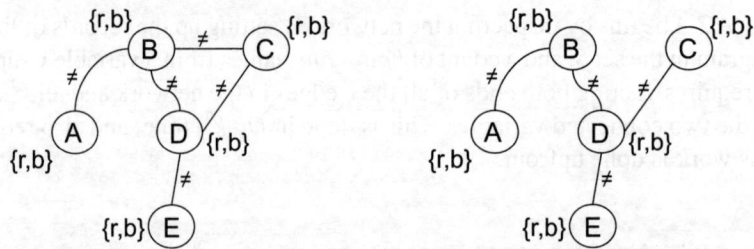

Figure 12.8 The CSP on the left is arc consistent but does not have a solution. The network on the right is similar to the one on the left except that it has one edge (*B,C*) less which makes it a tree. This network has two solutions.

The reader is encouraged to try out different orderings of the network and verify that the min-width ordering for the tree can be solved in a backtrack free manner.

12.3.2 The Waltz algorithm

During the early 1970s, a bunch of students at MIT worked on the problem of *interpreting* line drawings. It started with Adolfo Guzman, a graduate student of Marvin Minsky's, who took up the problem of writing a program to look at a line drawing to ascertain how many objects were present in it. David Huffman then limited the line drawings to those of trihedral objects without any cracks and shadows, and where the viewpoint was such that a slight shift did not produce drastic changes in the image. A trihedral object is one in which exactly three straight line edges meet at every vertex. This also means that a vertex was created by three plane surfaces meeting at one point.

The objective was to label each line drawing with one of three kinds of labels. A convex edge is one where two faces are visible and the solid matter subtends an angle less than 180°, like the edge of a cube. Such an edge is to be labelled '+'. A concave edge is one where two faces are visible and the solid matter subtends an angle more than 180°, like the line where two walls in a room meet. This is labelled with '−'. The third kind of label is an arrow, when only one face is visible. The visible face is on the right when one travels along the direction of the arrow. Thus there are two different arrow labels. This task is also known as Huffman–Clowes labelling (Clowes, 1971; Huffman, 1971).

The visible vertices are of four kinds as shown in Figure 12.9.

Constraint Satisfaction | 409

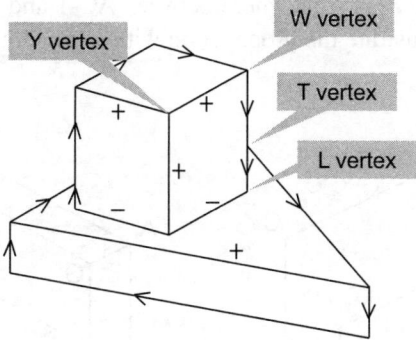

Figure 12.9 The three kinds of edge labels and four kinds of vertices in trihedral objects.

Each edge in a line drawing can be labelled in one of four ways: $+$, $-$, $\rightarrow$, and $\leftarrow$. Then a W or a Y or a T vertex can have $4^3 = 64$ different combined labels and an L vertex can have $4^2 = 16$. The interesting thing is that for trihedral objects without cracks or shadows, there are only 18 kinds of edge label combinations that are physically possible. These are shown in Figure 12.10.

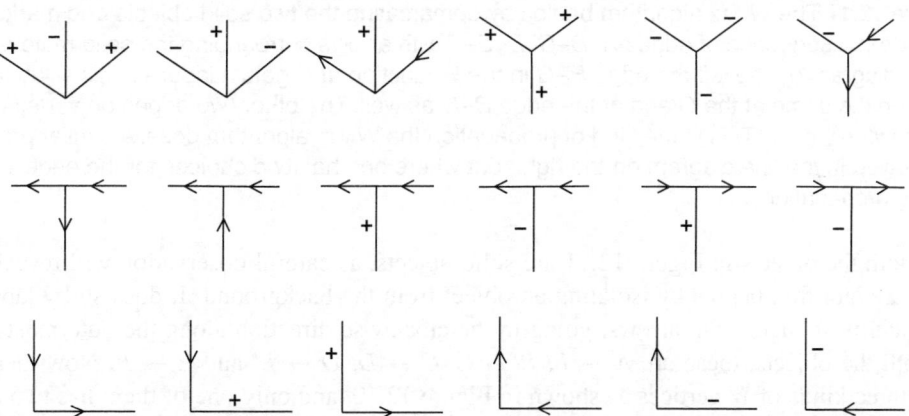

Figure 12.10 The 18 different kinds of vertices possible in line drawings for trihedral objects without cracks or shadows. Some texts leave out the middle two T vertices because that configuration comes from a non-normal viewpoint.

Every edge in a line drawing connects two vertices but it can have only one label. This lays the foundation of constraint propagation. If one knows the label at one end, then that label must be propagated to the other end as well. And at the other end, the other edges impinging on the vertex will be constrained by possibilities shown in Figure 12.10.

The constraint propagation algorithm was written by David Waltz who extended the scope of objects manifold (Waltz, 1975). The WALTZ algorithm, as it is now known, could handle objects with more than 3-edge vertices, objects with cracks, and images with light and shadows. The number of edge labels shot up from 4 to 50-plus, and the number of valid vertices shot up to thousands. The algorithm is somewhere between AC-1 and AC-3 and does propagation from vertex to vertex. We illustrate the propagation with a trihedral object shown on the left in Figure 12.11.

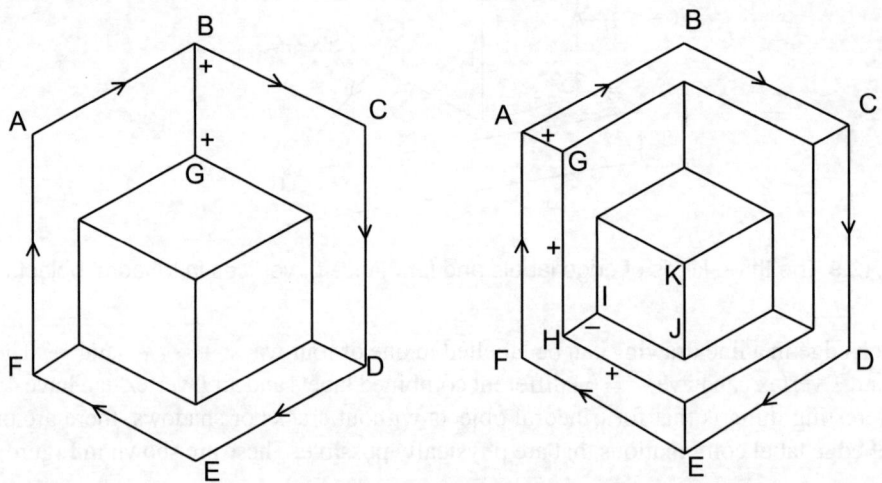

Figure 12.11 The WALTZ algorithm begins by demarcating the two solid objects and marks the external sequence of edges A–B–C–D–E–F with arrows surrounding the solid material. In the diagram on the left the edge B–G in the W junction at B gets a label +. This + label must be the same at the G end of the edge B–G as well. The other two edges on vertex G can now only be +. This is the kind of propagation the WALTZ algorithm does. A similar process is followed in the line diagram on the right, but where one has two choices for the edge 1–J during propagation.

Both the objects in Figure 12.11 are solid objects, as careful observation will reveal. The WALTZ algorithm begins by isolating an object from the background. It does so by labelling the outermost edges with arrows, going in the clockwise direction along the outermost lines. In both the objects, these are $A \to B$, $B \to C$, $C \to D$, $D \to E$, and $E \to F$. Now there are only three kinds of W vertices as shown in Figure 12.10, and only one of them has two arrow labels. Consequently, the third edge in these vertices *must* be convex edges and can be labelled with a '+'.

This is illustrated for the edge B–G for the object on the left. We have labelled it twice to emphasize the fact that the label is propagated from the W vertex B to the Y vertex G. Now there is only one kind of Y vertex that has '+' labels, and all three of the edges impinging on it must be labelled '+'. This label can now be propagated along the two edges emanating from G to the connected W edges. This process continues and the entire set of edges can be labelled unambiguously.

For the object on the right, the labelling may involve backtracking. The reader should verify that the edge H–I must be labelled with a '−'. Now there are two possibilities of labelling the other two edges at vertex *I*. Either both must be '−' or both must be arrows. If I–J is a '−', then the edge J–K can only be a '+' given the constraints on W vertices. This results in K being labelled with '+++'. If it is to be an arrow, then the direction must be $I \rightarrow J$. But then there is no label possible for the edge J–K. So if the algorithm were to select $I \rightarrow J$, it would have to backtrack and select the label '–'.

The reader is encouraged to complete the labelling process for both objects.

12.3.3 Path consistency

A network $\mathcal{R}$ is said to be *path consistent* (PC) or 3-consistent if a consistent assignment to any two variables can be extended to any third variable. Consider the simple map colouring problem with five regions on the left. As discussed earlier, the network is arc consistent but does not have backtrack free search. The assignment $\{A = r, B = b, C = r\}$ is consistent but cannot be extended to variable D. Making it path consistent involves adding a new constraint $R_{BC} = \{<r, r>, <b, b>\}$ to the network as shown on the right. When that happens the assignment $\{A = r, B = b, C = r\}$ is no longer consistent and BACKTRACKING has to choose $C = b$ instead, which allows $D = r$.

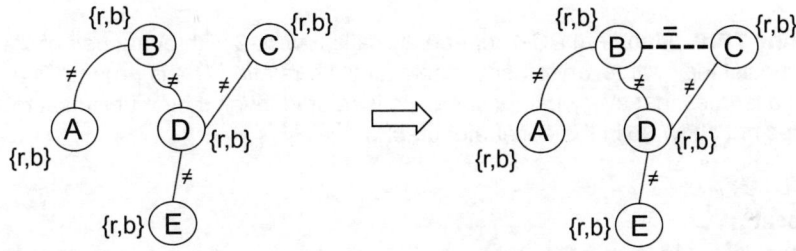

Figure 12.12 The network on the left is not path consistent because an assignment <B = r, C = b> cannot be extended to the variable D. Making it path consistent adds a new constraint $R_{BC} = \{<r, r>, <b, b>\}$ to the CSP. Now the variables B and C are related by the equality relation. Earlier, it was implicitly the universal relation.

The astute reader would have noticed that in the process of making the network 3-consistent we have introduced a new edge in the constraint graph for the relation $B = C$. The reader must also keep in mind that when the vertices B and C were not connected in the constraint graph, it meant that any value in B was locally consistent with any value in C. That is, no constraint between B and C was specified, and R_{BC} was a universal relation $\{<r, r>, <r, b>, <b, r>, <b, b>\}$. After the propagation this was pruned to $\{<r, r>, <b, b>\}$, and then an edge B–C was introduced in the constraint graph. This is done by the algorithm REVISE-3 which takes three variables X, Y, and Z, and removes any pair of values $<X = a, Y = b>$ when a and b are not connected to some value $c \in D_Z$ (Dechter, 2003). In other words, we are pruning the relation R_{XY}.

Algorithm 12.8. Algorithm REVISE-3 prunes the relation R_{XY}, removing any edge $<a, b>$ that does not have a matching value in the domain of variable Z.

REVISE-3((X,Y), Z)
1.   **for** every $<a, b> \in R_{XY}$
2.   **if** there is no $c \in D_Z$ s.t. $<a, c> \in R_{XZ}$ and $<b, c> \in R_{YZ}$
3.   **then** delete $<a, b>$ from R_{XY}

This is, in fact, an instance of the general case wherein making a network N-consistent induces a relation of arity $N - 1$, which essentially prunes an existing relation that could have been universal. This was the case also for arc consistency, because pruning the domain of a variable X is equivalent to inducing a relation R_X on the network. We will have more to say on this later.

The simplest algorithm to achieve path consistency is analogous to AC-1. It repeatedly considers *all* variable pairs X and Y and eliminates pairs of *values* $a \in D_X$ and $b \in D_Y$ that cannot be extended to a third variable Z. The algorithm is called *PC-1*.

Algorithm 12.9. Algorithm *PC-1* repeatedly calls REVISE-3 with *every pair* of variables for path consistency with *every* other variable, until no relation R_{YZ} is pruned. The algorithm assumes that every pair of variables is related, even if by a universal relation which does not show up in the constraint graph.

PC-i(X, D, C)
1.   **repeat**
2.   **for** each x in X
3.   **for** each y and z in X
4.   REVISE-3((y, z), x)
5.   **until** no relation changes in the cycle

Let there be n variables, each with domain of size k. The complexity of REVISE-3 is $\mathcal{O}(k^3)$ because the algorithm has to look at all values of the three variables. In each cycle the algorithm *PC-1* inspects $(n - 1)^2$ edges for each of the n variables, requiring $\mathcal{O}(k^3)$ computations for each call to REVISE-3. Therefore, in each cycle, the algorithm will do $\mathcal{O}(n^3 k^3)$ computations. In the worst case, PC-1 will remove one pair of values $<a, b>$ from some constraint R_{xy}. Then the number of cycles is $\mathcal{O}(n^2 k^2)$, because there are n^2 pairs of variables, each having k^2 elements. Thus in the worst case, algorithm PC-1 will require $\mathcal{O}(n^5 k^5)$ computations (Dechter, 2003).

Note that unlike AC-1, the algorithm PC-1 is not confined to working only with the edges in the constraint graph but considers all pairs of variables. It might be pertinent to remember that two variables in the constraint graph without an edge are related by the universal relation, which means that any combination of values is allowed. Achieving path consistency may delete some elements from the universal relation, as illustrated in Figure 12.13.

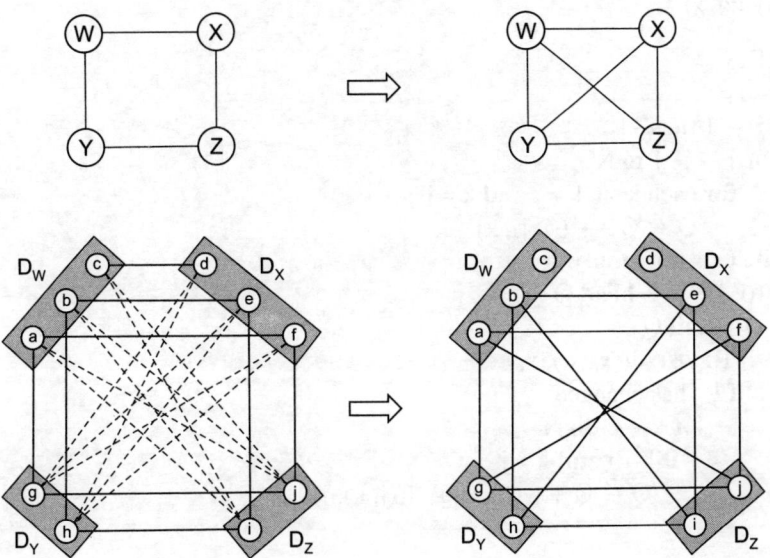

Figure 12.13 The two figures on the left are the constraint graph and the matching diagram for a network with four variables $W, X, Y,$ and Z. The dashed edges represent the implicit universal relations R_{WZ} and R_{XY}. On the right are the corresponding figures after the network is made path consistent. The relations R_{WZ} and R_{XY} are now non-universal and show up in the constraint graph. The edge c–d is deleted along with eight edges from R_{WZ} and R_{XY}.

It can be observed that after achieving path consistency every edge in the matching diagram is part of a triangle with all other variables. The edge c–d in Figure 12.13 on the left gets deleted because it is not part of any triangle with values in variables Y and Z. For the same reason, four edges from each of the two implicit universal relations R_{WZ} and R_{XY} are also deleted. In the network on the right, every edge is a part of two triangles.

Algorithm PC-1 looks at all triples of variables in every cycle. A better approach is to look only at variables where an edge deletion may have broken a triangle in the spirit of AC-3. Let the variables be an ordered set $X = \{x_1, x_2, ..., x_N\}$. Algorithm *PC-2* too tests every variable pair $<x_i, x_j>$ where $i < j$ against all other variables. It begins by enqueuing all such triples for calls to REVISE-3. Each pair of variables is added only once. If an edge $<a, b> \in R_{xy}$ is deleted by a call to REVISE-3, then PC-2 only checks for the triangles formed by all other variables with x and y. In the following algorithm, the indices $1, 2, ..., N$ of the variables $x_1, x_2, ..., x_N$ are stored in the queue Q to enable only one of $<x_i, x_j>$ and $<x_j, x_i>$ to be added.

Algorithm 12.10. Algorithm PC-2 begins by enqueuing all triples of distinct variables. It dequeues them one by one and calls REVISE-3$((x,y), z)$. If any edge in x–y is deleted, then for every other variable w, PC-2 sets up calls to REVISE-3$((x, w), y)$ and REVISE-3$((y,w), x)$.

PC-2 (X, D, C)
1. $Q \leftarrow [\,]$
2. **for** $i \leftarrow 1$ **to** $N-1$
3.   **for** $j \leftarrow i+1$ **to** N
4.   **for** each k s.t. $k \neq i$ **and** $k \neq j$
5. $Q \leftarrow Q$ ++ $((i, j), k)$
6. **while** Q is not empty
7. $((i, j), k) \leftarrow$ **head** Q
8. $Q \leftarrow$ **tail** Q
9.   REVISE-3$((x_i, x_j), x_k)$
10.   **if** R_{ij} has changed
11.   **for** $k \leftarrow 1$ **to** N
12.   **if** $k \neq i$ **and** $k \neq j$
13. $Q \leftarrow Q$ ++ $((i, k), j): [((j, k), i)]$

Each call to REVISE-3 is $\mathcal{O}(k^3)$. The minimum number of calls is $\mathcal{O}(n^3)$, which is the number of distinct calls that can be made initially. In the worst case, one pair of values is deleted in each call to REVISE-3 in the while loop. In the worst case, n^3 calls to REVISE-3 are made. In each call to REVISE-3, at most k^2 edges can be removed. Hence the while loop can be executed at most $\mathcal{O}(n^3 k^2)$ times and with REVISE-3 being $\mathcal{O}(k^3)$, the complexity of PC-2 is $\mathcal{O}(n^3 k^5)$.

Like AC-1 and AC-3, both PC-1 and PC-2 rely on calls to REVISE-3, which can be a little bit of overkill. Like in AC-4, one can work at the value level, but we will not pursue those edges in the matching diagram. Mohr and Henderson (1986) have devised such an algorithm *PC-4* with complexity $\mathcal{O}(n^3 k^3)$.

It is worth noting that path consistency does not automatically imply arc consistency. This is evident in Figure 12.13. The *CSP* is path consistent, but it is not arc consistent. In general, if a *CSP* is i-consistent it does not mean that it is $(i - 1)$-consistent as well.

12.3.4 i-Consistency

The concept of consistency can be applied to any number of variables. Without going into the details we observe that the process involves defining a function R_{EVISE}-i in which a tuple t_S from a set S of $(i - 1)$ variables is checked with one variable X for consistency. If there is no value v in X that is consistent with the tuple then t_S is deleted. This is equivalent to introducing a relation R_S of arity $(i - 1)$. In general, the complexity of REVISE-i is $\mathcal{O}(k^i)$, and algorithms for enforcing *i*-consistency have the worst case time complexity $\mathcal{O}((nk)^{2i} 2^i)$ and space complexity of $O(n^i k^i)$ as described in (Dechter, 2003).

A network is said to be *strongly i-consistent* if it is *i-consistent* and is *j-consistent* for all $j < i$. If a network is strongly *i*-consistent, then the algorithm BACKTRACKING will find the solution in the first shot and will be backtrack free. However, the cost of achieving this can be prohibitive, and there are other approaches that one can employ to chip away at the complexity of depth first search. We first look at directional consistency.

12.3.5 Directional consistency

When a network is *i-consistent*, then a consistent assignment to *any* $(i - 1)$ variables can be extended to *i* variables. In practice, though, we often have an order in which the algorithm BACKTRACKING searches. One only needs to ensure that the variables that *precede* a given variable X have values that admit a consistent value for X. This brings in the notion of *directional consistency* in which consistency is *only* in the direction from parents to children, and not for *any* subset of variables, as illustrated in Figure 12.14.

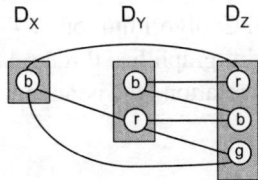

Figure 12.14 The above network is neither arc consistent nor path consistent, but is both directionally arc consistent and directionally path consistent. Given the order X, Y, Z, BACKTRACKING finds a solution without backtracking.

Given an ordering $X = (x_1, x_2, ..., x_N)$, a network $<X, D, C>$ is said to be *directionally arc consistent* (DAC) if for every edge $<x_i, x_j>$ in the constraint graph such that $i < j$, variable x_i is arc consistent with respect to variable x_j. DAC arc consistency can be achieved in a single pass, processing variables from the last to the first.

Algorithm 12.11. Algorithm DAC scans the variables from the last to first calling REVISE with all parents in the constraint graph.

DAC(X = [$x_1, x_2, ..., x_n$], D, C)
1.   **for** i ← N **downto** 2
2.   **for** j ← i−1 **downto** 1
3.   **if** $R_{ij} \in C$
4.   REVISE((x_j), x_i)

Directional path consistency (DPC) is similar, except that it prunes binary relations and looks at all triples without any heed to the constraint graph. When it prunes an edge in the matching diagram, it adds a relation to the constraint graph. In the following algorithm, we have included DAC as well.

Algorithm 12.12. Algorithm DPC does one pass from the last variable down to the first one. For each variable, it calls REVISE-3 with all the preceding variables, and then it also calls REVISE.

DPC(X = [$x_1, x_2, ..., x_N$], D, C)
1.   for i ← N **downto** 3
2.   for j ← i−1 **downto** 2
3.   for k ← j−1 **downto** 1
4.   REVISE-3((x_k, x_j), x_i)
5.   add R_{kj} to C
6.   for j ← i−1 **downto** 1
7.   **if** R_{ij} ∈ C
8.   Revise((x_j), x_i)

Figure 12.15 illustrates the DPC algorithm on a 4-variable 2-colour map colouring problem. To begin with, the constraint graph has three relations R_{WY}, R_{XZ}, and R_{YZ}. After the call REVISE-3((X, Y), Z) an induced relation R_{XY} is added, and after REVISE-3((W, X), Z) the relation R_{WX} is added.

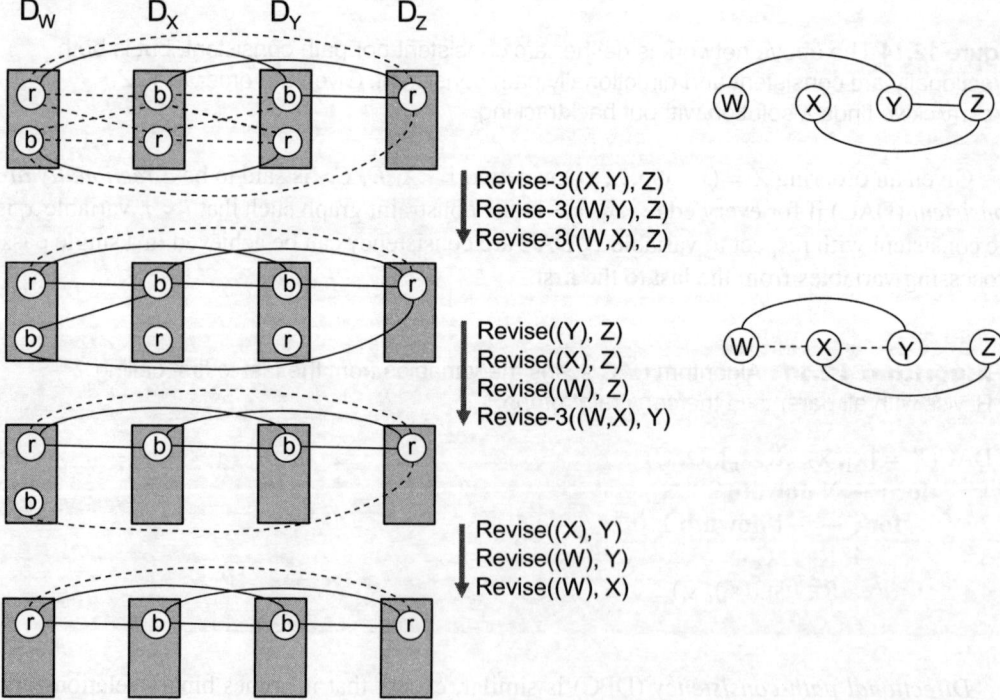

Figure 12.15 DPC and DAC process the network in one pass from the last to the first node. The original matching diagram and the constraint graph are on the top. The dashed edges are from the universal relations. The revised versions are shown progressively below. The final network is strongly path consistent and backtrack free.

The resultant network has an induced width 2. Observe that the edge <r, r> between variables W and Z is a remnant of the universal relation, and not a member of an induced relation. The induced width of the graph is 2, and for that DPC is sufficient for search to be backtrack free. If the induced width were to be higher, then a higher level of consistency would be required. This is neatly arrived at by the algorithm ADAPTIVECONSISTENCY, which also processes the variables from the last to the first, but for each variable the degree of consistency is tailor-made based on the number of parents the node has. One must keep in mind that as the algorithm achieves the requisite consistency for a variable, it induces new relations on the parents, which may increase the width of some nodes. The algorithm is described below. In the literature a variation, called *bucket elimination*, that focuses on the relations explicitly is also popular.

Algorithm 12.13. Algorithm ADAPTIVECONSISTENCY looks at the number t of parents of a variable x, and calls REVISE-t to filter out value combinations from the parents S. It then adds edges between all parents and augments the constraint graph. It also induces a relation R_s by deleting those sets of values in the parents that do not have a matching value in D_x.

AdaptiveConsistency($X = [x_1, x_2, ..., x_N]$, D, C)
1. $E \leftarrow$ set of edges in the constraint graph
2.   **for** $i \leftarrow N$ **downto** 2
3. $S \leftarrow$ Parents(x_i)
4. $t \leftarrow |S|$
5. $R_s \leftarrow$ REVISE-$t(S, x_i)$
6.   **for all** $x_j, x_k \in S$
7. $E \leftarrow \{<x_j, x_k>\} \cup E$
8. $C \leftarrow \{R_s\} \cup C$

The propagation techniques for combating combinatorial explosion seen so far are largely static and precede the search for solutions. Now we turn our attention to how some of these can be carried forward to the search phase itself. We have already mentioned dynamic variable ordering earlier. In the next section we look at ways for constraint propagation during search. Before picking a value for a variable, can we compute the impact on the domains of future variables?

Look before you leap.

12.4 Lookahead Search

Solving a CSP is often a mix of search and reasoning. Search tries out various assignments choosing variables and trying out values from their domains. Reasoning aims to compress the search space to minimize the work to be done by search. The consistency enforcement approaches described earlier process the CSP before BACKTRACKING takes over. While looking for a value for a variable, BACKTRACKING checks each value for consistency with the values assigned to earlier variables. The algorithms in this section look ahead at future variables in addition to the ones in the past. Of course, as before, there is a cost to be paid for the extra reasoning one does.

Consider trying to solve an *N*-queens problem on a real chessboard or one drawn on a piece of paper. Every queen one places rules out all the squares it attacks for the other queens to be placed. Imagine marking those squares with a cross. In Figure 12.16 we illustrate how placing six queens row by row, this marking process can help narrow down search and backtrack even before a dead end is reached.

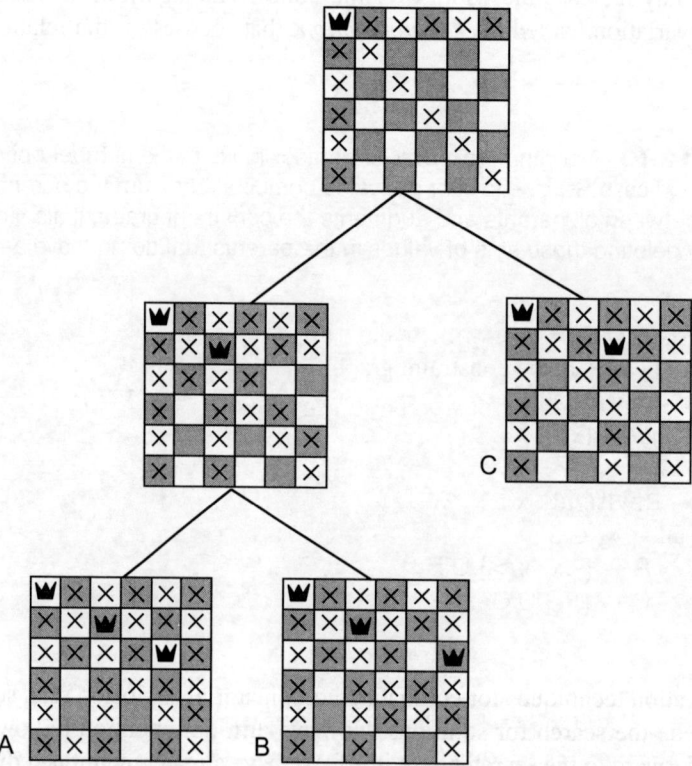

Figure 12.16 After placing a queen in the corner of the top row, the crosses mark the squares no longer available. By the time search places the third queen in board position A, many squares in the bottom half are already marked. At this point, placing a queen in the fourth row would block the entire sixth row. The algorithm backtracks and tries position B with similar effect. It next goes back to trying a new value for the second queen in board position C.

Placing queens row by row in the first available position, one finds oneself in board position *A* after placing three queens. There is one unmarked square in row 4, but if one were to place a queen there, row 6 would be completely blocked. The next, and last, option, marked *B*, for the third queen would have a similar impact with row 5 being ruled out this time. Without even placing the fourth queen one, can backtrack and try another square for queen number 2 in board *C*. This is in essence the algorithm *forward checking* (FC).

12.4.1 Algorithm forward checking

Algorithm FC is a variation of BACKTRACKING in which the function called for selecting a value for the next variable does some lookahead. While considering a value a_i for the variable x_i, the algorithm looks at all values in all future variables (Lines 4–7 of SELECTVALUE-FC in Algorithm 12.14) and deletes values that are not consistent with the proposed extension of the assignment $\mathcal{A}$. Only if no future domain is empty does it return the value a_i, else it undoes the deletions done with respect to this value (Lines 10–11). It is still in the while loop (Lines 1–11) and if there is another value available in the domain of x_i it considers that next. If it emerges from the while loop without success, it returns *null* which triggers the parent algorithm to go back and look for another value for variable x_{i-1} (Lines 7–10 in algorithm FC).

Algorithm 12.14. Given an ordering of the variables $x_1, \ldots, x_N$ algorithm FC is similar to algorithm BACKTRACKING except that the function SELECTVALUE-FC does more work pruning values from each future domain that is not consistent with the assignment $\mathcal{A}$ and the value a_i being considered for variable x_i.

FORWARDCHECKING(X, D, C)
1. $\mathcal{A} \leftarrow []$
2.   for $k \leftarrow 1$ to N
3. $D'_k \leftarrow D_k$
4. $i \leftarrow 1$
5.   while $1 \leq i \leq N$
6. $a_i \leftarrow$ SELECTVALUE-FC($D'_i, \mathcal{A}, C$)
7.   if $a_i = $ null
8.   then   Undo lookahead pruning done while choosing a_{i-1}
9. $i \leftarrow i - 1$ /* look for new value */
10. $\mathcal{A} \leftarrow$ tail $\mathcal{A}$
11.   else $\mathcal{A} \leftarrow a_i : \mathcal{A}$
12. $i \leftarrow i + 1$
13. return REVERSE($\mathcal{A}$)

SELECTVALUE-FC($D'_i, \mathcal{A}, C$)
1.   while D'_i is not empty
2. $a_i \leftarrow$ head D'_i
3. $D'_i \leftarrow$ tail D_i
4.   for $k \leftarrow i + 1$ to N
5.   for each b in D'_k
6.   if not CONSISTENT(b : $a_i : \mathcal{A}$)
7.   delete b from D'_k
8.   if no D'_k is empty
9.   then return a_i
10.   else for $k \leftarrow i + 1$ to N
11.   undo deletes in D'_k
12. return null

Algorithm FC does one pass over the future variables deleting values that are not going to be consistent with the current assignment. We illustrate the algorithm by following its progress on the matching diagram of a tiny example with five variables $x_1, ..., x_5$ processed in the given order. Figure 12.17 shows the constraint graph and the matching diagram at the start.

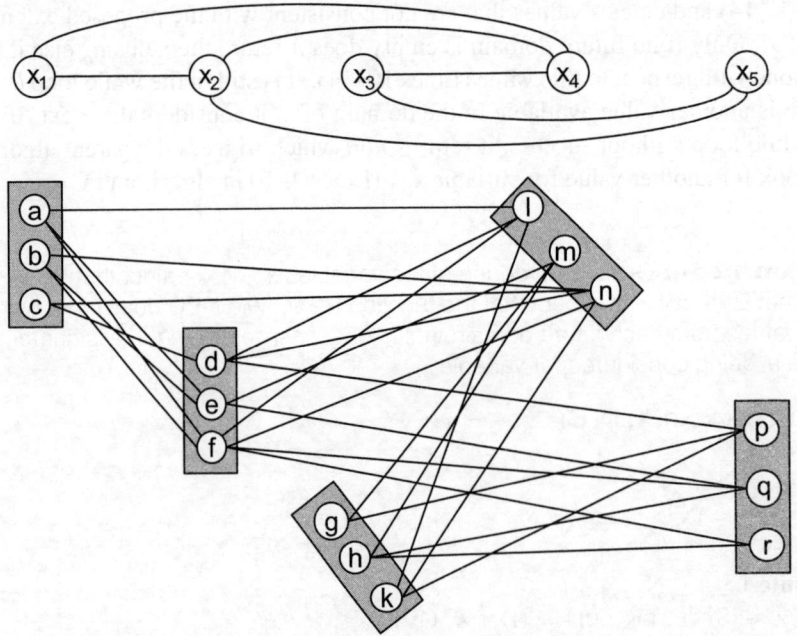

Figure 12.17 A tiny CSP with five variables processed in the order $x_1, x_2, x_3, x_4,$ and x_5. The constraint graph is shown at the top and the matching diagram at the bottom. Each domain has three values, selected in alphabetical order.

FC begins by calling for a value for variable x_1. SELECTVALUE-FC picks the first value a from x_1. This value is connected to values e and f in x_2 but not connected to value d. Consequently, SELECTVALUE-FC removes d from D_{x2}. Values deleted by SELECTVALUE-FC are shown with shaded circles in the figures that follow. In a similar manner, it also deletes values m and n from the domain of x_4. These are the only two variables which are related to x_1. SELECTVALUE-FC does no other pruning while considering value a. Then FC moves to x_2 and SELECTVALUE-FC tries the next *available* value e. This in turn deletes l from x_4 and p and r from x_5. The situation at this point is shown in Figure 12.18.

At this point, the domain of variable x_4 has become empty and the algorithm undoes the deletions done with respect to $x_2 = e$ and backtracks to try another value.

When SELECTVALUE-FC tries the next value $x_2 = f$, it deletes p from variable x_5 but that still has q and r. It returns $x_2 = f$ to FC, which calls it again looking for a value for x_3. SELECTVALUE-FC tries $x_3 = g$ and $x_3 = h$ but both delete l from D_{x4}. The next value k does not, but it deletes q and r from the domain of x_5, which now becomes empty. The situation is shown in Figure 12.19. It has assigned values to the first three variables but does not even try to for the fourth.

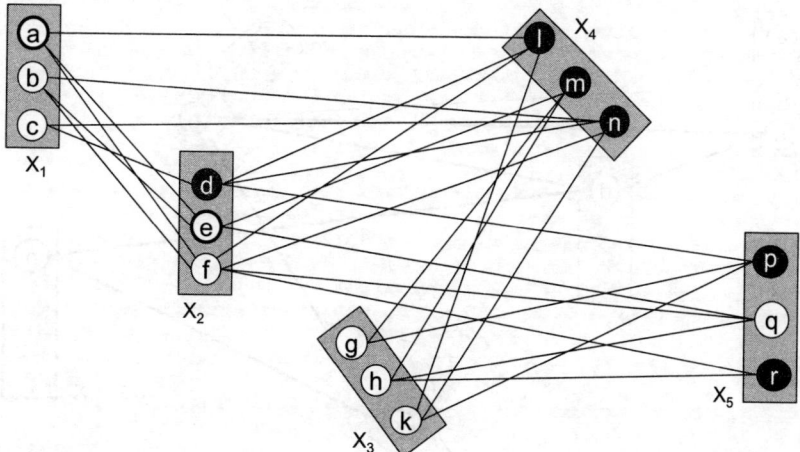

Figure 12.18 When FC tries $x_1 = a$ and the first available value $x_2 = e$, it discovers that the domain of variable x_4 has become empty, because all three values are not consistent with tries $x_1 = a$ and $x_2 = e$. It will now undo deletion of values $l, p,$ and r done while assigning $x_2 = e$ and will try the next value $x_2 = f$.

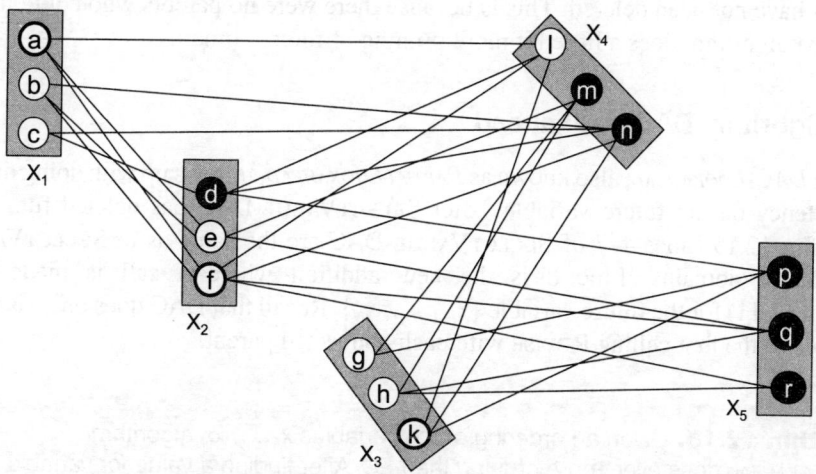

Figure 12.19 When FC tries the next value $x_2 = f$ after undeleting values $l, p,$ and r. This deletes p from x_5. It next tries values g and h for x_3 but both delete l in x_4. SELECTVALUE-FC next tries $x_3 = k$ but that deletes q and r from x_5, which becomes empty. There are no more values to backtrack to in x_2 and x_3 and it backtracks to x_1 and tries the value b.

SELECTVALUE-FC reports failure to find a value for x_3 and backtracks to x_2 but there is no other value available. It will next try $x_1 = b$. The reader is encouraged to verify that FC will backtrack because the D_{x5} will again become empty after assigning the last possible value to x_3. The algorithm next tries $x_1 = c$ and eventually finds a solution with matching diagram as shown in Figure 12.20.

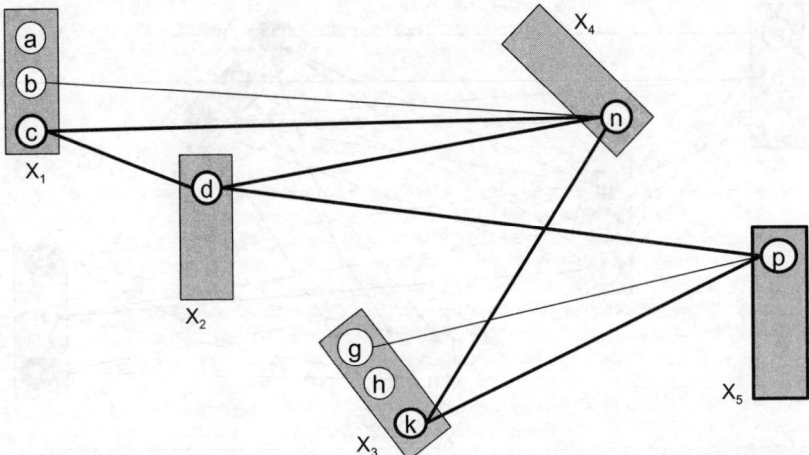

Figure 12.20 The matching diagram at the point when FC finds the solution $<c, d, k, n, p>$. Note that values a, b, g, and h were not deleted because variables x_1 and x_3 do not have any parents in the given ordering.

In the diagram in Figure 12.20 there are still some unconnected values in the domains of x_1 and x_3 that have not been deleted. This is because there were no parents who could have done so. The next algorithm does a little bit more pruning of future domains.

12.4.2 Algorithm DAC-Lookahead

Algorithm *DAC-LOOKAHEAD*, also known as *PARTIALLOOKAHEAD*, follows up with doing directional arc consistency on the future variables after SELECTVALUE-DAC has deleted future values. In Algorithm 12.15 Lines 4–7 of SELECTVALUE-DAC are the same as in SELECTVALUE-FC, pruning future domains. Line 8 is the new addition where a call is made to DAC (Algorithm 12.11) for the future variable $\{x_{i+1}, ..., x_N\}$. Recall that DAC does one pass from the last variable to the first calling REVISE with each connected parent.

Algorithm 12.15. Given an ordering of the variables $x_1, ..., x_N$ algorithm DAC-LOOKAHEAD does even more pruning than FC. After finding a value for variable X_i consistent with the assignment $\mathcal{A}$, and deleting future nodes like in Algorithm 12.5, SELECTVALUE-DAC does one pass of DAC on the future variables. If the current value a_i does not work, it undoes the deletion done before trying the next value.

DAC-LOOKHEAD(X, D, C)
1. $\mathcal{A} \leftarrow []$
2. **for** $k \leftarrow 1$ to N
3. $D'_k \leftarrow D_k$
4. $i \leftarrow 1$
5. **while** $1 \leq i \leq N$

```
6.        a_i ← SELECTVALUE-FC(D'_k, A, C)
7.        if a_i = null
8.           then      Undo lookahead pruning done while choosing a_{i-1}
9.                     i ← i − 1           /* look for new value */
10.                    A ← tail A
11.          else      A ← a_i : A
12.                    i ← i + 1
13.   return REVERSE(A)

SELECTVALUE-DAC(D'_i, A, C)
1.    while D'_i is not empty
2.       a_i ← head D'_i
3.       D'_i ← tail D'_i
4.       for k ← i + 1 to N
5.          for each b in D'_k
6.             if not CONSISTENT(b: a_i : A)
7.                delete b from D'_k
8.          DAC ({x_{i+1}.. x_N}, D', C)
9.          if no domain is empty
10.            return a_i
11.         else for k ← i + 1 to N
12.            undo deletes in D'_k
13.   return null
```

The extra work done in DAC-LOOKAHEAD are these calls to DAC. We illustrate the effect of these on the tiny problem in Figure 12.17. DAC-LOOKAHEAD too begins by selecting $x_1 = a$ and deleting d from x_2 and m, n from x_4. As in the diagrams previously discussed, we show these as shaded circles in Figure 12.21, but we have deleted the edges emanating from them for clarity. Now DAC is called with the future variables x_2, x_3, x_4, and x_5 shown inside the dashed oval. Both x_2 and x_3 are arc consistent with respect to x_5 and no deletions happen. But values e in x_2 and g, h in x_3 do not have supporting values in x_4 and are deleted. The deletions by DAC are shown with cross marks, and the situation is as shown in Figure 12.21.

In the situation in Figure 12.21, DAC-LOOKAHEAD next tries the value $x_2 = f$. The future variables are now only x_3, x_4, and x_5 as shown inside the dashed oval in Figure 12.22. Forward checking deletes the value p from the domain of x_5. This has a cascading effect when DAC kicks in with the value k being deleted from x_3. The domain of x_3 is now empty and DAC-LOOKAHEAD retreats to x_1 and will try the value b.

Algorithm FC had looked at all values in the domain of x_3 before backtracking to x_1 to try the next value. Algorithm DAC-LOOKAHEAD retreated because it could not find a consistent value for x_2 without going to x_3. The next algorithm AC-LOOKAHEAD finds that it is unable to even assign $x_1 = a$.

424 | Search Methods in Artificial Intelligence

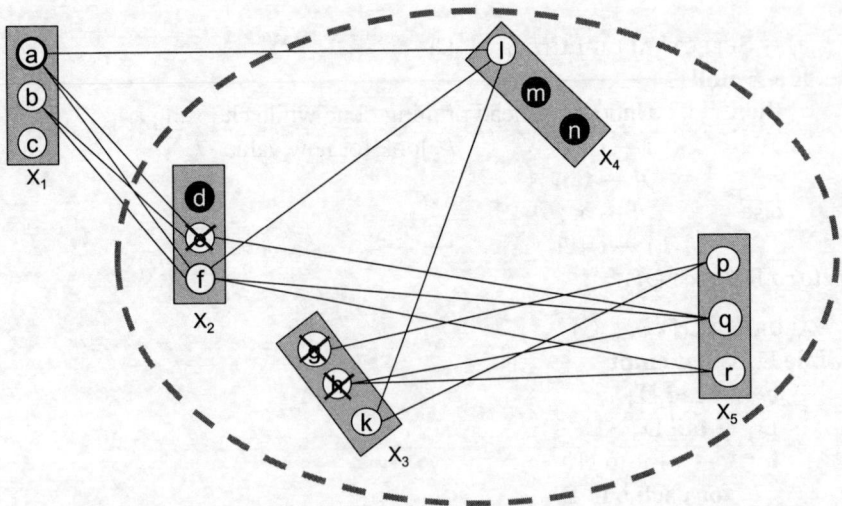

Figure 12.21 Algorithm DAC-LOOKAHEAD begins like FC with $x_1 = a$ deleting values d, m, and n. The DAC component in SELECTVALUE-DAC kicks in for the remaining four variables x_2, x_3, x_4, and x_5 shown in the dashed oval resulting in e being deleted from x_2 and g, h being deleted from x_3. DAC-LOOKAHEAD will try $x_2 = f$ next.

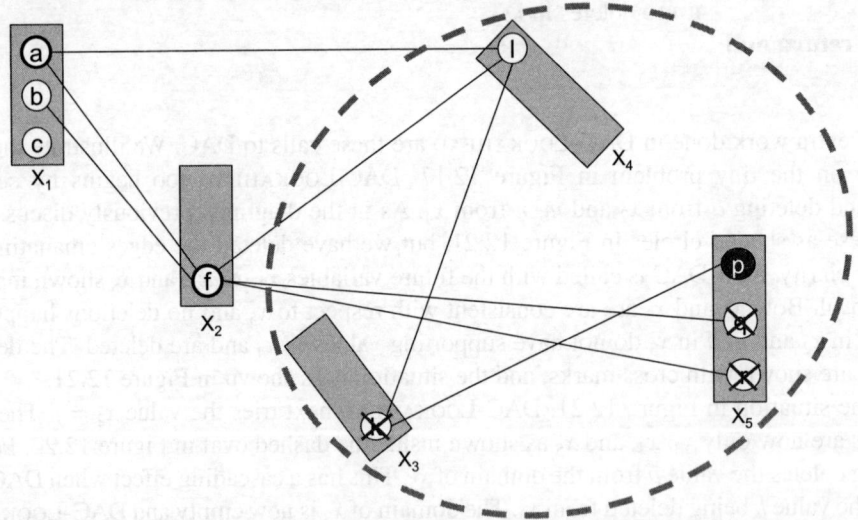

Figure 12.22 When algorithm DAC-LOOKAHEAD picks value f in x_2, forward checking deletes the value p in x_5. The DAC component in SELECTVALUE-DAC kicks in for the remaining three variables x_4, x_5, and x_5, shown in the dashed oval resulting in D_{x3} becoming empty. DAC-LOOKAHEAD retreats to variable x_1 without looking at x_3.

12.4.3 Algorithm AC-Lookahead

The algorithm *AC-LOOKAHEAD* is similar to DAC-LOOKAHEAD except that in Line 8 of SELECTVALUE-AC the algorithm calls for doing a full arc consistency of the future variables. This is clearly more work and also results in more pruning of the search space. We look at how the algorithm performs on the tiny problem from Figure 12.17.

Figure 12.23 shows the first part of the pruning phase when AC-LOOKAHEAD tries the first value a from the domain of x_1. Value e in x_2 has no support from x_4 and is deleted, as are values g and h in x_3. Likewise, the value p is deleted from the domain of x_5 because it does not have support from x_2.

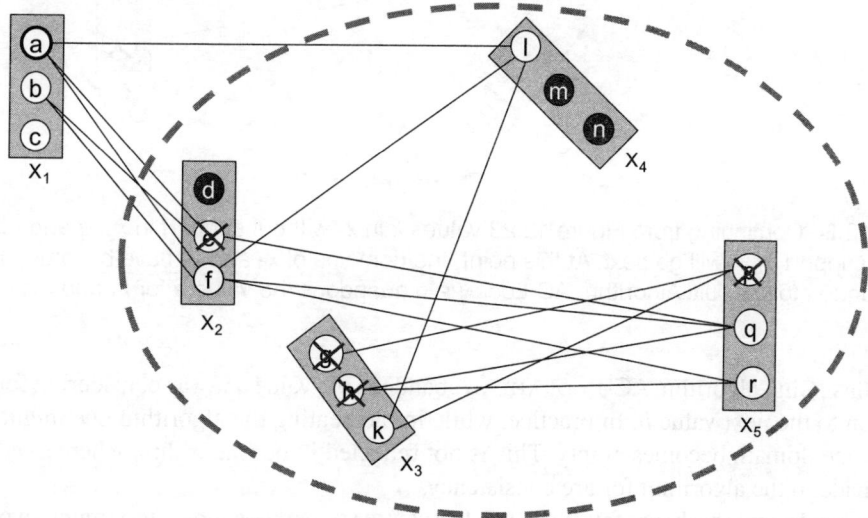

Figure 12.23 Algorithm AC-LOOKAHEAD begins like FC with $x_1 = a$ deleting values d, m, and n. After this variables x_2, x_3, x_4, and x_5 are made arc consistent. Values e in x_2 without support in x_4, g and h in x_3 also without support in x_4, and p in x_5 without support in x_2 are the first to go, shown by cross marks.

The matching diagram at this stage is shown in Figure 12.24 where the pruning process continues after we have removed the pruned nodes from the figure. At this point, there are only five values remaining in the four future variables. Value k in variable x_3 is deleted because it has no support in x_5, and this results in l in x_4 and p, q in x_5 also being deleted, after which f goes from x_2.

426 | Search Methods in Artificial Intelligence

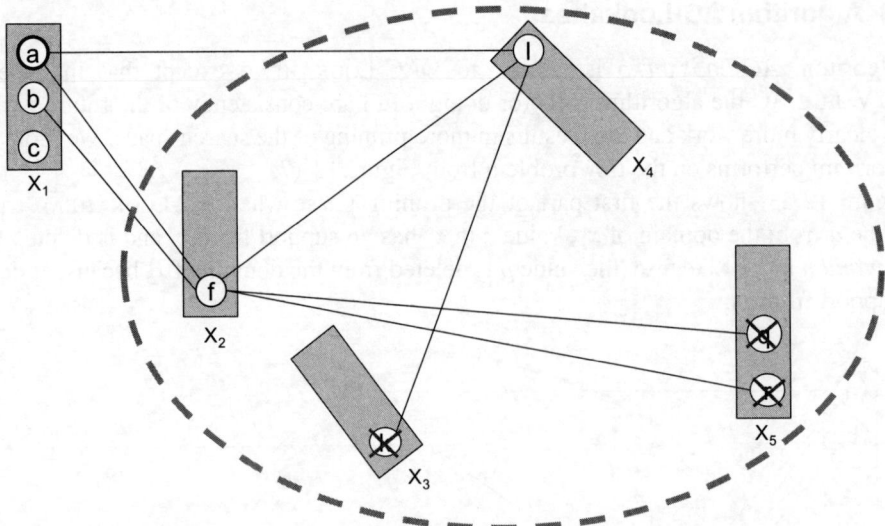

Figure 12.24 Continuing from Figure 12.23 values k in x_3 without support in x_5, q and r in x_5 without support in x_3 will go next. At this point, the domains of x_3 and x_5 have become empty, and x_2 and x_4 follow suit. Algorithm AC-LOOKAHEAD abandons the value a for x_1 and moves on to b.

At this point, algorithm AC-LOOKAHEAD abandons the value a it was considering for x_1 and moves on to the next value b. In practice, while implementing the algorithm one might exit as soon as one domain becomes empty. This is not reflected in our algorithm, where one blanket call is made to the algorithm for arc consistency.

The reader might have felt that AC-LOOKAHEAD perhaps does too much work. An algorithm we have not mentioned here is FULLLOOKAHEAD, which does a little bit less. This is like AC-LOOKAHEAD except that it does only one pass of calling REVISE for every pair of future variables.

We now turn our attention to informed or intelligent backtracking.

12.5 Informed Backtracking

Given an ordering of variables, a search algorithm builds an assignment incrementally, looking for a consistent value for each variable. We take up the action when the algorithm is looking for a value for x_i and has a partial assignment $\mathcal{A} = <a_1, a_2, ..., a_{i-1}>$. This assignment is consistent, which means that it satisfies all the constraints whose scope lies in the set $\{x_1, x_2, ..., x_{i-1}\}$. Now the algorithm seeks a value for x_i that is consistent with $\mathcal{A}$. If it cannot find one, then it has reached a dead end. The search must retreat and try options other than $\mathcal{A} = <a_1, a_2, ..., a_{i-1}>$.

Algorithm BACKTRACKING takes one step back and tries another value for x_{i-1}. The algorithm systematically tries different values for x_{i-1} and will not miss a solution if one of them were to lead to one. This is called *chronological* backtracking, because the *last* variable that was assigned a value is looked at again. But the real reason behind the dead end may lie elsewhere, and the work done trying different values for x_{i-1} may be futile. Informed backtracking aims to reduce such unnecessary search and *jump back* to a variable where a different value may allow some value in x_i.

We say that the assignment $\mathcal{A} = <a_1, a_2, ..., a_{i-1}>$ is a *conflict set* with respect to x_i if we cannot find a value b in D_i such that $<a_1, a_2, ..., a_{i-1}, b>$ is consistent. If no subset of $\mathcal{A} = <a_1, a_2, ..., a_{i-1}>$ is a conflict set with respect to x_i we say that $\mathcal{A}$ is a minimal conflict set. We say that $<a_1, a_2, ..., a_{i-1}>$ is a *dead end* with respect to x_i, and x_i is a *leaf dead end* variable. If in addition $<a_1, a_2, ..., a_{i-1}>$ cannot appear in *any* solution, we say that it is a *no-good*. It is possible for an assignment to be a no-good but not be a dead-end for any single variable. A minimal no-good is one which does not have any subset that is a no-good.

When $<a_1, a_2, ..., a_{i-1}>$ is a *conflict set* with respect to x_i, then search can jump back to *any* variable x_j such that $j < i - 1$ in the quest for a solution. This process is called *backjumping*. We say that the jump back is *safe* if there is no k between j and $i - 1$ such that a new value for x_k leads to a solution. Jumping back to a safe variable will thus not preclude any solution and affect the completeness of the algorithm. We say that a safe backjump to a variable x_j is *maximal* if there is no $m < j$ such that a backjump to x_m is safe.

The question is: given an assignment $<a_1, a_2, ..., a_{i-1}>$ that is a dead end for a variable, what is a safe and maximal backjump? We look at three well known algorithms for backjumping. Each collects differing kinds of data, based on which it decides the variable that is safe to jump back to. Each of them, however, arrives at a different answer to what is a maximal backjump that is safe.

12.5.1 Gaschnig's backjumping

Gaschnig's backjumping (GBJ) algorithm looks carefully at the current assignment $\mathcal{A} = <a_1, a_2, ..., a_{i-1}>$ while searching for a value for x_i. Before GBJ calls SELECTVALUE-GBJ for a value for x_i, it sets a variable $latest_i$ to 0. SELECTVALUE-GBJ starts with the first value $b \in D_i$ and scans $\mathcal{A}$ incrementally starting from a_1 to identify the index k in $\mathcal{A}$ where the sub-tuple first becomes inconsistent with b. If this k is greater than $latest_i$, its sets $latest_i$ to k. Then it moves on to the next value in D_i, where it could possibly increase the value of $latest_i$ further. If a value in D_i were to be consistent with $\mathcal{A}$, the $latest_i$ would end up with the value $i - 1$.

If the call to SELECTVALUE-GBJ were to return no value, then $latest_i$ would identify the *culprit* variable that GBJ needs to jump back to. We illustrate this process with the 8-queens example shown in Figure 12.25 in which five queens have been placed in rows 1–5, and GBJ is unable to place a queen in row 6.

When SELECTVALUE-GBJ tried placing the sixth queen on square $a6$, it was being attacked by the queen in row 1. The variable $latest_6$ is set to 1. Then it tries square $b6$ which is attacked by queens 3 and 5. If the culprit were one of these, then undoing the placement of queen 5 would not help, because queen 3 would still be attacking the square $b6$. So $latest_6$ is updated to 3. For square $c6$ the earliest queen attacking is queen 2. But it would *not* be safe to jump back to queen 2 because a solution *might* have been possible by trying a new square for queen 3. Therefore, the latest that it is safe to jump back to is still 3, as reflected in the value of $latest_6$. It cannot jump back farther than 3, so it would be a *maximal* jump that is safe. Then looking at square $d6$ this value is further increased to 4, where it stays over the next four squares in row 6.

428 | Search Methods in Artificial Intelligence

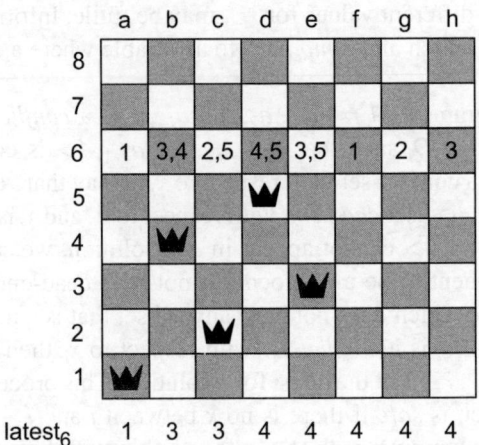

Figure 12.25 SELECTVALUE-GBJ is unable to find a value (column name) for the sixth queen. The numbers in row 6 are the numbers of the queens attacking that square. In each of these, the earliest counts for each square. The value of $latest_6$ begins with 1 for square a6, becomes 3, the earlier queen attacking b6, and so on. The largest value is 4 from the square d6. GBJ would backtrack to the fourth queen.

The assignment <a, c, e, b> for the first four queens is a no-good. One of the queens must be relocated. It can only be queen 4 because a solution by relocating that could still be possible. Skipping queen 4 and relocating an earlier queen might miss a solution that relocating 4 might yield. So jumping back to queen 4 is both maximal and safe, and queen 4 is the culprit.

The algorithm GBJ is described below. Observe that the variable $latest_i$ is a global variable, initialized in the main program, set in the call to SELECTVALUE-GBJ, and used in the main program for jumping back when a *null* value is returned.

Algorithm 12.16. Algorithm GBJ is similar to algorithm BACKTRACKING except that it can jump back more than one step from leaf dead ends. The function SELECTVALUE-GBJ operates a global ratchet variable that identifies the maximal safe variable to jump back to when a value for x_i cannot be found. For each value a_i in D_i it sets $latest_i$ to the index of the latest sub-tuple that is consistent with a_i if that index is higher than the current value.

GBJ(X, D, C)
1. $\mathcal{A} \leftarrow []$
2. $i \leftarrow 1$
3. $D'_i \leftarrow D_i$
4. **while** $1 \leq i \leq N$
5. $latest_i \leftarrow 0$
6. $a_i \leftarrow$ SELECTVALUE-GBJ($D'_i, \mathcal{A}, C$)
7.   **if** $a_i =$ **null**
8.   **then**
9.   **while** $i > latest_i$

```
10.            i ← i − 1
11.            A ← tail A
12.        else
13.            A ← a_i : A
14.            i ← i + 1
15.            D'_i ← D_i
16.    return REVERSE(A)
```

SELECTVALUE-GBJ(D'_i, A, C)
```
1.     while D'_i is not empty
2.         a_i ← head D'_i
3.         D'_i ← tail D'_i
4.         consistent ← true
5.         k ← 1
6.         while k < i and consistent
7.             A_k ← take k A
8.             if k > latest_i
9.                 latest_i ← k
10.            if not CONSISTENT(a_i : A_k)
11.                consistent ← false
12.            else
13.                k ← k + 1
14.        if consistent
15.            return a_i
16.    return null
```

When SELECTVALUE-GBJ does return a value for x_i, the variable $latest_i$ has a value $i - 1$ and GBJ moves on to x_{i+1}. If at a later point GBJ were to backtrack to x_i, and if that had no value left in its domain, where would it backtrack to? This is known as an *internal dead end*. The value of $latest_i$ is $i - 1$, and hence GBJ would just move one step back. The algorithm GBJ thus does a safe and maximal backjump from a leaf dead end, but just moves one step back from an internal dead end.

12.5.2 Graph based backjumping

While GBJ pays no heed to the constraint graph when deciding where to jump back, *Graph based backjumping* (GBBJ) relies only on the relations between variables. So much so that when a dead end is reached, it concludes that a parent *must* be the culprit.

The algorithm defines and utilizes the following information relating to the constraint graph with a given ordering of variables. The set of ancestors $anc(x)$ of a node x are all the nodes *connected* to x which precede it in the ordering. Of these nodes, the parent of x, $parent(x)$, is the most recent ancestor. When the algorithm GBBJ encounters a leaf dead end, it jumps back to its parent, *assuming* that the parent is the source of the conflict.

Figure 12.26 shows the graph from Figure 12.6 with the alphabetic ordering. Of the four nodes connected to node *E*, three are its ancestors in the given ordering. Of *A*, *B*, and *D*, the last one is the parent, and if node *E* were to be a leaf dead end, then it would try a new value of its parent *D*. This happens to be the last node visited, but that is not the case for nodes *C* and *F* who have only one ancestor and who is the parent. If *C* were to be a leaf dead end, GBBJ would try *A* next, and if *F* were to be a leaf dead end, GBBJ would try *D* next.

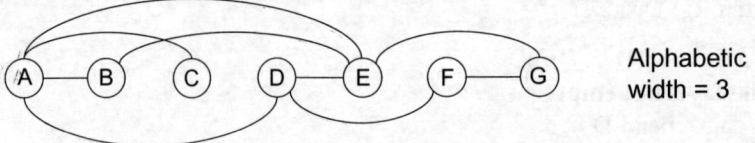

Figure 12.26 On the alphabetic ordering in the graph from Figure 12.6 node *E* has three ancestors *A*, *B*, and *D* of which *D* is the parent. Nodes *C* and *F* have only one ancestor each which is the parent.

When GBBJ jumps back from a leaf dead end, it may encounter an *internal dead end*, which is the node it has jumped back to but which does not have any consistent value left. Where does it go next? Consider the case when node *G* were to be a leaf dead end and GBBJ jumped back to its parent *F*. If there is no value left in *F*, should it jump to its parent *D*? That would not be safe, because the original conflict in *G* might have been caused by *E*, its other ancestor. Bypassing *E* would not be safe. GBBJ handles this and similar cases as follows.

We say that GBBJ *invisits* a node *x* when it visits it in the forward direction, that is, after a node preceding it in the ordering. Starting from there, the *current session* of *x* includes all the nodes it has tried after invisiting *x* till the time it finds *x* to be an internal dead end.

We define the set of *relevant* dead ends *r(x)* of a node *x* as follows:

- If *x* is a leaf dead end, then $r(x) = \{x\}$
- If *x* is an internal dead end after jumping back from node *y*, then $r(x) = r(x) \cup r(y)$

The set of *induced ancestors* of a node *x* is defined as follows. Let *Y* be the set of relevant dead ends in the current session of *x*. Then the set of induced ancestors $I_x(Y)$ of node *x* is the union of all ancestors of nodes in *Y* which precede *x* in the ordering. The induced parent $P_x(Y)$ of node *x* is the latest amongst the induced ancestors of *x*. When *x* is a dead end, algorithm GBBJ jumps to the induced parent of a node *x*.

In Figure 12.26, when *G* turns out to be a leaf dead end, GBBJ tries its induced parent *F* (which is also its parent). The induced parent of *F* is *E* because *E* is the latest induced ancestor of *F*, an ancestor of *G*, which is a relevant dead end for *F*. So if *F* is an internal dead end, GBBJ will try *E* next. If *E* is an internal dead end too, the algorithm will try *D*. The current session of *D* includes *E*, *F*, and *G*, and hence the induced parent of *D* is *B*, which is where GBBJ will jump back to if *D* were to be a dead end too.

Contrast this with the case when *F* is a leaf dead end. GBBJ will jump back to *D* the parent of *F*. The relevant dead ends of *D* are *F* and *D* itself. The only induced ancestor of *D* is *A*. Remember that *E* is not a relevant dead end. If *D* were to be an internal dead end, the algorithm will now jump back to *A*.

The algorithm GBBJ shown below begins by computing the set of ancestors $anc(x)$ for each variable, given the ordering $x_1, x_2, \ldots, x_n$. Whenever it moves forward to a node x_i, it initializes the set of induced ancestors l_i of x_i to $anc(x_i)$ in Lines 6, 7 and 22, 23 of Algorithm 12.17. The index p of the induced parent of the current node is always the latest node in the set of induced ancestors l_i (Lines 7, 17, and 23). When it encounters a dead end x_i, which is when there is no consistent value for the current variable, it jumps back to induced parent x_j. It updates the induced ancestors of x_j and identifies the induced parent to which it would jump back from x_j if needed (Lines 12–17).

Algorithm 12.17. Algorithm GBBJ begins by computing the ancestors of each node. It keeps track of the induced ancestors when it jumps back to a node. It always jumps back to the induced parent on reaching a dead-end. The SELECTVALUE function is the simple one used in BACKTRACKING.

GBBJ(X, D, C)
1. $\mathcal{A} \leftarrow [\,]$
2.   **for** $k \leftarrow 1$ to N
3.   compute $anc(k)$ the set of ancestors of x_k
4. $i \leftarrow 1$
5. $D'_i \leftarrow D_i$
6. $I_i \leftarrow anc(i)$
7. $p \leftarrow$ latest node in I_i
8.   **while** $1 \leq i \leq N$
9. $a_i \leftarrow$ SELECT VALUE($D'_i, \mathcal{A}, C$)
10.   **if** a_i = null
11.   **then**
12.   iprev $\leftarrow i$
13.   **while** $i > p$
14. $i \leftarrow i - 1$
15. $\mathcal{A} \leftarrow$ **tail** $\mathcal{A}$
16. $I_i \leftarrow I_i \cup \{l_{iprev} - \{x_i\}\}$
17. $p \leftarrow$ latest node in I_i
18.   **else**
19. $\mathcal{A} \leftarrow a_i : \mathcal{A}$
20. $i \leftarrow i + 1$
21. $D'_i \leftarrow D_i$
22. $I_i \leftarrow anc(i)$
23. $p \leftarrow$ latest node in I_i
24.   **return** REVERSE($\mathcal{A}$)

SELECTVALUE($D'_i, \mathcal{A}, C$)
1.   **while** D'_i **is not empty**
2. $a_i \leftarrow$ **head** D'_i

3. $D'_i \leftarrow$ **tail** D'_i
4.   **if** CONSISTENT($a_i : \mathcal{A}$)
5.   **then return** a_i
6. **return null**

The jumping back behaviour of GBBJ is the same whether it does so from an internal dead end or a leaf dead end. This is an improvement over GBJ, which can jump back only from leaf dead ends. But GBBJ is conservative and assumes that if a node is a parent in the constraint graph, it *must* be the cause of the dead end, and jumps to that as an insurance. It will also not miss out on any solution and its jumps are safe. But it may jump back less than it needs to because it does not look at the *values* in the domains that lead to the conflict. The next algorithm makes use of both kinds of information, based on the values that conflict, and also the graph topology.

12.5.3 Conflict directed backjumping

The reason why search has to backtrack is that it cannot find a value for the next variable x_i that is consistent with the current assignment $\mathcal{A}$ being constructed. A value a_i of x_i being considered conflicts with $\mathcal{A}$ because it conflicts with some constraint R_S whose scope S is within the set of variables $x_1 \ldots x_{i-1}$ already instantiated. As soon as the select value function in algorithm *conflict directed backjumping* (CDBJ) spots a conflict, it identifies the earliest constraint that conflicts with a_i.

Given an ordering of variables $x_1, x_2, \ldots, x_n$, a constraint R_i is said to be *earlier* than a constraint R_k if the latest variable in R_i which is not in R_k is earlier than the latest variable in R_k which is not in R_i. That is, the latest in $S_i - S_k$ is earlier that the latest in $S_k - S_i$ where S_i and S_k are the scopes respectively of R_i and R_k. For example, given $S_i = \{x_6, x_{10}\}$ and $S_k = \{x_1, x_3, x_7, x_8, x_{10}\}$, the constraint R_i is earlier than R_k because x_6 is earlier than x_8.

Algorithm CDBJ works with a set $emc(x_i)$ or $emc(\mathcal{A})$ of conflicting values called the *earliest minimal conflict set* associated with the variable x_i that the algorithm is seeking to pick a value from. The variables $var\text{-}emc(x_i)$ in this set define the jumpback set J_i for variable x_i. The jumpback set serves the same purpose that the ancestor set $anc(x_i)$ did in algorithm GBBJ, which is that the algorithm jumps back to the latest variable in this set. The difference is that GBBJ constructs the set based on the graph topology while CDBJ, like GBJ, does so based on an actual conflict of values. Both CDBJ and GBJ would jump back to the same variable from a leaf dead end, but CDBJ can jump back even from internal dead ends because it maintains an induced jumpback set J_i when it jumps back based on all relevant dead ends, like in GBBJ. At the same time, the induced jumpback set in CDBG can be a subset of the induced ancestor set of GBBJ because CDBG only adds variables when it detects a real conflict, whereas GBBJ conservatively assumes that if the variables are connected, one of them must be the culprit. Thus GBBJ may jump back to a variable that CDBJ knows is not the actual culprit.

When algorithm CDBJ visits a new variable (Lines 2–4 and 17–19 in Algorithm 12.18), it initializes the jumpback set to the empty set. Then it calls *SELECTVALUE-CDBJ*, which, like *SELECTVALUE-GBJ*, scans the values in the assignment $\mathcal{A}$ incrementally checking for consistency

(Lines 5–13 of SELECTVALUE-CDBJ). The moment the value a_i from the domain of x_i conflicts with $\mathcal{A}_k$, it adds the variables in the earliest conflict to the jumpback set J_i. Note that more than one constraint may simultaneously conflict with a_i for a particular value of k. That is why one needs to select the earliest one amongst them (Lines 11–13). Having done that, it moves on to the next value in the domain D'_i to test for consistency. For every value in D'_i it finds a conflict, it adds the earliest conflict to the jumpback set J_i. If SELECTVALUE-CDBJ cannot find a consistent value, then x_i would be a leaf dead end and the parent program would jump back to the latest variable in the jumpback set J_i. Observe that like in algorithm GBJ we have assumed that J_i is global data structure.

Algorithm 12.18. Algorithm CDBJ looks at actual conflicts of values a little bit like GBJ, but constructs the earliest minimal conflict set in SELECTVALUE-CDBJ when it spots a conflict based on the earliest constraint that conflicts with a value in x_i. Like GBBJ it can combine the data gleaned from relevant dead ends in the main algorithm to be able to jump back from internal dead ends as well.

CDBJ(X, D, C)
1. $\mathcal{A} \leftarrow []$
2. $i \leftarrow 1$
3. $D'_i \leftarrow D_i$
4. $J_i \leftarrow \{\}$
5. while $1 \leq i \leq N$
6. $a_i \leftarrow$ SELECTVALUE($D'_i, \mathcal{A}, C$)
7.   if $a_i =$ **null**
8.   then
9.   iprev $\leftarrow i$
10. $p \leftarrow$ latest node in J_i
11.   while $i > p$
12. $i \leftarrow i - 1$
13. $\mathcal{A} \leftarrow$ **tail** $\mathcal{A}$
14. $J_i \leftarrow J_i \cup \{J_{iprev} - \{x_i\}\}$
15.   else
16. $\mathcal{A} \leftarrow a_i : \mathcal{A}$
17. $i \leftarrow i + 1$
18. $D'_i \leftarrow D_i$
19. $J_i \leftarrow \{\}$
20. **return** REVERSE($\mathcal{A}$)

SELECT VALUE-CDBJ($D'_i, \mathcal{A}, C$)
1. while D'_i is not empty
2. $a_i \leftarrow$ **head** D'_i
3. $D'_i \leftarrow$ **tail** D'_i
4.   consistent $\leftarrow$ *true*
5. $k \leftarrow 1$

```
6.      while k < i and consistent
7.          A_k ← take kA
8.          if CONSISTENT(a_i : A_k)
9.              k ← k + 1
10.         else
11.             R_s ← earliest constraint with scope S causing the conflict
12.             J_i ← J_i ∪ {S − {x_i}}
13.             consistent ← false
14.     if consistent
15.         return a_i
16. return null
```

Observe that when CDBJ jumps back to variable x_i (Lines 10–14 of CDBJ), it is still in the current session of the variable, not yet having retreated from there. It might find a value for this variable and go forth to the next variable and onwards, till it strikes another dead end and again jumps back to x_i from a relevant dead end. The merging of jumpack sets in Line 14 of CDBJ is similar to the process of computing the induced ancestors in GBBJ which enables the algorithm to jump back safely and maximally from the internal dead end as well.

Summary

Constraints offer a uniform formalism for representing what an agent knows about the world. The world is represented as a set of variables each with its own domain of values, along with local constraints between subsets of variables. The fact that constraints are local obfuscates the world view. It is not clear what combination of values for each variable is globally consistent. The task of solving the CSP is to elucidate these values, which can be thought of as unearthing the solution relation that prescribes all consistent combinations of values.

There are two approaches to strive for this clarity, and constraint processing allows for the interleaving of both. On the one hand, there is constraint propagation or reasoning that eliminates infeasible combinations of values, in the process adding new constraints to the network. On the other is search, the basic strategy of problem solving by first principles.

We explored various combinations of techniques. This includes various levels of consistency that can be enforced. We looked at algorithms for arc consistency and path consistency. We also looked at the advantages of directional consistency and a little bit on ordering variables for search. Then we started with the basic search algorithm BACKTRACKING which essentially searches through combinations of values for variables. This can be augmented with look ahead methods that prune future variables, and by look back methods that make an informed choice of which variable to jump back to when a dead end is encountered. In all cases a dead end forces the search algorithm to retreat and undo some instantiations to try new combinations. One aspect we have not studied is no-good learning. Here, every time an algorithm jumps back to a culprit variable, the combination of conflicting values can be marked to be avoided in the future. Clearly, no-good learning would be meaningful in large problems with huge search trees that can benefit with such pruning.

We have not looked at many applications despite having said that the CSPs present a very attractive formulation in which all kinds of problems can be posed, and then solved using some of the methods we have studied. We did mention in the chapter on planning that planning can be posed as a CSP, and illustrated the idea by posing it as satisfiability. Another frequent application is classroom scheduling and timetable generation which has its own group of interested researchers. Also, we have confined ourselves to finite domain CSPs with the general methods that they admit. We have not looked at specialized constraint solving problems and methods like linear and integer programming that have evolved as areas of study in themselves, beyond the scope of this book.

Exercises

1. Which of the following statements are true regarding solving a CSP?
   a. Values must be assigned to ALL variables such that ALL constraints are satisfied.
   b. Values must be assigned to at least SOME variables such that ALL constraints are satisfied.
   c. Values must be assigned to ALL variables such that at least SOME constraints are satisfied.
   d. Values must be assigned to at least SOME variables such that at least SOME constraints are satisfied.
2. Pose the following cryptarithmetic problems as CSP:
   TWO + TWO = ONE
   JAB + JAB = FREE
   SEND + MORE = MONEY
3. Consider the following constraint network $R = <\{x_1, x_2, x_3\}, \{D_1, D_2, D_3\}, \{C\}>$ where $D_1 = D_2 = D_3 = \{a, b, c\}$ and $C = <\{x_1, x_2, x_3\}, \{<a, a, b>, <a, b, b>, <b, a, c>, <b, b, b>\}$. How many solutions exist?
4. Given a constraint satisfaction problem with two variables x and y whose domains are $D_x = \{1,2,3\}$, $D_y = \{1,2,3\}$, and constraint $x < y$, what happens to D_x and D_y after the REVISE(x,y) algorithm is called?
   a. Both D_x and D_y remain the same as before
   b. $D_x = \{1,2\}$ and $D_y = \{1,2,3\}$
   c. $D_x = \{2,3\}$ and $D_y = \{1,2\}$
   d. $D_x = \{\}$ and $D_y = \{1,2,3\}$
5. Draw the search tree explored by algorithm BACKTRACKING for the 5-queens problem till it finds the first solution.
6. What is the best case complexity of REVISE($(X), Y$) when the size of each domain is k?
7. Draw the matching diagram for the network in Figure 12.1 after it has been made arc consistent. Has the network become backtrack free?
8. What does one conclude when the domain of some variable X while computing arc consistency becomes empty?

436 | Search Methods in Artificial Intelligence

9. Try out different orderings for the two networks in Figure 2.8 and investigate how BACKTRACKING performs.
10. Is the following object a trihedral object? Label the edges, and explain your answer.

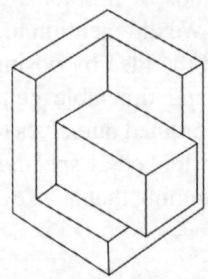

11. Draw trihedral objects to illustrate all the vertex labels shown in Figure 12.10.
12. [Baskaran] Label the edges in the following figure and identify each vertex type.

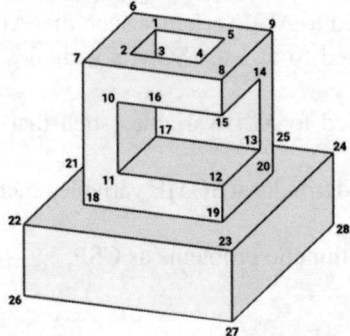

13. Given the CSP on variables $X = \{x, y, z\}$ and the relations R_{xy}, R_{xz}, R_{yz} depicted in the matching diagram below, draw the matching diagram after the CSP has been made arc consistent. State the resulting CSP.

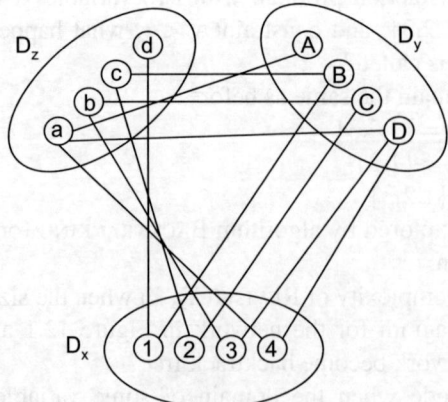

14. Consider the following CSP for a map colouring problem. Answer the questions that follow.

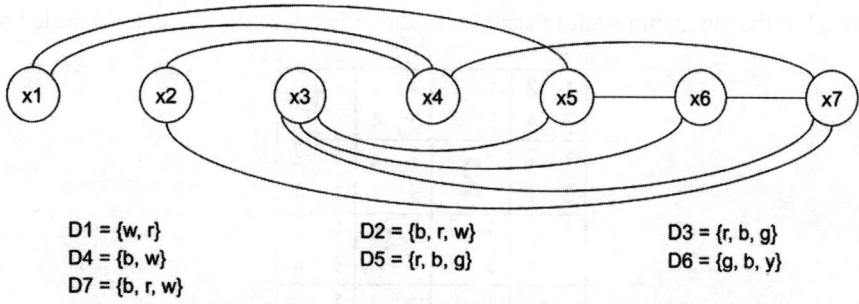

D1 = {w, r}   D2 = {b, r, w}   D3 = {r, b, g}
D4 = {b, w}   D5 = {r, b, g}   D6 = {g, b, y}
D7 = {b, r, w}

A. The GENERALIZEDLOOKAHEAD algorithm with SELECTVALUE-FC is applied to the CSP in the figure, for the ordering $(x_1, x_2, x_3, x_4, x_5, x_6, x_7)$. If the algorithm chooses the assignments $x_1 = r$, $x_2 = b$, and $x_3 = b$, how many total values are pruned from the domains of the variables x_4, x_5, x_6, and x_7?

B. The GENERALIZEDLOOKAHEAD algorithm with SELECTVALUE-AC is applied to the CSP in the figure, for the ordering $(x_1, x_2, x_3, x_4, x_5, x_6, x_7)$. If the algorithm chooses the assignments $x_1 = r$, $x_2 = b$, and $x_3 = b$, how many total values are pruned from the domains of the variables x_4, x_5, x_6, and x_7?

C. The GENERALIZEDLOOKAHEAD algorithm with SELECTVALUE-PARTIALLOOKAHEAD is applied to the CSP in the figure, for the ordering $(x_1, x_2, x_3, x_4, x_5, x_6, x_7)$. If the algorithm chooses the assignments $x_1 = r$, $x_2 = b$, and $x_3 = b$, how many total values are pruned from the domains of the variables x_4, x_5, x_6, and x_7?

D. The GENERALIZEDLOOKAHEAD algorithm with SELECTVALUE-FULLLOOKAHEAD is applied to the CSP in the figure, for the ordering $(x_1, x_2, x_3, x_4, x_5, x_6, x_7)$. If the algorithm chooses the assignments $x_1 = r$, $x_2 = b$, and $x_3 = b$, how many total values are pruned from the domains of the variables x_4, x_5, x_6, and x_7?

15. The objective of a 4 × 4 Sudoku puzzle is to fill a 4 × 4 grid so that each column, each row, and each of the four disjoint 2 × 2 subgrids contains all of the digits from 1 to 4.

x1	x2	x3	x4
x5	x6	x7	x8
x9	x10	x11	x12
x13	x14	x15	x16

The following figure depicts the domains of the variables for the given 4 × 4 Sudoku problem. Note that some cells have only one value in their domain. Show the order in which *dynamic variable ordering with forward checking* (DVFC) algorithm will fill in the values. Let the algorithm prefer variables in the order $(x_1, x_2, \ldots, x_{16})$ at each tie break.

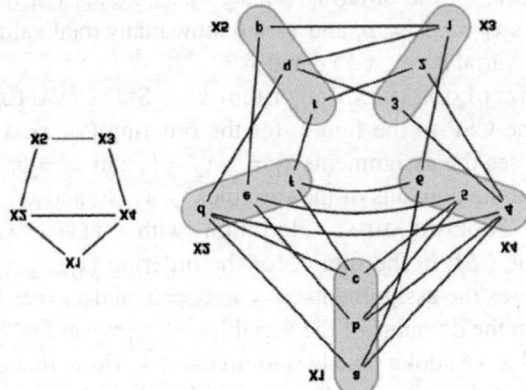

16. [Baskaran] The following figure shows a constraint graph of a binary CSP and a part of the matching diagram. When a pair of variables (like X_1 and X_3) do not have a constraint in the constraint graph then assume a *universal relation* in the matching diagram.

The FC algorithm begins by assigning $X_1 = a$. What are the *next* four values assigned to variables by the FC algorithm? List the values as a comma separated list in the order they are assigned.

17. What is the first solution found by the FC algorithm for the above problem?
18. The following figure shows the constraint graph of a binary CSP on the left and a part of the matching diagram on the right. Please assume a *universal relation* in the matching diagram where there is no constraint between variables in the constraint graph. The variables, and their values, are to be considered in *alphabetical* order.

Algorithm FC is about to begin by assigning $X_1 = a$. What are the next six values assigned to variables? Draw the matching diagram at this point.
What is the first solution found by the algorithm?

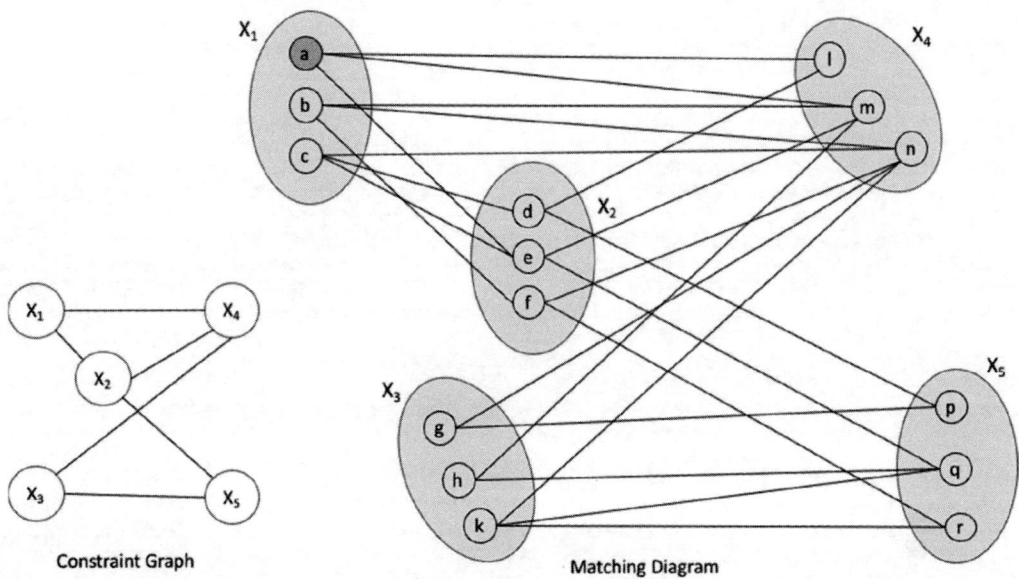

Constraint Graph   Matching Diagram

19. Repeat the above question for the following CSP:

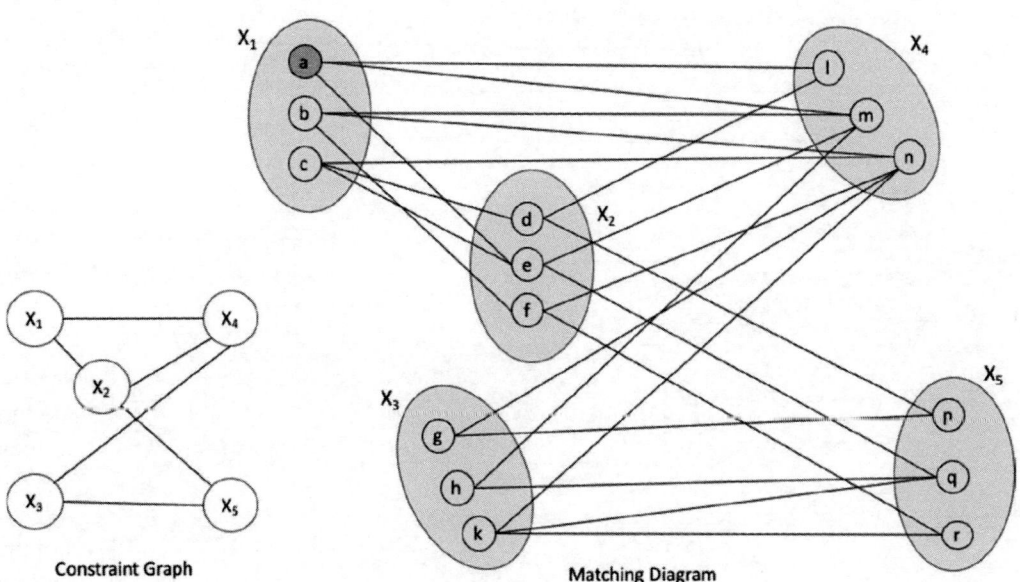

Constraint Graph   Matching Diagram

APPENDIX

Algorithm and Pseudocode Conventions

S. Baskaran

The algorithms presented in this book assume eager evaluation. The values of primitive types (integers, reals, strings) are passed by value, and tuples, lists, arrays, sets, stacks, queues, etc., are passed by reference, similar to how Java treats primitive values and objects.

The data structures (container types) like sets, arrays, stacks and queues, and the operations on those structures carry their usual meaning, and their usages in the algorithms are self explanatory.

Tuple

A tuple is an ordered collection of fixed number of elements, where each element may be of a different type. A tuple is represented as a comma separated sequence of elements, surrounded by parenthesis.

$$tuple \to (ELEMENT_1, \ ELEMENT_2, \ \ldots, \ ELEMENT_k)$$

A tuple of two elements is called a pair, for example, (S, **null**), ((A, S), 1), (S, [A, B]) are pairs. And a tuple of three elements is called a triple, for example, (S, **null**, 0), (A, S, 1), (S, A, B) are triples. A tuple of k elements is called a k-tuple, for example, (S, MAX, $-\infty$, ∞), (A, MIN, LIVE, ∞, 42).

Note: parenthesis is also used to indicate precedence, like in (3+1) * 4 or in (1 : (4 : [ ])), its usage will be clear from the context.

List

A list is an ordered collection of an arbitrary number of elements of the same type. A list is read from left to right and new elements are added at the left end. Lists are constructed recursively like in Haskell.

$$list \to ELEMENT : list$$
$$list \to [\ ]$$

The ':' operator is a list constructor; it takes an element (HEAD) and a list (TAIL) and constructs a new list (HEAD : TAIL) similar to **cons**(HEAD, TAIL) in LISP. Using head:tail notation, a list such as [3, 1, 4] is recursively constructed from (3 : (1 : (4 : [ ]))), similar to **cons**(3, **cons**(1, **cons**(4, **nil**))) in LISP. The empty list [ ] has no head or tail.

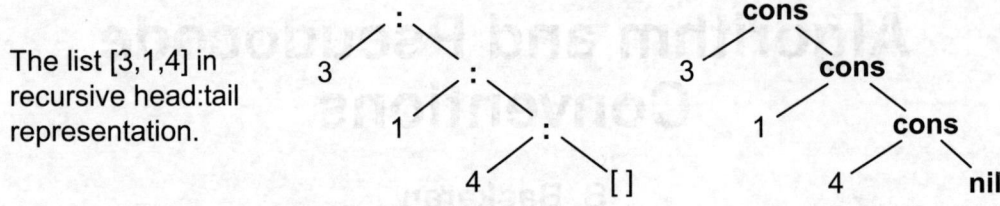

The list [3,1,4] in recursive head:tail representation.

In the head:tail representation, elements are always added to and removed from the head (left end) of a list and in that respect a list behaves like a stack.

To reduce clutter, we allow the list (3 : (1 : (4 : [ ]))) to be expressed in any of the following equivalent forms:

(3 : (1 : (4 : [ ])))   3 : 1 : 4 : [ ]   3 : 1 : [4]   3 : [1, 4]   [3, 1, 4]

Assignment

The assignment statements take the general form

PATTERN ← EXPRESSION

where the values contained in the EXPRESSION are assigned to the variables contained in the PATTERN. A PATTERN is an expression that is constructed out of variables and underscores. An underscore is a placeholder for an unnamed variable, whose value is of no interest. Such values can be called 'don't care values'. A few examples of assignment statements and the resulting assignments are shown in the table below.

$x \leftarrow 3^2 + 4^2$ $x \leftarrow 25$ because of eager evaluation.

$x : y \leftarrow 3 : 1 : 4 : [\,]$ $x \leftarrow 3; y \leftarrow 1 : 4 : [\,]$

$x : y \leftarrow [3, 1, 4]$ $x \leftarrow 3; y \leftarrow [1, 4]$. Same as the above.

$x : y : z \leftarrow [3, 1, 4]$ $x \leftarrow 3; y \leftarrow 1; z \leftarrow [4]$

$(x, y, \_) \leftarrow (S, LIVE, 125)$ $x \leftarrow S; y \leftarrow LIVE$ and the value 125 is ignored.

$(x, \_, \_) \leftarrow (S, LIVE, 125)$ $x \leftarrow S$ and the remaining values are ignored.

Tests

The equality, inequality, and '**is**' tests used in the algorithms are standard ones; for the most part they are self explanatory. Here, we describe equality testing on structures and context dependent tests.

The equality tests are of the form

$$\text{EXPRESSION}_1 = \text{EXPRESSION}_2$$

where two expressions are equal if their structures match and the corresponding values in the respective structures also match, for example:

$5^2 = 3^2 + 4^2$ is true because of eager evaluation.

$(3 : (1 : (4 : [\ ]))) = [3, 1, 4]$ is true because both lists are equivalent.

$(3, 1, 4) = [3, 1, 4]$ is false because a tuple is not a list.

$(A, S, 1) = (A, S, 2)$ is false because the third element fails to match.

$(A, S, \_) = (A, S, 2)$ is true because the structures match, the elements match and the underscore matches any value.

$3 : 1 : \_ = [3, 1, 4, 1, 5]$ is true because the underscore matches any value.

In a few cases, we use context dependent tests like

$$\text{EXPRESSION}_1 \textbf{ is better than } \text{EXPRESSION}_2$$

which tests whether the value generated by one expression is better than the value generated by another expression. The notion of what is better will be clear from the context. For maximization problems, '**better than**' means '**greater than**' and for minimization problems, it means '**less than**'.

Built-In Functions

The standard functions on sets, tuples, arrays, lists, stacks, and queues are treated as built-in functions and are typeset like keywords and invoked like commands, for example:

    $list_1$ **++** $list_2$ returns a new list that is a concatenation of two lists.
                          [3, 1, 4] **++** [1, 5] returns [3, 1, 4, 1, 5].

    head list   returns the head element of a list.
                    head [3, 1, 4] returns 3.

    tail list   returns the tail of a list.
                    tail [3, 1, 4] returns [1, 4].

take *n* list   returns the first *n* elements of a list, at most *n* elements.
                       take 2 [3, 1, 4] returns [3, 1].
                       take 2 [3] returns [3].
                       take 2 [ ] returns [ ].

reverse list   returns a new list with elements in reverse order.
                       reverse [3, 1, 4] returns [4, 1, 3].

sort$_h$ list   sorts the elements of a list by their **h** values.
                       sort$_h$ [(D, 4), (A, 1), (N, 3)] returns [(A, 1), (N, 3), (D, 4)],
                       where the list elements are of the form (__, **h**).

first tuple*   returns the first element of a tuple.
                       first (A, S, 1) returns A.

second tuple   returns the second element of a tuple.
                       second (A, S, 1) returns S.

In some cases, a chain of built-in functions may be applied to a value; such chain of calls are evaluated from right to left, for example, "**second head tail** [(A, 1), (N, 3), (D, 4)]" is evaluated in the following manner:

second head tail [(A, 1), (N, 3), (D, 4)]
(**second** (**head** (**tail** [(A, 1), (N, 3), (D, 4)])))
(**second** (**head** [(N, 3), (D, 4)]))
(**second** (N, 3))
(3)
3

Procedures

We use indentation to capture the scope (body, block, block structure) of procedures and control statements.

A procedure has a name followed by zero or more parameters and a body, and, optionally, one or more nested procedures. A procedure may accept zero or more inputs and may produce an output.

RECONSTRUCTPATH(nodePair, CLOSED)

1   SKIPTO(parent, nodePairs)
2   **if** parent = **first head** nodePairs
3   **then return head** nodePairs
4   **else return** SKIPTO(parent, **tail** nodePairs)

```
5    (node, parent) ← nodePair
6    path ← node : [ ]
7    while parent is not null
8        path ← parent: path
9        nodePair ← SkipTo (parent, CLOSED)
10       (_, parent) ← nodePair
11   return path
```

RECONSTRUCTPATH is a procedure that takes two inputs and returns one output; its scope spans lines 1 through 11; it contains a nested procedure (SKIPTO, lines 1 through 4) and the body (lines 5 through 11). RECONSTRUCTPATH will start executing from line 5.

The level of abstraction (the amount of detail) in a procedure can vastly vary, for example, the GAMEPLAY procedure describes the general idea of game playing at a much higher abstraction than what RECONSTRUCTPATH does for path reconstruction.

GAMEPLAY lacks detail; it does not tell us what a move is or how to make a move or when the play ends or how the winner is decided, etc. GAMEPLAY is a high-level algorithm.

```
GAMEPLAY(MAX)
1    while game not over
2        call k-ply search
3        make move
4        get MIN'S move
```

Control Statements

The control statements such as if-then, if-then-else, for loop, for-each loop, while loop, repeat loop, and repeat-until loop carry their usual meaning. And we use indentation to capture the scope (body, block, block structure) of control statements.

```
1    if test
2        block
```

Execute the *block* only if the *test* succeeds, otherwise skip that block.

```
1    if test
2    then block₁
3    else block₂
```

Execute *block₁* if the *test* succeeds, otherwise execute *block₂*. In each pass, only one of the two blocks is executed.

```
1   while test                    if test succeeds then execute body
2       body                      if test succeeds then execute body
                                  ...
                                  if test succeeds then execute body
```

A while loop represents a computation that essentially executes an arbitrarily long sequence of identical if-then statements, and the computation (the loop) ends when the *test* fails. In a while loop, the *body* may not execute, or execute one or more times.

```
1   repeat                        body
2       body                      if test fails then execute body
3   until test                    if test fails then execute body
                                  ...
                                  if test fails then execute body
```

A repeat-until loop represents a computation that essentially executes the *body* followed by an arbitrarily long sequence of identical if-then statements, and the computation (the loop) ends when the *test* succeeds. In a repeat-until loop, the *body* will execute one or more times.

```
1   repeat N times                body; body; ...; body;
2       body
```

Execute the *body* many times (N times).

```
1   for i ← 1 to N                i ← 1; body; i ← 2; body; ...; i ← N; body;
2       body
```

A for loop represents a sequence of computations that essentially sets the loop variable to a number and executes the *body*, and does so for each integer from 1 to N.

Algorithm and Pseudocode Conventions | 447

```
1  for each x in list                x ← e₁; body; x ← e₂; body; …; x ← eₖ; body;
2      body                          where, list = [e₁, e₂, …, eₖ]
```

A for-each loop represents a sequence of computations that essentially sets the loop variable to an element in the *list* and executes the *body*, and does so for each element in the *list*.

The for-each loop can iterate over lists, sets, arrays, stacks, and queues.

```
1  for i ← 1 to N, if test           1  for i ← 1 to N
2      body                          2      if test
                                     3          body
```

A loop with a guarded body is a special case; it iterates like a regular loop, but in each iteration the *body* is executed only if the *test* succeeds. The guarded body may occur in a for-each loop, while loop, repeat loop, and repeat-until loop.

References

Agarwala, R, D. L. Applegate, D. Maglott, G. D. Schuler, and A. A. Schaffler. 2000. 'A Fast and Scalable Radiation Hybrid Map Construction and Integration Strategy'. *Genome Research* 10: 350–64.

Aho, Alfred V., John E. Hopcroft, and Jeffrey D. Ullman. 1974. *The Design and Analysis of Computer Algorithms, Addison-Wesley Series. Computer Science and Information Processing*. Boston: Addison-Wesley.

Allen, James F. 1983. 'Maintaining Knowledge about Temporal Intervals'. *Communications of the ACM* 26 (11): 832–43.

———. 1991. 'Temporal Reasoning and Planning'. In *Reasoning about Plans*, edited by James F. Allen, Henry A. Kautz, Richard N. Pelavin, and Josh D. Tenenberg. San Mateo: Morgan Kaufmann.

Andrews, Robin. 2019. 'Volcano Space Robots Are Prepping for a Wild Mission to Jupiter'. *The Wired, Science*. Accessed October 2022. https://www.wired.co.uk/article/nasa-submarines-searching.

Antoniou, Grigoris and Frank van Harmelen. 2008. *A Semantic Web Primer*. 2nd ed. Cambridge: MIT Press.

Applegate, D., W. Cook, and A. Rohe. 2003. 'Chained Lin-Kernighan for Large Traveling Salesman Problems'. *INFORMS Journal of Computing* 15 (1): 82–92.

Applegate, David L., Robert E. Bixby, Vasek Chvatal, and William J. Cook. 2007. *The Traveling Salesman Problem: A Computational Study*. Princeton Series in Applied Mathematics. Princeton: Princeton University Press.

Arora, S. 1998. 'Polynomial Time Approximation Schemes for Euclidean Traveling Salesman and Other Geometric Problems'. *Journal of ACM* 45: 753–82.

Baader, Franz and Ulrike Sattler. 2001. 'An Overview of Tableau Algorithms for Description Logics'. *Studia Logica: An International Journal for Symbolic Logic* 69 (1): 5–40. Analytic Tableaux and Related Methods. Part 1: Modal Logics.

Belgum, Erik, Curtis Roads, Joel Chadabe, T. Emile Tobenfeld and Laurie Spiegel. 1988. 'A Turing Test for "Musical Intelligence"?' *Computer Music Journal* 12 (4): 7–9. doi:10.2307/3680146.

Berliner, Hans J. 1987. 'Pennsylvania Chess Championship Report-HITECH Wins Chess Tourney'. *AI Magazine* 8 (4): 101–02.

Bland, R.G., and D.F. Shallcross. 1984. 'Large Traveling Salesman Problems Arising from Experiments in X-Ray Crystallography: A Preliminary Report on Computation'. *Operation Research Letters* 8: 125–28.

Blum, A., and Furst, M. 1997. 'Fast Planning through Planning Graph Analysis." *Journal of Artificial Intelligence* 90 (1–2): 281–300.

Bolander, Thomas. 2017. "A Gentle Introduction to Epistemic Planning: The DEL Approach'. In *Proceedings of M4M@ICLA 2017*, 1–22.

Bonet, Blai, and Hector Geffner. 2001a. 'Planning as Heuristic Search'. *Artificial Intelligence* 129 (1–2): 5–33.

———. 2001b. 'Heuristic Search Planner 2.0'. *AI Magazine* 22 (3): 77–80.

Brachman, Ronald J. and Hector J. Levesque. 2004. *Knowledge Representation and Reasoning*. Burlington: Morgan Kaufmann.

Browne, Cameron B., Edward Powley, Daniel Whitehouse, Simon M. Lucas, Peter I. Cowling, Philipp Rohlfshagen, Stephen Tavener, Diego Perez, Spyridon Samothrakis, and Simon Colton. 2012. 'A Survey of Monte Carlo Tree Search Methods'. *IEEE Transactions on Computational Intelligence and AI in Games* 4 (1): 1–43. doi:10.1109/TCIAIG.2012.2186810.

Brownston, Lee, Robert Farrell, Elaine Kant, and Nancy Martin. 1985. *Programming Expert Systems in OPS5*. Boston: Addison-Wesley.

Bryce, Daniel and Subbarao Kambhampati. 2007. 'A Tutorial on Planning Graph Based Reachability Heuristics'. *AI Magazine* 28 (1): 4.

Buchanan, Bruce G. and Edward H. Shortliffe. 1984. *Rule Based Expert Systems: The Mycin Experiments of the Stanford Heuristic Programming Project*. Addison-Wesley Series in Artificial Intelligence. Boston: Addison-Wesley.

Bylander, Tom. 1994. 'The Computational Complexity of Propositional STRIPS Planning'. *Artificial Intelligence* 69 (1–2): 165–204.

Cain, Stephanie. 2022. 'A.I.-Driven Robots are Cooking your Dinner'. *Fortune*, October. Accessed December 2022. https://fortune.com/2022/10/18/tech-forward-everyday-ai-robots-pizza/.

Caltabiano, Daniele and Giovanni Muscato. 2005. 'A Robotic System for Volcano Exploration'. In *Cutting Edge Robotics*, edited by Vedran Kordic, Aleksandar Lazinica and Munir Merdan. I-Tech.

Campbell, A. N., V. F. Hollister, R. O. Duda, and P. E. Hart. 1982. 'Recognition of a Hidden Mineral Deposit by an Artificial Intelligence Program'. *Science* 217: 927–29.

Campbell, Murray, A. Joseph Hoane Jr., and Feng-hsiung Hsu. 2002. 'Deep Blue'. *Artificial Intelligence* 134 (1–2): 57–83.

Charniak, Eugene and Drew McDermott. 1985. *Introduction to Artificial Intelligence*. Boston. Addison-Wesley.

Chien, Steve, Anthony Barrett, Tara Estlin, and Gregg Rabideau. 2000. 'A Comparison of Coordinated Planning Methods for Cooperating Rovers'. In *Proceedings of the Fourth International Conference on Autonomous Agents* (Agents 2000), Barcelona, Spain, 100101, June 2000. doi:10.1145/336595.337057.

Clowes, M. B. 1971. 'On Seeing Things'. *Artificial Intelligence* 2: 79–116.

Cobb, William S. 1997. 'The Game of Go: An Unexpected Path to Enlightenment'. *The Eastern Buddhist* (New Series) 30 (2): 199–213. https://www.jstor.org/stable/44362178.

Cohen, Paul. 2016. 'Harold Cohen and AARON'. *AI Magazine* 37 (4).

Coles, Andrew, Maria Fox, Derek Long, and Amanda Smith. 2008. 'Planning with Problems Requiring Temporal Coordination'. In *Proceedings of the Twenty-Third AAAI Conference on Artificial Intelligence, AAAI 2008*, edited by Dieter Fox, Carla P. Gomes, 892–97. Palo Alto: AAAI Press.

Colorni, A., M. Dorigo, and V. Maniezzo. 1991. 'Distributed Optimisation by Ant Colonies'. In *Proceedings of ECAL'91, European Conference on Artificial Life*. Amsterdam: Elsevier Publishing.

Cook, S. and D. Mitchell. 1997. 'Finding Hard Instances of the Satisfiability Problem: A Survey'. In *Proceedings of the DIMACS Workshop on Satisfiability Problems*, 11–13. Providence: American Mathematical Society.

Cope, David. 2004. *Virtual Music: Computer Synthesis of Musical Style*. Cambridge: MIT Press.

Cormen, Thomas H., Charles E. Leiserson, Ronald L. Rivest, and Clifford Stein. 2001. *Introduction to Algorithms*. 2nd ed. Cambridge: The MIT Press.

Cushing, William, Subbarao Kambhampati, Mausam, and Daniel S. Weld. 2007. 'When Is Temporal Planning Really Temporal?' In *IJCAI 2007, Proceedings of the 20th International Joint Conference on Artificial Intelligence*, edited by Manuela M. Veloso, 1852–59. San Francisco, CA: Morgan Kaufmann Publishers Inc.

Cushing, William Albemarle. 2012. 'When Is Temporal Planning *Really* Temporal?' PhD diss., Arizona State University. Accessed October 2022. https://rakaposhi.eas.asu.edu/cushing-dissertation.pdf.

Dantzig, G. B., R. Fulkerson, and S. M. Johnson. 1954. 'Solution of a Large-scale Traveling Salesman Problem'. *Operations Research* 2: 393–410.

Darwiche, Adnan. 2018. 'Human-Level Intelligence or Animal-Like Abilities'. *CACM* 61 (10): 56–57.

Davis, Ernest, Leora Morgenstern, and Charles L. Ortiz, Jr. 2017. 'The First Winograd Schema Challenge at IJCAI-16'. *AI Magazine* 38 (3). doi:10.1609/aimag.v38i4.2734.

Dawkins, Richard. 1996. *The Blind Watchmaker: Why the Evidence of Evolution Reveals a Universe Without Design*. New York: W. W. Norton & Company.

Dawkins, Richard. 2006. *Climbing Mount Improbable*. London: Penguin UK.

Dechter, Rina. 2003. *Constraint Processing*. Burlington: Morgan Kaufmann.

Dechter, Rina and Robert Mateescu. 2007. 'And/or Search Spaces for Graphical Models'. *Artificial Intelligence* 171 (2): 73–106.

Dempster, A. P., N. M. Laird, and D. B. Rubin. 1977. 'Maximum Likelihood from Incomplete Data via the EM Algorithm'. *Journal of the Royal Statistical Society*, Series B 39 (1): 1–38. JSTOR 2984875. MR 0501537.

Dijkstra, E. W. 1959. 'A Note on Two Problems in Connexion with Graphs'. *Numerische Mathematik* 1: 269–71. https://link.springer.com/article/10.1007/BF01386390.

Do, Minh Binh, and Subbarao Kambhampati. 2001. 'Planning as Constraint Satisfaction: Solving the Planning Graph by Compiling it into CSP'. *Artificial Intelligence* 132 (2): 151–82.

———. 2003. 'SAPA: A Multi-objective Metric Temporal Planner'. *Journal of Artificial Intelligence Research* 20: 155–94.

Dorigo, Marco and Thomas Stutzle. 2004. *Ant Colony Optimisation*. Cambridge: Bradford Books, MIT Press.

Dudani, S. A. 1976. 'The Distance-weighted k-Nearest Neighbour Rule'. *IEEE Transactions on System, Man, and Cybernetics* SMC-6: 325–27.

Edelkamp, S., and J. Hoffmann. 2004. '*PDDL2.2*: The Language for the Classical Part of IPC-4'. In *Competition Proceeding Hand-Out, 14th International Conference on Automated Planning and Scheduling*, June 3-7, 2004, Whistler, British Columbia, Canada. Menlo Park, CA: The AAAI Press.

Engesser, Thorsten, Thomas Bolander, Robert Mattmuller, and Bernhard Nebel. 2017. 'Cooperative Epistemic Multi-Agent Planning for Implicit Coordination'. In *Proceedings of M4M@ICLA 2017*, 75–90. https://icla.cse.iitk.ac.in/M4M/.

Fagin, Ronald, Joseph Y. Halpern, Yoram Moses, and Moshe Vardi. 2004. *Reasoning About Knowledge*. Cambridge: MIT Press.

Fikes, R. E., and N. Nilsson. 1971. 'STRIPS: A New Approach to the Application of Theorem Proving to Problem Solving'. *Artificial Intelligence* 5 (2): 189–208.

Fitting, Melvin. 2013. *First-Order Logic and Automated Theorem Proving*. 2nd ed. Springer.

Fix, Evelyn and J. L. Hodges, Jr. 1989. 'Discriminatory Analysis Nonparametric Discrimination: Consistency Properties'. *International Statistical Review/Revue Internationale de Statistique* 57 (3): 238–47.

Forgy, Charles. 1981. *OPS5 User's Manual, Technical Report CMU-CS-81-135*. Pittsburgh: Carnegie Mellon University.

———. 1982. 'Rete: A Fast Algorithm for the Many Patterns/Many Objects Match Problem'. *Artificial Intelligence* 19 (1): 17–37.

Fox, Douglas. 2016. 'What Sparked the Cambrian Explosion?' *Nature* 530: 268–70.

Fox, M. and D. Long. 'PDDL2.1: An Extension to PDDL for Expressing Temporal Planning Domains'. *Journal of Artificial Intelligence Research* (special issue on the *3rd International Planning Competition*) 20: 61–124.

Frege, Gottlob. 1967. *Begriffsschrift, eine der arithmetischen nachgebildete Formelsprache des reinen Denkens*, Halle a. S.: Louis Nebert; translated as *Concept Script, a Formal Language of Pure Thought Modelled upon That of Arithmetic* by S. Bauer-Mengelberg. In *From Frege to Godel: A Source Book in Mathematical Logic, 1879–1931*, edited by J. van Heijenoort. Cambridge, MA: Harvard University Press.

Freuder, E. C. 1997. 'In Pursuit of the Holy Grail'. *Constraints* 2: 57–61. Accessed December 2022. doi:10.1023/A:1009749006768.

Gardner, Martin. 1970. 'The Fantastic Combinations of John Conway's New Solitaire Game "Life".' *Scientific American* 223: 120–23.

Gamow, George. 1989. *One, Two, Three ... Infinity: Facts and Speculations of Science* (Dover Books on Mathematics). New ed. Mineola: Dover Publications.

Gerevini, A. and D. Long. 2005. *Plan Constraints and Preferences in PDDL3. Technical Report, Department of Electronics for Automation*. Brescia: University of Brescia.

Gerrig, Richard J. and Mahzarin R. Banaji. 1994. 'Language and Thought'. In *Thinking and Problem Solving* (2nd ed.), edited by Robert J. Sternberg, vol. 2 in Handbook of Perception and Cognition. Academic Press.

Ghallab, M., D. Nau, and P. Traverso. 2004. *Automated Planning, Theory and Practice*. Amsterdam; Burlington: Elsevier, Morgan Kaufmann Publishers.

Ginsberg, M. L. 1999. 'GIB: Steps Toward an Expert-Level Bridge-Playing Program'. In *Proceedings of the Sixteenth International Joint Conference on Artificial Intelligence*, 584–93. Stockholm: Morgan Kaufmann.

Gleick, James. 1987. *Chaos: Making a New Science*. New York: Viking.

Gleiser, Marcelo. 2022. 'Not Just Light: Everything Is a Wave, Including You'. *Big Think* 13 (8). Accessed December 2022. https://bigthink.com/13-8/wave-particle-duality-matter/.

Glover, Fred. 1986. 'Future Paths for Integer Programming and Links to Artificial Intelligence'. *Computers and Operations Research* 13 (5): 533–49. doi:10.1016/0305- 0548(86)90048-1.

Goldberg, David E. 1989. *Genetic Algorithms in Search, Optimisation, and Machine Learning*. Reading, MA: Addison-Wesley.

Gonzalez, Rafael C. and Michael G. Thomason. 1978. *Syntactic Pattern Recognition*. Reading, MA: Addison-Wesley Pub. Co., Advanced Book Program.

Grand, Steve. 2001. *Creation: Life and How to Make It*. Cambridge, MA: Harvard University Press.

Green, Cordell. 1969. 'Application of Theorem Proving to Problem Solving'. In *IJCAI'69: Proceedings of the 1st International Joint Conference on Artificial Intelligence*, 219–39, May 7–9, 1969, Washington, DC. San Francisco, CA: Morgan Kaufmann Publishers Inc.

Gruber, T. R. 1993. 'A Translation Approach to Portable Ontologies'. *Knowledge Acquisition* 5 (2): 199–220.

Guarino, N., D. Oberle, and S. Staab. 2009. 'What Is an Ontology?' In *Handbook on Ontologies. International Handbooks on Information Systems*, edited by International Handbooks on Information Systems. Berlin, Heidelberg: Springer. doi:10.1007/978-3-540-92673-3_0.

Haas, A. 1987. 'The Case for Domain-Specific Frame Axioms'. In *The Frame Problem in Artificial Intelligence, Proceedings of the 1987 Workshop*, edited by F.M. Brown. Burlington: Morgan Kaufmann Publishers.

Hadamard, Jacques. 1945. *The Mathematician's Mind: The Psychology of Invention in the Mathematical Field*. Princeton, NJ: Princeton University Press. Reprinted, Dover Publications Inc., 2003.

Haken, Armin. 1985. 'The Intractability of Resolution'. *Theoretical Computer Science* 39: 297–308.

Hart, P., N. Nilsson, and B. Raphael. 1968. 'A Formal Basis of Heuristic Determination of Minimum Cost Paths'. *IEEE Trans. System Science and Cybernetics, SSC* 4(2): 100107.

Haugeland, John. 1985. *Artificial Intelligence: The Very Idea, A Bradford Book*. Cambridge: The MIT Press.

Hawking, Stephen. 2003. *On the Shoulders of Giants: The Great Works of Physics and Astronomy*. London: Penguin UK.

Hayes, Patrick. 'What the Frame Problem Is and Isn't'. In *The Robot's Dilemma: The Frame Problem in Artificial Intelligence*, edited by Z. W. Pylyshyn. Norwood, NJ: Ablex Publ.

Hayes-Roth, Frederick, Donald A. Waterman, and Douglas B. Lenat. 1983. *Building Expert Systems*. Boston, MA: Addison-Wesley Longman Publishing Co.

Hoffmann, Jorg and Bernhard Nebel. 2001. 'The FF Planning System: Fast Plan Generation Through Heuristic Search'. *Journal of Artificial Intelligence Research (JAIR)* 14: 253–302. doi:10.1613/jair.855.

Hofstadter, Douglas. 1979. *Godel, Escher, Bach: An Eternal Golden Braid*. New York: Basic Books.

———. 1996. 'Number Numbness.' In *Metamagical Themas: Questing for the Essence of Mind and Pattern*, ch. 6. Revised ed. New York: Basic Books.

———. 2009. 'Essay in the Style of Douglas Hofstadter'. *AI Magazine* 30 (3): 82–88. doi:10.1609/aimag.v30i3.2256.

Holland, John H. 1975. *Adaptation in Natural and Artificial Systems: An Introductory Analysis with Applications to Biology, Control and Artificial Intelligence*. Ann Arbor: The University of Michigan Press.

———. 1999. *Emergence: From Chaos to Order*. New York: Basic Books.

Hoos, Holger H. and Thomas Stutzle. 2004. *Stochastic Local Search: Foundations and Applications*. The Morgan Kaufmann Series in Artificial Intelligence. Burlington: Morgan Kaufmann.

Horn, Alfred. 1951. 'On Sentences Which Are True of Direct Unions of Algebras'. *J. Symbolic Logic* 16(1): 14–21.

Huffman D. A. 1971. 'Impossible Object as Nonsense Sentences'. In *Machine Intelligence*, vol. 6, edited by B. Meltzer and B. Michie, 295–323. New York: American Elsevier.

Ito, Joi and Jeff Howe. 2017. *Emergent Systems Are Changing the Way We Think*. Blogpost on the Aspen Institute website, 30 January. Accessed November 2021. https://www.aspeninstitute.org/blog-posts/emergent-systems-changing-way-think/.

Jackson, Peter. 1986. *Introduction to Expert Systems*. 1st ed. Boston: Addison-Wesley.

Jimenez, Sergio, Tomas De la Rosa, Susana Fernandez, Fernando Fernandez, and Daniel Borrajo. 2012. 'A Review of Machine Learning for Automated Planning'. *The Knowledge Engineering Review* 27 (4): 433–67. doi:10.1017/S026988891200001X.

Johnson, David S. 1990. 'Local Optimization and the Traveling Salesman Problem'. In *Automata, Languages and Programming, 17th International Colloquium, ICALP90, Proceedings*, Lecture Notes in Computer Science 443, edited by Mike Paterson, 446–61. Berlin: Springer.

Johnson, Steven. 2002. *Emergence: The Connected Lives of Ants, Brains, Cities, and Software*. New York: Scribner. Reprint edition.

Kambhampati, S. 1993. 'On the Utility of Systematicity: Understanding the Tradeoffs Between Redundancy and Commitment in Partial-order Planning'. In *Proceedings of IJCAI-93*, 1380–85. San Francisco, CA: Morgan Kaufmann Publishers Inc.

Kautz, Henry and Bart Selman. 1992. 'Planning as Satisfiability'. In *Proceedings of the 10th European Conference on Artificial Intelligence*, 359–63. New York: Wiley.

Kautz, Henry, David McAllester, and Bart Selman. 1996. 'Encoding Plans in Propositional Logic'. In *Proceedings of the 4th International Conference on Knowledge Representation and Reasoning (KR-96)*, 374–85. Burlington: Morgan Kaufmann.

Kautz, Henry A. and Bart Selman. 1999. 'Unifying SAT-based and Graph-based Planning'. In *Proceedings of the Sixteenth International Joint Conference on Artificial Intelligence, IJCAI99*, edited by Thomas Dean, vol. 2, 318–25. Burlington: Morgan Kaufmann.

Kelsey, Hugh. 1995. *Logical Bridge Play*. 2nd ed. London: Orion.

Khemani, Deepak. 1989. 'Theme Based Planning in an Uncertain Environment'. PhD thesis, Department of CS&E, IIT Bombay.

———. 1994. 'Planning with Thematic Actions'. In *Proceedings of the Second International Conference on Artificial Intelligence Planning Systems*, June 13–15, University of Chicago, Chicago, Illinois, 287–92. AAAI, AIPS.

———. 2013. *A First Course in Artificial Intelligence*. New York: McGraw-Hill Education India.

Khemani, Deepak, Radhika B. Selvamani, Ananda Rabi Dhar, and S. M. Michael. 2002. 'InfoFrax: CBR in Fused Cast Refractory Manufacture'. In *Advances in Case-Based Reasoning, 6th European Conference, ECCBR 2002*, edited by S. Craw and A. Preece, 560. Aberdeen, Scotland, UK, 4–7 September. Proceedings, Springer LNAI 2416.

Khemani, D. and S. Singh. 2018. 'Contract Bridge: Multi-Agent Adversarial Planning in an Uncertain Environment'. In *Sixth Annual Conference on Advances in Cognitive Systems*, 161–80. Accessed September 2018. www.cogsys.org/papers/ACSvol6/posters/Khemani.pdf.

Korf, Richard E. 1985. *Learning to Solve Problems by Searching for Macro-Operators*. Research Notes in Artificial Intelligence, vol. 5. London: Pitman Publishing.

Koehler, Jana, Bernhard Nebel, Jorg Hoffmann, and Yannis Dimopoulos. 1997. 'Extending Planning Graphs to an ADL Subset'. In *Recent Advances in AI Planning, 4th European Conference on Planning, ECP'97, September 24-26, 1997, Toulouse, France, Proceedings*, Lecture Notes in Computer Science 1348, edited by Sam Steel and Rachid Alam, 273–85. Springer.

Korf, Richard. 1985a. 'Depth First Iterative Deepening: An Optimal Admissible Tree Search Algorithm'. *Artificial Intelligence* 27: 97–109.

———. 1985b. *Learning to Solve Problems by Searching for Macro-Operators*. Research Notes in Artificial Intelligence, vol. 5. Pitman Publishing, 1985.

———. 1993. 'Linear-Space Best-First Search'. *Artificial Intelligence* 62: 41–78.

Korf, R. and W. Zhang. 2000. 'Divide-and-Conquer Frontier Search Applied to Optimal Sequence Alignment'. In *Proceedings of the 17th National Conference on Artificial Intelligence (AAAI-00)*, July 30–August 3, 2000, Austin, TX, 910–16. AAAI Press.

Korf, R., W. Zhang, I. Thayer, and H. Hohwald. 2005. 'Frontier Search'. *Journal of the ACM* 52 (5): 715–48.

Korf, Richard E. and Peter Schultze. 2005. 'Large-Scale Parallel Breadth-First Search'. In *AAAI'05: Proceedings of the 20th National Conference on Artificial Intelligence*, vol. 3, July 9–13, 2005, 1380–85. AAAI Press.

Korte, B. 1990. 'Applications of Combinatorial Optimization in the Design, Layout and Production of Computers'. In *Modelling the Innovation 1990*, edited by H.-J. Sebastian and K. Tammer, 517–38. Springer.

Krizhevsky, Alex, Ilya Sutskever, and Geoffrey E. Hinton. 2017. 'ImageNet Classification with Deep Convolutional Neural Networks'. *Communications of the ACM* 60 (6): 84–90. doi:10.1145/3065386. ISSN 0001-0782. S2CID 195908774.

Kumashi, Praveen Kumar and Deepak Khemani. 2002. 'State Space Regression Planning Using Forward Heuristic Construction Mechanism'. In *Artificial Intelligence: Theory and Practice, proceedings of the Int. Conf. on Knowledge Based Computer Systems, KBCS 2002*, edited by M. Sasikumar, J. J. Hegde, and M. Kavitha, 489–99. Noida: Vikas Publishing House.

Laird, John E., Paul S. Rosenbloom, and Allen Newell. 1986. 'Chunking in Soar: The Anatomy of a General Learning Mechanism'. *Machine Learning* 1: 11–46.

Laird, John E., Allen Newell, and S. Paul. 'Rosenbloom: SOAR: An Architecture for General Intelligence'. *Artificial Intelligence* 33 (1): 1–64.

Larranaga, P., C. M. H. Kuijpers, R. H. Murga, I. Inza, and S. Dizdarevic. 1999. 'Genetic Algorithms for the Travelling Salesman Problem: A Review of Representations and Operators'. *Artificial Intelligence Review* 13: 129–70.

Larson, Eric J. 2021. *The Myth of Artificial Intelligence: Why Computers Can't Think the Way We Do*. Cambridge: The Belknap Press.

Leslie, A. M. 2001. 'Theory of Mind'. In *International Encyclopedia of the Social and Behavioral Sciences*, edited by Neil J. Smelser and Paul B. Baltes. Elsevier.

Levesque, Hector J., Ernest Davis, and Leora Morgenstern. 2012. 'The Winograd Schema Challenge'. In *Principles of Knowledge Representation and Reasoning: Proceedings of the Thirteenth International Conference, KR 2012*, edited by Gerhard Brewka, Thomas Eiter, and Sheila A. McIlraith, Rome, Italy, 10–14 June. AAAI Press.

Levesque, Hector J. 2017. *Common Sense, the Turing Test, and the Quest for Real AI*. 1st ed. Cambridge: MIT Press.

Levy, David. 2008. *Love and Sex with Robots: The Evolution of Human–Robot Relationships*. London: Gerald Duckworth & Co Ltd.

Lewis, D. D. 1998. 'Naive (Bayes) at Forty: The Independence Assumption in Information Retrieval'. In *Machine Learning: ECML-98: 10th European Conference on Machine Learning*, Chemnitz, Germany, April, 21–23, 1998, edited by C. N'edellec and C. Rouveirol, 4–15. Berlin/Heidelberg: Springer.

Lifschitz, V. 1994. 'Circumscription'. In *Handbook of Logic in Artificial Intelligence and Logic Programming*, vol. 3, 297–352. Oxford: Oxford University Press.

Lindsay, Robert K., Bruce G. Buchanan, Edward A. Feigenbaum, and Joshua Lederberg. 1980. *Applications of Artificial Intelligence for Organic Chemistry: The Dendral Project*. New York: McGraw-Hill Book Company.

Lindsay, Robert K., Bruce G. Buchanan, and Edward A. Feigenbaum. 1993. 'DENDRAL: A Case Study of the First Expert System for Scientific Hypothesis Formation'. *Artificial Intelligence* 61 (2): 209–61.

Litke, John D. 1984. 'An Improved Solution to the Traveling Salesman Problem with Thousands of Nodes'. *Communications of the ACM* 27 (12): 1227–36.

Long, Derek and Maria Fox. 2003. 'The 3rd International Planning Competition: Results and Analysis'. *Journal of Artificial Intelligence Research*. Special issue on the *3rd International Planning Competition*, 20: 1–59.

Lorenz, Edward N. 1993. *The Essence of Chaos*. Seattle: University of Washington Press.

Love, Clyde C. 2010. *Bridge Squeezes Complete: Winning End Play*. 2nd updated, revised ed. Toronto: Master Point Press.

Lucy, J. A. 2001. 'Sapir–Whorf Hypothesis'. In *International Encyclopedia of the Social and Behavioral Sciences*, edited by Neil J. Smelser and Paul B. Baltes. Elsevier.

Mackworth, Alan K. 1977. 'Consistency in Networks of Relations'. *Artificial Intelligence* 8 (1): 99–118.

Mackworth, Alan K. and Eugene C. Freuder. 1985. 'The Complexity of Some Polynomial Network Consistency Algorithms for Constraint Satisfaction Problems'. *Artificial Intelligence* 25 (1): 65–74.

MacLean, P. D. 1990. *The Triune Brain in Evolution: Role in Paleocerebral Functions*. Berlin: Springer.

MacQueen, J. B. 1967. 'Some Methods for classification and Analysis of Multivariate Observations'. In *Proceedings of 5th Berkeley Symposium on Mathematical Statistics and Probability*, vol. 1, 281–97. Berkeley: University of California Press.

Mainzer, Klaus. 2003. *Thinking in Complexity: The Computational Dynamics of Matter, Mind, and Mankind*. 4th ed. Berlin: Springer.

Manna, Zohar. 1974.*Mathematical Theory of Computation*. McGraw Hill.

Martelli, Alberto and Ugo Montanari. 1978. 'Optimising Decision Trees Through Heuristically Guided Search'. *Communications of the ACM* 21 (12): 1025–039.

McAllester, D. and D. Rosenblitt. 1991. 'Systematic Nonlinear Planning'. In *AAAI'91: Proceedings of the Ninth National Conference on Artificial Intelligence*, vol. 2, July 14–19, 1991, Anaheim, CA, 634–39. AAAI Press.

McCarthy, J. 1980. 'Circumscription: A Form of Non-monotonic Reasoning'. *Artificial Intelligence* 13: 27–39.

———. 1986. 'Applications of Circumscription to Formalizing Common-sense Knowledge'. *Artificial Intelligence* 28: 89–116.

McCarthy, J. and P. J. Hayes. 1969. 'Some Philosophical Problems from the Standpoint of Artificial Intelligence'. In *Machine Intelligence 4*, edited by D. Michie and B. Meltzer, 463–502. Edinburgh: Edinburgh University Press.

McCorduck, Pamela. 2004. *Machines Who Think: A Personal Inquiry into the History and Prospects of Artificial Intelligence*. 2nd ed. Boca Raton: A K Peters/CRC Press.

McCulloch, Warrant. 1961. 'What Is a Number, That a Man May Know It, and a Man, That He May Know a Number'. *General Semantics Bulletin Nos* 26, 27: 7–18.

McCulloch, Warren S., and Walter Pitts. 1943. 'A Logical Calculus of Ideas Immanent in Nervous Activity'. *Bulletin of Mathematical Biophysics* 5: 115–33.

McDermott, John P. 1980. 'RI: An Expert in the Computer Systems Domain'. In *Proceedings of the First National Conference on Artificial Intelligence (AAAI-80)*, August 18–21, 1980, Stanford University, Stanford, CA, 267–71. Palo Alto, CA: AAAI Press.

———. 1980. 'R1: The Formative Years'. *AI Magazine* 2 (2): 21–29.

McDermott, D. 1987. 'We've Been Framed: Or Why AI Is Innocent of the Frame Problem'. In *The Robot's Dilemma: The Frame Problem in Artificial Intelligence*, edited by Z. W. Pylyshyn. Norwood, NJ: Ablex Publ.

———. 1998. *PDDL - The Planning Domain Definition Language Version 1.2*. New Haven: Yale Center for Computational Vision and Control, Yale, October.

McGann, Conor, Frederic Py, Kanna Rajan, Hans Thomas, Richard Henthorn, and Rob McEwen. 2008. 'A Deliberative Architecture for AUV Control'. In *Proceedings of IEEE International Conference on Robotics and Automation, ICRA 2008, IEEE*, Pasadena, CA, 1049–54.

McClelland, James L. and David E. Rumelhart. 1986. *Parallel Distributed Processing: Explorations in the Microstructure of Cognition*. Vol. 1: *Foundations*. Cambridge, MA: MIT Press.

———. 1986. *Parallel Distributed Processing: Explorations in the Microstructure of Cognition*. Vol. 2: *Psychological and Biological Models*. Cambridge, MA: MIT Press.

Minsky M. L. and S. A. Papert. 1969. *Perceptrons*. Cambridge, MA: MIT Press.
Minsky, Marvin. 1975. 'A Framework for Representing Knowledge'. In *The Psychology of Computer Vision*, edited by P. Winston. New York: McGraw-Hill.
———. 1986. *The Society of Mind*. New York: Simon & Schuster.
Mitchell, Tom M. 1997. *Machine Learning*. Singapore: McGraw Hill.
Mohr, R. and T. C. Henderson. 1986. 'Arc and Path Consistency Revisited'. *Artificial Intelligence* 28: 225–33.
Morrison, Philip, Phylis Morrison, Charles Eames, and Ray Eames. 1986. *Powers of Ten: About the Relative Size of Things in the Universe*. New ed. New York: W.H. Freeman & Co Ltd.
Moses, Joel. 1967. *Symbolic Integration. Technical report MAC-TR-47*. Cambridge, MA: MIT.
———. 2008. 'Macsyma: A Personal History'. *Journal of Symbolic Computation* 47 (2): 123–30. doi:10.1016/j.jsc.2010.08.018.
Murphy, Robin R., Satoshi Tadokoro, Daniele Nardi, Adam Jacoff, Paolo Fiorini, Howie Choset, and Aydan M. Erkmen. 2008. 'Search and Rescue Robotics'. In *Springer Handbook of Robotics*, edited by B. Siciliano and O. Khatib. Berlin, Heidelberg: Springer. doi:10.1007/978-3-540-30301-5_51.
Murray, H. J. R. 2015 [1985]. *A History of Chess: The Original 1913 Edition*. New York: Skyhorse.
Needleman, Saul B. and Christian D. Wunsch. 1970. 'A General Method Applicable to the Search for Similarities in the Amino Acid Sequence of Two Proteins'. *Journal of Molecular Biology* 48 (3): 443–53. doi:10.1016/0022-2836(70)90057-4. PMID 5420325.
Newell A. 1973. 'Production System Models of Control Structures'. In *Visual Information Processing*, edited by W. G. Chase. New York: Academic Press.
Newell, A. and H. A. Simon. 1963. 'GPS: A Program That Simulates Human Thought'. In *Computers and Thought*, edited by E. A. Feigenbaum and J. Feldman. New York: McGraw-Hill.
———. 1976. 'Computer Science as Empirical Inquiry: Symbols and Search'. *Communications of the ACM* 19 (3): 113–26. doi:10.1145/360018.360022.
Nilsson, N. 1971. *Problem-Solving Methods in Artificial Intelligence*. New York: McGraw-Hill.
———. 1980. *Principles of Artificial Intelligence*. Burlington: Morgan Kaufmann.
———. 1998. *Artificial Intelligence: A New Synthesis*. 1st ed. Burlington: Morgan Kaufmann.
Ottenheiomer, Harriet. 2009. *The Anthropology of Language: An Introduction to Linguistic Anthropology*. 2nd ed. Belmont, CA: Wadsworth.
Ottlik, Geza and Hugh Kelsey. 1983. *Adventures in Card Play*. New ed. London: Gollancz.
Payne, Matt. 2021. *What Is Beam Search? Explaining the Beam Search Algorithm*. Blog on Width.AI. Accessed September 2022. https://www.width.ai/post/what-is-beam-search.
Pearl, Judea. 1984. *Heuristics: Intelligent Search Strategies for Computer Problem Solving*. The Addison-Wesley Series in Artificial Intelligence. Boston: Addison-Wesley Pub
Pearl, Judea. 2019. *The Book of Why: The New Science of Cause and Effect*. London: Penguin.
Poundstone, William. 1993. *Prisoner's Dilemma: John Von Neumann, Game Theory and the Puzzle of the Bomb*. Anchor, Reprint edition.
Premack, David and Guy Woodruff. 1978. 'Does the Chimpanzee Have a Theory of Mind?' *Behavioral and Brain Sciences* 1 (4): 515–26. doi:10.1017/S0140525X00076512.
Premchand, Munshi. 2020. *Shatranj Ke Khiladi* (in Hindi). New Delhi: Diamond Books.
Quillian, M. R. 1967. 'Word Concepts: A Theory and Simulation of Some Basic Semantic Capabilities'. *Behavioral Science* 12 (5): 410–30.
———. 1968. 'Semantic Memory'. In *Semantic Information Processing*, edited by M. Minsky, 227–70. Cambridge: MIT Press.

Rajan, Kanna, Frederic Py, Conor Mcgann, John Ryan, and Thom Maughan. 2009. 'Onboard Adaptive Control of AUVs using Automated Planning and Execution'. In *Proceedings of the 16th Annual International Symposium on Unmanned Untethered Submersible Technology 2009 (UUST 09)*, August, 23–26, 2009, Durham, NH, 1–13. Autonomous Undersea Systems Institute (AUSI).

Reifsteck, D., T. Engesser, R. Mattmuller, and B. Nebel. 2019. 'Epistemic Multi-Agent Planning using Monte-Carlo Tree Search'. In *Joint German/Austrian Conference on Artificial Intelligence*, 277–89. Kunstliche Intelligenz.

Reinelt G. 1995. 'TSPLIB: Discrete and Combinatorial Optimization'. Accessed August 7 2022. http://comopt.ifi.uni-heidelberg.de/software/TSPLIB95/.

Reiter, Raymond. 1980. 'A Logic for Default Reasoning'. *Artificial Intelligence* 13: 81132.

Rich, Elaine. 1983. *Artificial Intelligence*. New York: McGraw Hill.

Rich, Elaine and Kevin Knight. 1991. *Artificial Intelligence*. New York: Tata McGraw Hill.

Richards, Mark and Eyal Amir. 2007. 'Opponent Modeling in Scrabble'. In *Proceedings of the 20th International Joint Conference on Artificial Intelligence*, January, 6–12, 2007, Hyderabad, India, edited by Manuela M. Veloso, 1482–87.

Robinson, J. Alan. 1965. 'A Machine-Oriented Logic Based on the Resolution Principle'. *Journal of the ACM (JACM)* 12 (1): 23–41.

Rosenblatt, Frank. 1958. 'The Perceptron: A Probabilistic Model for Information Storage and Organization in the Brain'. *Psychological Review* 65 (6): 386–408.

Rumelhart, David E., Geoffrey E. Hinton, and Ronald J. Williams. 1986. 'Learning Representations by Back-Propagating Errors'. *Nature* 323: 533–36. doi:10.1038/323533a0.

Schaeffer, J., N. Burch, Y. Bjornsson, A. Kishimoto, M. Muller, R. Lake, P. Lu, and S. Sutphen. 2007. 'Checkers Is Solved'. *Science* 317: 1518–522.

Schank, Roger C. and Robert P. Abelson. 1977. *Scripts, Plans, Goals, and Understanding: An Inquiry into Human Knowledge Structures*. Hillsdale, NJ: Lawrence Erlbaum.

Schank, R. C. and C. K. Riesbeck. 1981. *Inside Computer Understanding: Five Programs Plus Miniatures*. Mahwah: Lawrence Erlbaum.

Schirber, Michael. 2005. *Dancing Bees Speak in Code*. Live Science, 27 May. Accessed November 2021. https://www.livescience.com/3812-dancing-bees-speak-code.html.

Schrittwieser, Julian, Ioannis Antonoglou, Thomas Hubert, Karen Simonyan, Laurent Sifre, Simon Schmitt, Arthur Guez, Edward Lockhart, Demis Hassabis, Thore Graepel, Timothy Lillicrap, and David Silver. 2020. 'Mastering Atari, Go, Chess and Shogi by Planning with a Learned Model'. *Nature* 588: 604–09. doi:10.1038/s41586-020- 03051-4.

Shanahan, M.P. 1997. *Solving the Frame Problem: A Mathematical Investigation of the Common Sense Law of Inertia*. Cambridge: MIT Press.

Shanahan, Murray. 1999. 'The Event Calculus Explained'. *Artificial Intelligence Today* 1999: 409–30.

Shannon, Claude E. 1950. 'Programming a Computer for Playing Chess'. *Philosophical Magazine* (Ser.7) 41 (314): 256–75.

Sheppard, Brian. 2002. 'World-championship-caliber Scrabble'. *Artificial Intelligence* 134: 241–275.

Shortliffe E. H. 1976. *Computer Based Medical Consultation: MYCIN*. New York: Elsevier.

Silver, David, Aja Huang, Chris J. Maddison, Arthur Guez, Laurent Sifre, George van den Driessche, Julian Schrittwieser, Ioannis Antonoglou, Veda Panneershelvam, Marc Lanctot, Sander Dieleman, Dominik Grewe, John Nham, Nal Kalchbrenner, Ilya Sutskever, Timothy Lillicrap, Madeleine Leach, Koray Kavukcuoglu, Thore Graepel, and Demis Hassabis. 2016. 'Mastering the Game of Go with Deep Neural Networks and Tree Search'. *Nature* 529: 484–89. doi:10.1038/nature16961.

Silver, David, Julian Schrittwieser, Karen Simonyan, Ioannis Antonoglou, Aja Huang, Arthur Guez, Thomas Hubert, Lucas Baker, Matthew Lai, Adrian Bolton, Yutian Chen, Timothy Lillicrap, Fan Hui, Laurent Sifre, George van den Driessche, Thore Graepel, and Demis Hassabis. 2017. 'Mastering the Game of Go without Human Knowledge'. *Nature* 550: 354–59. doi:10.1038/nature24270.

Silver, David, Thomas Hubert, Julian Schrittwieser, Ioannis Antonoglou, Matthew Lai, Arthur Guez, Marc Lanctot, Laurent Sifre, Dharshan Kumaran, Thore Graepel, Timothy Lillicrap, Karen Simonyan, and Demis Hassabis. 2018. 'A General Reinforcement Learning Algorithm That Masters Chess, Shogi, and Go through Self-Play'. *Science* 362 (6419): 1140–44. doi:10.1126/science.aar640.

Singhal, Amit. 2012. 'Introducing the Knowledge Graph: Things, Not Strings'. Google Official Blog, May 2012. Accessed December 2022. https://blog.google/products/search/introducing-knowledge-graph-things-not/.

Skolem, Thoralf A. 1970. *Selected Works in Logic*. Edited by J. E. Fenstad. Oslo: Scandinavian University Books.

Slagle J. 1961. 'A Heuristic Program that Solves Symbolic Integration Problems in Freshman Calculus'. Ph.D. diss., MIT, May.

———. 1963. 'A Heuristic Program That Solves Symbolic Integration Problems in Freshman Calculus'. *JACM* 10 (4): 507–20.

Smith, David E. and Daniel S. Weld. 1998. 'Conformant Graphplan'. In *Proceedings of the Fifteenth National Conference on Artificial Intelligence and Tenth Innovative Applications of Artificial Intelligence Conference, AAAI 98, IAAI98*, 889–96. AAAI Press.

Smullyan, Raymond M. 2009. *Logical Labyrinths*. Natick, MA: A K Peters.

Sterling, Leon and Ehud Shapiro. 1994. *The Art of Prolog, Second Edition: Advanced Programming Techniques (Logic Programming)*. Cambridge, MA: The MIT Press.

Stieger, Allison. 2014. 'Myth and Creativity: Ariadne's Thread and a Path through the Labyrinth'. *The Creativity Post*, 16 June. Accessed 6 April 2020. https://www.creativitypost.com/article/myth_and_creativity_ariadnes_thread_and_a_path_through_the_labyrinth.

Stockman, George C. 1979. 'A Minimax Algorithm Better than Alpha-Beta?' *Artificial Intelligence* 12 (2): 179–96.

Stoll, Robert R. 1979. *Set Theory and Logic*. New ed. Dover Publications Inc.

Sussman, G. 1975. *A Computer Model of Skill Acquisition*. Amsterdam: Elsevier/North Holland.

Sutton, R. S. 1988. 'Learning to Predict by the Methods of Temporal Differences'. *Machine Learning* 3: 9–44.

Sutton, Richard S. and Andrew G. Barto. 1998. *Reinforcement Learning: An Introduction*. Cambridge, MA: MIT Press.

Tate, A., B. Drabble, and R. Kirby. 1994. *O-Plan2: An Architecture Command Planning and Control*. Burlington: Morgan Kaufmann.

Tesauro, Gerald and Terrence J. Sejnowski. 1989. 'A Parallel Network That Learns to Play Backgammon'. *Artificial Intelligence* 39 (3): 357–90.

Tessauro, G. 1989. 'Neurogammon Wins Computer Olympiad'. *Neural Computation* 1: 321–23.

———. 1994. 'TD-Gammon, A Self-Teaching Backgammon Program Achieves Master-level Play'. *Neural Computation* 6 (2): 215–19.

———. 1995. 'Temporal Difference Learning and TD-Gammon'. *Communications of the ACM* 38 (3): 58–68.

———. 2002. 'Programming Backgammon using Self-teaching Neural Nets'. *Artificial Intelligence* 134 (1–2): 181–99.

Truscott, A. and D. Truscott. 2004. *The New York Times Bridge Book: An Anecdotal History of the Development, Personalities, and Strategies of the World's Most Popular Card Game*. Basingstoke: Macmillan.

Turing, A. M. 1950. 'Computing Machinery and Intelligence'. *Mind* 59: 433–60. http://loebner.net/Prizef/TuringArticle.html.

van Beek, Peter and Xinguang Chen. 1999. 'CPlan: A Constraint Programming Approach to Planning'. In *Proceedings of the Sixteenth National Conference on Artificial Intelligence and the Eleventh innovative Applications of Artificial intelligence*, American Association for Artificial Intelligence, Menlo Park, CA, 585–90.

van Ditmarsch, H., W. van Der Hoek, and B. Kooi, 2007. *Dynamic Epistemic Logic*. Berlin: Springer.

van Gelder, A. and Y. K. Tsuji. 1996. 'Satisfiability Testing with More Reasoning and Less Guessing'. In *Cliques, Coloring, and Satisfiability: Second DIMACS Implementation Challenge*, edited by D. S. Johnson and M. Trick, 559–86. DIMACS Series in Discrete Mathematics and Theoretical Computer Science. Providence, RI: American Mathematical Society.

von Neumann, John and Oskar Morgenstern. 1944. *Theory of Games and Economic Behavior*. Princeton: Princeton University Press.

Waltz, D. L. 1975. 'Understanding Line Drawings of Scenes with Shadows'. In *The Psychology of Computer Vision*, edited by P. H. Winston, 19–91. New York: McGraw-Hill.

Waterman, D. A. and Frederick Hayes-Roth. 1978. *Pattern-Directed Inference Systems*. New York: Academic Press.

Watson, Ian. 1997. *Applying Case-Based Reasoning: Techniques for Enterprise Systems*. The Morgan Kaufmann Series in Artificial Intelligence. Burlington: Morgan Kaufmann.

———. 2002. *Applying Knowledge Management: Techniques for Building Corporate Memories*. The Morgan Kaufmann Series in Artificial Intelligence. Burlington: Morgan Kaufmann.

Weizenbaum, Joseph. 1966. 'ELIZA: A Computer Program for the Study of Natural Language Communication Between Man and Machine'. *Communications of the ACM* 9 (1): 36–45.

Weld, Daniel S. 1994. 'An Introduction to Least Commitment Planning'. *AI Magazine* 15 (4): 27–61.

———. 1999. 'Recent Advances in AI Planning'. *AI Magazine* 20 (2): 93. doi:10.1609/aimag.v20i2.1459.

Winograd, T. 1972. *Understanding Natural Language*. New York: Academic Press.

Winston, Patrick Henry. 1977. *Artificial Intelligence*. Boston: Addison-Wesley Pub. Co.

Woolsey, Kit. 2000. *Computers and Rollouts, GammOnline*. http://gammonline.com/members/lan00/articles/roll.htm.

Xie, Lingyun and Du Limin, Du. 2004. 'Efficient Viterbi Beam Search Algorithm using Dynamic Pruning'. In *International Conference on Signal Processing Proceedings, ICSP. 1*, vol. 1, 699–702. doi:10.1109/ICOSP.2004.1452759.

Zhou, R. and E. A. Hansen. 2003. 'Sparse-Memory Graph Search'. In *Proceedings of the 18th International Joint Conference on Artificial Intelligence (IJCAI-03)*, 1259–66. San Francisco, CA: Morgan Kaufmann Publishers Inc.

———. 2004. 'Breadth-First Heuristic Search'. In *Proceedings of the 14th International Conference on Automated Planning and Scheduling*, June 3–7, 2004, Whistler, British Columbia, Canada, 92–100. AAAI Press.

———. 2005. 'Beam Stack Search: Integrating Backtracking with Beam Search'. In *Proceedings of the 15th International Conference on Automated Planning and Scheduling*, edited by Susanne Biundo, Karen Myers, Kanna Rajan, 90–98. Palo Alto: AAAI Press.

Index

2-city-exchange 96
2-edge-exchange 98
2-SAT 35
3-edge-exchange 98
3-SAT 35
8-puzzle 36–7, 77
$\mathcal{ALC}$ 358
α-cutoff 237
β-cutoff 237

A* 147, 156–68
AARON 3
abduction 21, 319, 369
abnormal 359
ABox 358
abstraction 11
action description language 310
action layer 296
action variables 303
actions 192, 273, 295
activation function 385
Add list 267
addition 345
additive heuristic 275
adjacency representation 134
admissibility 163
admissible 162–3
Al-Khalili, Jim 139
AlexNet 8
algorithm 406, 418
algorithm AC-1 404
algorithm AC-3 406
algorithm AC-4 407
algorithm AC-LOOKAHEAD 425, 426
algorithm ADAPTIVECONSISTENCY 417
algorithm AO* 211
algorithm CDBJ 433

algorithm BACKTRACKING 398
algorithm DAC 415
algorithm DAC-LOOKAHEAD 422
algorithm DPC 416
algorithm FC 419
algorithm GBJ 428
algorithm Minimax 235–6
algorithm PC-1 412
algorithm PC-2 413
algorithm SSS* 239
Allen's interval algebra 309
alpha cutoff 241
Alpha nodes 198, 236
ALPHABETA algorithm 236, 238
AlphaBeta pruning 236–9
AlphaGo 4, 223, 246
AlphaGo Zero 246
AlphaZero 246
alternating edges crossover 135
ancestors $anc(x)$ 429
And-Or (AO) 205
And-Or graph 343
And-Or tree 211
ANN 6, 382
annealing 119
ant colony optimization 142
anthropic cosmological principle 125
AO 205
applicable 273
arc consistent 404
Archimedes 75
Ariadne's thread 51
Aristotelian logic 328
artificial neural networks 6, 382
aspiration criterion 106
assertion box 358
assignment 394, 398

assignment statements 442
association 7
atomic formulas 330, 331, 349
atomic sentence 321
axiomatization 326

B&B 148, 150
backgammon 247
backjumping 427
BACKPROPAGATION 6, 388
backtrack free 403
backtracking 57–8
backward chaining 341, 342, 343
backward propagation 212
backward state space planning 277
Baskaran, S. 441
Bayesian classifier 378
Bayesian reasoning 381
BCN 394
beam search 101–3
beam stack 177, 178
beam stack search (BSS) 176–8
beam width 101
BEAMSEARCH 102
Belgum, Erik 3
beliefs 314, 369
Bengio, Yoshua 8
Berliner, Hans 235
Berlioz, Carole-Ann 205
Bernstein, Alex 222
best first 239
best first search 81–6
BESTFIRSTSEARCH 83
beta nodes 198, 236
BFHS 176
BFS 58
bias 370
bidding 251
bidding system 253
binary constraint networks 394
binary constraints 394
binary predicates 357
binding constraints 286
biosemiotics 10, 141
blind search 47
block structure 445
blocks world 90–1, 186
blocks world domain 268

Blum and Furst 295
board games 227–47
Bolander 313
Bonet and Geffner 274, 279, 280
Boolean formula 95
Boolean satisfiability 35
Bottom 320
bound variable 333
BOUNDARY 175
branch & bound (B&B) 147, 148
breadth first heuristic search (BFHS) 176
breadth first search 58–60
Brownston 189
BSS 176
BSSP 278
Buchanan and Shortliffe 195
bucket elimination 417
business rule management systems 188

Caltabiano and Muscato 264
Cambrian explosion 124
Campbell 195
Campbell et al. 232
Campbell, Hoane, and Hsu 223
case based reasoning 15
categories 19, 357
category 330
causal links 286
CBR 15
cellular automaton 139
chain of thought 197
Chaos theory 139
Charniak and McDermott 341
ChatGPT 2, 9
chaturanga 221
Chef Watson 3
chess 221, 227, 246
Chien 313
Chinese room 15
chromosomes 125, 128
chronological backtracking 426
circumscription 359
classical logic 320
classification 377
classification problem 388
class name 200
clause form 348
clauses 304

Clavier 16
clobbered 291
CLOSED 51, 56, 68, 82, 159, 175
closed world assumption 359
cluster 240, 241
clustering algorithm 389
CNF 35, 304
Cobb 246
Coco-Pope, David 3
Cope, David 3
cognition 13–14, 369
Cohen, Harold 3
CombEx 44, 47, 67, 180
competing needs 298
competition 123
complete information game 254
completeness 32, 64, 69–70, 85, 89, 152, 167, 326
concepts 357
conditional effects 310–11
conditional probabilities 379
configuration problems 29, 33–5
conflict 432
conflict directed backjumping 432
conflict resolution 195
conflict set 427
CONFORMANTGRAPHPLAN 311
CONGEN 207
conjunctive normal form 304
connectives 332
consistency condition 166
consistency enforcement 398, 403
consistent 394, 427
consistent arguments 323
constant symbol 330
constrained generator (CONGEN) 207
constraint graph 395
constraint processing 393
constraint propagation 398, 403, 405, 417
constraints 393, 403
constraint satisfaction 393
constructive methods 42–3
content 328
contingency 322
contingent planning 311
contract 251
contract bridge 196, 250, 314
contradiction 322

control statements 445
convolutional neural network 8
cooling rate 120
coordinated planning 313
creation 11, 123
CRIKEY3 309
crossover 125, 127, 128
crossover operators 130
cryptarithmetic 74
cryptarithmetic problems 435
CSP 302, 393, 403
culprit variable 427
Cuneiform 12
current session 430
Cushing, Subbarao Kambhampati, and Weld 309
cutoff 237
CWA 359
CX 132
cycle crossover 131

DALL-E 3, 9
Dartmouth conference 222
Darwiche 9
Darwin, Charles 123
database 341
DAWG 249
Dawkins, Richard 116
DB-DFS 65
DCBFS 178
DCBS 178
DCBSS 178
DCFS 173
dead end 427
deception 256, 315
decision boundaries 377
decision trees 17, 371
De Corpore 13
deduction 21, 319, 324, 359, 369
deduction system 346
Deduction theorem 327, 349
deductive retrieval 341, 341–3
Deep Blue 4, 223, 232
DeepMind 4, 223, 246
deep neural networks 8–9
default logic 362
default reasoning 196, 359
defeasible 359

DEL 313
Delete list 267
delta rule 384
Demotion 288
DENDRAL 206
denotational 328
depth bounded depth first search (DB-DFS) 65
depth first iterative deepening
    (DFID) 64–70, 236
depth first search 49–58, 343, 398
Descartes, René 13
description logics 357, 358
DFID 66, 67, 236
DFS 50, 55, 57–8, 235
Dijkstra's algorithm 156
directed acyclic word graph (DAWG) 249
directional consistency 415–17
directional path consistency 415
directionally arc consistent 415
discriminate 384
discrimination tree 198
disjunction 349
disjunctive refinement 296
divide and conquer beam search (DCBS) 178
divide and conquer beam stack search
    (DCBSS) 178
divide and conquer breadth first search
    (DCBFS) 178
divide and conquer frontier search (DCFS) 173
DL 357
DNA 171
domain 329
domain independent heuristics 274
domain independent planning 264
domains 403
Dorigo, Marco 142
Doxastic logic 362
durative actions 308
dynamic epistemic logic 313
dynamic variable ordering 403
dynamic variable ordering with forward
    checking 438

earliest minimal conflict 432
ecosystem 124
edge exchange 97
effective branching factor 86
Eliza 2

emergent systems 138
EM 390
EMI 3
empty clause 349
ENFORCEDHILLCLIMBING 277
entailed 323
entailment 323, 359
entities 328
entropy 373, 374
epistemic actions 361
epistemic coordinated actions 314
epistemic logic 313, 361
epistemic planning 313–15
epistemic reasoning 361
equality tests 443
equivalent 349
error surface 385
escaping local optima 103–8
E step 390
estimated cost 147, 154
Euclidean distance 79
eureka 75
EVAL(N) 118
EVAL(J) 231
evaluation function 116, 231, 232
event calculus 265, 360
evidence 378
evolution 122–38
evolutionary algorithms 125, 126
EWI 3
execute 194
existential generalization 334
existential instantiation 340, 334
existential quantifier 329, 331
existential query 347
existential variable 335
expectation 390
expert systems 185
explanations 369
explanatory frame axioms 305
exploitation 104, 116
exploration 104, 115, 116

FASTFORWARD 276
feature space 387
feedforward artificial neural network 7
FF 276
Fikes and Nilsson 266

finite branching 162
first-in-first-out 49
first order logic 328
first principles 22
first tuple 444
fitness 127
fitness function 126
flaws 287
flip-1-bit 105
Flood, Merrill 224
fluents 265
$f(n)$ 159
$f^*(n)$ 163
FOL 328, 329, 339, 351–5, 356
FOL with equality 351–5
for loop 445
Forgy, Charles 194, 197
form 320, 328
formal 328
formal logic 328
formulas 333, 337
forward chaining 340
forward checking 418
forward phase 211
forward reasoning 339
forward state space planning (FSSP) 273
FORWARDCHECKING 403
Fox and Long 310
frame axioms 305
frame problem 265
frames 19
free variables 333
Frege 326
Frege's axiomatic system 327
Freuder 404
FSSP 273
FULLLOOKAHEAD 426
functions 329, 331
function symbol 329
future variables 403

Galilei, Galileo 13
Game of Life 139
game theory 223
game tree 227
Gardner, Martin 139
Gaschnig's backjumping 427
GBBJ 430

gene 125
generalization 22, 334
generalize 370
generate and test 31
generative model 381
generative neural networks 9
GENETICALGORITHM 125
genetic algorithms 125–30
genotype 125
Gleick 139
Gleiser, Marcelo 11
$g(n)$ 163
$g^*(n)$ 163
go 4, 223, 246
goal description 269, 304
goals 48, 263, 277, 341
goal stack planning 280
goal state 30
GOALTEST 23, 30
goal trees 185
Goldberg 127
Google 9
Gosper's gliding gun 140
Grand, Steve 11, 123
gradient descent 8, 384, 388
graph based backjumping 429
GRAPHPLAN 276, 295, 295–302, 306
GREEDY 80
greedy algorithm 88
Gruber 19
Guzman, Adolfo 408
GSP 280, 281

Hamilton, W.R. 39
Hamiltonian 39
Hannabi 315
Hansen and Mladenovic 100
Haugeland, John 1, 13
Hayes-Roth, Waterman, and Lenat 205
head list 443
helpful actions 276
Hercules 44
heuristic 76
heuristic crossover 136
heuristic function 81, 91
heuristic regression planner 279
heuristic search 75, 81
heuristic search planner (HSP) 274

Heuristic search terrains 90–6
heuristic terrain 99
hidden layer 6
hill climbing 88–9
Hinton, Geoffrey 8
Hitech 235
HILLCLIMBING 88, 96
$h(n)$ 76
$h*(n)$ 163
Hobbes, Thomas 13
Hoffmann and Nebel 276
Hofstadter, Douglas 3
HoldsAt 265
Holland, John 125, 139
horizon 232
horizon effect 233
Horn clause logic 356–7
Horn clauses 356
HSP 274
HSPr 279
Huffman, David 408
Huffman–Clowes 408
Hume, David 14
Hydra 44
hypothesis 370
hypothesis space 371, 373, 377
hypothetical syllogism 327

i-consistency 404, 414
IBM 3, 4
ID3 17
IDA* 169
IE 189, 194
if-then-else 445
IHC 107
Imagen 9
imagination 369
imagine 23–6
imitation game 1, 14
implicit facts 341
implicit quantifier form 340, 347
IMPLIES 321
incomplete information games 227
incompleteness 346–7
inconsistent effects 298
indel 171
index of cities 136
induced ancestors 430

induced jumpback set 432
induced parent 430
induced width 401
induction 21, 319, 369
inference 319, 381
inference engine 185, 189, 194
Inference Player 249
inferences 255
information gain 373, 374
informed backtracking 427
initiates 265
input layer 6
input node 383
instance based learning 376
instantiation 394
intelligent agent 263
intelligent backtracking 426
internal dead end 430
interpretation 303, 330
invisits 430
iterated hill climbing (IHC) 107
iterative deepening A*(IDA*) 169

Jackson 195
Java 441
Jeopardy 3
Julia sets 139

k plies 232
Kasparov, Garry 4, 223
Kautz and Selman 302
KB 322
Kelsey, Hugh 4, 255
KERNEL 175
Khemani 196, 311
Khemani and Singh 314
K-MEANS clustering 389
k-nearest neighbour 376
kNN 376
knowledge 10–2, 75, 314
knowledge acquisition
    bottleneck 205
knowledge base 322
knowledge graphs 19
Korf, Richard 37, 103, 170
Korf and Zhang 173
Kowalski, Robert 345
Krizhevsky, Alex 8

Laird 189
language 10, 10–2
Larson, Eric 21
last-in-first-out 49
latest$_i$ 427
lazy learner 377
leaf dead end 427
learning 15
least commitment planning 286
LeCun, Yann 8
Leela Chess Zero 246
left hand side 189
level off 301
Levesque, Hector 2
Levy, David 223
lexical order 196, 203
LHS 189, 191
Lifschitz 196
likelihood 378, 379
linear decision boundary 384
linearly separable 384
linear planning 281
linear regression 370
line drawing 409
Lisp 40
list 441
list$_1$ ++ list$_2$ 443
literal 349
live 211
live node 243
local maximum 93
local optimum 89
local search 86–90, 93
logical connectives 320, 332
logic of propositions 320
logic programming 345
long term memory (LTM) 189
lookahead 232
Lorenz, Edward 139
lower bound 153
Loyd, Sam 36
LTM 189
Ludo 247

machine learning 367
Mackworth 404
MacLean, Paul 5

macro-operators 103
Macsyma 209
Make 192
MAKEPAIRS 55
MAKEPATHS 148
makespan 276, 294, 306
Mandelbrot, Benoit 139
Mandelbrot sets 139
Manhattan distance 79, 83
map colouring problem 33, 77, 394–5
Martelli and Montanari 211
mass spectrometer 207
match 189
Match 194
Match-Resolve-Execute 194–5
matching diagram 395
matching patterns 337
Mathematica 209
mathematical logics 319, 360
matrix 224
Maven 248
Max 227, 233
max heuristic 275, 300
maximal jump 427
maximal backjump 427
maximization 390
maze 51
McCarthy, John 4
McCarthy and Hayes 305
McCorduck, Pamela 2
McCulloch 15
McCulloch and Pitts 6
McDermott, Drew 195, 311, 316
McGann et al., 264
means-ends analysis (MEA) 197, 204–5
mean squared error 369
meaning 328
Melvin 224
memoization 302
memory based reasoning 15
Mesopotamia 12
metric domains 310
MGU 337, 349
Midjourney 3
Min 227, 233
mind-body dualism 13

MINFILL 401
minimal conflict set 427
minimal models 360
Minimax rule 232
minimax value 228
MININDUCEDWIDTH 401
Minkowski norm 79
Minsky, Marvin 4, 13, 19, 408
Minsky and Papert 6
min-width ordering 400
ML 367
MMP 339
modal operator 361
model 303
model based reasoning 29
modified modus ponens 339
Modify 192
modus ponens 323, 405
monotone condition 166–7
Monte Carlo 246
more informed 165
Morgenstern, Oskar 223
Moses, Joel 209
most general unifier 337
MOVEGEN 23, 30, 37, 48
MSE 370
M step 390
multilayer network 388
multi-layer perceptron 6
multiplayer games 226
Murphy et al., 264
Murray, H.J.R. 221
musical intelligence 3
mutation 123, 128
mutex free 301
mutex pairs 279
mutex relations 296, 299
MYCIN 195, 206
mystery domain 316

N-queens 395, 418
NAND 321
Nash, John 225
Nash Equilibrium 225
natural deduction 339
natural numbers 345
Naïve Bayes classifier 379
NEARESTNEIGHBOUR 80, 96

negation by failure 359
negative effect edges 296
negative sum games 226
neighbourhood function 29, 96–103
network 394, 411
neural networks 6–9, 368, 387
Neurogammon 247
neurons 5
Newell, Alan 12, 189
Newell and Simon 197
Nilsson, Nils 36
NN 345
no-good 427
nodePair 53, 82
node structure 244
non-decreasing 301
non-linear planning 286
nonlinear regression 370
non-mutex 301
non-serializable 36, 286
No-op action 297
NOR 321
noughts and crosses 187
NP-complete 35
NP-hard 41
N-queens 34

OALN 386
objective function 116, 369
observations 368
OCCURSIN 56
ontology 19
OPEN 31, 48, 68, 76, 82, 159
open goal 287
OPS5 189, 193, 195, 203
optimal solutions 147
optimal strategy 240
optimization 41, 87–8
optimization problem 389
order crossover 134
ordering links 286
ordinal representation 136
Othello 227
Ottlik, Geza 4, 255
output layer 6
output node 383
overfitting 370
OWL 359

partial assignment 398
PARTIALLOOKAHEAD 422
partially mapped crossover (PMX) 133
partial order planning 286
partial plan 286, 290
partial strategy 240
path consistency 411–4
pathPair 148
path representation 131
pattern 186
pattern directed inference systems 185
payoff 224, 253
PDDL 186, 264, 270
penetrance 86
perceptron 6, 382, 383
perturbation 97
perturbative methods 43, 81
phenotype 125
pheromone 141, 142
physical symbol system 13
physical symbol system hypothesis 12
Pickup 269
PL 320
plan 35
planning 263
planning actions 272
planning as satisfiability 302
planning domain definition languages (PDDL) 186, 264
planning graph 295, 300
planning problems 29, 35, 264
plan space planning 286
PMX 133
POP 286
populations of agents 141
positive definite clause 356
positive effect edges 296
positive sum games 226
possible worlds 252
posterior 381
posterior probability 378
precondition edges 296
preconditions 267, 305
predicate logic 328
predicates 331
prediction 369, 381
preferences 312
primitive nodes 215

prior 378
priority queue 244
prisoner's dilemma 224–5
probability 121, 127
problem decomposition 185, 205–16
problem solving 4–6
procedures 444
production systems 189
programming language 193
progresses 273
Prolog 343, 356–7
Promotion 288
proof 323, 339
proof procedure 358
property 330
propositional calculus 326
propositional logic 320
propositional variables 303, 321
proposition layer 296
propositions 269, 295
PROSPECTOR 195
pruning 178
pseudocode conventions 441
PSP 286
Putdown 269

quality 67–9, 84–5, 89, 152, 167
quality of solution 33, 64
quantified formulas 332
quantifiers 331
queries 341
queue 49
Quinlan, Ross 19

R1 195
rack leave 249
Rajan et al. 264
randomness 115
random walk 117
RANDOMWALK 117
Ray, Satyajit 221
RBFS 170
RBox 358
reachability heuristics 295
reachable 296
reasoning 20
recency 197, 203
reconstructing 53–5

RECONSTRUCTPATH 53
recursive best first search 170
recursive rules 345
refinement 287
reflexivity 352
refractoriness 196, 203
regression 278
reification 11
Reinelt, Gerhard 130
reinforcement learning 8, 371
Reiter 196
relations 330, 357, 394
relaxed plan 276
RELAXEDPLANEXTRACTION 300
relay layer 174
relevant dead ends 430
relevant actions 277
Remove 192
REMOVESEEN 55, 56
repeat loop 445
representation 20
Reproduction 127, 128
required concurrency 309
resolution refutation method 347–51, 350
resolution rule 347, 349
RESOLVE 194, 288
resolvent 349
Rete algorithm 185
RETEALGORITHM 197
Rete net 197–201
Rete-NT™ 205
reverse list 444
REVISE$((X), Y)$ 404
REVISE-i 414
REVISE-3 412
reward 371
RHS 191
Rich, Elaine 91
right hand side (RHS) 191
river crossing puzzles 37–8
RNA 171
Robinson, Alan 347
role axioms 358
roles 357
route finding 78–80
RR-F 280
Rubik's cube 36

rule based systems 189
rule of inference 323, 337
rules 188, 189, 371
Rumelhart, Hinton, and Williams 6, 389

safe backjump 427
SAINT 209
Samuel, Arthur 222, 367
SAPA 309
Sapir–Whorf hypothesis 130
SAT 29, 35, 77, 95, 115, 306
satisfiable 322
satisfies 394
SATPLAN 302
SAVINGS heuristic 81
Schank, Roger 19, 22
Schank and Abelson 19
Schank and Riesbeck 19
Scrabble 247
search 342, 373, 398
searchable dictionary 249
search space 32
search tree 48, 59
Searle, John 15
second tuple 444
Sedol, Lee 4, 246
SELECTVALUE 398
SELECTVALUE-DAC 422
SELECTVALUE-FC 420
SELECTVALUE-GBJ 427
semantic networks 19
semantic notion 323
semantics 328, 332
Semantic Web 19, 359
semiotics 10
sense-deliberate-act 4, 263
sentence 321, 333
Separation 288
sequence alignment 171, 172
set of support strategy 353
Sheppard, Brian 248
Shakey 266
Shannon, Claude 222
shogi 246
shortest path 147
Shortliffe 195
short term memory (STM) 189
Sierpiński triangle 139

sigmoid function 118, 386
signalling 255
Silver et al. 246
similarity 377
Simon, Herbert 12, 189
SIMPLESEARCH 31, 48
simplification 325
simulated annealing 119–22
SIMULATEDANNEALING 119
SIN 209
single point crossover 125
Skolem, Thoralf 348
Skolem constants 340, 351
Skolemization 335, 348, 351
Slagle, James 209
SLD resolution 353, 356
smart memory graph search (SMGS) 175
Smith, David 311
SOAR 189
Socratic argument 328
solution 395
solution plan 287
solution space 81, 42–3
solution space search 98
solved 211, 243
sort$_h$ list 444
sound 324
soundness 324–6
space complexity 33, 61, 67, 85, 90, 152, 167
space saving variations 169
spam 368
species 123
specificity 196, 203
Stable Diffusion 3
Stack 49, 269
Stanford Research Institute Problem Solver (STRIPS) 266
state space 29, 30
state space planning 272
state transition function 273
static 400–2
statistical hypothesis test 370
steepest descent 385
steepest gradient 87
stereotypical situations 22
Sterling and Shapiro 196, 345
stimulus–response 4

STM 189
stochastic 115
stochastic hill climbing 117–9
STOCHASTICHILLCLIMBING 118
stochastic local search 42
strategy 195, 229, 240
STRIPS 266
strong interference 298
structure matching 358
subgoal 36, 345
substitution 337, 341, 349
substitution for functions 352
substitution for predicates 352
subsumption 359
sub-symbolic 8
SUBUNIFY 338
successor 345
successor function 345
Sudoku 437
Sumerian 12
sum heuristic 300
supervised 390
supervised learning 8, 371
survival of the fittest 122–38
Sussman, Gerald 284
Sussman anomaly 284, 305
swarm intelligence 138–43
syllogisms 328
symbolic AI 10–4
symbolic integration 209
symbols 10–2
symmetry 352
syntactic notion 323
syntax 328
systematic 32

tableau method 358
taboo 105
tabu 105
tabu search 104
tabu tenure 105
tail list 443
take n list 444
TALN 388
tautological implication 324
tautology 322, 326–7, 349
taxonomy 19
TBox 357

TD-Gammon 247
team games 226
temperature 120
temporal difference 247
temporal planning 308
Tensorflow 9
terminal node 235
terminological box 357
terms 329
Tesauro, Gerald 247
test data 370
test instance 376
The Assayer 13
the h-value 81
thematic techniques 254
theorem prover 265
theorem proving 323
'Theory of Mind' 255, 258
The Powers of Ten 11
The Society of Mind 13
threat 288
threshold 368, 383, 386
tic-tac-toe 187
time complexity 33, 62, 67, 86, 89, 152, 167
Top 320
training 8
training data 370, 376
trajectory constraints 312
Transitivity 352
travelling salesman problem (TSP) 29, 39–42, 80, 115, 126, 130–8, 153
T-REX 264
trie 249
trihedral objects 408
triples 19
Triune Brain 5
true sentences 323
truth functional 328
truth tables 321
TSPLIB 41, 131
tuple 441
Turing, Alan 1, 222
Turing Test 1
two armed robot 293–5
two player games 226

UAV 264
UI 333
unary predicates 357
underestimate 162
underfitting 370
unification 337–9
UNIFICATION algorithm 338
unifier 337
unifies 337
uniform formalism 434
UNIFY 338
uninformed 47, 70
unit clause strategy 353
universal generalization 334
universal instantiation 333
universal quantifier 329, 331
unsatisfiable 322, 349
Unstack 269
unsupervised 390
unsupervised learning 8, 371
upper bound 241
utilities 224

Valéry, Paul 125
valid 324
valuation function 321, 331
value 240
value of a strategy 229
VARIABLENEIGHBOURHOODDESCENT 96, 100
Variable neighbourhood descent 99–101
variable ordering 400–2
variance 370
VARUNIFY 338
visibility 142
VITERBI 101
von Neumann, John 222–3, 231

wA* 168
Waltz, David 409
WALTZ algorithm 408, 409
water jug problem 38
Waterman and Hayes-Roth 185
Watson 3
weak interference 298
Web Ontology Language 359
weighted A* 167
weights 377, 383
Weizenbaum, Edward 2
Weld, Daniel 311
while loop 445

width 400
winning strategy 229
Winograd, Terry 2
Winograd Schema Challenge 2
Winston, Patrick Henry 82, 86
WolframAlpha 209
working memory (WM) 189–90
working memory elements (WMEs) 189–90, 192, 198

XCON 195
XOR 321, 384

Yale shooting problem 360

zero-sum 247
zero sum games 226–7
Zhou and Hansen 175–6